CO-CEN 179

Wool

Polymer Interfaces

Richard P. Wool

Polymer Interfaces

Structure and Strength

Hanser Publishers, Munich Vienna New York

Hanser/Gardner Publications, Inc., Cincinnati

Prof. Dr. Richard P. Wool, Director, Center for Composite Materials, University of Delaware, Newark, DE 19716-3144, USA

Distributed in the USA and in Canada by
Hanser/Gardner Publications, Inc.
6600 Clough Pike, Cincinnati, Ohio 45244-4090, USA
Fax: (513) 527-8950
Phone: (513) 527-8977 or (800) 950-8977

Distributed in all other countries by
Carl Hanser Verlag
Postfach 86 04 20, 81631 München, Germany
Fax: +49 (89) 98 12 64

Library of Congress Cataloging-in-Publication Data
Wool, Richard P.
Polymer interfaces: structure and strength / Richard P. Wool.
p. cm.
Includes bibliographical references and index.
ISBN 1-56990-133-3
1. Polymers--Surfaces. 2. Polymers--Mechanical properties.
3. Micromechanics. 4. Microstructure. I. Title.
TA455.P58W66 1995
620.1'92--dc20 94-29396

Die Deutsche Bibliothek - CIP-Einheitsaufnahme
Wool, Richard P.:
Polymer interfaces : structure and strength / Richard P. Wool.
- Munich ; Vienna ; New York : Hanser ; Cincinnati : Hanser/
Gardner, 1995
ISBN 3-446-16140-6

Printed and bound in Germany by Kösel, Kempten

Special Dedication

With love

To

Deborah

Sorcha, Meghan and Breeda

Preface

Polymer interfaces are ubiquitous and play a critical role in controlling the properties and function of a broad range of materials. In this book, we discuss model polymer interfaces in terms of their structure and strength, and apply the results to practical applications, including melt processing, weld lines, injection molding, composite lamination, particle sintering, rubber tack, blends, compatibilizers, adhesion, and fracture. The strength is related to the interface structure in terms of microscopic deformation mechanisms involving disentanglement and bond rupture of polymer chains. The subject is very rich if one considers the wide variety of interface pairs involving polymers and materials with different chemical composition, crystalline and amorphous content, molecular weight distributions, additives, and surface chemistry. The structure of welding interfaces is described in terms of the chain dynamics, interface roughness, and the thermodynamics of chemically interacting monomers. In all cases, the fundamental formulae for strength are compared with suitable spectroscopic and fracture experiments on polymer interfaces.

The book should appeal to a wide range of readers with experimental and theoretical backgrounds in chemistry, physics, materials, and engineering. It will be highly useful to readers with interests in material surfaces, interfaces, manufacturing, melt rheology, polymer dynamics, fractals, percolation, material design, fracture, fatigue, crack healing, welding, smart materials, electronic materials, and high performance composites. It is primarily intended as a research resource but can also be used as a textbook or reference for a class on polymer surfaces and interfaces. In this regard, it contains many solved problems, exercises, comments, and research suggestions.

Most of the ideas in this book were developed in collaboration with a delightful set of graduate students and postdocs in my laboratory at the University of Illinois during the past 17 years. I had the great privilege to share the joy, anxiety, and fun in this research with my unforgettable colleagues (alphabetically) Gaurav Agrawal, David Bailey, Yves Bernaert, Serge Billieux, Robert Bretzlaff, Michael Daley, Pre Desai, Lou Dolmon, Ken Foster, Anthony Friend, Susanna Goheen, Stuart Jackson, Brian Joss, Young-Hwa Kim, David Kline, Don Klosterman, Andre Lee, Yueh-Ling Lee, Marisa Lohse, John Long, Owen McGarel, John McGonigle, Kevin O'Connor, Karen Paulson, John Peanasky, Raj Raghavan, Alvin Rockhill, Karla Salin, Suresh Upadyayula, Mark VanLandingham, Jerry Wagner, Wanda Walczak, Scott Whitlow, J. L. Willett, Baoling Yuan, and Huanzhi Zhang. I also express my appreciation to the many high school students and undergraduates who contributed substantially to our research efforts.

Each chapter starts with a special dedication to the individuals who have strongly influenced the direction of my research and personal development as a scientist.

Urbana, Illinois *Richard P. Wool*

Contents

1 Introduction to Polymers and Interfaces*

1.1 Introductory Remarks

Polymer interfaces are ubiquitous. They play a critical role in determining the properties, reliability, and function of a broad range of materials. For example, polymer melt processing via injection molding (for plastic auto parts) and extrusion (for plastic pipe) creates many interfaces in the form of internal weld lines where the fluid fronts coalesce and weld. Compression molding and sintering (of artificial hip joints) involve coalescence when the surfaces of pellets or powder make contact in the mold. Drying of latex paints and coatings entails a very large number of interfaces per unit volume, as the relatively tiny (about 1000 Å) latex particles interdiffuse to form a continuous film. Construction of composites with thermoplastic matrices (for aircraft bodies) requires the fiber-filled laminates to weld by an interdiffusion process at the interface. Welding of two pieces of polymer by thermal or solvent bonding is a commonly encountered example of strength development at a polymer–polymer interface; tack between uncured rubber sheets during auto tire manufacture is an important example of this kind of welding.

Numerous examples of *symmetric interfaces* exist, where the same polymer occurs on both sides of the interface. In contrast, rubber toughening of glassy polymers (for bullet-proof glass) requires that the interface formed by the rubber particle with the glass (Plexiglas™) matrix be sufficiently strong to promote stable energy dissipation, and the rubber/Plexiglas is an example of an *asymmetric interface*, where two different polymers meet. Asymmetric or dissimilar interfaces come about when polymers are blended, when dissimilar materials are laminated (in electronics applications), when coupling agents are used in composites, when plastic mixtures from municipal solid waste are recycled, when different materials are coextruded, and when photographs are developed. As our knowledge of material interfaces expands, so does our ability to fabricate more sophisticated material systems, be they artificial organs, biomedical implants, rocket motors, high speed integrated circuits, the Super Auto of 2000, advanced composite space vehicles, or, perhaps, a better paint.

We find the subject of interfaces is very rich if we consider the wide variety of possible pairs comprising polymers and materials with different chemical composition, crystalline and amorphous content, compatibility, incompatibility, molecular weight

---

* Dedicated to P. J. Flory

distributions, additives, surface chemistry, etc. In this book, we examine how interfaces form, describe their structure, and provide an understanding of the relationship between interface structure and strength. The vast number of possible interface pairs is reduced to a few important examples. These include the *symmetric amorphous*, the *incompatible amorphous*, the *incompatible crystalline*, and the *polymer–nonpolymer* interfaces. We first examine the structure of these interfaces and then relate the structure to fracture strength through microscopic deformation mechanisms involving chain disentanglement and bond rupture. The evolution of structure at diffuse interfaces is controlled by the dynamics of the chains and the thermodynamics of chemically interacting species. Once the structure is known, a *connectivity relation* is required, to relate the structure to mechanical properties. The connectivity relation for amorphous polymers is developed in terms of an entanglement model. We can then understand what is required to break the connectivity at the interface by disentanglement, by bond rupture, or both. These two microscopic deformation mechanisms consume a known amount of energy, from which the strength of the interface can be determined.

Chapter 1 is a general review of the polymer science and interface concepts that are used throughout the book. A classification of interfaces is provided, with a brief description of each interface. In Chapters 2 and 3, the structure of amorphous symmetric interfaces during welding is presented in terms of molecular properties $H(t)$. At this stage, we have a "tool kit" in the form of $H(t)$, which is used in later chapters to examine the properties of a wide range of interfaces; the Rouse and reptation dynamics models are used to derive $H(t)$. The generic fractal-like structure of diffuse interfaces is examined in Chapter 4, with several important applications to polymer–metal interfaces. Chapters 5 and 6 give the experimental analyses of interface structure and validation of the $H(t)$ tool kit. Microscopic deformation and fracture mechanisms are presented in Chapter 7, and the remaining chapters, 8–12, focus on specific structure–strength relations for the model interfaces. The healing of microscopic damage and cracks in polymers and composites is treated in Chapters 11 and 12.

1.2 Polymer Molecules

Polymers or *macromolecules*, commonly referred to as *plastics*, are typically long-chain molecules with high molecular weights, M, in the range 10^4–10^6. *Polyethylene (PE)* consists of strings of monomers with a molecular structure $-CH_2-CH_2-$ and monomer molecular weight, M_0, of 28. The monomers can be strung together like beads in a necklace, where the number of beads, $N = M/M_0$, is the *degree of polymerization*. Polymers can be made with different architectures, resulting in plastic materials with a wide range of microstructures and physical properties. The most useful properties are found with the high-molecular-weight polymers. Mechanical properties (such as fracture energy) and rheological properties (such as viscosity) are functions of molecular weight, and generally M must exceed a threshold or critical molecular weight, M_c, before useful properties are obtained. As the molecular weight of a polymer increases above M_c, the fracture properties typically increase to an asymptotic limit while the viscosity continues to increase without limit.

The architecture of individual polymer chains is usually described in terms of *linear, branched*, or *cross-linked* structures. The chain architecture can have profound effects on properties. For example, linear PE chains like strings can pack closely in the solid state and have a relatively high degree of crystallinity, such as in *high density polyethylene* (*HDPE*). Typical examples of items made of HDPE are gallon milk jugs and large chemical containers, where rigidity and strength are important properties. Strings with multiple branches of varying lengths do not pack as closely together or crystallize as readily, and the result is *low density polyethylene* (*LDPE*). LDPE plastic is used to make baby bottles, butter tubs, and other objects that have flexibility as well as strength. PE chains with regularly spaced small branches (four to nine carbon atoms) can pack more closely; the resulting plastic is *linear low density polyethylene* (*LLDPE*). This material is excellent for high-strength plastic bags. Linear PE chains that are very long (M of about 10^7) may have difficulty crystallizing due to self-entanglement and high viscosity; the result is called *ultra high molecular weight polyethylene* (*UHMWPE*). This material has excellent fatigue and wear resistance and is often used in artificial hip joints. HDPE that has been lightly cross-linked in the melt exhibits better creep resistance at high temperatures and behaves like UHMWPE. Such materials are excellent for wire coatings. The linear and lightly branched polymer materials constitute the vast bulk of commodity plastics. The abbreviated designation or acronym for each plastic is that prescribed by ASTM D 1600 [49], and all these abbreviations can be found in the Appendix.

Polymers are made by chain reaction and condensation polymerization mechanisms. The vinyl family, represented by P (the polymer) with a molecular structure $-(CH_2-CHX)_n-$, is a very important example of polymers formed by chain reaction kinetics in the presence of free radical or ionic initiators. The substituent –X determines the character of the polymer. For example, when –X is –H, $-CH_3$, –Cl, $-C_6H_5$, –OH, or –CN, then P is polyethylene, polypropylene, poly(vinyl chloride), polystyrene, poly(vinyl alcohol), or polyacrylonitrile, respectively. Free radical reactions are initiated when an initiator species, I, is added to the monomer, usually in liquid form. The initiator decomposes into active radical fragments and attacks the monomer. The activated monomer adds other monomers in a chain reaction that continues to propagate until it is terminated, which can occur when two active radicals combine to form a stable polymer. The more initiator present, the greater the number of polymer chains created and the lower the molecular weight, M. The molecular weight is related to the initiator concentration by an inverse square relationship, $M \sim [I]^{-1/2}$. For example, 0.5 g benzoyl peroxide initiator in 100 ml styrene monomer ([I] of 0.5 g/100 ml) produces polystyrene with a viscosity average molecular weight, M_v, of about 10^5, whereas [I] of 0.02 g/100 ml gives an M_v of about 1.7×10^6. Cross-linking can be induced with tetrafunctional monomers such as divinyl benzene.

Polyamides (nylons), polyesters, polyurethanes, polyureas, and other polymers can be made by condensation reactions. The monomers (ARB) have reactive end groups, A and B, such as –OH, –COOH, or $–NH_2$, which react with each other. The group R in the ARB monomer is not involved in the reaction process and typically consists of methylene sequences, $–(CH_2)_n–$, or phenyl rings, $–C_6H_5$, in the chain backbone, or other chemical species. The individual family (polyamide, polyester, etc.), is identified by the functional group linkage resulting from the condensation reactions. For example, acids and alcohol end groups combine to form polyesters with –COO– links; $–NH_2$ and –COOH give the nylon family with –NHCO– links in the chain. The monomers can be bifunctional (ARB), bi–bifunctional (ARA + BRB), or polyfunctional (ARA + BRB + RB_3). The polyfunctional monomers give cross-linked structures (gels) if present in sufficient quantity. Flory's theory of condensation reactions can be used to predict whether linear, branched, or cross-linked structures will develop.

Example

Problem: *In a polyfunctional condensation reaction*

$$RA_2 + RA_3 + xRB_2 = gel$$

determine the value of x such that the system gels by forming a three-dimensional cross-linked network when all the B groups have reacted.

Solution: *According to Flory, the critical branching probability α_c at the gel point is given in terms of the functional group ratio, $r = A/B$; the branch density, $\beta = \Sigma A(\text{in branch})/\Sigma A(\text{total})$; and the extent of reaction of B groups, p_B; as*

$$\alpha_c = p_B^2 \beta/[r - p_B^2 (1 - \beta)]$$

This equation gives the probability that a growing chain segment that begins at a branch point in the network will reach another branch point. For trifunctional branch points in RA_3, $\alpha_c = 1/(3-1) = 1/2$; $p_B = 1$ when all the B groups have reacted; $r = (2 + 3)/2x = 5/2x$; and the branch density $\beta = 3/(3 + 2) = 3/5$. Substituting for all values, we obtain $x = 25/16$ and the stoichiometric ratio $r = 8/5$.

In the absence of cross-linking monomers RA_3, $\beta = 0$ and only linear chains form; their molecular weight is determined by $M = M_0/(1-p_B)$, for the case of $r = 1$. In the presence of RA_3 groups, branched chains and partially cross-linked clusters of finite molecular weight develop before gelation, at p_B less than 1. At gelation, the clusters become critically connected at the percolation threshold and M becomes infinite. This problem is relevant to interfaces formed during reaction injection molding (RIM) and to transesterification reactions at composite interfaces with polyester matrices. Cross-linked polymers are important in composites, electronics materials, foams, solid rocket engines, auto tires, and applications requiring high-temperature dimensional stability.

1.2.1 Polymer Entanglements

Figure 1.1 shows a computer simulation of polyethylene chains in the melt, by R. H. Boyd [9]; only the C–C bond axes are shown for clarity. Polymer chains in the melt

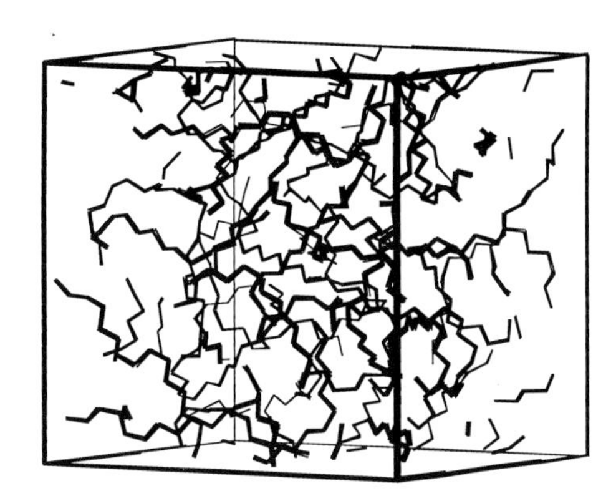

Figure 1.1 Computer simulation showing a periodic box containing 32 chains of 24 methylene units, representing a dense polymer melt (courtesy of R. H. Boyd).

and concentrated solutions are often described as a "bowl of entangled spaghetti noodles". This analogy presents some interesting food for thought, if you will, and it is worth noting that the creation of new surface area in this system involves the formation and breakdown of an oriented deformation zone by disentanglement or fracture, provided the noodles are long enough. Fracture of linear polymers involves similar mechanisms whereby a crack propagates through a deformation zone in which the chains align, disentangle, or rupture. The extent of chain rupture versus disentanglement depends on the molecular weight, deformation rate, plasticizer content, temperature, and other aspects of the problem, such as the stress field at the crack tip (details in Chapter 7).

Several properties of chains in the melt (Figure 1.1) are dependent on the chain length M, and change rapidly at a critical length M_c, known as the *critical entanglement molecular weight*. For example, the zero-shear viscosity, η_0, behaves as $\eta_0 \sim M$ when M is less than M_c, but $\eta_0 \sim M^{3.4}$ at M greater than M_c. Below M_c, the molecules behave as a simple Rouse fluid whose resistance to motion is determined only by the chain length (Chapter 3). Above M_c, it is assumed that an entanglement network exists that further restricts the motion of the molecules and results in a much higher viscosity. Other properties related to molecular mobility, such as the self-diffusion coefficient, D, behave as $D \sim M^{-1}$ at M less than M_c, and $D \sim M^{-2}$ at M greater than M_c. The stress relaxation time, τ, changes from $\tau \sim M^2$ to $\tau \sim M^{3.4}$ at M_c. The fracture energy of glassy polymers, G_{1c}, which involves craze formation and viscous flow at the crack tip, exhibits similar behavior: at M less than M_c, $G_{1c} \sim M$, and at M greater than M_c, $G_{1c} \sim M^x$, where the exponent x is in the range 2–4 (reviewed in Chapter 8). The onset of rubber-elastic–like behavior in polymer melts is manifest by the development of the plateau modulus, G_N^0, in the terminal relaxation zone at an entanglement molecular weight $M_e \approx M_c/2$. Ferry [3] provides an excellent review of entanglements and their effects on specific properties.

For polyethylene, the crossover from a nonentangled to an entangled state occurs at M_c = 4,000, which corresponds to an aspect ratio (length/width) of about 30 statistical (random walk) units. Polystyrene (PS) has a monomer structure $-CH_2-CH\phi-$, where ϕ is the phenyl group $-C_6H_5$. M_c for this polymer is 30,000, so the molecular weight must be increased to 30,000 to develop an entangled state, but the aspect ratio at this M_c is also about 30, because the monomer is larger. I take up the subject of entanglements in more detail in Chapter 7.

1.2.2 Configuration and Crystallization

When a PE chain crystallizes from a melt of entangled chains, about half the chain (on average) is contained in thin lamellar crystals, typically 100–200 Å thick. The other half is distributed in amorphous regions interspersed between the crystals. The details of the distribution of the chain segments in crystalline or amorphous regions and the fraction thereof (degree of crystallinity) depend sensitively on the processing and crystallization conditions. The crystallization mechanism requires that about 50% of the chain disentangle from the melt and order itself in a crystalline lattice.

Polystyrene when cooled from the melt fails to crystallize but forms a glass at the glass transition temperature, T_g, of 100 °C. Crystallization is impeded because of the atactic chain configuration. However, the intermolecular driving force for compaction is great and the chains aggregate into a vitrified glassy state at temperatures below T_g. The nonuniform placement of the phenyl group on the carbon atom (*atacticity*) does not allow the chains to adopt conformations that pack uniformly in a crystal lattice. A chain's configuration is determined by the method of synthesis and does not change once formed. During free radical synthesis of PS, the addition of the next monomer to the growing chain end occurs randomly with respect to the phenyl group placement on the chain. Although the configuration is fixed by synthesis, torsional rotations about carbon–carbon bonds allow the molecule to take on a variety of different spatial conformations. The most probable conformations are those that minimize the intramolecular energy associated with steric, polar, and nonbonded interactions.

Stereoregularity is imparted to the chain by stereospecific catalysts during synthesis and remains unchanged thereafter. A regular structure allows the chain to achieve optimum symmetry by adopting helical conformations that pack in crystal arrays. The stereoregularity of vinyl polymers, $-CH_2-CHX-$, determines whether they are rubber-like, glassy, or semicrystalline materials. For example, polypropylene (in which $-X$ is $-CH_3$) in its atactic form is a rubbery gummy material at room temperature and forms a glass at a T_g of −5 °C, but is of little commercial value. However, the isotactic configuration adopts helical conformations with threefold symmetry, resulting in a semicrystalline material at room temperature. Varying the fraction of stereoregular sequences affects the degree of crystallinity and results in materials with intermediate properties.

The glass transition temperature of the amorphous fraction of isotactic PP occurs at a T_g of −5 °C, and the melting point at a T_m of 170 °C. Thus, at room temperature the material is a composite of stiff crystalline regions and rubbery amorphous regions. The structure of PE can be similarly explained. Even though the degree of crystallinity can be quite low, the crystals may be considered as acting as effective cross-links, holding the rubbery components in place and reinforcing the material. Thus, PE can be a reasonably stiff but ductile material with elongations to break in uniaxial strain, ϵ, of about 1000%. Other polymers, such as poly(ethylene terephthalate) (PET), polycarbonate (PC), and poly(vinyl chloride) (PVC), have glass transition temperatures above room temperature, resulting in properties that are a combination of glassy and semicrystalline, depending on the degree of crystallinity.

1.2.3 Conformation and Random Walks

When the PE chain melts at a temperature of about 132 °C, it adopts a "random walk" shape which on average occupies a space with spherical dimensions, as shown in Figure 1.2. The instantaneous real shape of the chain does not resemble a sphere. Computer simulations at our laboratory show that real random walks (phantom, self-avoiding, and truly or absolutely self-avoiding) adopt irregular ellipsoidal forms, which in the melt constantly change shape due to the Brownian motion of all the chains. As the chain segments move around, the random walk explores many different conformations, which taken on average approximate a most probable spherical envelope, shown in Figure 1.2.

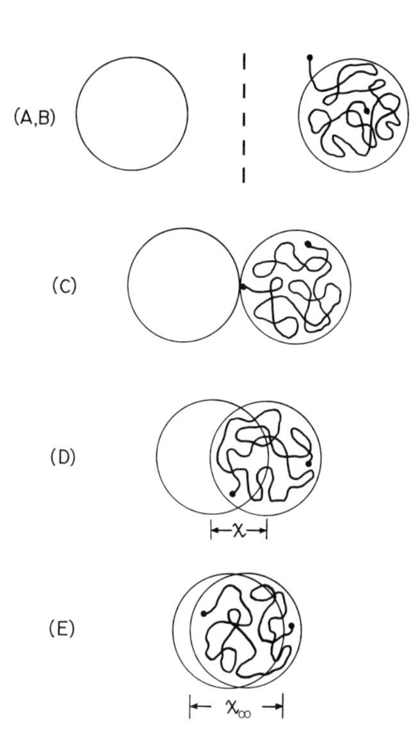

Figure 1.2 Interpenetration of random-walk chains (one chain shown for clarity). X is the overlap distance. (A,B) No interpenetration; (C) $X = 0$; (D) $0 < X < X_\infty$; (E) $X = X_\infty$, the equilibrium value.

The radius of gyration of the average sphere, R_g, is determined from the root mean square of the ensemble average, $R_g = \langle R_g^2 \rangle^{1/2}$, and is given by [1, 2]

$$R_g = \left(\frac{C_\infty Mj}{6 M_0} \right)^{1/2} b_0 \tag{1.2.1}$$

where C_∞, M_0, j, and b_0 are the characteristic ratio, monomer molecular weight, number of backbone bonds per monomer, and bond length, respectively. This corresponds to a random walk of N steps of length b_0, for which $R_g = N^{1/2} b_0 / \sqrt{6}$. The number of C–C steps, N, taken by a polymer chain is jM/M_0. However, since the bonds have biased directions in space determined by fixed valence angles and preferred torsional angles, the real chain is more expanded in space compared to the freely jointed random walk. The number of steps in the equivalent freely jointed random walk needs to be increased by the factor C_∞ (so that $N = C_\infty jM/M_0$).

The distance that a random walker advances from its origin is known as the *end-to-end vector*, $R = \langle R^2 \rangle^{1/2}$. R and R_g are simply related by

$$R = \sqrt{6}\, R_g \tag{1.2.2}$$

or $R^2 = N b_0^2$. This corresponds to the distance X that a randomly diffusing particle travels in a time t. If it moves a distance b_0 in a hopping time t_0, then $X^2 = 2Dt = N b_0^2$, and the diffusion coefficient $D = b_0^2/(2t_0)$. The use of these equations is illustrated in the following example.

Problem (a): *Determine the radius of gyration and end-to-end vector of high density polyethylene with a molecular weight* $M = 10^6$ *in the melt at 150 °C.*

Problem (b): *What molecular weight of polystyrene would be required to give the same radius of gyration as this polyethylene?*

Solution (a): *At 150 °C,* $C_\infty \approx 6.7$ *(the expansion factor can change slightly with temperature) [1],* $M_0 = 28$ *for the* $-CH_2-CH_2-$ *monomer,* $j = 2$ *backbone C–C bonds per monomer, and* $b_0 = 1.54$ *Å for the C–C bond length. Eqs 1.2.1 and 1.2.2 give the radius of gyration as* $R_g = 435$ *Å and the end-to-end vector as* $R = 1065$ *Å, respectively.*

Solution (b): *Let us use the subscripts 1 for PS and 2 for PE. We know PS and PE have the same backbone structure, so* $j_1 = j_2$ *and* $b_{01} = b_{02}$*. If we set* $R_{g1} = R_{g2}$*, Eq 1.2.1 gives*

$$M_1 = M_2 \, (M_{01}/M_{02}) \, (C_{\infty 2}/C_{\infty 1})$$

For PS, $M_{01} = 104$ *and* $C_{\infty 1} = 10$ *and hence* $M_1 = 2.5 \times 10^6$*.*

The radius of gyration, R_g, is the most important static property for polymer melt interfaces and varies with molecular weight M as $R_g \sim M^{1/2}$. The proportionality factor can readily be determined from Eq 1.2.1 since all the constants are known. The exponent of ½ in Eq 1.2.1 comes about from the classical random-walk analysis, where the direction of the next step in the walk is independent of all the previous steps. Happily, this situation appears to describe most polymers in a homogeneous melt, where excluded volume effects are screened out. Excluded volume refers to the fact that real polymer chains occupy space and the next step in the walk may have to avoid previously occupied space. Thus, some walks can be *self-avoiding walks* (*SAW*), which in three dimensions have exponents of the order of 0.6 instead of 0.5. SAW exponents are used to describe the static properties of real polymers in good solvents at temperatures different from the Θ temperature (Θ is the temperature at which the exponent is ½). Also, the interdiffusion of dissimilar chains that are partially miscible could result in static properties other than those described above.

The static properties of chains at a surface and interface are different from those in the bulk melt. The interface is formed by contact of two polymer surfaces. The chains whose centers are at a distance x less than R_g from the surface are restrained from adopting average random-coil configurations and have distorted dimensions. The latter can be described by reflecting boundary conditions: a hypothetical mirror (surface plane) is placed through a random walk with spherical dimensions, and all segments on one side of the mirror are reflected (physically placed) into the other side. More complex rearrangements of the molecule are also possible, particularly at smaller length scales. The confinement effect of the surface may induce local segmental rearrangements that are partially ordered conformational sequences of monomers. The local ordering effect is expected to decay rapidly once diffusion commences at the interface, but it may have particular significance for the properties of chains near a surface.

1.2.4 Interpenetrated Chains

To develop an understanding of entanglements and topological constraints to chain motion at interfaces, it is useful to consider the extent to which chains in the melt are interpenetrated with each other. Consider the random-coil PE chain with M of 10^6, as depicted in Figure 1.2. In a dense PE melt with density ρ, the number of chains per unit volume, N_v, is

$$N_v = \rho N_a / M \quad (1.2.3)$$

where N_a is Avogadro's number, 6.02×10^{23} molecules/mol. If we place the center of each random-coil sphere on a point of a cubic lattice, we obtain the average distance, d, between chain centers as

$$d = (M / \rho N_a)^{1/3} \quad \text{or} \quad d = N_v^{-1/3} \quad (1.2.4)$$

With a typical PE melt density, ρ, of 0.8 g/cm³ and M of 10^6, we find $d = 127$ Å and $R_g = 435$ Å. Since R_g is much greater than d (Figure 1.2), the random-coil sphere of any one chain is highly interpenetrated with those of other chains. A single random-coil sphere intersects Ω other spherical envelopes, where Ω is given by

$$\Omega = (4/3)\pi(2R_g/d)^3 \tag{1.2.5}$$

The molecular weight dependence of Ω is

$$\Omega \sim M^{1/2} \tag{1.2.6}$$

In this example, when $R_g = 435$ Å and $d = 127$ Å, the PE chain intersects $\Omega = 1347$ other random-coil spheres.

If we extract this PE chain from the melt with a mechanical stress σ, we might expect that a resistance would be created by the other Ω entangled chains. The relation between mechanical stress σ and Ω is unknown but we might guess that $\sigma \sim \Omega^\beta$, where β is an unknown exponent. From Eq 1.2.6, we might then deduce a "first guess" structure–property relation. If the guess is correct, a necessary but not sufficient condition for the structure–property relation would be that the molecular weight dependence of σ should be

$$\sigma \sim M^{\beta/2} \tag{1.2.7}$$

Additional relations can be developed from the time dependence of interdiffusion of chains at interfaces, so that Ω, and hence σ, are also functions of both time and molecular weight, M. Relations similar to Eq 1.2.7 were found to have relevance for determining the strength of polymer–polymer interfaces in terms of microscopic deformation mechanisms. For example, tack and green strength of uncured linear elastomers can be described by such relations (Chapters 7 and 8).

The nature of the mechanical interaction of polymer chains in the melt and in concentrated solutions has received considerable attention and has resulted in the concept of entanglements and entanglement networks. Ferry [3] and Graessley [4] have presented extensive reviews of this topic. The entanglement concept plays a central role in all modern theories of melt dynamics and fracture micromechanics. It is considered by some as a device to bridge a large gap in the understanding of the structure and dynamics of the complex many-body problems presented by dense polymer melts, and it has become an elegant tool for developing reasonable models for the mechanical properties of polymer melts.

1.3 Strength of Polymer Interfaces

1.3.1 Polymer Welding

At a polymer–polymer interface, the chains have been physically separate, and diffusion to a distance of R_g is necessary to reestablish the interpenetrated structure of the virgin state of the melt, as shown in Figure 1.2. The time to achieve this level of interpenetration can be designated as the welding time, t_w. The dynamic properties of polymer chains in dilute and concentrated states have been analyzed in detail by Rouse [5], de Gennes [6, 8], Doi and Edwards [7], and many others.

We can estimate the welding time from the diffusion behavior in the bulk or in solution. For example, one-dimensional Fickian diffusion of polymer chains gives the center-of-mass motion, $<X>^2$, as

$$<X>^2 = 2\,Dt \tag{1.3.1}$$

where D is the self-diffusion coefficient. If we assume for polymer welding that $<X>^2$ is approximately equal to R_g^2 when t is equal to t_w, it follows that

$$t_w \sim R_g^2 / D \tag{1.3.2}$$

From the classical work of Rouse, we find that for both dilute chains (non-entangled) and for melt chains with molecular weights less than the critical entanglement molecular weight, M_c, we have the molecular weight dependence of the self-diffusion coefficient, D_{RO}, as

$$D_{RO} \sim M^{-1} \tag{1.3.3}$$

Chains with Rouse dynamics therefore have a welding time given by

$$t_w(\text{Rouse}) \sim M^2 \tag{1.3.4}$$

For highly entangled polymer melts with M greater than M_c, the diffusion coefficient is given by de Gennes [6] and Doi–Edwards [7] as

$$D_r \sim M^{-2} \tag{1.3.5}$$

and the welding time can be estimated as

$$t_w(\text{reptation}) \sim M^3 \tag{1.3.6}$$

Thus, small differences in molecular weight can result in large differences in the time to achieve optimal weld conditions.

The time dependence of welding can be deduced from the dynamics of the chains. The short-time dynamic properties involving distances less than or equal to R_g are particularly important since the weld is complete when the molecules have diffused a distance equal to the radius of gyration. The time dependence cannot be readily determined from properties at distances greater than or equal to R_g. The short-time dynamic properties of polymer chains with both Rouse and reptation dynamics are dominated by correlated motion of monomers due to the connectivity of monomers in the chain. The correlated motion effects on the properties disappear at times greater than the characteristic relaxation time of the chains. For example, the center-of-mass motion described by Eq 1.3.1 above is valid for $<X>^2 \geq R_g^2$, but is not valid for interfaces at distances less than or equal to R_g. It is valid for all times, and distances in the melt far from the surface. The fact that the diffusion relation is valid at a distance equal to R_g for interfaces allows us to determine t_w, as outlined above, but cannot be used to calculate the time dependence of welding directly. Thus, polymer welding provides a new method of exploring correlated motion effects on the dynamic relations, with the important assumption that the relation between structure and strength is known. In a later section, the details of the dynamic properties and their influence on the structure and strength development of polymer interfaces will be investigated.

1.3.2 Defining the Problem

The problem of evaluating the fracture energy, G_{1c}, of polymer interfaces is represented in Figure 1.3. Material A is brought into contact with material B to form an A/B interface; the weld is fractured, and the strength is related to the structure of the interface through microscopic deformation mechanisms. Typically, a crack propagates through the interface region, preceded by a deformation zone at the crack tip.

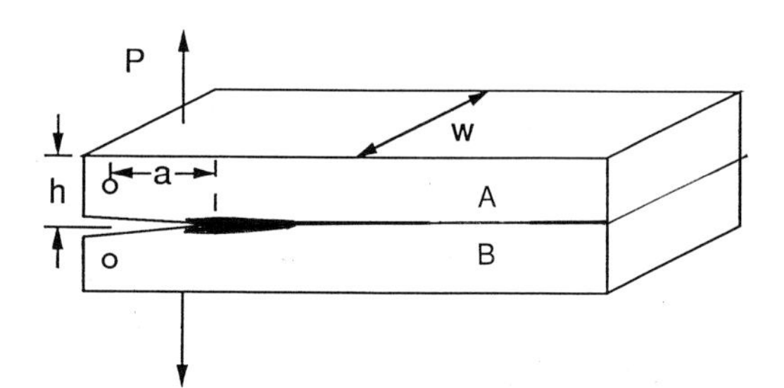

Figure 1.3 A crack with a deformation zone at its tip propagates through an interface formed from materials A and B.

The development of the relation between interface structure and strength proceeds as follows:

Step 1

The time-dependent structure of the interface is determined. Relevant properties may be characterized by a general function $H(t)$, which for the case of polymer melts can usually be described in terms of the static and dynamic properties of the polymer chains. For example, with symmetric (A = B) amorphous melt interfaces, $H(t)$ describes the average molecular properties developed at the interface by the interdiffusion of random-coil chains as [10]

$$H(t) = H_\infty (t/\tau)^{r/4} \tag{1.3.7}$$

where

$$H_\infty \sim M^{(3r-s)/4} \tag{1.3.8}$$

in which τ is a characteristic relaxation time, H_∞ is the equilibrium value of the property $H(t)$ at $t = \tau$, and r and s are integers whose value (1, 2, or 3) depends on the specific molecular property. $H(t)$ can be measured from the concentration depth profile $C(x,t)$ for symmetric interfaces; it could be a measure of the extent of diffusion across the interface, for example, the average monomer interpenetration distance or the number of chains crossing the interface plane (Chapter 2). With incompatible amorphous interfaces, limited interdiffusion occurs up to an equilibrium depth d_∞, which becomes the important descriptor of the interface structure (Chapter 9).

In the case of polymer–metal interfaces, the function $H(t)$ could be a measure of the interface roughness, diffusion depth, fractal properties, etc. For example, when metal atoms are vapor deposited on polymers during the fabrication of a material for an electronics application, a diffuse polymer–metal interface can form which is highly ramified [10]. The time-dependent roughness, $N_f(t)$, of the interface depends on the average diffusion depth $<X(t)>$, according to [11]

$$N_f(t) \sim <X(t)>^{1-1/D_f} \tag{1.3.9}$$

where D_f is the fractal dimension of the interface (Chapter 4). Thus, for each interface we examine the structure in terms of relevant descriptors that could affect the mechanical, thermal, optical, or electrical properties.

The interface properties can usually be independently measured by a number of spectroscopic and surface analysis techniques such as secondary ion mass spectroscopy (SIMS) (Chapter 5), X-ray photoelectron spectroscopy (XPS) (Chapter 9), specular neutron reflection (SNR) (Chapter 6), forward recoil spectroscopy (FRES), scanning (SEM) and transmission electron microscopy (TEM), infrared (IR), and several other methods. Theoretical and computer simulation methods can also be used to evaluate $H(t)$ (Chapters 2 and 3). Thus, we assume for each interface that we have the ability to measure $H(t)$ at different welding times and that the function is well defined.

Step 2

After a contact time t, the material is fractured or fatigued and the mechanical properties, $G(t)$, are determined. The measured properties are functions of the test configuration, rate of testing, temperature, etc., and include the *critical strain energy release rate* G_{1c}, the *critical stress intensity factor* K_{1c}, the *critical crack-opening displacement* δ_c, the *critical fracture stress* σ_c, the *fatigue crack propagation rate* da/dN (the incremental increase in crack length a per fatigue cycle), and others. Related properties can be measured when other modes of fracture (torsion and shear) are used.

Step 3

The fracture properties $G(t)$ are related to the interface structure $H(t)$ through suitable deformation mechanisms deduced from the micromechanics of fracture. This is the most difficult part of the problem, but the analysis of the fracture process *in situ* can lead to valuable information on the microscopic deformation mechanisms. SEM, optical, and XPS analysis of the fractured interface usually determine the mode of fracture (cohesive, adhesive, or mixed) and details of the fracture micromechanics. However, considerable modeling may be required with entanglement and chain fracture mechanisms to realize useful solutions (Chapter 7). We then obtain a solution to the problem,

$$G(t) = f[H(t)] \tag{1.3.10}$$

where f is a known function of $H(t)$ at constant temperature and pressure. In the simplest case, $G(t)$ is proportional to $H(t)$, as found for many polymer welding problems [10].

1.3.3 Structure and Strength of Interfaces

When the interrelation of the structure and strength of the interface is determined, several unique predictions for the mechanical properties can usually be made. In the case of polymer melts, the strength predictions are made in terms of the static and dynamic scaling laws of the polymer chains. Thus, we can investigate scaling laws for fracture energy G_{1c}, fatigue crack propagation rate da/dN, etc., as functions of time and molecular weight. For example, if $G_{1c} \sim H(t)$, where $H(t)$ is given by Eq 1.3.7, then we expect solutions of the form

$$G_{1c}(t) \sim t^{r/4} M^{-s/4} \tag{1.3.11}$$

$$G_{1c}(\tau) \sim M^{(3r-s)/4} \tag{1.3.12}$$

The fracture energy is then predicted to have a precise time dependence, a molecular weight dependence of the rate of welding, and a molecular weight dependence of the virgin or fully welded state. Experimental determination of the constants r and s should also indicate which of the molecular properties is important in controlling the fracture energy. We have argued for the case of interface fracture dominated by chain disentanglement that $r = s = 2$ [10]. This suggests that the average interpenetration chain length plays a major role in controlling the strength of polymer melt interfaces. G_{1c} is related to $H(t)$ by

$$G_{1c}/G_{\infty} = [H(t)/H_{\infty}]^{2/r} \tag{1.3.13}$$

This is a very interesting problem, discussed also by Kausch [12], Prager and Tirrell [13], and de Gennes [14]. Many other investigators have studied the diffuse interface in polymer melts; their contributions will be presented in later chapters.

The understanding of the interrelation of structure and strength from such a fundamental vantage point permits us to tailor the properties of a wide variety of interfaces. From a practical point of view, the solution to this problem also allows us to optimize the strength of specific interfaces and design complex material systems with greater reliability.

The problem of the relation between structure and strength of polymer interfaces is very rich if one considers the multitude of symmetric and asymmetric A/B interface pairs that can be formed with different chemical compositions, molecular weights, molecular weight distributions, and molecular architectures, in addition to numerous polymer–nonpolymer interfaces. The different types of interfaces are presented in the next section.

1.4 Symmetric Polymer–Polymer Interfaces

1.4.1 Classification of Polymer Interfaces

Polymer interfaces can be categorized in four broad groups as follows:

1. Symmetric polymer–polymer interfaces
2. Asymmetric polymer–polymer interfaces
3. Polymer–nonpolymer interfaces
4. Multicomponent polymer interfaces

The different classes within each group are listed in Table 1.1.

The symmetric interfaces are most commonly encountered in high-volume applications with melt processing of plastics. Manufacturing methods involving extrusion, injection molding, compression molding, powder and pellet sintering, welding, and lamination all involve amorphous polymer–polymer interfaces.

When a melt is cooled, the interface of polymer glasses (PS, PMMA, PVC, PSAN, PC, etc.) or rubbers (PB, PI) may remain in the amorphous state; that of semicrystalline materials (PE, PP, PET, PA, etc.) may crystallize. The crystallization process can significantly change the structure that existed in the melt.

1.4.2 Comment on Physical Aging

For glassy polymers, the interface can be quenched, essentially retaining the same structure as in the melt. However, subtle effects on the interface structure (such as physical aging) may need to be considered [16]. Physical aging of glasses involves relatively small changes in both the free volume [16] and internal energy [17] of the chains, at temperatures below the glass transition temperature, T_g. However, minor

Table 1.1 Polymer Interfaces

1. Symmetric (A/A) polymer–polymer interfaces
amorphous materials
crystalline materials
liquid crystals
2. Asymmetric polymer–polymer interfaces
amorphous structurally asymmetric materials
amorphous dissimilar compatible materials
amorphous dissimilar incompatible materials
crystalline similar materials
crystalline dissimilar compatible materials
crystalline dissimilar incompatible materials
3. Asymmetric polymer–nonpolymer interfaces
polymer–metal interfaces
polymer–ceramic interfaces
polymer–biological interfaces
4. Multicomponent polymer interfaces
blends
compatibilizers
interactive components
noninteractive components

changes in the volume can have dramatic effects on the mechanical properties of the glass. Typically, as the glass ages, it becomes more brittle, stress-relaxes more slowly, creeps at a lower rate, and changes its volume in a complex manner dependent on the thermal and pressure history. These property changes have in common that the molecular mobility is decreasing with physical aging. These effects are not unique to polymers but occur in all glassy materials. A decrease in free volume provides a convenient first explanation for a decrease in mobility. Free volume change alone cannot explain the physical aging process, but it is a useful intuitive tool to gain a foothold on this complex topic.

Investigators of interfaces in glassy materials should enquire how the physical aging process in the interface behaves compared to that of the bulk material. It is not obvious that the two aging processes should be the same. In particular, molecular weight effects should be considered. If the interface ages more rapidly than material in the bulk, it becomes more brittle and the likelihood of fracture increases. If the interface ages more slowly, or can be tailored to age more slowly by molecular engineering or by additives, then this could be a very favorable situation for the stability of the material.

1.4.3 Interfaces in Liquid Crystals

Liquid crystal polymers (LCP) form symmetric interfaces that can have complex structures depending on the thermal and flow history. They exist in several phases, smectic, nematic, amorphous, and crystalline. Because of their ability to form ordered regions in the melt, orientation effects at LCP interfaces are unusually important, if not unique to these materials. The axes of the chains tend to align parallel with surfaces, and this causes considerable difficulty with weld-line strength. When the chains behave as rigid rods, thermal motion causes them to diffuse parallel to, rather than perpendicular to, the interface. Consequently, very little interdiffusion occurs and the weld remains weak in general. This problem can be solved by designing LCP chains that align perpendicular to the surface and promote facile interdiffusion. The relation between structure and strength for the latter case is described in Chapter 2.

1.4.4 Orientation Effects

Orientation effects are also important for the more flexible polymer melts discussed above, especially when internal weld lines are involved. For example, when a polymer melt flows into a mold, the leading edge of the flow develops a biaxial orientation, known as the *fountain effect*. In a double- or multi-gated mold, the internal weld lines involve the coalescence of oriented fluid layers followed by diffusion of chains across the plane of contact. The surface orientation has to be considered in the analysis of the development of time-dependent interface structure. The extent to which the oriented layer perturbs the diffusion process is a complex problem. In rubbery materials, small degrees of orientation are known to result in considerable loss of tack or stickiness [15]. Thus, the local orientation in the interface may influence the entanglement structure as well as the rate of interdiffusion. The fountain effect at weld lines is particularly deleterious with fiber-filled resins, since the fibers orient themselves parallel to the weld line.

As molecules become oriented and lose their random-coil configuration, their ability to participate in entanglements with other chains decreases. An exact quantitative analysis of this phenomenon is lacking, primarily due to the absence of a suitable model for polymer entanglements. It is perhaps constructive to recall the maximum draw ratio, λ, required to fully extend a random-coil chain with radius of gyration $R_g \sim M^{1/2}$, to a length $L \sim M$, as

$$\lambda \sim L/R_g \tag{1.4.1}$$

and

$$\lambda \sim M^{1/2} \tag{1.4.2}$$

When the chain is fully extended it is no longer considered to be entangled and can be readily extracted from the melt with minimum perturbation of the other chains. Thus, orientation of chains in a polymer interface should be considered to have deleterious effects on the mechanical properties.

Conversely, the path to fracture could involve a molecular orientation step that would readily lead to chain separation via disentanglement. This argument is applicable to the formation and disintegration of the oriented fibrillar material in the deformation zone ahead of an advancing crack; an example is the advance of a crack through the fibrillar material of a craze in glassy polymers. We use such arguments later to relate interface structure to strength via disentanglement models (Chapters 7 and 8).

1.4.5 Polydispersity and Chain Ends

Many interesting details remain to be addressed about the molecular properties of symmetric amorphous interfaces. Consider the amorphous interface in Figure 1.4. The following features influence both the structure and the strength of the interface:

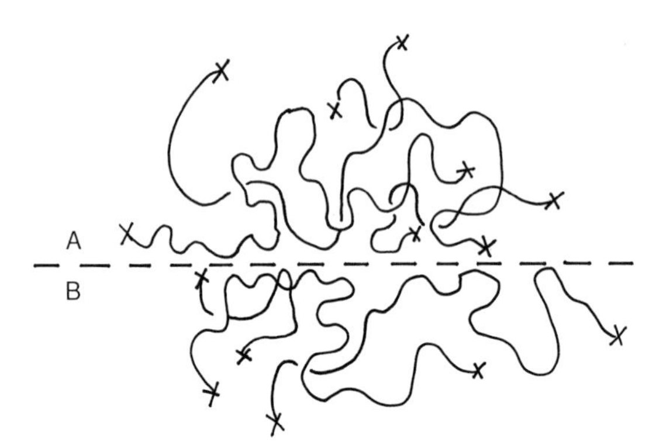

Figure 1.4 Polymer chains at an A/B interface.

Polydispersity

Polymer melts, particularly commercial plastics, are rarely monodisperse in molecular weight, and the spatial distribution of the molecular weight distribution, $\phi(M,x)$, with respect to the distance x from the interface should be considered. Since the rate of diffusion in polymer melts has a strong molecular weight dependence ($D \sim M^{-2}$) [6, 7, 18], we need to be concerned with whether the surface layer of a polymer slab is rich in high- or low-molecular-weight species. The thickness of the affected layer may be only of the order of the radius of gyration of the number average molecular weight (M_n) species, that is, a few hundred angstroms. However, this is the layer that dominates the strength of the interface. One can argue that the large chains suffer a severe conformational entropy penalty due to reflecting boundary conditions by being placed at the surface, which favors a layer rich in low-molecular-weight chains. On the other hand, segregating the low-molecular-weight chains on the surface also incurs a penalty in the free energy of mixing. Which factor controls depends on the molecular weight and the chemical nature of the species, but, as a first guess, we expect the low-molecular-weight species to segregate to the surface.

Chain Ends

For a given spatial distribution of molecular weights in a polydisperse melt, the location of the chain ends is of importance. If the chain ends are preferentially on the surface, the rate of interdiffusion is greater than if they are randomly distributed. The segregation of the chain ends on the surface is determined by (a) the loss of entropy due to ordering the chain ends on the surface and (b) the effect of the chemical structure of the chain end on the surface tension. The latter effect can be very strong and dominate the rather weak entropy effect. This fact could be used to advantage in the chemical design of surfaces and interfaces. In Chapter 5, we show that the entropic contribution to surface enrichment by chain ends is essentially negligible for linear polymers with M greater than M_c.

Using Langmuir's principle of independent surface action (discussed in reference [20]), one can make an estimate based on chemical interactions as to whether the chain end prefers to reside on the surface. For example, one may ask whether the ethanol molecule, CH_3–CH_2–OH, is oriented with the –CH_3 or the –OH group preferentially on the air/liquid surface at ambient conditions. The surface comprising mainly –OH groups would be expected to have a surface energy, Γ, of about 57 erg/cm^2, extrapolated from half a water molecule (with Γ of 72.9 erg/cm^2); the surface composed of the organic CH_3–CH_2– group might be expected to resemble that of another organic molecule such as octane (with Γ of 21.8 erg/cm^2). The surface tension of ethanol is found to be 22.75 erg/cm^2, which suggests that the orientation with the organic group on the surface is favored. Therefore, with hydroxy-terminated polybutadiene elastomers, one can expect that the chain ends do not sit on the surface but diffuse into the interior, to reduce the surface tension. If the chain ends are moved to the surface by mechanical action, the history of the surface becomes important with respect to the subsequent rate of diffusion at a joined interface. This problem occurs, for example, in crack healing of elastomeric materials (Chapter 12), which fracture primarily by a disentanglement process, leaving the chain ends on the surface [19]. If the surfaces are immediately joined the healing rate is faster compared to the healing rate for aged surfaces, where the chain ends have diffused away from the surface (Chapter 11). Langmuir's principle is considered primitive by many but provides an excellent first guess as to the location of the chain ends. Useful discussions of the chain end problem have been presented by Adamson [20] and others [21].

1.4.6 Surface Ordering

The restricted configurations of the polymer chain segments near a surface can induce ordering of segments compared to the bulk. To investigate subtle changes in local conformations induced by the additional surface kinks on the chain would require details of the balance of the conformational energy function. This effect could be more important in asymmetric interfaces where the chemical potential of the molecules on one side could further induce ordering on the other. This is especially true for polymer–metal and polymer–ceramic interfaces, where the effect would be considered in terms of adsorption.

Wattenbarger *et al.* [22] used a computer to simulate the conformations of small polymer molecules (the degree of polymerization, N, was 16) near an interface. They found for noninteracting chains that (a) the number of available conformations decreases with distance from the surface as the entropy of the chains decreases; (b) the radius of gyration, R_g, becomes anisotropic as the normal random-coil structure becomes influenced by the presence of the surface (the parallel radius becomes larger while the perpendicular radius becomes much smaller near the surface); (c) the conformational energy loss required to bring the center of mass of the chain from the bulk to the surface is about 7 kT (k is Boltzmann's constant); and (d) steric constraints imposed by the interface eliminate some conformations but favor other more compact conformations such as helices and sheets. Their main conclusion is to predict that the surface or interface induces and enhances internal structure in polymers. With strong interactions of the chain with the surface, the chain can adsorb on the surface and assume an essentially flat configuration with significant internal structure. The subject of polymer adsorption at surfaces and interfaces has been reviewed [23, 24].

1.5 Crystallization at Interfaces

When a polymer interface is cooled to temperatures T_c below the melting point, T_m, crystallization commences, as with PP, PE, PET, etc. Crystallization occurs via homogeneous or heterogeneous nucleation, followed by growth of crystalline lamellae, which involves a secondary nucleation process. The growth rate and thickness L of the lamellae are strong functions of the degree of supercooling ($\Delta T = T_m - T_c$). The method of incorporating new chains on the growing crystal surface via secondary nucleation is affected by the growth rate and thickness of the lamellae as well as by the disposition of the amorphous component between the lamellae. The lamellae, with dimensions, L, on the order of 100 Å, are the building blocks for the more complex spherulitic microstructures, which typically are several micrometers in diameter, depending on their number per unit volume.

We may enquire how the crystallization processes in the vicinity of the interface affect interface structure and strength. Several points are of interest, especially for partially diffused interfaces (interfaces with average monomer interdiffusion distance $X(t) < X_\infty$, where $X_\infty \approx R_g$). Let us consider two melt pieces whose surfaces have been equilibrated in air (without oxidation), and then brought into good contact for a time t that is less than the relaxation time, τ, and rapidly cooled below T_m to T_c.

1.5.1 Homogeneous Nucleation

In the absence of heterogeneous nucleation particles (impurities or additives), homogeneous nucleation begins. Chain segments in the melt aggregate as they experience favorable intermolecular interactions (with driving force Δf) leading to crystallization. The thermal energy (kT), local fluctuations, and the surface energy (σ_e) of the newly formed crystals all oppose the aggregation or clustering process. One obtains an equilibrium population of subcritical-sized embryos that are constantly forming and decaying. This amounts to a partial ordering in the melt and represents an interesting competition between segmental Rouse dynamics and intramolecular conformational ordering.

With the nucleation theory developed by Turnbull and Fisher for general phase transitions [25], Hoffman and Lauritzen examined crystal nucleation and growth in polymers [26, 28]. An embryo crystal of critical size has the length dimension L^* (Figure 1.5):

$$L^* = 4\sigma_e/\Delta f \tag{1.5.1}$$

where the driving force is determined by

$$\Delta f = \Delta H \Delta T / T_m^0 \tag{1.5.2}$$

in which ΔH is the heat of fusion per unit volume of crystal, ΔT is the degree of supercooling, and $T_m{}^0$ is the melting point of an infinitely large crystal. Then we have

$$L^* = 4\sigma_e T_m^0/(\Delta H \, \Delta T) \tag{1.5.3}$$

At low supercooling, the critical dimensions are large but the rate of crystal growth is small and approaches zero at T_m. As ΔT increases, the rate of crystallization increases and the critical dimensions decrease, typically to about 50–100 Å.

For polymers, the critical embryo is considered to be cylindrical, with a volume v^*:

$$v^* = \pi L^* a^{*2}/4 \tag{1.5.4}$$

where a^* is the diameter. The surface containing the chain stems and folds has a surface energy σ_e, which is typically ten times larger than the side surface energy σ_s. By a derivation similar to that for L^*, the critical diameter is

$$a^* = 4\sigma_s T_m^0/(\pi \Delta H \Delta T) \tag{1.5.6}$$

and it follows that the critical volume is

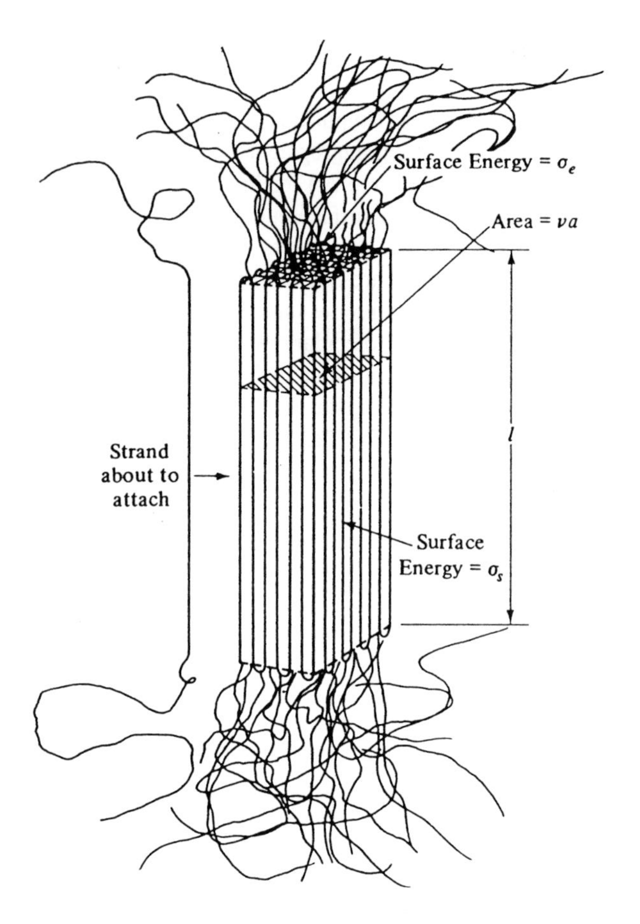

Figure 1.5 Critical primary polymer nucleus (courtesy of J. M. Schultz [50]).

$$v^* = 16\sigma_s^2\sigma_e/(\pi\,\Delta f^3) \tag{1.5.5}$$

The relevance of the embryonic dimensions to the interface structure can be deduced from the following problem.

Problem

Polyethylene with a molecular weight, M, of 10^5 is being injection-molded into a double-gated cold mold. An internal weld line forms at a local temperature of 122 °C.

(a) *Determine the dimensions L^* and a^* of the critical embryo in the weld interface.*

(b) *How does the critical length L^* compare with the radius of gyration R_g of the chains in the melt? The necessary data for polyethylene are given below.*

(a) Critical Dimensions:

Data *[from Table 3, J. D. Hoffman, Polymer,* ***Vol. 23****, 656 (1982)].*

$\sigma_e = 90\ erg/cm^2$ *(fold surface)*

$\sigma_s = 11.4\ erg/cm^2$ *(side surface)*

$\Delta H = 2.8 \times 10^9\ erg/cm^3$ *(heat of fusion)*

$T_m^0 = 145\ °C = 418.2\ K$

$\Delta T = 145 - 122 = 23\ K$

Solution: *We substitute the given values into Eq 1.5.3:*

$$L^* = \frac{4\,(90\,erg/cm^2)\,(418.2\,K)\,(10^8\,Å/cm)}{(2.8 \times 10^9\,erg/cm^3)\,(23\,K)}$$

$= 234\ Å$ *(critical length)*

The ratio $L^*/a^* = \pi\ \sigma_e/\sigma_s = 24.8$ *can be derived from Eqs 1.5.3 and 1.5.6.*

Therefore, $a^* = 9.4\ Å$ *(critical diameter)*

and $v^* = 16{,}241\ Å^3$ *(critical volume).*

(b) Radius of Gyration:

Data:

$C_\infty = 6.7$ *(characteristic ratio)*

$M_0 = 28\ g/g\ mol$ *(monomer molecular weight)*

$b_0 = 1.54\ Å$ *(C–C bond length)*

$j = 2$ *(C–C bonds per monomer)*

$M = 10^5$

Solution: *The radius of gyration is determined by Eq 1.2.1:*

$R_g = [C_\infty Mj/(6\ M_0)]^{1/2} b_0$

$= 137\ Å.$

The critical embryo dimension L^* (234 Å) is of the same magnitude as the optimal interface thickness ($X_\infty \approx R_g \approx 137$ Å). As a consequence of the formation of liquid-crystal–like embryos, considerable disentanglement of chain segments from the equilibrated melt structure occurs. The subsequent growth of crystals near the interface can have positive and negative effects on the strength of the interface.

There are several interesting differences between crystallization in the bulk and at interfaces. The chains near the surface have a reduced entropy and a more highly ordered state, with potentially partially oriented regions. Therefore, at times immediately after contact before extensive diffusion has occurred, the surface layer of chains should have a higher probability for the formation of critical-sized embryos. In effect, the surface layer acts as pre-embryos for the normal bulk homogeneous nucleation mechanism. Also, the orientation of the embryos on the surface is such that their principal director or *c*-axis orientation direction lies parallel to the surface. Crystal growth on both sides of the interface with this orientation would result in an interface region with poor mechanical properties. The oriented embryo problem is further aggravated by orientation due to melt flow in a mold. In the above problem, the impinging melt fronts forming the weld line in a double-gated (two injection points) mold have the chains in a biaxially oriented state.

The orientation of chains at a surface is of paramount importance for liquid crystal interfaces where, for example, the parallel alignment of two smectic phases, one on each side of the interface, would result in very little entanglement and, consequently, a very weak interface. Liquid crystals with smectic and nematic phases are in many respects like stable embryos and could provide useful models for the transient crystallization problem and the structure of interfaces.

The preferred crystallization of chains on the melt surface layer is well known for asymmetric interfaces, particularly polymer–metal and polymer–ceramic interfaces. However, this may be considered as heterogeneous nucleation, usually involving highly aligned nuclei. The high concentration of aligned nuclei (such as on a fiber surface) can result in simultaneous growth into the melt to produce an epitaxial layer with a microstructure considerably different from that in the bulk. The extent to which this happens for symmetric interfaces is expected to be magnified at short contact times before the diffusion process destroys most of the ordered surface layer. As the interface broadens, the orientation of nuclei should become progressively more random as the virgin state is approached.

1.5.2 Heterogeneous Nucleation

Heterogeneous nucleation involves the formation of crystals from a foreign substrate. Most commercial thermoplastics contain high levels of impurities, in addition to nucleating agents. The number of such foreign nuclei controls the number of spherulites per unit volume and hence their size. For polymers as well as metals, better mechanical properties are usually obtained with fine-grained microstructures. For metals, which are good heat conductors, the rate of nucleation is enhanced by rapid solidification processing with deep quench methods; for polymers such as PE, which are poor heat conductors, relatively high concentrations of additives are used to promote nucleation and rapid solidification.

At a symmetric polymer–polymer interface, the spatial distribution of nucleating agents is an interesting problem and depends on the history of the melt. Before the interface was formed in the melt, the polymer may have been crystallized several times as a normal part of the processing and pelletizing operations. During crystallization, many of the foreign particles that are not participating in nucleation can be zone refined, or exuded to the boundaries of the local crystallization front. These impurities may collect in pockets that could lie in the interstices of the packed spherulitic structure. Their redistribution in the melt depends on the melt flow and mixing history. Compression molding and sintering of powder and pellet resin should produce more predictable distributions than injection molding or coextrusion, for example.

We find that the distribution of nuclei near the interface affects both the structure and the strength [29] (Chapter 10). However, the influence of the crystallization processes at an interface should be viewed more in terms of first-order transitional phenomena than of changes in entanglement density. This important feature is discussed below. Before we leave this topic, one more point is worth noting: nucleation agents typically are not designed with interfaces in mind. It should be possible to design additives with properties optimally suited to strength development at interfaces. For example, this could involve a modification to the chemical structure, which for steric or polar reasons would allow preferential segregation of the nucleating agent to, or away from, the weld line.

1.5.3 Crystal Growth

The subject of crystal growth at an interface [30] is very rich in that it requires information of the mechanistic details at a very intimate level, currently bordering on the edge of our knowledge. In fact, most studies of interfaces are demanding of considerable skill, largely due to the small-scale nature of the parameters of the problem ($X \leq R_g$). While debate continues on the exact mechanism of crystallization in the bulk [35], relatively little is known about surface crystallization mechanisms. Although surface effects are likely to be quite specific, we tend to seek solutions to polymer interface problems that are logical extrapolations of bulk phenomena. This may not always be valid but represents an excellent starting point until more information becomes available.

A useful approach to crystal growth in the bulk is that presented by Hoffman [28]. He considers the growth of crystals in terms of three regimes, which are defined with respect to the magnitude of the degree of supercooling, ΔT, and the associated microstructures. Two of the three regimes are illustrated in Figure 1.6.

Regime I

Regime I occurs at small ΔT. The lamellar growth vector, G_I, is given by

$$G_I = b\,i\,L_0 \tag{1.5.7}$$

where b is the width of the chain deposited (about 4 Å for PE), L_0 is the width of the growing crystal front (about 1 μm) and i is the secondary nucleation rate. The rate-controlling step is the deposition of a new chain segment of length L^* to the completed crystal surface. The lamellar thickness in secondary nucleation is predicted to be half that for primary nucleation as given by Eq 1.5.1. The addition of a new chain segment on the completed crystal surface involves the creation of new surface area, the energetics of which controls the secondary nucleation rate, i. After the first chain is deposited, the lateral deposition rate is rapid because the completion of the substrate layer involves no additional surface area per chain segment.

The rapid completion of the crystal surface advances the crystal by a distance b. The growth rate, G_I, is proportional to the substrate length, L_0, because the secondary nucleation rate is defined with respect to the surface area of the crystal front (number

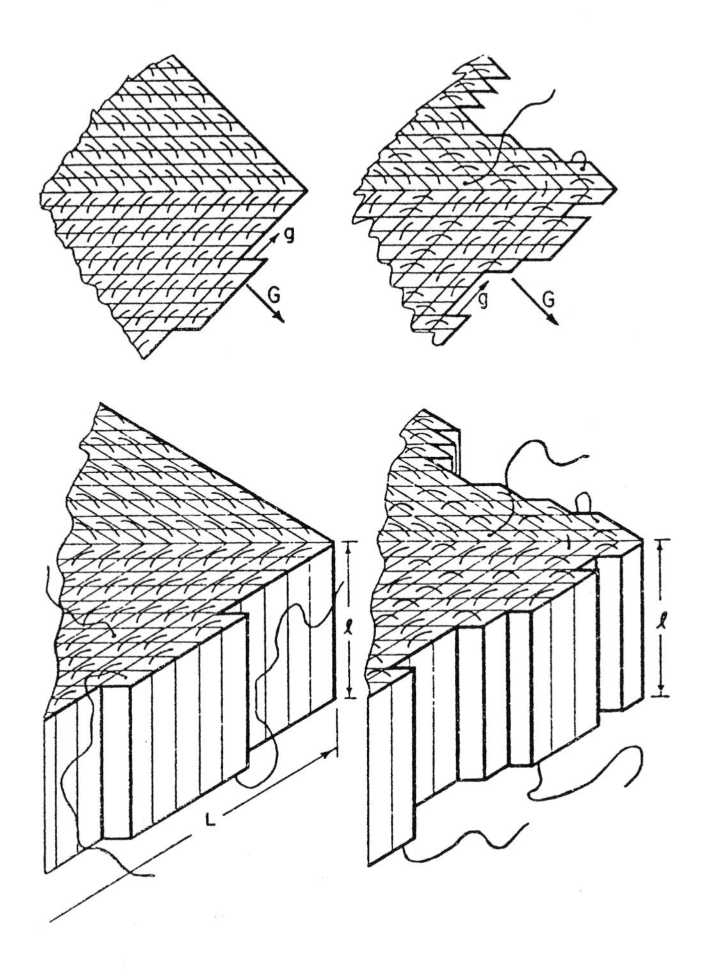

Figure 1.6 Growth front morphology for Regime I and Regime II growth (schematic). Regime I (diagrams at left): single nucleus forms on surface, rapidly completes new layer; folds are all parallel to edge of crystal. Regime II (diagrams at right): many new surface nuclei form before previous layer is complete, leading to reentrant or crenellated growth front; some folds are now parallel to direction of overall growth. (Courtesy of J. D. Hoffman).

of chain stems per unit area). The lateral completion rate is much faster than the nucleation rate, i. The crystal microstructure developing during Regime I is considered to be highly regular, with few crystal defects and a high degree of adjacent reentry of chain stems from the same chain. Single crystals, for example, are considered to form by this mechanism [28]. Adjacent reentry results in chain folding, with a hairpin bend and change of direction, as depicted in Figure 1.6. The growth face in polyethylene is the <110> plane of the orthorhombic unit cell.

The rate of crystal growth in Regime I is much slower than the interdiffusion rate at an interface. Therefore, the interface is expected to reach equilibrium before the

Regime I growth process advances the crystal by a distance R_g. The "reeling in" of chains from the melt as they align themselves on the crystal surface involves a disentanglement process where the dynamics of the chain causes frictional forces that compete with the driving force for crystallization. If the chain were very long, the crystallization driving force might not be sufficient to drag the chain (overcome the frictional force) through the melt and deposit it on the crystal surface.

If primary nucleation and crystal growth occur at the interface plane, then chains may be pulled across the interface and provide a strengthening mechanism at early contact times. Otherwise, Regime I kinetics primarily involves the fully healed interface. At small degrees of supercooling, homogeneous nucleation is particularly influenced by the surface orientation, which could result in the initial growth vector G_I being normal to the interface contact plane.

Regime II

The transition from Regime I to Regime II occurs in polyethylene at a supercooling ΔT of about 16 °C. Regime II crystal growth occurs when multiple nucleation events involve the deposition of many chain segments on a crystal growth surface. This can occur at increasing supercooling, where the energy barrier for secondary nucleation decreases. In this case, the crystal surface is very rough or crenellated, as shown in Figure 1.6, and the growth rate is given by

$$G_{II} = b(iL_0)^{1/2} \tag{1.5.8}$$

The crossover in growth kinetics from Regime I to II has been observed by Hoffman.

Due to multiple nucleation events involving the deposition of chain segments from several different chains, the crystal structure is not as well ordered as in Regime I. Again, the maximum growth rate is limited by the ability of the chain to pull itself through the melt onto the crystal surface. In a fully healed interface, the crystal growth is the same as in the bulk. With increasing molecular weight, the growth rate exceeds the relaxation rate of the chains and the crystal can propagate through the partially healed interface. In this case the propagating crystal structure is expected to be altered by the deposition of a chain that has a lower entropy due to reflection at the surface. The effect of such growth through nonequilibrated melts at interfaces on the mechanical properties is an unsolved problem.

Regime III

Most polymer melt processing occurs at high degrees of supercooling (ΔT is greater than 20 °C) in Regime III, which results in the characteristic formation of spherulites. Spherulites consist of lamellar ribbons with noncrystallographic branching emanating radially from a central nucleus to form a sphere-like object of uniform density. Because the ribbons are optically anisotropic, the spherulites typically exhibit maltese crosses and other patterns when viewed between the crossed polars of an optical microscope. In the case of PE, which has an orthorhombic unit cell, the ribbons are *b*-axis–oriented and sometimes twist along their propagation direction. Growth microstructures have been considered by Keith and Padden [31, 32] and Bassett [33].

The kinetics of crystal growth in Regime III involves a cascade of secondary nuclei on the crystal surface, with lateral propagation being small. In this case, the secondary nucleation rate, i, is of the same order of magnitude, or greater than, the lateral completion rate. The growth rate is given by

$$G_{III} = b\,i\,L' \tag{1.5.9}$$

where L' is the effective substrate length [28]. Rapid deposition of chains on a growing crystal creates a highly disordered crystal that may have few regular chain folds.

The microstructure of the chains in quenched PE was investigated by Flory and Yoon using neutron scattering on deuterated PE in the melt and in the crystalline state [34]. They made the interesting discovery that the radius of gyration of the deuterated PE chain in the melt was approximately the same as in the semicrystalline state, R_{gc}. They argued that the chain was enveloped by the rapidly growing crystal in a time that was far less than the conformational relaxation time of the chains in the melt. Therefore, the chain effectively condensed in place on the growing crystal along with the other chains, thus the equality of R_g and R_{gc}. The resulting crystalline microstructure appears to preclude the possibility of a regular chain-folded lamellar structure, which has existed as the conceptual cornerstone of polymer morphology for many years. Regime III proposed by Hoffman (only Regimes I and II existed at the time of the Flory–Yoon study) seems to be a compromise between the 100% and 0% chain-folded perspectives; Regime III allows for varying percentages of chain-folded structure. The exact amount remains as a problem whose solution depends on molecular weight, crystallization temperature, pressure, and annealing treatments.

Future studies should elucidate the microstructure of semicrystalline polymers, and, in particular, address attendant issues relative to the chain-folding percentage, such as the constitution and connectivity of the intervening amorphous layer known to exist between crystalline lamellae. We need to know more about the fraction of loops (loose chain folds), bridges or "tie-chains" connecting crystalline lamellae, cilia (long chain ends emanating from one lamellar surface), and unattached chains. We expect the details of the connectivity between lamellae to strongly influence the mechanical properties, especially the fracture properties. Many models have been proposed for the deformation of semicrystalline polymers as affected by morphology, constitution, molecular orientation of both the amorphous and crystalline phases, and deformation conditions. Kausch has presented comprehensive reviews [12, 36].

Consider the propagation of a crack through a deformation zone at an A/B interface, as in Figure 1.3. For present purposes we assume that the amorphous phase ideally straddles the interface plane, with a crystal lamella on each side of the interface. Let us assume that the stress is acting normal to the interface plane. The deformation zone initially involves the softer amorphous phase, which elongates by several hundred percent to an extent dependent on the rate and entanglement structure. We can expect strain hardening of the amorphous phase to involve extension of the slack between entanglement points. With continued straining, chain slippage and bond rupture occur in the oriented amorphous layer as preludes to fracture. The mechanical resistance to fracture now depends sensitively on the constitution of the amorphous phase and details of the individual components, such as the length distribution of cilia and of the tie-chains between lamellae. Additional straining results in a breakdown of the amorphous strained layer, with the prospect for additional damage in the crystalline lamellae due to chain pullout, lamellar unravelling, etc. The latter damage may propagate into adjoining regions, resulting in very large crack-opening displacements. At semicrystalline interfaces, the amorphous phase may have a constitution that depends on the state of interdiffusion existing at the time crystallization occurred.

The mechanics of semicrystalline deformation processes are complex (reviewed in [12]). In fact, relatively few solutions to this problem have been attempted, largely due to inadequate knowledge of the initial conditions in the bulk or interface. The problem is further complicated by the participation of many components in the spherulitic microstructure and the nature of the stress field in the deformation zone. In our discussion we have examined issues related to mechanisms and deformation of structures with dimensions of the order of 200 Å. However, the deformation zone in semicrystalline polymers typically involves micrometer scales, and the deformation processes are not isolated in the interface region. We will also see in Chapter 10 that for interfaces the volume contraction accompanying the first-order transition of crystallization of spherulites has a dramatic effect on the large-scale structure of the interface.

1.6 Asymmetric Polymer–Polymer Interfaces

While the number of symmetric A/A interfaces can be counted on one hand, the number and complexity of asymmetric A/B interfaces is far greater, as shown before in Table 1.1.

1.6.1 Amorphous Structurally Asymmetric Materials

This interface is essentially chemically symmetric but involves an A/B couple whose architectural structures are different (Figure 1.7). An important example is that of the amorphous interface of linear polymer chains with different molecular weights, M_i and M_j, and molecular weight distributions. Such interfaces are commonly encountered in coextrusion of different grades of the same polymer, in lamination of composites with varying reactivity ratios in the matrices, in coatings, in the recycling of plastics, and more generally in high-shear-rate extrusion processes with homopolymers, where the residence time affects the molecular weight distributions.

More exotic examples exist when one considers A/B combinations of linear, branched, cyclic, star, comb, cross-linked, and other structures. For example, the melt processing of the PE family (HDPE, LDPE, LLDPE, and UHMWPE) with different

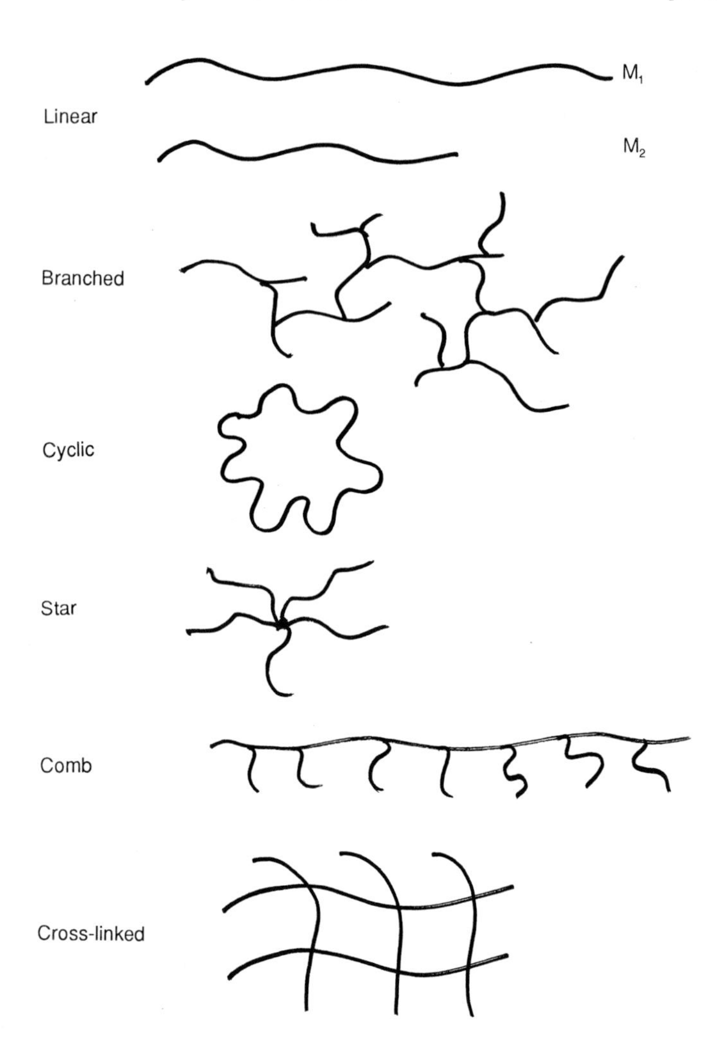

Figure 1.7 Molecular architectures.

molecular weights involves a matrix of four (diagonal) M_1/M_2 interfaces and six (off-diagonal) asymmetric interfaces, as shown in Table 1.2. In general, if we have n different structures within a given family of the same chemical composition, we have n symmetric (A/A) interfaces and $N_{A/B}$ asymmetric (A/B) interfaces, given by

$$N_{A/B} = n(n-1)/2 \tag{1.6.1}$$

One might conjecture that the study of interfaces formed from the same chemical species but different structure represents a simple extension of that of the symmetric

Table 1.2 Interface A/B Pairs for Polyethylene

A/B	HDPE	LDPE	LLDPE	UHMWPE
HDPE	HDPE/HDPE	LDPE/HDPE	LLDPE/HDPE	UHMWPE/HDPE
LDPE		LDPE/LDPE	LLDPE/LDPE	UHMWPE/LDPE
LLDPE			LLDPE/LLDPE	UHMWPE/LLDPE
UHMWPE				UHMWPE/UHMWPE

A/A interface. However, this is not the case, and the structure of these interfaces in terms of the static and dynamic properties of the chains can be exceedingly complex. For example, with a high- and low-molecular-weight couple of linear chains, the flux of molecules across the interface plane is uneven due to the more rapid motion of the low-molecular-weight species. As a result, the interface plane moves (in the direction of the low-molecular-weight side), the high-molecular-weight side swells with the low-molecular-weight species, and the motion of the chains alters with time and position during diffusion. This specific problem is discussed in detail in Chapter 6.

Current technologies provide many practical examples of interfaces with structural asymmetry. Fabrication of composites with epoxies often occurs in a sequential process, with the first laminate, A, more cured (cross-linked) than the second, B. For example, hundreds of laminates may be formed by tape winding in the manufacture of one-foot-thick submarine hulls of fiber-reinforced epoxy. Similar technology is used for rocket motors. If sufficient diffusion does not occur at the A/B interface before the reaction takes place, the composite can be weak. Thus, the diffusion rate must exceed the chemical reaction rate. One can therefore use additives that promote the interdiffusion but do not alter the reaction rate significantly. The fabrication of materials for electronics applications from polyimides (PI) and metallized layers uses a similar process, where a fresh PI layer is placed on an older, more cured PI layer. I could cite many other important examples of interfaces with similar chemistry but different structure, involving conformation, chirality, and metastable phases (in liquid crystals).

1.6.2 Incompatible Amorphous Interfaces

The second most important interface, after the symmetric amorphous, is the incompatible amorphous interface. Most polymer pairs are not miscible to the extent that they can form homogeneous solutions. Immiscible polymers can be forced to form pseudo-homogeneous mixtures by rapid pressure-drop quenching from supercritical solvents, but annealing induces phase separation. Incompatible polymer–polymer interfaces are commonly encountered in semicrystalline polymer blends (PP particles dispersed in a PE matrix), toughened glass (rubber particles in a PMMA matrix), recycled plastics (commingled plastics from municipal solid waste streams), coextruded articles (PSAN in PC), and many other items [37, 38]. The major practical difficulty with incompatible interfaces is that they are very weak compared to the virgin strength of either component. For example, the PMMA/PS interface has a strength of about 40 J/m^2, much lower than the virgin strength (PS) of about 1000 J/m^2 [39, 40]. The strength can be considerably improved with *compatibilizers*, typically diblock, triblock, grafted, or reactively blended copolymers that have miscible groups capable of diffusing into one side and anchoring the other by straddling the interface (Chapter 9).

The cornerstone for the treatment of miscibility in polymer blends and interfaces is the Flory–Huggins thermodynamic theory [1]. For a binary A/B blend, the Helmholtz free energy per monomer, ΔA, of mixing chains with degree of polymerization N_A and N_B and monomer volume fractions ϕ and $(1 - \phi)$, respectively, is

$$\Delta A / kT = [\chi \phi (1-\phi)]_{\text{enthalpy}} + [\phi \ln\phi / N_A + (1-\phi) \ln(1-\phi)/N_B]_{\text{entropy}} \tag{1.6.2}$$

where T is the temperature, k is Boltzmann's constant, and χ is the Flory–Huggins interaction parameter. The entropy term on the right side, known as the combinatorial entropy of mixing, is always negative but becomes smaller and approaches zero with increasing N_A and N_B. This theory was developed for a lattice with a coordination number z. While the entropic factor favors mixing as in ideal solutions, the monomer interactions expressed through nearest neighbor attractive energies, ε, may oppose mixing, and this effect is quantified through the χ parameter

$$\chi = -(z-2)(\varepsilon_{AA} + \varepsilon_{BB} - 2\varepsilon_{AB})/2kT \tag{1.6.3}$$

where ε_{AA}, ε_{BB} and ε_{AB} are the interaction energies between AA, BB, and AB monomers, respectively. Thus, the χ parameter examines specific monomer interactions and determines the balance of energy required to break the homopolymer bonds and form two new AB bonds. Ideally, for any A/B pair, each χ parameter could be decomposed into additive contributions from polar, steric, electronic, nearest neighbor, architecture, and other effects. The analysis can be complex. Three situations are important:

1. If $\varepsilon_{AA} + \varepsilon_{BB} = 2\varepsilon_{AB}$, χ is 0 and the blend is *athermal*; this case is similar to that of the symmetric A/A interface. Strength through interdiffusion can occur.
2. If $\varepsilon_{AA} + \varepsilon_{BB} < 2\varepsilon_{AB}$, AB bond formation is favored, χ is negative, and the blend is *compatible*. Strength develops more rapidly than in an athermal blend, since the molecules experience an additional driving force for diffusion across the interface.
3. If $\varepsilon_{AA} + \varepsilon_{BB} > 2\varepsilon_{AB}$, χ is positive and mixing is not favored. Under these conditions, A and B components do not mix, and mechanical mixtures attempt to phase-separate. At an A/B interface, diffusion is limited to a narrow region whose thickness is less than the radius of gyration of the chains.

When $N_A = N_B$, the miscibility of A in B depends only on χ_c, the critical χ value:

$$\chi_c = 2/N \tag{1.6.4}$$

When χ is greater than χ_c, the blend is immiscible; when χ is less than χ_c, mixing occurs. The critical value χ_c is independent of temperature; the temperature dependence of χ is determined by

$$\chi(T) = a/T + b \tag{1.6.5}$$

where a and b are constants and T is in degrees Kelvin. With the PS/PMMA pair, for example, the temperature dependence of χ was determined by Russell *et al.* [41] as

$$\chi(T) = (3.902/T) + 0.0284 \tag{1.6.6}$$

As the temperature increases, χ decreases and compatibilization is enhanced. The temperature at which compatibility is obtained is known as the *upper critical solution temperature, UCST*. The following example illustrates the use of the above equations.

Example

Problems:

(a) *Determine if the PS/PMMA pair is compatible at 140 °C, given that the molecular weight, M, of both polymers is 200,000.*

(b) *What temperature, T_{cr}, would be required to produce mixing at this molecular weight?*

(c) *What molecular weight would be required to produce compatibility at 140 °C?*

Solutions:

(a) *The χ_c value for the PS/PMMA pair is determined by $\chi_c = 2M_0/M$, where $N_A = N_B = N = M/M_0$. Since the monomer molecular weights, M_0, are approximately equal, with PS(104) ≈ PMMA(100), we have $\chi_c \approx 2 \times 100/200{,}000$, or $\chi_c = +0.001$. At 140 °C = 413 K, Eq 1.6.6 gives $\chi(T) = 0.038$. Since $\chi(T) > \chi_c$, the pair is immiscible at this temperature and molecular weight.*

(b) *Letting $\chi_c = \chi(T)$ and solving for $T_{cr} = 3.902/(0.001 - 0.0284)$, we have $T_{cr} = -142$ K. This impossible result is due to the finite value of b in Eq 1.6.6, which remains as (a/T) goes to zero with increasing T. Thus, the PS/PMMA pair is incompatible at all temperatures at this molecular weight.*

(c) *At 140 °C, when $\chi_c = \chi(T) = 0.038$, $N = 52$ and $M \approx 5{,}300$. This molecular weight is considered to be very low, since $M_c = 30{,}000$ for PS. Such a material would have little practical value. We expect that most commercially available PS/PMMA blends are incompatible.*

Polymers such as PS/PMMA have great difficulty in forming homogeneous blends, but the limited mixing that can occur at the interface merits study. Helfand and co-workers have given this considerable attention [42–45]. They examined the incompatible interface between two immiscible polymers (A and B) of infinite molecular weight, and described the diffusion of A in B by solving the diffusion equation of a random walk in a potential field created by the incompatible monomer. In the simple case a balance of two forces determines the extent of interfacial mixing. The compressed configurations of the chains in the surface layer provide an entropic driving force to diffuse across the interface by a distance d, and expand the random coil dimensions. This is counterbalanced by the unfavorable mixing of incompatible monomers.

The dominant entropic force for this limited diffusion is not combinatorial entropy but rather the configurational entropy change at the surface. The entropy change for a random-coil chain of n steps that relaxes from a surface-reflected configuration to a completely random one (with end-to-end vector d_∞), as shown in Figure 1.8, is

$$\Delta S = k \ln n \tag{1.6.7}$$

However, the enthalpy change for mixing this chain of length n with incompatible monomers is

$$\Delta H \sim \chi n \tag{1.6.8}$$

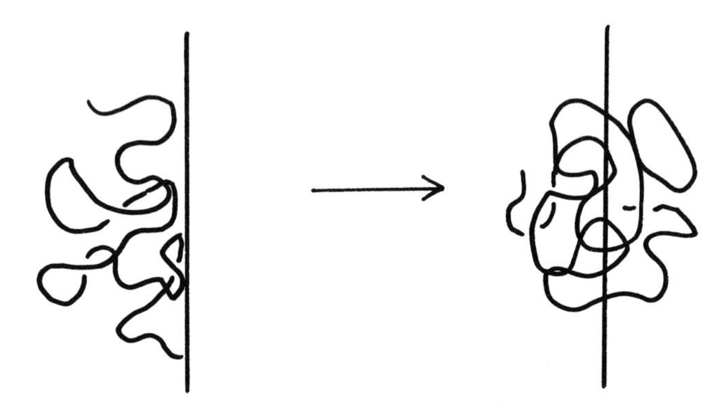

Figure 1.8 Conformational relaxation of a chain at an interface.

Minimizing the free energy, $\Delta A = \Delta H - T\Delta S$, with respect to n and solving for the equilibrium value of $d_\infty \sim \sqrt{n}$, Helfand obtained the solution for the equilibrium interface thickness, d_∞, as

$$d_\infty = 2\,b/(6\,\chi)^{1/2} \tag{1.6.9}$$

where b is the statistical segment length, which is about 6.5 Å for PS and PMMA. For the PS/PMMA interface at 140 °C, Eq 1.6.9 gives $d_\infty = 27$ Å. The significance of this layer thickness on mechanical properties of the interface is discussed in Chapter 9.

The Flory–Huggins theory has been used with considerable success to describe incompatible blends and interfaces but suffers from several serious deficiencies. These include the incompressibility of the lattice and a lack of correlation in concentration and connectivity. Schweizer [46] recently addressed these problems using the *reference interaction site model* (*RISM*) of Chandler [47], which had been successfully applied to small-molecule fluids. For the case of athermal blends involving differences in local structure (topological asymmetry), Curro and Schweizer [48] derived the Flory–Huggins χ parameter in terms of composition, density, monomer size, and aspect ratio. In contrast to the mean field theory, for which $\chi = 0$, they find that structural asymmetry results in a negative χ value with an approximately linear dependence on composition. Both the local structure at the monomer level and the global structure differences at the radius of gyration level lead to significant entropies of mixing that are noncombinatorial.

1.7 Multicomponent Polymer Interfaces

Several important multicomponent interfaces are obtained with polymer blends and composites. With rubber-toughened epoxies and thermoplastic matrices, one has a mixture of phase-separated (typically spherical), small rubber particles in a higher modulus matrix. For example, the tough ABS plastic consists of a terpolymer matrix of acrylonitrile–butadiene–styrene with a small volume fraction of rubber particles. The rubber particles can have a complex microstructure, with inner cores of concentric layers of varying moduli and an outer skin with a grafted molecular layer to enhance adhesion with the matrix.

In a uniaxial stress field, the lower modulus particles cause a stress concentration in the brittle matrix at the equatorial poles (perpendicular to the stress field) of the particle. The local high stress initiates a craze, and the rubber particle deforms to accommodate the craze opening. If the adhesion of the rubber particle to the matrix is sufficient to prevent the craze from reaching the critical crack-opening displacement, which could result in catastrophic failure, the energy adsorption mechanism has successfully accomplished its purpose. The craze propagates safely and usually terminates in another particle. Other toughening mechanisms in these composites involve cavitation, crack arrest or blunting, and shear yielding of the matrix ligaments between particles.

The blending of plastics with rubber particles can result in a substantial enhancement of properties, for example, the creation of bullet-proof glass from brittle PMMA and PC. The process is delicate in that one is deliberately inducing extensive microscopic damage as the energy-absorbing mechanism, while attempting to prevent the propagation of a single fatal crack. Considerable attention is being paid to the design of these composites, with emphasis on the particle microstructure and adhesion characteristics. The success of the composite design is typically evaluated from the bulk properties. However, in larger injection-molded parts (such as auto body parts) and compression-molded laminates, the interface properties may be radically different from the bulk properties. In the following sections, we discuss several cases involving multicomponent interfaces.

1.7.1 Interfaces of Two-Component Blends with Particles

Consider the two-component blend consisting of matrix A, with volume fraction ϕ_B of spherical particles (of radius R), as shown in Figure 1.9, where PS spheres are dispersed in a PE matrix [27]. The number of particles per unit volume, N_v, is

$$N_v = 3\phi_B/(4\pi R^3) \tag{1.7.1}$$

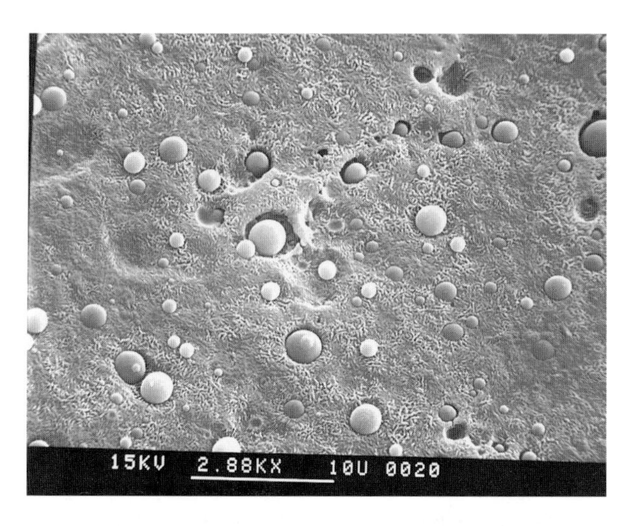

Figure 1.9 Surface of a blend containing polystyrene spheres dispersed in a polyethylene matrix. The indicated bar is 10 μm (courtesy of P. H. Geil).

When the particles are *randomly* distributed, the number touching a unit area of free surface, N_B, is determined from the number of particles whose centers are within a depth of radius R from the surface as

$$N_B = 3\phi_B/(4\pi R^2) \tag{1.7.2}$$

with $0 < \phi_B << 1$. Touching in this case allows protrusion of spherical particles through the matrix surface (Figure 1.9) to a distance not greater than $2R$. The average surface area of a sphere is $\frac{2}{3}\pi R^2$, so the surface area fraction, F_B, occupied by particles is

$$F_B = \phi_B/2 \tag{1.7.3}$$

When two of these surfaces are brought into contact, the resulting interface is composed of three component interfaces, namely A/A, B/B, and A/B. Thus we have two symmetric and one asymmetric interface; their area fractions add up:

$$F_{AA} + F_{BB} + F_{AB} = 1 \tag{1.7.4}$$

The area fraction of each component interface can be estimated as follows. At the interface we now have $2N_B$ particles that contribute to either A/B or B/B interfaces only. Therefore, the fraction of A/A interfaces is $1 - (F_{AB} + F_{BB})$, or

$$F_{AA} = 1 - \phi_B \tag{1.7.5}$$

The fraction of B/B contacts is proportional to $\phi_B{}^2$, with a proportionality constant of order unity, so that

$$F_{BB} \approx \phi_B^2 \tag{1.7.6}$$

and

$$F_{AB} \approx \phi_B - \phi_B^2 \tag{1.7.7}$$

Thus, for a typical blend containing 10% by volume of particles, the composite interface consists of 90% pure matrix/matrix, 9% matrix/particle, and 1% particle/particle interfaces. To an excellent first approximation, one can ignore the $\phi_B{}^2$ term and consider the influence of the particle/matrix interface strength on the pure matrix interface. When the pure matrix is healed to a strength comparable to the virgin state, the healed composite strength should be identical to the bulk virgin strength. However, two situations can arise that adversely affect the interface strength; these involve either a matrix-rich skin on the surface or a particle-rich surface.

A matrix-rich skin typically develops during compression and injection molding where the blend fluid comes in contact with a hard wall and the random distribution of particles protruding from the surface is not allowed. The matrix forms a skin when the particles bounce back from the hard wall. In that case, a crack may propagate continuously through this particle-depleted layer when the interface is formed from two such surfaces.

Particle-rich surfaces can form at internal weld lines, due to the fountain effect inducing particle separation; at semicrystalline surfaces, by a type of zone refining, when crystallization occurs in a temperature gradient; in single- and twin-screw extruders, due to high shear of the fluid between the screw and the barrel wall; and for other reasons usually associated with the melt flow. When this situation develops, the resulting interface can be very weak. In complex parts, this problem can be minimized to a certain extent. If the mold design is controlled via location of the injection gates, the weld lines can usually be located at planes that are not orthogonal to anticipated service stresses.

Exercise

Determine an expression for the fraction of component interfaces in a three-component (ABC) composite interface.

1.8 References

1. P. J. Flory, *Principles of Polymer Chemistry*; Cornell University Press, Ithaca, NY; 1953.
2. P. J. Flory, *Statistical Mechanics of Chain Molecules*; John Wiley & Sons, NY; 1969.
3. J. D. Ferry, *Viscoelastic Properties of Polymers*; John Wiley & Sons, New York; 3rd ed., 1980.
4. W. W. Graessley, *The Entanglement Concept in Polymer Rheology*; Vol. 16 in series *Advances in Polymer Science*, H. J. Cantow *et al.*, Eds.; 1982, Springer-Verlag, Berlin 1974.
5. P. E. Rouse, *J. Chem. Phys.* ***21***, 1272 (1953).
6. P.-G. de Gennes, *Scaling Concepts in Polymer Physics*; Cornell University Press, Ithaca, NY; 1979.
7. M. Doi and S. F. Edwards, *The Theory of Polymer Dynamics*; Clarendon Press, Oxford; 1986.
8. P.-G. de Gennes, *J. Chem. Phys.* ***55***, 572 (1971).
9. R. H. Boyd, *Macromolecules* ***22***, 2477 (1989).
10. R. P. Wool, "Dynamics and Fractal Structure of Polymer Interfaces"; chapter in *New Trends in Physics and Physical Chemistry of Polymers*, L.-H. Lee, Ed.; Plenum Press, New York; 1989; p 129.
11. B. Sapoval, M. Rosso, and J.-F. Gouyet, *J. Phys. Lett.* ***46***, L149 (1985).
12. H. H. Kausch, *Polymer Fracture*; Springer-Verlag, Berlin; 2nd ed., 1987.
13. S. Prager and M. Tirrell, *J. Chem. Phys.* ***75***, 5194 (1981).
14. P.-G. de Gennes, *C. R. Seances Acad. Sci., Ser. 2* ***292***, 1505 (1981).
15. R. P. Wool, conclusions based on discussions with H. Schonhorn.
16. L. C. E. Struik, *Physical Aging in Amorphous Polymers and Other Materials*; Elsevier Scientific Publ., Amsterdam; 1978.
17. B. L. Joss, R. S. Bretzlaff, and R. P. Wool, *Polym. Eng. Sci.* ***24***, 1130 (1984).
18. J. Klein and B. J. Briscoe, *Proc. R. Soc. London, A* ***365***, 53 (1979).
19. R. P. Wool and K. M. O'Connor, *J. Appl. Phys.* ***52***, 5194 (1981).
20. A. W. Adamson, *Physical Chemistry of Surfaces*; Wiley Interscience, New York; 5th ed., 1990.
21. W. J. Feast and H. S. Munro, Eds., *Polymer Surfaces and Interfaces*; John Wiley & Sons, New York; 1987.
22. M. R. Wattenbarger, H. S. Chan, D. F. Evans, V. A. Bloomfield, and K. A. Dill, *J. Chem. Phys.* ***93***, 8343 (1990).
23. L.-H. Lee, Ed., *Adhesion and Adsorption of Polymers*; Vol. 12A in series *Polymer Science and Technology*; Plenum Press, New York; 1980.
24. I. C. Sanchez, *Physics of Polymer Surfaces and Interfaces*; Butterworth–Heinemann, Stoneham, MA; 1992.
25. D. Turnbull and J. C. Fischer, *J. Chem. Phys.* ***17***, 71 (1949).
26. J. D. Hoffman and J. I. Lauritzen, *J. Appl. Phys.* ***44***, 4340 (1973).
27. P. H. Geil and S. Kent, *IUPAC Working Party 4.2.1 Report on Polymer Blend/Alloys*; 1988.
28. J. D. Hoffman, *Polymer* ***24***, 3 (1983).
29. B.-L. Yuan and R. P. Wool, *Polym. Eng. Sci.* ***30***, 1454 (1990).
30. P. H. Geil, *Polymer Single Crystals*; Vol. 5 in series *Polymer Reviews*, H. F. Mark and E. H. Immergut, Eds.; Wiley–Interscience, New York; 1963.
31. F. J. Padden, Jr., and H. D. Keith, *J. Appl. Phys.* ***30***, 1479 (1959).

32. H. D. Keith and F. J. Padden, Jr., *J. Appl. Phys.* ***35***, 1270 (1964).
33. D. C. Bassett, F. C. Frank, and A. Keller, *Philos. Mag.* ***8***, 1739 (1963).
34. P. J. Flory and D. Y. Yoon, *Nature (London)* ***272***, 226 (March 16, 1978).
35. F. C. Frank, *Faraday Discussions of the Royal Society of Chemistry* ***68***, 7 (1979).
36. H. H. Kausch, J. A. Hassell, and R. I. Jaffee, Eds., *Deformation and Fracture of High Polymers*; Plenum Press, New York; 1973.
37. *Engineering Plastics*; Vol. 2 in *Engineered Materials Handbook*; ASM International, Metals Park, OH; 1988.
38. S. S. Schwartz and S. H. Goodman, Eds., *Plastics Materials and Processes*; Van Nostrand Reinhold Company, New York; 1982; Chapter 18, "Fastening and Joining Techniques", p 757.
39. K. L. Foster and R. P. Wool, *Macromolecules* ***24***, 1397 (1991).
40. J. L. Willett and R. P. Wool, *Macromolecules* ***26***, 5336 (1993).
41. T. P. Russell, A. Menelle, W. A. Hamilton, G. S. Smith, S. K. Satija, and C. F. Majkrzak, *Macromolecules* ***24***, 5721 (1991).
42. E. Helfand and Y. Tagami, *J. Chem. Phys.* ***56***, 3592 (1971).
43. E. Helfand and A. Sapse, *J. Chem. Phys.* ***62***, 1327 (1975).
44. E. Helfand, *Macromolecules* ***25***, 1676 (1992).
45. E. Helfand, "Polymer Interfaces"; chapter in *Polymer Compatibility and Incompatibility; Principles and Practices*, K. Šolc, Ed.; MMI Press Symposium Series; Harwood Academic Publishers, New York; 1982; also E. Helfand, *J. Chem. Phys.* ***62***, 999 (1975).
46. K. S. Schweizer, *J. Chem Phys.* ***91***, 5802 (1989).
47. D. Chandler, "Equilibrium Theory of Polyatomic Fluids"; In *The Liquid State of Matter: Fluids, Simple and Complex,* E. W. Montroll and J. L. Lebowitz, Eds.; Vol. VIII in series *Studies in Statistical Mechanics*; E. W. Montroll and J. L. Lebowitz, Eds.; North–Holland, Amsterdam; 1982; p 275.
48. J. Curro and K. S. Schweizer, *J. Chem Phys.* ***87***, 1842 (1987).
49. ASTM D 1600 – 92a, Standard Terminology for Abbreviated Terms Relating to Plastics. In *Vol. 08.01, Plastics (I), 1993 Annual Book of ASTM Standards*; American Society for Testing and Materials, Philadelphia, PA, 1993.
50. J. M. Schultz, *Polymer Materials Science*; Prentice Hall, Englewood Cliffs, NJ; 1984.

2 Structure of Symmetric Amorphous Interfaces*

2.1 Introduction

The strength of a symmetric amorphous polymer–polymer interface (A/A) depends on the structure that develops during welding. The average structure is uniquely related to the dynamics and shape (statics) of the chains attempting to diffuse across the interface. The detailed ramified structure of diffuse interfaces with fractal characteristics can be described by gradient percolation and will be presented in detail in Chapter 4. The dynamics of chains at a surface or interface is complex and is divided into five time regions, namely, (1) short-range Fickian diffusion of monomers, (2) Rouse relaxation of entanglements, (3) Rouse relaxation of the whole chain, (4) reptation, and (5) long-range Fickian diffusion. These regions are discussed more fully in Chapter 3.

For the evaluation of strength development during welding of A/A interfaces, the reptation region involving diffusion of chain segments to depths of order R_g is the most important. The word *reptation* has as its root the Latin word meaning "to creep", which well describes the snake-like motion of interdiffusing chains. In this chapter we first elucidate the development of structure brought about by the dynamics of reptation. Secondary contributions from segmental motion and Rouse dynamics are treated in Chapter 3. The average structure of the interface can be described by the concentration profile, $C(x,t)$, from which a set of molecular properties, $H(t)$, can be deduced. The latter are important in developing microscopic deformation models for the strength of interfaces. SIMS and neutron reflection experiments are used to investigate the concentration profiles at distances less than R, the end-to-end vector (Chapters 5 and 6). Relations between the interface fracture energy G_{1c} and the molecular properties $H(t)$ are presented here and further examined in Chapters 7–12. The results presented in this chapter have application to a wide range of polymer and nonpolymer interfaces.

2.1.1 Welding and Crack Healing

Many researchers in recent years have investigated welding and crack healing in polymers [1–16]. When two similar pieces of bulk polymer are brought into contact

---

* Dedicated to P.-G. de Gennes

at a temperature above the glass transition temperature T_g, the interface gradually disappears and mechanical strength develops as the crack or weld heals. Crack healing and welding are related problems but are fundamentally very different. In *crack healing* we are concerned with the healing of two fractured surfaces that contain the remnants of the deformation zone through which the crack propagated. We expect the molecular weight distribution to be altered depending on the microscopic deformation mechanisms preceding and leading to fracture. If chain disentanglement occurs, as with uncured linear elastomers, the molecular weight distribution is relatively unaffected, but the location of the chain ends may be preferentially biased towards the surface. If significant chain fracture occurs, as during the fracture of high-molecular-weight glasses, the molecular weight in the surface layer is substantially altered. We have used GPC methods to estimate the change in molecular weight; I discuss this work in Chapter 8.

Welding involves the contact of two melt surfaces that had not previously been in contact; the molecular weight at each surface is expected to be the same as in the corresponding bulk. Allowances for molecular weight segregation can be made for polydisperse molecular weight distributions. In this chapter we focus our discussion on the monodisperse case, and present general solutions for the polydisperse result, which is of interest in most practical plastic welding applications.

Healing, implying the recovery of mechanical strength with time, is primarily the result of the diffusion of chains across the interface. The chain diffusion is a special type of mass transfer that cannot be described by the conventional diffusion equation. Wool and O'Connor have studied healing of interfaces in terms of the following stages: (1) surface rearrangement, (2) surface approach, (3) wetting, (4) diffusion, and (5) randomization. By the end of the wetting stage, potential barriers associated with inhomogeneities in the interface disappear and chains are free to move across the interface in the subsequent stages of diffusion and randomization, which are the most important because that is when the characteristic strength of the polymer material appears.

The essential features of chain motion in a polymer consisting of entangled random-coil chains are well described by the reptation model, which was developed by de Gennes [17] for evaluating the dynamics of polymer melts, and by Doi and Edwards [18–24] with particular application to the rheological properties of polymer melts. The reptation model has been applied to the healing problem at a polymer–polymer interface by de Gennes [25–27], Prager and Tirrell [13], and Wool and co-workers [14–16, 28, 29]. In the healing problem, the basic question to be answered is, "How does one relate the microscopic description of motion of chains to the macroscopic measurements, such as fracture energy?" Also, the initial conformations of chains and the applicability of the reptation model in the case of highly oriented conformations of chains on the surface must be considered. We do not yet have exact solutions to these questions, and therefore we must make a number of assumptions in order to apply the reptation model to the problem of healing at an interface. The validity of these assumptions is examined in the course of the development of this chapter.

2.2 Chain Dynamics in Amorphous Polymer Materials

In a bulk amorphous polymeric material, the motion of a chain is greatly restricted by the entanglements of neighboring chains. The reptation model is suitable for describing the motion of a chain that is entangled with many other chains. According to the model, the constraints imposed by the entanglements from other chains effectively confine the chain to a tube-like region (Figure 2.1). A chain is conceived of as wriggling around in the tube because of thermal fluctuations. According to de Gennes, the wriggling motions occur rapidly, their magnitudes are small (compared to R_g), and, on a time scale greater than that of the wriggling motion, the chain moves coherently back and forth along the center line of the tube with a certain diffusion constant, keeping the arc length constant. Therefore, the chain as a whole moves one-dimensionally but randomly along the tube. The chain conformation at a given moment is the same as the shape of the tube that confines the chain at that instant. For a chain to change its conformation, it must disengage itself from the tube that was defined at an earlier moment. When a chain moves "forward", the "head" of the chain chooses its direction randomly, and when it moves "backward", the "tail" chooses its direction randomly. This particular feature is also the essential algorithm in developing a computer simulation of the reptation model, as we demonstrate later.

The microscopic details of polymer melt dynamics have been explored by many investigators using experimental, computer simulation, and theoretical approaches. Infrared studies of stress relaxation in selectively deuterated polystyrene chains (typically labeled at the chain center, or at the ends) demonstrated that the chain ends relaxed faster than the chain center when the probe chain was placed in a higher molecular weight matrix [30–37]. These results were in agreement with stress relaxation studies evaluated by neutron scattering [38]. However, birefringence [39] and IR dichroism studies [33] of centrally labeled homopolymers (no matrix) demonstrated little difference between the chain-end and chain-center relaxation rates. Computer simulation work by several groups provided both support and further controversy [40–43], and it is generally considered that more powerful computers are needed to provide a detailed molecular dynamics analysis of high-molecular-weight chains in the melt. Tracer diffusion studies of polymers in matrices of different molecular weights have given considerable support to the concept of topological constraints (and constraint release) and have demonstrated the effect of these constraints on center-of-mass diffusion coefficients D [44–48]. Significant insight on polymer dynamics has been obtained from SIMS (Chapter 5) and neutron reflection (Chapter 6) experiments, which probe characteristic correlated motion effects while the molecules are diffusing over distances on the order of R_g [49–53]. Several theoretical models involving constraint release, chain end fluctuation, and nematic orientation have been advanced to make corrections to the reptation model [54–57], or suggest new approaches to interpreting melt dynamics [55]. The experimental and theoretical arguments up to 1990 are reviewed by Lodge *et al.* [58]. However, the most

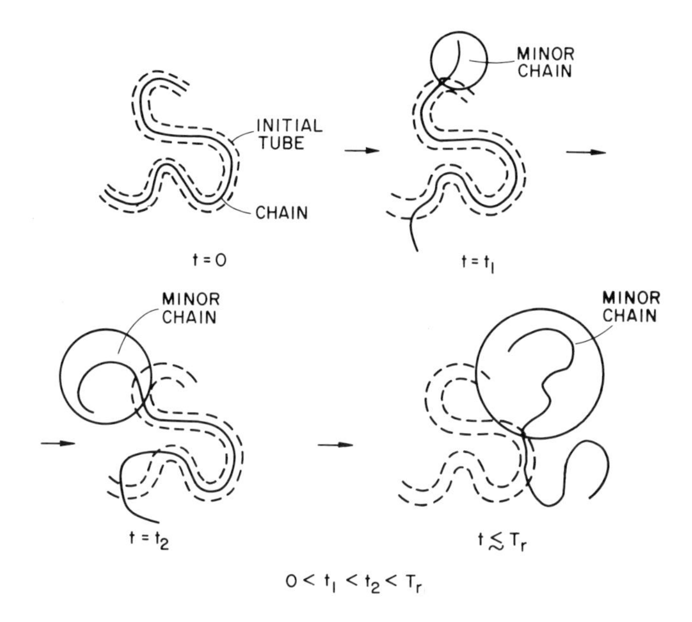

Figure 2.1 Minor chain reptation model; disengagement of a chain from its initial tube. The emergence and growth of minor chains are shown at times up to the tube renewal time, T_r (Kim and Wool).

compelling evidence for the reptation model was recently obtained from SIMS [59] and neutron reflection [60] studies of interdiffusion with matching pairs of centrally labeled polymers. The "ripple" experiment, using HDH/DHD matching pairs, is described in Chapters 5 and 6; it supplies significant support for the minor chain reptation model, which is described in the next section.

2.2.1 Minor Chain Reptation Model

The way a chain disengages itself from the initial tube is illustrated by the chain conformations at different times in Figure 2.1 [29]. The tube defined at time 0 (the initial tube) is shown as two dotted lines at various times. At t_1, some end lengths of the chains have already escaped from the initial tube. The portions of the chains that are no longer in the initial tube lengthen with time and are called the *minor chains* (Kim and Wool [29]). The random-coil (most probable) spherical envelopes enclosing the minor chains at different times are shown as circles. A minor chain has length $l(t)$, which is an increasing function of time and has the conformation of a random Gaussian chain. We will see in Chapters 7 and 8 that $l(t)$ is the most important molecular property controlling strength development. The minor chains are the portions of the chains that have lost the memory of their initial conformations. As the minor chains become longer, the initial conformations are gradually replaced by new conformations.

There are three fundamental time scales in the reptation model. The first time, τ_e, describes the *Rouse relaxation time* between entanglements and is a local characteristic of the wriggling motion. The second time, τ_{RO}, describes the propagation of wriggle motions along the contour of the chain and is related to the Rouse relaxation time of the whole chain. The important time for the welding problem is the *reptation time*, T_r, which describes the time required for the chain to escape from its initial tube. T_r is often referred to as the *tube renewal time*. It is the characteristic relaxation time for the chain and properly refers to the time when about 70% of the chain has escaped from the tube. The *complete escape time*, T_d, is related to T_r by $T_d = 1.94\ T_r$ [30], so it is about twice T_r. The molecular weight dependences of the three relaxation times are $\tau_e \sim M_e^2$, $\tau_{RO} \sim M^2$, and $T_r \sim M^3$, where M is the molecular weight of the linear chains and M_e is the entanglement molecular weight. Since M is typically large, the reptation time, T_r, is much greater than the Rouse relaxation time, τ_{RO}. For example, with polystyrene with M of 245,000 and welded to itself at 118 °C, the characteristic relaxation times are τ_e of about 10 s, τ_{RO} of about 21 min, and T_r of about 1860 min [60].

The time scale of importance in welding is in the region between τ_{RO} and T_r. Welding at times less than τ_{RO} is of importance for high-rate fracture processes and for incompatible interfaces, which will be discussed later. Thus, for now we will ignore times less than τ_{RO}; if we say t is greater than zero, we are implicitly saying it is greater than τ_{RO}. We will return to this issue in Chapters 3 and 7. In Figure 2.2, the minor chains and their spherical envelopes at different times are shown. At $t = 0$, only the locations of the end segments are shown, because minor chains start to grow at the end segments. The concept of the minor chain turns out to be very useful for evaluating molecular properties at an interface.

2.3 Interdiffusion at a Polymer–Polymer Interface

We now use the minor chain model to examine the interdiffusion behavior and resulting molecular properties (summarized in Table 2.1) at a symmetric polymer–polymer interface.

2.3.1 Characterization of Chain Motion at an Interface

The conformations of two chains at the interface before and after the stages of diffusion and randomization are shown in Figure 2.3. At $t = 0$, the chains are separated by the interface plane and their shapes (most probable envelopes) are nonspherical and quite flat. However, at $t \approx T_r$, the chains have crossed the interface, adopted more spherical configurations, and interpenetrated and made new entanglements with chains

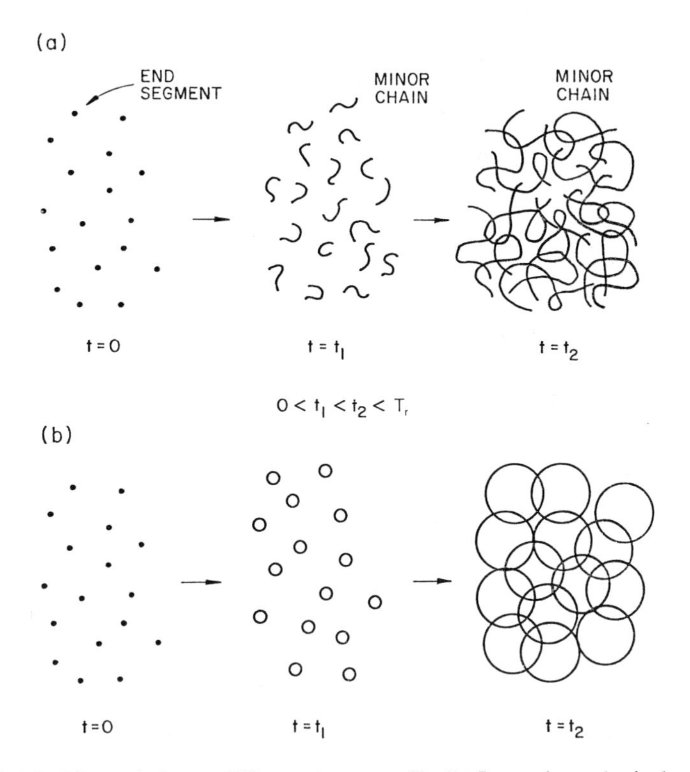

Figure 2.2 (a) Minor chains at different times, $t < T_r$. *(b)* Increasing spherical envelopes of minor chains at different times.

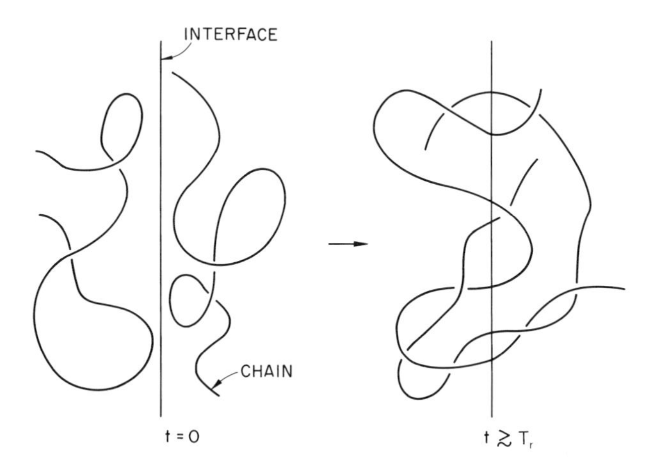

Figure 2.3 Conformations of two chains at an interface before and after diffusion.

Table 2.1 Molecular Aspects of Interdiffusion at a Polymer-Polymer Interface

Molecular Aspect	Symbol	Dynamic relation, $H(t)$	Static relation, H_∞	r	s
General property	$H(t)$	$t^{r/4} M^{-s/4}$	$M^{(3r-s)/4}$	r	s
Number of chains	$n(t)$	$t^{1/4} M^{-5/4}$	$M^{-1/2}$	1	5
Number of bridges	$p(t)$	$t^{1/2} M^{-3/2}$	M^0	2	6
Average monomer depth	$X(t)$	$t^{1/4} M^{-1/4}$	$M^{1/2}$	1	1
Total monomer depth	$X_0(t)$	$t^{1/2} M^{-3/2}$	M^0	2	6
Average contour length	$l(t)$	$t^{1/2} M^{-1/2}$	M	2	2
Total number of monomers	$N(t)$	$t^{3/4} M^{-7/4}$	$M^{1/2}$	3	7
Average bridge length	$l_p(t)$	$t^{1/4} M^{-1/4}$	$M^{1/2}$	1	1
Center of mass depth[a]	X_{cm}	$t^{1/2} M^{-1}$	$M^{1/2}$	2	4
Diffusion front length	N_f	$t^{1/2} M^{-3/2}$	M^0	2	6

[a] This equation applies to chains in the bulk and does not apply to chains whose center of mass is within a radius of gyration of the surface.

on the other side. Let us consider the motion of a chain from the point of view of the reptation model. The conformations at different healing times of a chain at an interface are shown in Figure 2.4. At $t = 0$, a chain and its initial tube are shown at the interface. At $t = t_1$, the minor chains are shown having escaped from their tubes. The minor chains assume random-coil conformations as they cross the interface. Only the portion of the initial tube that still confines part of the chain is shown. At a later time ($t = t_2$), the minor chains are much longer. They may cross the interface many times and penetrate into the other side more deeply. A small portion of the chain is still confined in the initial tube.

Let us now consider the minor chains at the interface. We assume that the chain ends are randomly distributed in the bulk layer near the surface. (Segregation of chain ends can occur at the surface and this important case will also be evaluated, since it has relevance to many asymmetric interfaces.) We further assume that the chain dynamics is unperturbed by the non–Gaussian conformations at the surface. If this is the case, then the interdiffusion problem can be solved in principle with any spatial distribution function for the chain ends and molecular weights. In Figure 2.5, the minor chains that poked through from one side of the interface (the vertical lines) are shown at different times. It is important to note that only minor chains can cross the interface. Therefore, Figure 2.5 shows only those portions of the chains that are responsible for mass transfer of chain segments across the interface. At $t = 0$, the chain conformations at or near the interface are not in equilibrium but the minor chains are now free to cross the interface. The minor chains grow out of the initial nonequilibrium conformations, they do not encounter obstacles at the interface, and they assume

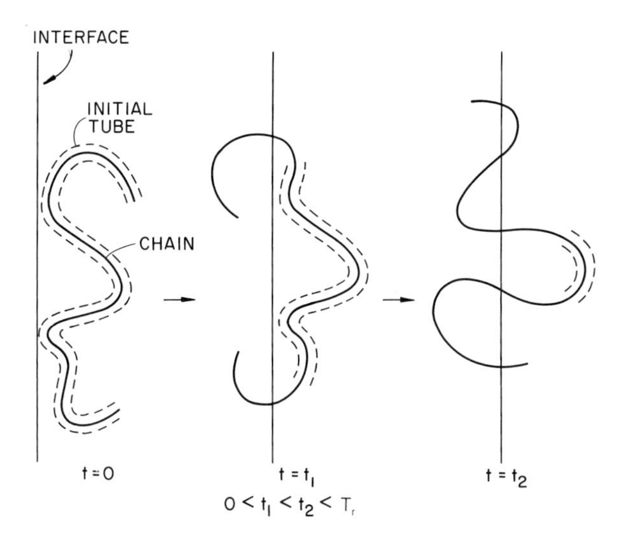

Figure 2.4 Disengagement of a chain from its initial tube near the interface. Only the portion of the initial tube that still confines part of the chain is shown.

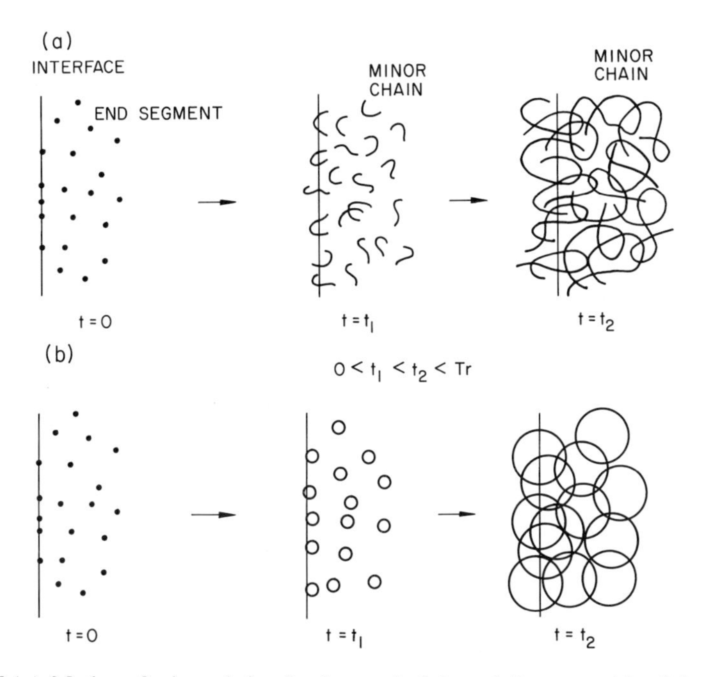

Figure 2.5 (a) Motion of minor chains that have poked through from one side of the interface; *(b)* growth of the minor chain spherical envelopes that have emerged from one side of the interface.

spherically symmetrical Gaussian conformations as they interpenetrate the interface (Figure 2.5b).

The growth and random conformations of the minor chains (Fig 2.5) constitute the diffusion and randomization stages of healing. This view of the motion of chains at an interface may be compared with an earlier view proposed by Voyutskii [1–5], shown in Figure 2.6. This is the situation one pictures if one attempts to keep track of the motion of entire chains during diffusion at an interface. Figure 2.5, on the other hand, shows only those parts of the chains (that is, the minor chains) that contribute to mass transfer across the interface. The behavior of the minor chains is easy to grasp once the length $l(t)$ is obtained as a function of time t and molecular weight M.

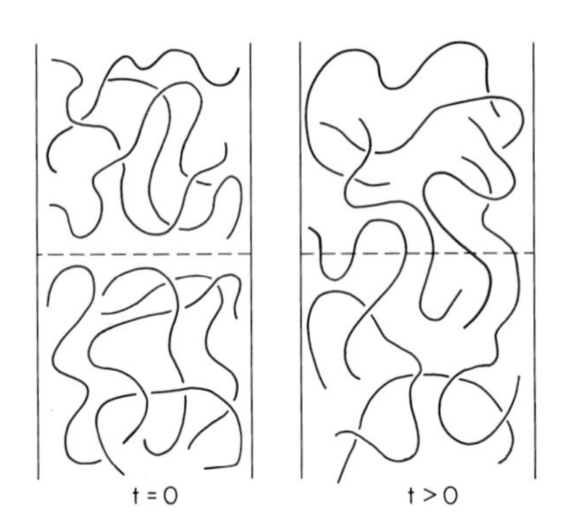

Figure 2.6 Voyutskii model for interdiffusion of chains at an interface.

2.3.2 Chain Conformation at an Interface

Let us consider a chain near the interface and let ξ be the distance of the end segment from the interface plane. At $t = 0$, the conformations of a chain with end segments at ξ_1 and ξ_2 is determined by the partition function $Z(\xi_1, \xi_2)$ as

$$Z(\xi_1, \xi_2) = (g/\pi)^{1/2} \{\exp[-g(\xi_1 + \xi_2)^2] + \exp[-g(\xi_1 - \xi_2)^2]\} \tag{2.3.1}$$

where $g = ½\, a^2N$, in which a is the length of one segment and N is the number of segments in a chain. The relaxed conformation of the chain at $t > T_r$ is determined by the Gaussian function $Z_e(\xi_1, \xi_2)$,

$$Z_e(\xi_1, \xi_2) = (g/\pi)^{1/2} \exp[-g(\xi_1 - \xi_2)^2] \tag{2.3.2}$$

The stages of diffusion and randomization are characterized by the transition of chain conformations from the nonequilibrium $Z(\xi_1, \xi_2)$ to the equilibrium $Z_e(\xi_1, \xi_2)$. This transition is shown in Figure 2.7, where the average shape and size of a chain at the

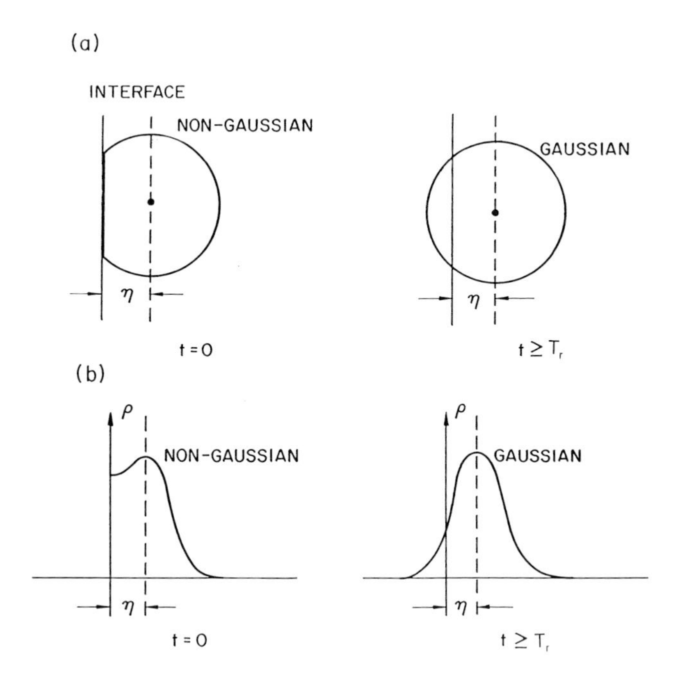

Figure 2.7 (a) Average shape and size of a chain's most probable envelope at the interface before and after healing. *(b)* Segment density, ρ, of the chain in *(a)* before and after healing.

interface plane (solid vertical line) before and after the transition are depicted. The distance η from the location of the highest segment density profile of the chain before and after the transition is shown. This transition occurs during the reptation time T_r, which marks the end of the diffusion and randomization stages. During the transition, the chains on the two sides of the interface begin to entangle with each other as the minor chains diffuse. At $t > T_r$, chains with spherical symmetry continue to diffuse across the interface. However, this type of diffusion is irrelevant to the healing problem in the symmetric interface. Therefore, the time dependence of the interpenetration of minor chains should strongly influence the time dependence of welding.

2.4 Analysis of Chain Motion at an Interface

We wish to derive an expression for the minor chain contour length, $l(t)$, during interdiffusion at the interface. We first consider the time dependence of the average minor chain length, $\langle l \rangle$, as it escapes from its tube by reptation, and we then evaluate contributions from many minor chains diffusing across the interface.

2.4.1 Minor Chain Length Evolution from Tube

In order to derive $l(t)$, let us consider the one-dimensional random walk of a chain in its tube, as shown in Figure 2.8. The ξ axis is the curvilinear distance a chain travels, and the tube is shown as two dotted lines. At $t = 0$, the end segments A and B are located at $\xi = -L$ and $\xi = 0$, respectively. The location of B after τ random walk steps is denoted by ξ_τ. There are two distances, b and c, to consider, where $b + c = L$ is required. In this section, we derive <*l*>, the average of $l(t)$. The reader may wish to skip the following details and resume at Section 2.4.2.

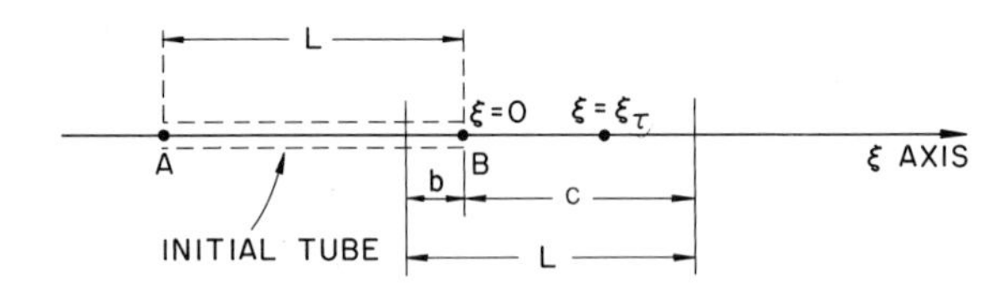

Figure 2.8 Coordinate and variables for the one-dimensional random walk of a chain in its tube for the derivation of the escape length, <*l*>.

Let $G(b,c,\xi_\tau)$ be the fraction of the number of walks of τ steps from $\xi = 0$ to $\xi = \xi_\tau$, with two absorbing walls at $\xi = -b$ and $\xi = c$. The end segment B is not allowed to "touch" the absorbing walls. This is given by

$$\begin{aligned} G(b,c,\xi_\tau) = (\beta/\pi)^{1/2} \sum_{m=0}^{\infty} \Big(& \exp\{-\beta[2m(b+c)+\xi]^2\} \\ & + \exp\{-\beta[2m(b+c)-\xi]^2\} \\ & - \exp\{-\beta[2m(b+c)+2b+\xi]^2\} \\ & - \exp\{-\beta[2m(b+c)+2c-\xi]^2\}\Big) \\ & - (\beta/\pi)^{1/2}\exp(-\beta\,\xi^2) \end{aligned} \tag{2.4.1}$$

where $\beta = ½\, a^2\tau$, a is the length of one step, and m is a summation variable.

For the derivation of <*l*>, it is necessary to obtain $J(b,c,\xi_\tau)$, which is the fraction of the number of random walks of τ steps from $\xi = 0$ to $\xi = \xi_\tau$ for $-b \le \xi_\tau < c$ with at least one "touch" at $\xi \equiv -b$, where $b + c = L$. This is because when the end segment B arrives at $\xi = -b$, the part of the initial tube from $\xi = -b$ to $\xi = 0$ is destroyed. The length of the minor chain that has escaped at end segment B is $b + \xi_\tau$. The same rule applies to the end segment A. This is related to $G(b,c,\xi_\tau)$ as follows:

$$J(b,c,\xi_\tau) = a\frac{\partial}{\partial b} G(b,c,\xi_\tau) \tag{2.4.2}$$

The average of l, that is, $<l>$, is equal to $<b>$, because of the symmetry of the chain motion with respect to the "head" and "tail" of the chain. Therefore, we have

$$<l> = \frac{\int_0^L \mathrm{d}b\, b \int_{-b}^{L-b} \mathrm{d}\xi_\tau\, J(b, L-b, \xi_\tau)}{\int_{-b}^{L-b} \mathrm{d}\xi_\tau\, J(b, L-b, \xi_\tau)} \tag{2.4.3}$$

From these three equations, $<l>$ is given by

$$<l> = k_2(x) / k_1(x) \tag{2.4.4}$$

where we have defined

$$k_1(x) \equiv a\sum_{m=0}^{\infty} [(6m+2)\,\mathrm{erf}(2mx+x) - (6m+1)\,\mathrm{erf}(2mx) - (2m+1)\,\mathrm{erf}(2mx+2x) + 2m\,\mathrm{erf}(2mx-x)] \tag{2.4.5}$$

and

$$k_2(x) \equiv \frac{aL}{\pi^{1/2}x}\sum_{m=0}^{\infty} \{\exp[-(2mx)^2] - \exp[-(2mx+2x)^2]\} + aL\sum_{m=0}^{\infty} [(m+1)\,\mathrm{erf}(2mx+2x) - (m+1)\,\mathrm{erf}(2mx+x) - m\,\mathrm{erf}(2mx) + m\,\mathrm{erf}(2mx-x)] \tag{2.4.6}$$

Also, $x \equiv L/(2a^2\tau)^{1/2}$, and erf (x) is the error function of variable x.

The asymptotic form of $<l>$ at early times, that is, when $a^2\tau < L^2$, is obtained as follows.

$$\lim_{x\to\infty} [\mathrm{erf}(x)] \simeq 1 - \exp(-x^2) / (\pi^{1/2} x) \tag{2.4.7}$$

From Eqs 2.4.3–2.4.7, we have

$$<l> \approx a(2\tau/\pi)^{1/2}\, \{1 - \exp(-L^2/2a^2\tau)\} \tag{2.4.8}$$

when $a^2\tau < L^2$. We write τ in terms of t as

$$\tau = 2Dt/a^2 \tag{2.4.9}$$

where D is the curvilinear diffusion constant, which is proportional to $1/M$. Therefore, we have

$$<l> \approx 2(Dt/\pi)^{1/2}[1 - \exp(-L^2/2Dt)] \tag{2.4.10}$$

From Eq 2.4.10 we find that

$$<l> \approx 2(Dt/\pi)^{1/2} \tag{2.4.11}$$

until t becomes comparable to T_r, since $T_r \approx L^2/(2D)$. This is an important result and is the basis for the reptation model derived by de Gennes in 1971.

The relation $<l> \sim t^{1/2}$ applies to molecular weights greater than M_c and times less than T_r. Once $<l>$ is obtained, the average interpenetration distance of chains or the number of links (or bridges) and other molecular properties $H(t)$ can be readily calculated.

So far, only $<l>$, the average of $l(t)$, has been calculated. In reality, $l(t)$ has a distribution around $<l>$. Also, $<l>$ should approach $L/2$ as t approaches infinity. These aspects are considered now. Let α_t be the fraction of the number of chains that have disengaged themselves completely from their initial tubes at time t. In order to derive α_t, let us define $f(\xi_1,L,t)$ as the probability that a one-dimensional random walk that starts at $\xi = \xi_1$ terminates at $\xi = 0$ for $t > 0$ when there are two absorbing walls at $\xi = 0$ and $\xi = L$ (see Figure 2.9). In the figure, the initial locations of the end segments are denoted as A and B, and the initial tube is shown as two dotted lines. We need to define another quantity $H(\xi_1,L,t)$ as the probability that a one-dimensional random walk that starts at $\xi = \xi_1$ terminates at $\xi = 0$ or $\xi = L$ at t when there are two absorbing walls at $\xi = 0$ and $\xi = L$. This is given by

$$H(\xi_1,L,t) = f(\xi_1,L,t) + f(L-\xi_1,L,t) \tag{2.4.12}$$

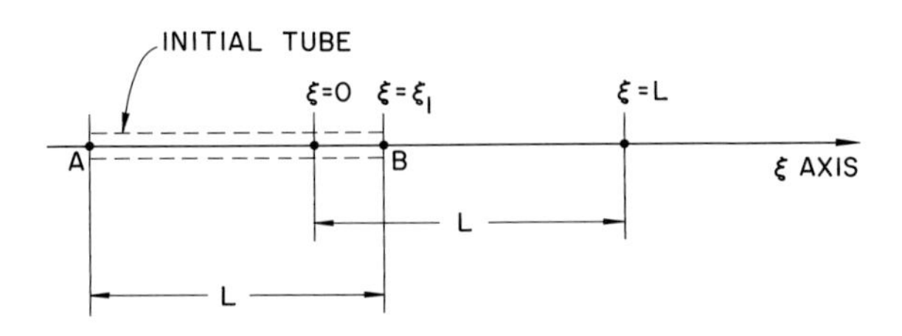

Figure 2.9 Coordinate and variables for the one-dimensional random walk of a chain in its tube for the derivation of α_t.

Then α_t is given by

$$\alpha_t = \frac{\int_0^t \mathrm{d}t \int_0^L \mathrm{d}\xi_1 \, H(\xi_1, L, t)}{\int_0^\infty \mathrm{d}t \int_0^L \mathrm{d}\xi_1 \, H(\xi_1, L, t)} = \frac{\sum_{\nu=1}^{\infty} \frac{[1-(-1)^\nu]}{\nu^2} \left[1 - \exp\left(-\frac{\pi^2 D \nu^2 t}{L^2}\right)\right]}{\sum_{\nu=1}^{\infty} \frac{[1-(-1)^\nu]}{\nu^2}} \tag{2.4.13}$$

where ν is a summation variable. From Eq 2.4.13, α_t approaches unity as t approaches infinity; also, $\alpha_t \approx 0$ until t becomes comparable to T_r, because of the exponential term and the factors of $1/\nu^2$ in the numerator. Therefore, for $0 < t < T_r$, it may be assumed that $\alpha_t = 0$. This result is also of importance since it permits many "back of the envelope" calculations of related molecular properties (summarized in Table 2.1) and demonstrated in Section 2.6.

Next, let us check whether $\langle l \rangle$ approaches $L/2$ as t approaches infinity. When t is very large, the following form of $G(b,c,\xi_\tau)$ is more useful than Eq 2.4.1 in the derivation of $\langle l \rangle$:

$$G(b, c, \xi_\tau) = \frac{2}{b+c} \sum_{m=1}^{\infty} \sin\left(\frac{\pi m b}{b+c}\right) \sin\left[\frac{\pi m (b+\xi)}{b+c}\right] \times \exp\left[-\frac{\pi^2 a^2 m^2 t}{2(b+c)^2}\right] \tag{2.4.14}$$

From Eqs 2.4.14, 2.4.2, and 2.4.3, $\langle l \rangle$ is given by

$$\langle l \rangle = (L/2) \frac{\sum_{m=1}^{\infty} \exp\left(-\frac{\pi^2 a^2 m^2 t}{2L^2}\right) \left[\frac{1-(-1)^m}{m^2}\right] \times \left(\frac{\pi^2 a^2 m^2 t}{L^2} + 1\right)}{\sum_{m=1}^{\infty} \exp\left(-\frac{\pi^2 a^2 m^2 t}{2L^2}\right) \left[\frac{1-(-1)^m}{m^2}\right] \times \left(\frac{\pi^2 a^2 m^2 t}{L^2} + 3\right)} \tag{2.4.15}$$

From this equation, we see that in the limit of t approaching infinity, $\langle l \rangle = L/2$.

The scaling law for the minor chains is given by

$$\langle l \rangle = L/2 \; (t/T_r)^{1/2} \tag{2.4.16}$$

where $L/2 \sim M$ is the equilibrium property.

2.4.2 Minor Chain Diffusion at an Interface

In the vicinity of the interface, the minor chains grow at different distances from the interface plane and diffuse across the interface, as shown in Figure 2.5. The evolution of the minor chain Gaussian segment density profiles is shown in Figure 2.10. Here, the minor chains are uniformly distributed and y is a coordinate perpendicular to the interface plane located at $y = 0$. The bell-shaped Gaussian curves of the minor chains eventually grow into the region $y < 0$ on the other side. The distance X is the average distance of these segments from the interface. According to this definition, the average monomer diffusion depth is given by

$$X = \frac{\int_0^\infty dy_0 \int_0^\infty dy\, y \times \exp\left[-\frac{(y_0+y)^2}{2a^2<l>}\right]}{\int_0^\infty dy_0 \int_0^\infty dy \times \exp\left[-\frac{(y_0+y)^2}{2a^2<l>}\right]} \quad (2.4.17)$$

From Eq 2.4.17, we determined

$$X \sim <l>^{1/2} \quad (2.4.18)$$

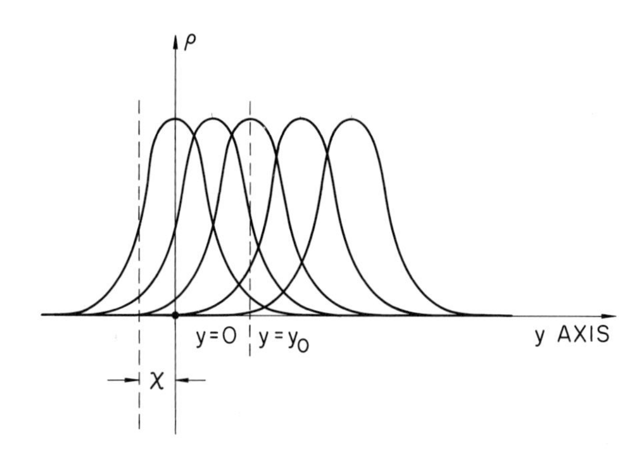

Figure 2.10 Segment density profiles, ρ, of minor chains and the average interpenetration distance, X.

The thickness of the interface, as determined by X, is shown schematically in Figure 2.11. Note that X is defined for only one side of the interface. The double interface width, d, for both sides of the interface is $2X$. The equilibrium monomer interdiffusion distance X_∞ of the interface at $t = T_r$ is

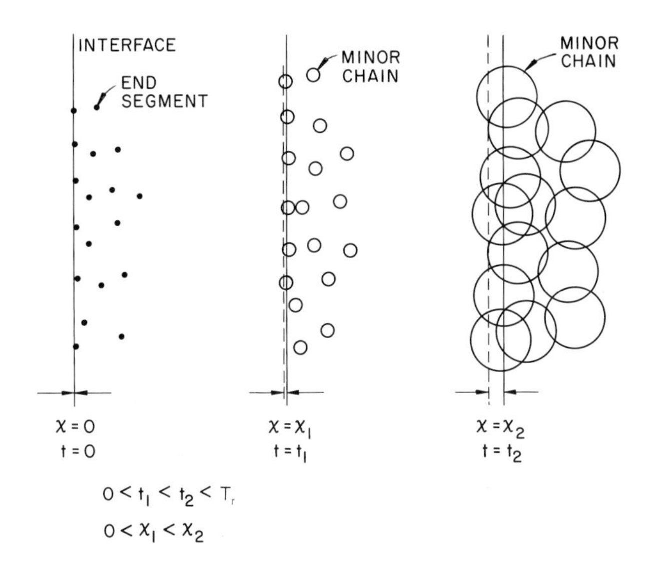

Figure 2.11 Average interpenetration distance, X, and the minor chains (one side of the interface) during diffusion.

$$X_\infty \approx 0.8\,R_g \tag{2.4.19}$$

where $R_g \sim M^{1/2}$ is the radius of gyration. Thus, the optimal interface width, d, is obtained from $d = 2X_\infty$, as $d \approx 1.6\,R_g$. In terms of the end-to-end vector $R = \sqrt{6}R_g$, $d = 0.66\,R$.

The scaling law for the interface thickness and the average monomer interpenetration distance as a function of time, $X(t)$, is obtained from the above equations as

$$X(t) = X_\infty\,(t/\,T_r)^{1/4} \tag{2.4.20}$$

Since $X_\infty \sim M^{1/2}$ and $T_r \sim M^3$, then $X(t) \sim t^{1/4}M^{-1/4}$, at $t < T_r$.

When $t > T_r$, all correlated motion effects are lost, normal Fickian diffusion resumes, and Eq 2.4.20 is replaced by

$$X(t) = X_\infty\,(t/\,T_r)^{1/2} \tag{2.4.21}$$

The exponent's transition from ¼ to ½ is a characteristic of the reptation model, which we demonstrate in Chapter 5. The normal Fickian diffusion result, $X^2 \sim Dt$, is evident from the molecular weight dependence of X_∞ and T_r in Eq 2.4.21, as $X^2 \sim t/M^2$, in which the center-of-mass diffusion coefficient is $D \sim M^{-2}$.

We have found from many welding studies that $X(t)$ and $l(t)$ provide excellent correlations and predictions for mechanical properties of interfaces (Chapters 7 and 8), with the fracture energy $G_{1c} \sim l(t)$ and the critical stress intensity factor $K_{1c} \sim X(t)$.

2.5 Concentration Profile at an Interface

The concentration depth profile $C(x,t)$ for a symmetric A/A polymer interface was analyzed in detail by Zhang and Wool [61, 62] and by Tirrell, Adolf, and Prager [63]. The concentration profiles can be used to extract molecular properties relevant to the dynamics, and in fact provide a useful vehicle for the investigation of the reptation model. The calculated concentration profiles can also be compared with experimental profiles determined from secondary ion mass spectroscopy (Chapter 5) and neutron reflection (Chapter 6). As discussed in Section 2.4, the concentration profiles should be particularly interesting if the reptation dynamics provides a non–Fickian diffusion region at $t < T_r$, which changes to a normal Fickian depth profile at longer times. The primary contribution to the depth profile from reptation dynamics is given below. Secondary contributions from Rouse relaxation are discussed in Chapter 3.

2.5.1 Concentration Depth Profile Due to Reptation

In this section we evaluate the concentration depth profile $C(x,t)$ due to the minor chains. We assume that the initial chain-end distribution function is flat, but the derivation can be performed in principle with any distribution function. As time goes by, the minor chains emerging from the ends of the initial tubes become longer and diffuse across the interface (Fig 2.5). In the one-dimensional span analysis in the last section, it was shown that the average minor chain length, $l(t)$, increases as

$$l(t) \approx 2(Dt/\pi)^{1/2} \tag{2.5.1}$$

where $D = L^2/\pi^2 T_r$ is the one-dimensional curvilinear diffusion coefficient, inversely proportional to molecular weight M.

If ρ is the uniform mass density of a bulk polymer in the melt state, the number of chain ends per unit volume, N_v, is $(2\rho N_a/M)$, where N_a is Avogadro's number. If n is the number of monomers of length a in a minor chain at healing time t, then

$$n = l(t)/a \tag{2.5.2}$$

The monomer concentration depth profile is calculated as follows for times T_r. Consider a minor chain of length $l(t)$ that emerged from its tube end at $x = -x_0$, as shown in Figure 2.12. The origin of the coordinates is chosen so that the interface plane is at $x = 0$ and the piece A is located on the region where $x < 0$. Since the conformation of the minor chain is random, the probability $p(s, x/x_0)$ that the sth monomer in the minor chain occurs at the interval distance $(x, x + \mathrm{d}x)$ from the interface (Figure 2.12(a)) is given by

$$p(s, x/x_0) = 1/(2\pi sa^2)^{1/2} \exp[-(x+x_0)^2/2sa^2]\,\mathrm{d}x \tag{2.5.3}$$

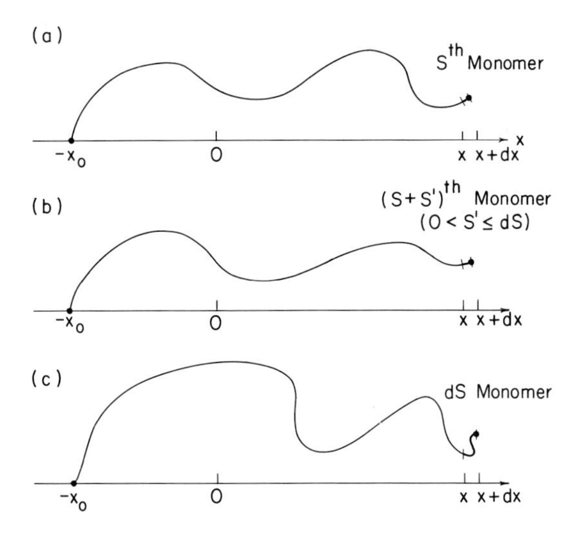

Figure 2.12 (a) Minor chain with s monomers. The last monomer (that is, the sth monomer) occurs in the distance interval $(x, x + dx)$. *(b)* Minor chain with $(s + s')$ monomers, where $0 < s' < ds$. The monomer $(s + s')$ occurs in the interval $(x, x + dx)$. *(c)* Minor chain with $(s + ds)$ monomers, where ds monomers occur in the interval $(x, x + dx)$.

Since ds is very small compared to s, Eq 2.5.3 also gives the probability that each monomer contained in the interval $(s, s + ds)$ of the minor chain occurs in the interval distance $(x, x + dx)$, as shown in Figure 2.12(b). Thus, the probability that the quantity ds monomers of the minor chain occur in the interval distance $(x, x + dx)$, as shown in Figure 2.12(c), is the same as $p(s, x/x_0)$. Therefore, the number of monomers $f(x, n)$ of the minor chain occurring at the interval distance $(x, x + dx)$ from the interface at time t is determined as

$$f(x, n) = \int_0^n (½\pi s a^2)^{1/2} \exp[-(x + x_0)^2/2sa^2] \, dx \, ds \tag{2.5.4}$$

For an interface with cross-sectional area A, there are $(2\rho N_a A\,dx_0/M)$ minor chain ends in the volume element Adx in the interval distance $(-x_0, -x_0 - dx_0)$. Among them there are $(2\rho N_a A\,dx_0/M)\,f(x, n)$ monomers that occur in the interval distance $(x, x + dx)$. The total number of the monomers $f_t(x, t)$ occurring in the interval distance $(x, x + dx)$ is given by

$$f_t(x, t) = \int_{x_0=0}^{x_0=\infty} (2\rho N_a A\,dx/M)\, dx_0 \int_{s=0}^{s=n} ½\pi s a^2)^{1/2} \exp[-(x + x_0)^2/2sa^2]\, ds \tag{2.5.5}$$

Therefore, the number of moles of monomers per unit volume at x (> 0) coming from the polymer piece A, that is, the monomer concentration profile, $C_A(x,t)$, occurring at the interval distance (x, x + dx) at time t is given by

$$C_A(x,t) = (2\rho/M)\int_{x_0=0}^{x_0=\infty} \mathrm{d}x_0 \int_{s=0}^{s=n} (\tfrac{1}{2}\pi s a^2)^{1/2} \exp[-(x+x_0)^2/2sa^2]\,\mathrm{d}s \tag{2.5.6}$$

If we exchange the order of integration and evaluate the second integral with respect to x_0, we find

$$C_A(x,t) = (\rho/M)\int_{s=0}^{s=n} \mathrm{erfc}[x/a\sqrt{2s})]\,\mathrm{d}s \tag{2.5.7}$$

where for $y = x/[a\sqrt{(2s)}]$, the complementary error function is

$$\mathrm{erfc}(y) \equiv (2/\sqrt{\pi})\int_{y}^{\infty} \exp(-z^2)\,\mathrm{d}z \tag{2.5.8}$$

with z a dummy variable. Integrating Eq 2.5.7 gives

$$\begin{aligned} C_A(x,t) = (\rho/M)\{n\ \ &\mathrm{erfc}[x/(a\sqrt{2n})] \\ &- [x/(a\sqrt{2})]\int_{s=0}^{s=n} (1/\sqrt{s})\exp[-x^2/(2sa^2)]\,\mathrm{d}s\} \end{aligned} \tag{2.5.9}$$

Upon evaluating the integral in Eq 2.5.9, we obtain the monomer concentration profile for the symmetric polymer–polymer interface as [61]

$$\begin{aligned} C(x,t) = (\rho/M_0)\{[&l(t)/L + x^2/aL]\ \mathrm{erfc}[x/(a\sqrt{2n})] \\ &- \sqrt{(2n/\pi)}(x/L)\exp[-(x^2/(2na^2)]\} \end{aligned} \tag{2.5.10}$$

with M_0 the monomer molecular weight, $l(t)$ and n defined in Eqs 2.5.1 and 2.5.2, and $L = Na$ the contour length of a chain of N monomers. Because of symmetry, $C_A(x,t)$ is identical to $C_B(-x,t)$, and the subscript A was dropped in Eq 2.5.10.

The calculated concentration profiles $C(x,t)$ given by Eq 2.5.10 are shown in Figure 2.13 at decimal time increments. Here, the time is reduced with respect to T_r, the diffusion distance is reduced with respect to R_g, and the concentration is reduced with respect to the bulk density, ρ/M_0, (moles of monomers per unit volume). The profile is peculiar in that it contains a gap in the concentration profile at the interface plane $x = 0$. The origin of the gap and its effect on mechanical properties of interfaces are discussed in the next section.

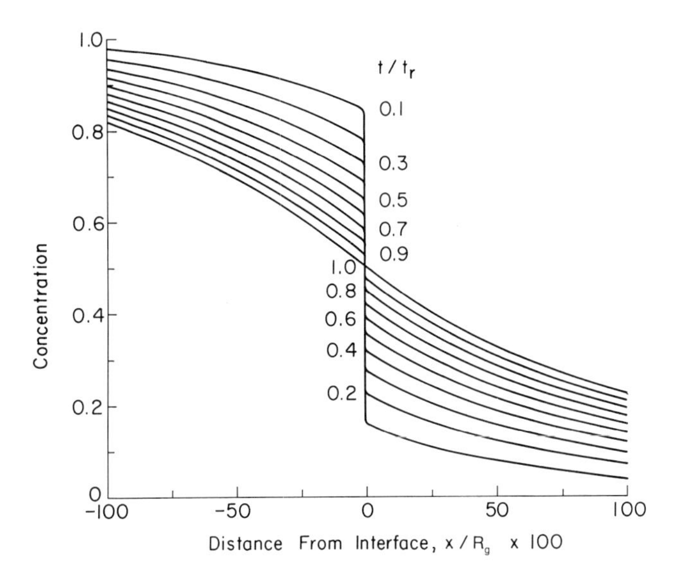

Figure 2.13 Normalized concentration depth profiles $[C(x,t)/(\rho/M_0)]$ at a symmetric polymer-polymer interface at decimal fractions of reduced time t/T_r. The depth coordinate is normalized with the radius of gyration, R_g (Zhang and Wool).

2.5.2 The Gap at the Interface Plane

Figure 2.13 clearly indicates that the concentration at the initially joined surfaces ($x = 0$) is discontinuous when the healing time is shorter than the reptation time. This gap does not disappear until the reptation time, as shown in Figure 2.14, where the reduced concentration at ($x = 0$) is plotted against the healing time. The gap represents regions of the interface plane that have not been threaded through by reptating chains. Simple Fickian diffusion of small molecules does not produce a gap, since after one molecular hop at the interface, half of the molecules have jumped to one side and half to the other, resulting in $C(0,t) = ½$. This also happens with polymer monomers during Rouse relaxation of entanglement segments over distances of the order of the radius of the tube (discussed in Chapter 3). However, for polymers with M much greater than M_c, the gap becomes more apparent with respect to the ratio R_g/R_{ge}, where R_{ge} is the entanglement radius of gyration.

The gap helps us understand the development of mechanical strength at an interface, since it provides a relative measure of the lack of bridges and entanglements. Regions of the interface that have been threaded through by the minor chain ends have new entanglements formed with chains from the other side, while regions that have only Rouse diffusion from lateral segmental motion remain relatively weak. The gap can also be understood in liquid crystal interfaces formed, for example, by the contact of aligned smectic sheets. Diffusion is significantly hindered by the alignment

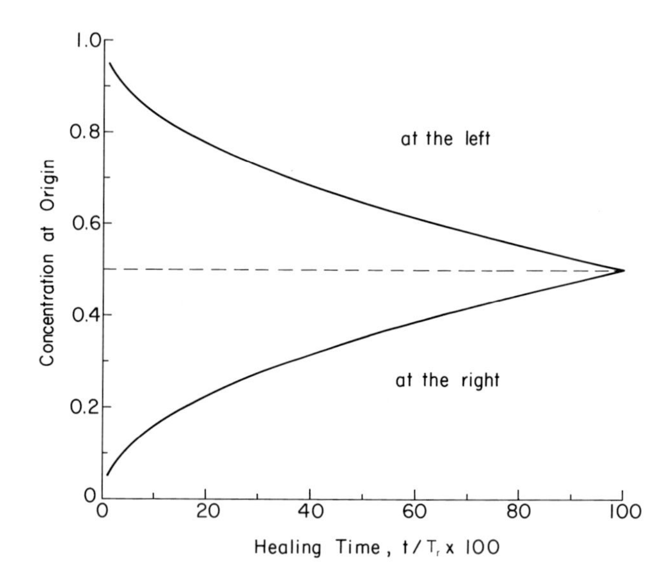

Figure 2.14 The behavior of the gap in the concentration profile at $x = 0$ as a function of time.

constraints. Gradual mixing at the $x = 0$ plane results in a gap that eventually decays to 0.5, consistent with normal diffusion. Other interfaces with gaps involve situations with development of a local resistance to diffusion. This could happen with impenetrable particles, or in nonwetted or noncontacted regions of the interface.

To determine the time dependence of the gap, we let $C_{12}(x,t)$ represent the reduced concentration of those monomers that are on side 1 of the interface at time $t = 0$ and on side 2 at time t. For the symmetric interface, from Eq 2.5.10 we have

$$C_{12}(x,t) = C_{21}(-x,t) = C_A(x,t)/(\rho/M_0) \tag{2.5.11}$$

and

$$C_{12}(x,t) + C_{22}(x,t) = C_{21}(-x,t) + C_{22}(-x,t) = 1 \tag{2.5.12}$$

Therefore, the change in the reduced concentration $J(t)$ from one side of the plane $x = 0$ to its other side is given by

$$J(t) = C_{11}(0^-,t) - C_{12}(0^+,t) = 1 - 2\,C_{12}(0^+,t) \tag{2.5.13}$$

Combining Eqs 2.5.10, 2.5.11 and 2.5.13 gives

$$J(t) = 1 - (4\sqrt{2}/\sqrt{\pi})\,(Dt)^{1/2}/L \tag{2.5.14}$$

or

$$[1 - J(t)] \sim t^{1/2} M^{-3/2} \quad (2.5.15)$$

Since the reduced concentration is always smaller than 0.5 when t is less than T_r and equal to 0.5 at t equal to T_r, then $J(t)$ is always positive if t is less than T_r and zero if t is equal to T_r. This means that there is a discontinuity in the profile at $x = 0$ when t is less than T_r and it does not disappear until the reptation time. The discontinuity and its decay rate are affected by the molecular weight. The larger the molecular weight, the more pronounced is the discontinuity and the more slowly it decays with time. The decay function of the gap has the same time dependence as the number of bridges formed at the interface [25] and the crossing density [13].

The gap at $C(0,t)$ may be difficult to observe experimentally because of the short-range segmental motion. Rouse motion of the chain in the tube allows for lateral motion to depths of about the tube radius (about 50 Å for polystyrene), which effectively broadens the depth profile. However, with high molecular weights, the gap effect should be more pronounced when the broadening due to Rouse relaxation of entanglement segments is small compared to R_g. This is observed to be the case, and will be discussed in Chapters 3 and 6.

2.6 Molecular Properties of a Polymer–Polymer Interface

For the purpose of modeling the development of strength at polymer interfaces, it is useful to have a set of molecular properties, $H(t)$, with their time- and molecular-weight-dependent functions. $H(t)$ is described by a simple scaling law of the form $H(t) = H_\infty (t/T_r)^{r/4}$, where H_∞ is the molecular-weight-dependent equilibrium or virgin property at $t = T_r$, and $r = 1$, 2, or 3, as shown in Table 2.1. These can then be used to examine proposed models and develop new insight into welding problems. For many applications, the strength of the interface is found to depend on $H(t)$ via $G_{1c} \sim H(t)^{2/r}$. The molecular properties are developed in terms of the static and dynamic properties of linear random-coil chains via the minor chain model and compared with computer simulation, depth profile analysis, and scaling relations.

2.6.1 Computer Simulation of Polymer–Polymer Interdiffusion

We did a computer simulation of reptating random-coil chains diffusing on a square lattice across an interface plane [66, 67]. We had several objectives in this study: (a) to investigate the theoretical scaling laws derived for polymer–polymer interdiffusion; (b) to evaluate the non–Fickian concentration profiles for polymer diffusion at times less than the reptation time; and (c) to examine fractal characteristics of polymer diffusion fronts (presented in Chapter 4).

Reptation dynamics was modeled by an algorithm that randomly chose a chain end and added a new monomer in one of three possible new directions and subtracted a monomer from the other end. The process was then repeated for each chain. The chain-end distribution function was chosen to be uniform behind the interface plane. The basic reptation dynamic results were checked for consistency, namely, that (1) the relaxation time behaved as $T_r \sim M^3$, (2) the center-of-mass diffusion coefficient behaved as $D \sim M^{-2}$ for chains whose conformations were not affected by the initial surface, and (3) the monomer displacement function, $r(t)$, behaved as

$$[\langle r(t) - r(0)\rangle^2]_n \sim t^{1/2} \qquad \text{for } t < T_r \tag{2.6.1}$$

and

$$[\langle r(t) - r(0)\rangle^2]_n \sim t \qquad \text{for } t > T_r \tag{2.6.2}$$

where $r(t)$ is the position vector of the nth monomer at time t. The monomer displacement average was determined over all monomers on each chain, and the ensemble average for about 1000 chains was calculated. The chain density was chosen to be inversely proportional to the molecular weight, and the monomer density in the virgin state was chosen to consist of an average of five monomers per lattice site, to account for interpenetration of random-coil chain statistical segments.

The reptating chains diffused across the interface plane and formed fractal-like structures at $t = T_r$, as shown in Figure 2.15. In this example, the molecular weight is expressed as $M/M_c = 80$. The polymer interface is seen to consist of features resembling lakes, islands, and seashore. The remaining holes near the $x = 0$ plane are due to incomplete filling of the gap, shown in Figure 2.13 and 2.14. The fractal aspects of the interface structure are very interesting and are treated separately in Chapter 4. The simulation results for symmetric interfaces are incorporated into the following sections and compared with theory.

2.6.2 Number of Chains Intersecting the Interface, *n*(*t*)

The number of random-coil chains intersecting a unit area of the interface as a function of contact time, t, and molecular weight, M, behaves as

$$n(t) = n_\infty (t/T_r)^{1/4} \qquad \text{for } t < T_r \tag{2.6.3}$$

and the equilibrium property is

$$n_\infty \sim M^{-1/2} \qquad \text{for } t < T_r \tag{2.6.4}$$

with $n(t) \sim t^{1/4} M^{-5/4}$.

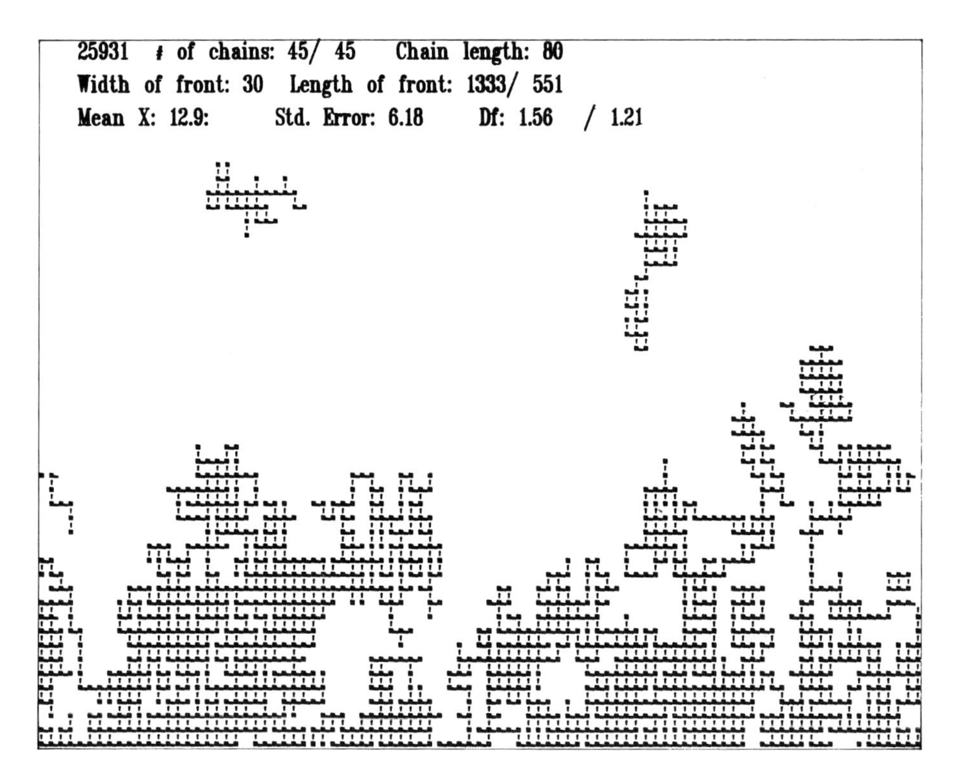

Figure 2.15 Computer simulated polymer interface at T_r for random-coil chains ($N = 80$) reptated on a 2-d lattice. The fractal nature of this result is discussed in Chapter 4 (Wool and Long).

A chain is considered to intersect the interface when one of its ends has crossed the interface contact plane defined at $t = 0$. The number of chains doing that, $n(t)$, also scales as X/M, where X is the average monomer segment interpenetration distance. Eq 2.6.3 is derived on the basis that all chains whose ends are within a distance of X from the interface will have intersected the interface at time t, and that the number of chains per unit volume varies as M^{-1}. The $t^{1/4}$ time dependence is atypical compared to the usual $t^{1/2}$ expected for atomic diffusion processes.

While we are primarily interested in chains in contact with a unit area of the interface plane, we can also determine the total number of chains that have crossed through the unit area to the other side at times greater than the reptation time. Since $n(t) \sim X/M$ and $X \sim t^{1/2} M^{-1}$, then we have at t much greater than T_r,

$$n(t) \sim t^{1/2} M^{-2} \tag{2.6.5}$$

or

$$n(t) = n_\infty (t/T_r)^{1/2} \tag{2.6.6}$$

The change in the time dependence ($t^{1/4}$ to $t^{1/2}$ at T_r) and the molecular weight dependence ($M^{-5/4}$ to M^{-2}) of this molecular property is quite characteristic of the reptation model and provides several opportunities for experimental investigation.

The equilibrium or virgin state solution at $t = T_r$ is obtained by substitution for the reptation time $T_r \approx M^3$ in the dynamic solutions. The equilibrium solutions are independent of the molecular dynamics model and can be derived from other considerations. Equation 2.6.4 was derived previously by Wool and Rockhill [68]. If we pass an imaginary plane through dense interpenetrated random-coil chains, the number of chains intersecting a unit area of this plane from both sides is determined by

$$n_\infty = 1.31\,\rho\,N_a\,(C_\infty\,j/M_0)^{1/2}\,b_0\,M^{-1/2} \tag{2.6.7}$$

with ρ, N_a, C_∞, b_0, and j the density, Avogadro's number, characteristic ratio, bond length, and number of backbone bonds per monomer, respectively. We can use Eq 2.6.7 to evaluate the number of bonds and chains broken in the fracture of glassy polymers (Chapter 8), and it is a key part of the entanglement analysis in Chapter 7.

When a molecular property is suggested as controlling the time dependence of welding, that property can be used to simultaneously predict both the molecular weight dependence of welding and the molecular weight dependence of fracture in the virgin state. This three-way concurrence is useful in the analysis of potential solutions to welding problems. For example, it is known from the experiments of Kausch and coworkers [6–12] and our work [14–16] that the fracture stress, σ, during crack healing and welding increases as $\sigma \sim t^{1/4}$. We could hypothesize upon inspection of Eq 2.6.3 that the number of chains intersecting the interface, $n(t) \sim t^{1/4}$, is responsible for the strength development. However, we reject this hypothesis when we examine the virgin state property, $\sigma_\infty \sim M^{-1/2}$. The fracture stress of polymers is known to increase with molecular weight to an asymptotic value. Further experimental study would show that the healing rate also does not depend on $M^{-5/4}$. Consider the case where the chain ends are initially segregated on the interface plane so that the number of chains, n_∞, remains constant during welding. As the chains interdiffuse the weld strength increases with time and it is intuitively obvious that some length scale, such as the contour length $l(t)$, is necessary to describe the strength development.

Conversely, molecular models used to describe the virgin state strength G_{1c} can also be evaluated by their ability to describe the time and molecular weight dependence of healing. If we hypothesize that $G_{1c} \sim M^2$, based on the energy to pull out a chain of length L behaving as $G_{1c} \sim L^2$, then we might expect that $G_{1c} \sim t/M$, which is incorrect. This argument applies when the same microscopic deformation mechanisms are operative during welding and fracture of the virgin state.

2.6.3 Number of Bridges Crossing the Interface, $p(t)$

Each time a piece of diffusing chain crosses once through the interface, it creates a molecular bridge. Intuitively, this may be visualized as a "sewing up" of the interface.

The number of these bridges, $p(t)$, is determined by $p(t) \sim n(t) \sqrt{l(t)}$, where $\sqrt{l(t)}$ is proportional to the number of bridges created by each minor chain (a random walk of N steps originating at a plane crosses the plane $\sqrt{N}$ times). Thus we obtain

$$p(t) \sim t^{1/2} M^{-3/2} \tag{2.6.8}$$

$$p_\infty \sim M^0 \tag{2.6.9}$$

This result was derived by de Gennes [25] using scaling methods, and is equivalent to the crossing density determined by Prager and Tirrell [13] from a more detailed calculation of the interdiffusion process.

Figure 2.16 shows the computer simulation result of the number of crossings $p(t)$ versus time on a log–log plot, with molecular weights $N = M/M_e$ ranging from $N = 60$ to $N = 100$. At t less than T_r, the simulation slopes are about 0.5, in agreement with Eq 2.6.8. At t equal to T_r, the number of crossings becomes saturated fairly abruptly on the log scale, and the slope becomes zero. The molecular weight dependence of the crossing rate, $M^{-3/2}$, is a strong dependence and denotes a clear distinction between the crossing density and other molecular properties such as $l(t)$, which has an $M^{-1/2}$ dependence on healing rate.

We can also derive the bridge relations from the concentration profile analysis in the last section. The number of bridges intersecting a unit area is the same as the

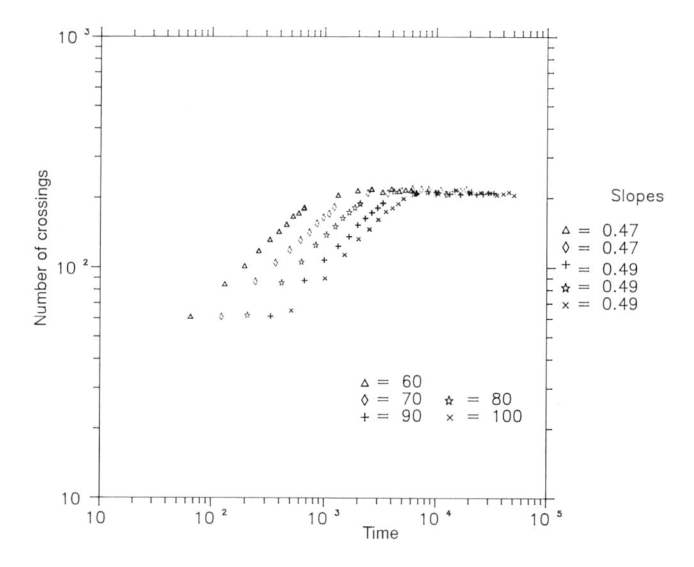

Figure 2.16 Number of interface crossings versus time for N of 60–100. The slopes for each molecular weight are given. The predicted slope is 0.5 (Wool and Long).

number of monomers at the interface plane. Therefore, when $x = 0$, Eq 2.5.10 for the concentration profile gives

$$p(t) = C(0,t) = (\rho / M_0)\, l(t) / L \tag{2.6.10}$$

Substituting for $l(t)$ and $L \sim M$ gives $p(t) \sim t^{1/2} M^{-3/2}$. The scaling law for $p(t)$ is then

$$p(t) = p_\infty (t / T_r)^{1/2} \tag{2.6.11}$$

To determine the crossing density of the virgin state, p_∞, we consider the local structure of the chain as it crosses the interface plane. For example, a section of chain of length c, with z monomers, has a volume $V = Ac$ (A is the cross-sectional area). The number of crossings per unit area is $p_\infty = 1/\sqrt{2}A$; the $\sqrt{2}$ accounts for random orientation of segments. The repeat unit volume is $V = M_0 z / \rho N_a$, and p_∞ is given by

$$p_\infty = c \rho N_a / (\sqrt{2} M_0 z) \tag{2.6.12}$$

This relation is important in the analysis of entanglement development and is discussed further in Chapter 7. The reader may find the following solved problem helpful.

Problem

Find (a) the number of bridges, (b) the number of chains per unit area, and (c) the number of bridges per chain, at a healed (equilibrium) interface in a polystyrene melt of density 0.9 g/cm³ and number average molecular weight, M, 250,000 g/g mol.

Solution

(a) Assume that the local structure of the chain is similar to the isotactic threefold helix. (This is not a critical assumption: an atactic structure of the monomers in the immediate vicinity of the plane gives a cross-sectional area essentially identical to either the syndiotactic or isotactic forms). Thus, c = 6.5 Å, z = 3 monomers per c-axis length, $M_0 = 104$, and Eq 2.6.12 gives

$p_\infty = (6.5\ \text{Å})(0.9 g/cm^3)(6.02 \cdot 10^{23}/g\ mol)/\ (\sqrt{2})(104\ g/g\ mol)(3)(10^8\ \text{Å}/cm)$
$\quad = 8 \times 10^{13}$ *crossings/cm²*

(b) We calculate the number of chains per unit area from Eq 2.6.6 with $C_\infty = 10$, $b_0 = 1.54$ Å, and $j = 2$ as

$$n_\infty = \frac{(1.31)(0.9 g/cm^3)(6.02 \cdot 10^{23} g/g mol)\{(10)(2)/(104 g/g mol)\}^{1/2}(1.54 \text{Å})(10^{-8} \text{Å}/cm)}{(250{,}000\ g/g\ mol)^{1/2}}$$

$\quad = 9.6 \times 10^{12}$ *chains/cm²*

(c) The number of crossings per chain, $p_c = p_\infty / n_\infty$
= 8.3 crossings per chain

Note that at $M_c \approx 32{,}000$, $p_c \approx 3$.

The virgin-state number of bridges is independent of molecular weight; hence, bridges alone cannot be used to describe strength development during welding in cases where the virgin strength exhibits a molecular-weight dependence over some range of molecular weight, as observed for most polymers. Modifications to the bridge models have been suggested by Adolf, Tirrell, and Prager [70], and by Mikos and Peppas [71], and involve a minimum interpenetration depth, X_e, to develop entanglements. These approaches introduced a molecular-weight dependence for the virgin state over a range of molecular weight, resolving the problem posed by Eq 2.6.9, and gave a more complex expression for the time dependence of healing. Extrapolation of long-time healing data to zero strength is predicted by this modification to intersect the time axis at a finite time t_e, required to diffuse to the minimum entanglement distance X_e.

2.6.4 Average Monomer Interpenetration Depth, $X(t)$

The average monomer interpenetration depth, $X(t)$, is obtained from an integration over the Gaussian segment density profiles of the interdiffused minor chain population, as shown in Section 2.4, or from the concentration profile $C(x,t)$. $X(t)$ is derived from the concentration profile as follows.

$$X(t) = \langle x \rangle = \int_0^\infty x\, C(x,t)\, \mathrm{d}x \Big/ \int_0^\infty C(x,t)\, \mathrm{d}x \tag{2.6.13}$$

This method is used in Chapter 5 to examine experimental depth profiles. As a check on the theoretical depth profile, substituting Eq 2.5.10 in this equation we obtain,

$$X(t) = 0.47\, a \sqrt{n} \tag{2.6.14}$$

Since $n/a = l(t)$, Eq 2.6.14 yields

$$X(t) \sim t^{1/4} M^{-1/4} \tag{2.6.15}$$

At T_r, the equilibrium monomer interpenetration distance, X_∞, is obtained from Eq 2.6.14, letting $n = N/2$ and the radius of gyration $R_g = N^{1/2}\, a/\sqrt{6}$, as $X_\infty = 0.81\, R_g$.

The computer simulation results are shown in Figure 2.17, where log $X(t)$ is plotted versus log $t^{1/4}$ for a range of molecular weights. At $t \leq T_r$, the data show excellent linearity with the $t^{1/4}$ plot, but because the dependence on molecular weight is weak ($M^{-1/4}$), the data for different molecular weights appear superposed. However, at $t > T_r$, $X(t)$ behaves as $X(t) \sim t^{1/2} M^{-1}$ and the stronger molecular weight dependence gives a wider spread of the data. We can understand the transition from $t^{1/4}$ to $t^{1/2}$ by considering the displacement, s, of an individual monomer along the contour of the tube as $s \sim t^{1/2}$ for $t < T_r$. However, the curvilinear motion is related to translational motion as $X \sim s^{1/2}$ because the chain is a random coil. This is analogous to a

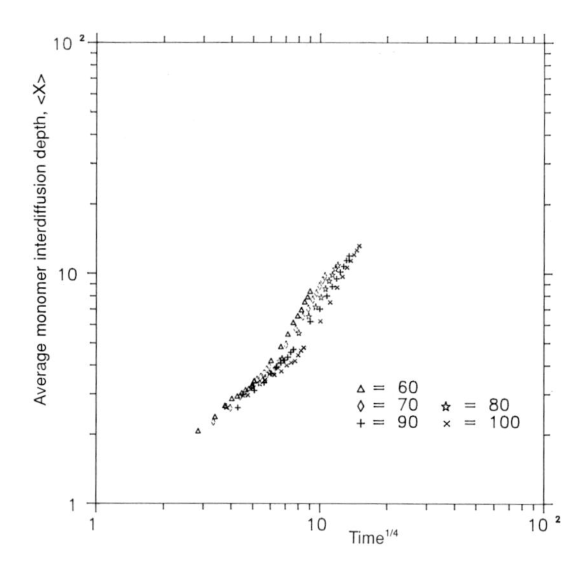

Figure 2.17 Average monomer diffusion depth as a function of the fourth power of the time on a double log plot for molecular weights in the range 60–100. When t is less than T_r, $X \sim t^{1/4}$, and when t is greater than T_r, $X \sim t^{1/2}$. The transition in slope occurs near $X(T_r) \approx 0.8\ R_g$, where $R_g = N^{1/2}/\sqrt{6}$.

"Brownian rabbit" in a burrow of length L twisted randomly in a hillside. The rabbit moves along the burrow at a pace determined by $s \sim t^{1/2}$ but is unaware that the burrow is a random walk. Therefore his translational progress into the hillside is determined by a double square root law, or $t^{1/4}$. For times greater than T_r, when the original tube has disappeared, the motion of the monomer becomes uncorrelated, since s is greater than the chain length and its displacement becomes similar to the center-of-mass displacement.

The crossover from $t^{1/4}$ to $t^{1/2}$ has been examined by secondary ion mass spectrometry (SIMS) [50, 51], and specular neutron reflectivity (NR) [52–53, 73–75]. These methods were used to probe the concentration profile at a PS/PS interface for $t < T_r$ and distances $X < R_g$. Strong evidence for the crossover was obtained and will be presented in detail in Chapters 5 and 6. Center-of-mass motion for $t > T_r$ and $X >> R_g$ is well established from self-diffusion studies [44–48]. For chains in the bulk, far away from the interface, the monomer motion is predicted to proceed as $t^{1/4}$ for $t < T_r$, while the center of mass proceeds as $t^{1/2}$ at **all** times. Evans and Edwards [72] showed that the center-of-mass motion is uncorrelated at all times by virtue of the random motion of the chain ends. Center-of-mass motion is different for chains whose random-coil conformations are altered by the interface, as shown in Figure 2.7. The chains initially have a non–Gaussian conformation due to the reflecting boundary condition at the surface. As the minor chains interdiffuse, the configurations of the affected chains in the surface layer relax to Gaussian configurations. Since the center of mass of each chain is close to the peak of the segment density profile, very little motion of the center of mass occurs for times t less than T_r.

2.6.5 Number of Monomers Crossing the Interface, $N(t)$

The number of monomers that have crossed from one side of the interface to the other is equivalent to the total interpenetration contour length, $N(t) \sim n(t)\ l(t)$, and is obtained by integration of the concentration profile $C(x,t)$ over the positive semi-infinite region

$$N(t) = \int_0^\infty C_A(x,t)\,\mathrm{d}x \tag{2.6.16}$$

Substituting Eq 2.5.10 in Eq 2.6.16 and evaluating the integral, we obtain the (non–Fickian) result

$$N(t) = 0.265\ l(t)^{3/2}\ a^{1/2}/L \tag{2.6.17}$$

Substituting for $l(t)$, we have,

$$N(t) \sim t^{3/4}\ M^{-7/4} \tag{2.6.18}$$

The scaling law is determined by

$$N(t) = N_\infty\ (t/\ T_r)^{3/4} \tag{2.6.19}$$

when $t \leq T_r$, and by $N_\infty(t) = N_\infty\ (t/T_r)^{1/2}$ when $t > T_r$. From Eq 2.6.17, N_∞ is related to R_g via $l(T_r) = 0.718\ L$, and $R_g = \sqrt{3}\ La^{1/2}$, so that

$$N_\infty \approx 0.28\, R_g \tag{2.6.20}$$

or $N_\infty \approx R_g/4$.

$N(t)$ and $X(t)$ are related to each other at $t < T_r$ via

$$N(t) \sim X(t)^3/M \tag{2.6.21}$$

and by $N(t) \sim X(t)$ at $t > T_r$. Figure 2.18 shows the computer simulation result where the number of monomers, $N(t)$, is plotted against the average monomer interpenetration depth, $X(t)$, for several molecular weights. The slope of 3.0 changes to 1.0 at $t = T_r$, and the data for different molecular weights are seen to collapse onto a single line at $t > T_r$, in excellent agreement with the above relations. The unusual $t^{3/4}$ dependence is

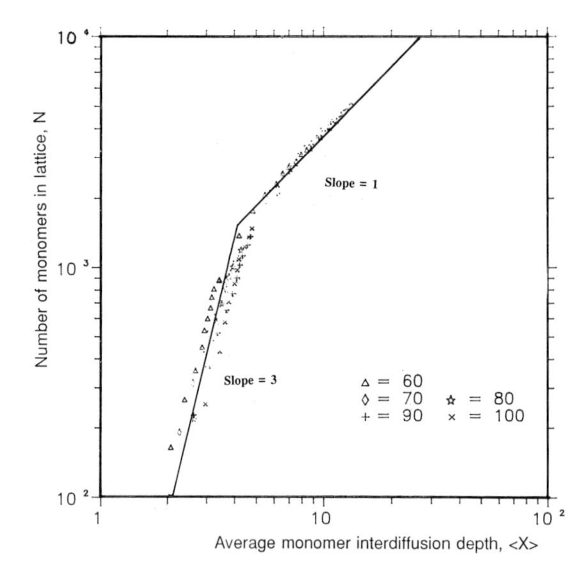

Figure 2.18 Number of monomers, *NL*, diffused versus average diffusion depth for molecular weights in the range 60–100. When t is less than T_r, $NL \sim X^3$, and when t is greater than T_r, $NL \sim X$ (Wool and Long).

again due to correlated motion effects and reverts to a normal $t^{1/2}$ diffusion dependence for times greater than T_r. This result is supported with SIMS experiments [51] in Chapter 5. However, $N(t)$ is sensitive to the initial position of the chain ends, which we discuss in the next section.

2.7 Interface Properties with Chain Ends on Surface

In the above discussions, we assumed that the chain-end concentration, ρ_e, was constant and was a function only of molecular weight via

$$\rho_e \sim 1/M \tag{2.7.1}$$

and that chain ends are uniformly distributed in space and show no preference for migration to the surface. Prager and Tirrell [13] and de Gennes [65, 76–78] considered the case where the chain ends preferentially segregate to the surface before the welding starts. This can be very important in healing cracks where chain disentanglement and fracture produce many chain ends near the surface.

2.7.1 Surface Segregation Effect on Molecular Properties

With surface segregation, the concentration of chain ends, ρ_s, at $t = 0$ is determined by Eq 2.6.4, so that

$$\rho_S \sim 2/M^{1/2} \tag{2.7.2}$$

The difference of $M^{-1/2}$ between the uniform ($\rho \sim M^{-1}$) and the surface-segregated case results in a much higher chain-end concentration. With reptation-dominated diffusion, the surface-segregated case would give enhanced diffusion and a more rapid gain in welding strength.

Surface chain-end segregation has the following effect on the molecular properties, $H(t)$: the average properties, such as the average contour length, $l(t)$, average monomer interpenetration distance, $X(t)$, and average bridge length, $l_p(t)$, are unaffected, but the number of chains, $n(t)$, number of bridges, $p(t)$, and number of monomers crossing the interface, $N(t)$, have different scaling laws. Using subscript "s" for the surface-segregated case, we have the following new results for $n_s(t)$, $p_s(t)$, and $N_s(t)$.

The number of chains crossing the interface is essentially constant, since all the chain ends at distances $x \leq R_g$ have their chain ends on the surface at $t = 0$ and no new additional chain ends contribute at $t \leq T_r$. Thus

$$n_S(t) \sim t^0 M^{-1/2} \tag{2.7.3}$$

The number of bridges crossing the interface is determined from the number of bridges per chain, $p_c(t)$, times the number of chains per unit area, $n_s(t)$. A random walker taking $l(t)$ steps that begin at $x = 0$ crosses the interface plane $\sqrt{l(t)}$ times. Hence we obtain from $p_s \sim p_c n_s$,

$$p_S(t) \sim t^{1/4} M^{-3/4} \tag{2.7.4}$$

$$p_S(T_r) \sim M^0 \tag{2.7.5}$$

where the virgin-state property at $t = T_r$ is the same as the uniform chain-end case but the time dependence has changed from $t^{1/2}$ to $t^{1/4}$. This change in exponent with surface segregation could be used to explore the effect of bridges on weld strength, since the exponents for the average molecular properties such as $l(t)$ remain unchanged.

The number of monomers crossing the interface is determined by $N_s(t) \sim l(t)\, n_s(t)$:

$$N_S(t) \sim t^{1/2} M^{-1} \tag{2.7.6}$$

$$N_S(T_r) \sim M^{1/2} \tag{2.7.7}$$

Note that when t is less than T_r, $N(t) \sim t^{1/2}M^{-1}$; we observe that surface segregation has eliminated the interesting $t^{3/4}$ to $t^{1/2}$ transition described previously for the uniform case.

In summary, the effect of surface segregation on $H(t)$ is to change the time and molecular weight dependence of the "number" properties (p, N, n), while the average properties (l, X) and the virgin state properties (H_∞), remain the same. For other chain-end distribution functions, which are intermediate between the uniform and the segregated cases, we can expect intermediate exponents for the scaling laws. Depending on the specific shape of the distribution function, the exponents may no longer be constant with time, or may have intermediate values. For example, with the number of monomers crossing the interface, $N(t)$, we could have time exponents between ¾ (uniform) and ½ (segregated). SIMS experiments could be used to examine this problem for specific cases.

2.7.2 Asymmetric Chain-End Segregation Effects on Interdiffusion

Brochard-Wyart and Pincus [80] considered the interesting case depicted in Figure 2.19, where a symmetric interface (in terms of the molecular weight and chemical species) has the chain ends segregated on one side at $t = 0$. This situation could arise if the chain ends had different chemistry or different rheological history. We could set up an example of this interface for welding two polymers if we placed a fractured (by disentanglement) surface in contact with a normal surface of the same material.

Interfaces with chain-end asymmetry have an unusual accordion-like feature; the position of the interface moves by a distance $u(t)$, then returns to its original position, like an accordion or concertina. This behavior is explained in Figure 2.20 as follows:

a. (Figure 2.20a) At times less than the Rouse relaxation time for the entanglements $\tau_e = \tau_0 N_e^2$, the position of the chain ends is not important and an equal flux of

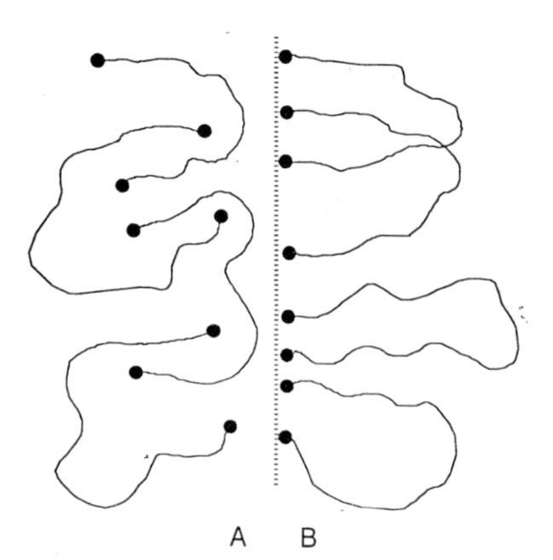

Figure 2.19 An interface of identical polymers with chain ends segregated on one side (courtesy of F. Brochard-Wyart and P. Pincus).

monomers to the A and B sides occurs. The resulting concentration profile is symmetric. This behavior exists only for $t < \tau_0 N_e^2$.

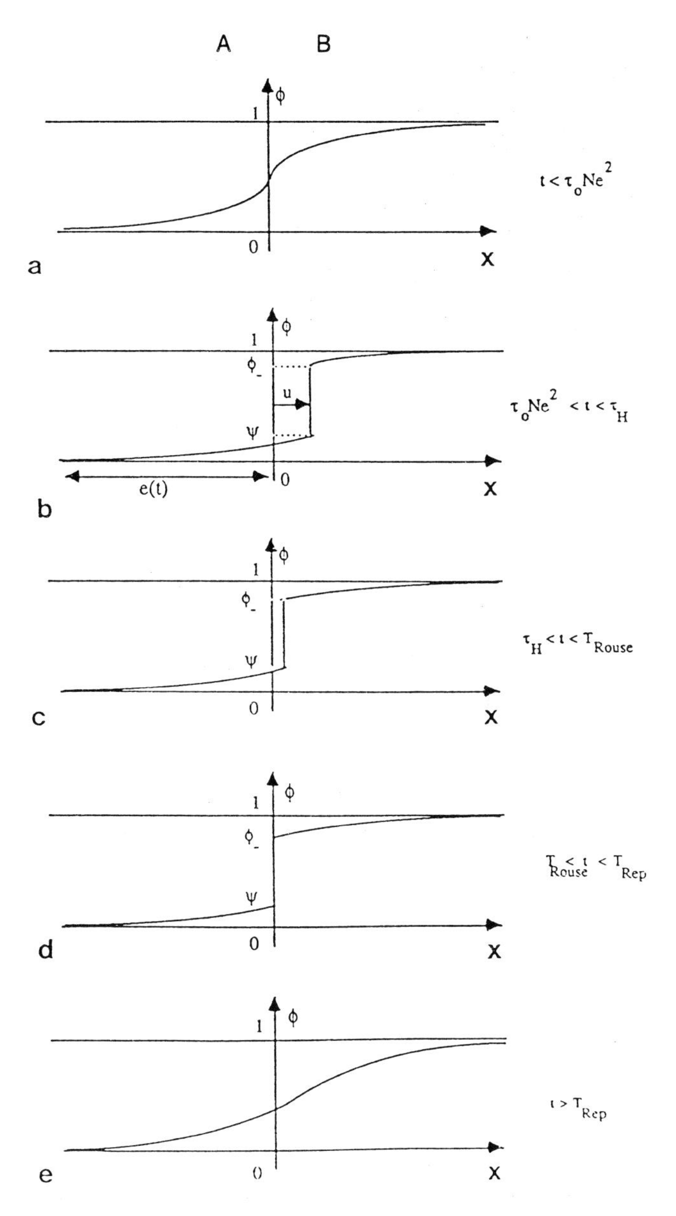

Figure 2.20 "Accordion" motion of the position of the interface (Figure 2.19) during interdiffusion. Details discussed in text (courtesy of F. Brochard-Wyart and P. Pincus).

b. (Figure 2.20b) For $t > \tau_e$, the chain-end asymmetry takes effect and we get more flux from the B monomers into the A side than the reverse flux of A to B. The excess of chain ends on the B surface initially results in more rapid diffusion into the A side. This causes the position of the interface to move by a distance $u(t)$ into the B side. The movement of the interface is in the opposite direction to the net flux (Chapters 3 and 6).
c. (Figure 2.20c) The unequal flux of monomers continues until the melt incompressibility factor generates a resistance stress, which slows down the rapid diffusion of B monomers into the A side. This is proposed to occur when the osmotic swelling pressure is balanced by the gel stress at the A/B interface. The resistance to further enhanced diffusion allows the flux from the A side to catch up and offset the imbalance. As a result, the interface position begins to move back to its original position, thus completing the accordion cycle. The time at which this occurs is between τ_e and the Rouse relaxation time of the whole chain, τ_{RO}.
d. (Figure 2.20d) Between τ_{RO} and the reptation time T_r, the chains weave back and forth through the interface with equal probability from both sides and normal reptation occurs, giving the gap in Figure 2.20d. In fact the gap has been eliminated by the Rouse diffusion that occurs in Figure 2.20a and is obvious only with very high molecular weights (Chapter 3). Thus, the unusual accordion effect is lost at times in the range between τ_e and τ_{RO}.
e. (Figure 2.20e) At t much greater than T_r, normal Fickian diffusion occurs and the profile broadens.

It should be possible to investigate the accordion phenomenon using neutron reflection. However, the trick in this experiment would be to ensure that the chains on the B side existed at the interface initially and that the experiments could be reliably conducted over a diffusion depth of about 30–50 Å.

2.8 Scaling Laws for Polymer–Polymer Interdiffusion

The molecular properties $H(t)$ described above and others listed in Table 2.1 are interrelated and have a convenient common scaling law. In this section, we show how this scaling law can be used to address complex issues with polydispersity and provide structure–strength interrelations for a range of model interfaces.

2.8.1 General Scaling Law

The dynamic properties, $H(t)$, for the symmetric interface are related to the static properties, H_∞, and the reduced time, t/T_r, by

$$H(t) = H_\infty (t / T_r)^{r/4} \tag{2.8.1}$$

$$H_\infty \sim M^{(3r-s)/4} \tag{2.8.2}$$

where r, $s = 1, 2, 3, \ldots$.

If the chain ends are segregated on the surface at the start of the welding as discussed in the last section, then the same general scaling law applies but with different values of r and s for some of the "number" properties. The average properties remain unaffected.

2.8.2 Polydispersity Effects on Interdiffusion

Polydispersity effects were considered by Tirrell *et al.* [63, 64] and by our group. Assuming independent chain diffusion, the contribution to the concentration profile from individual reptating chains can be determined for any molecular weight distribution. At short times, the weld strength is dominated by contributions from the faster diffusing shorter chains, followed later by the slower diffusing longer chains. In this section we provide a general solution for the effect of the molecular weight distribution, $f(M)$, on the molecular properties, $H(t)$.

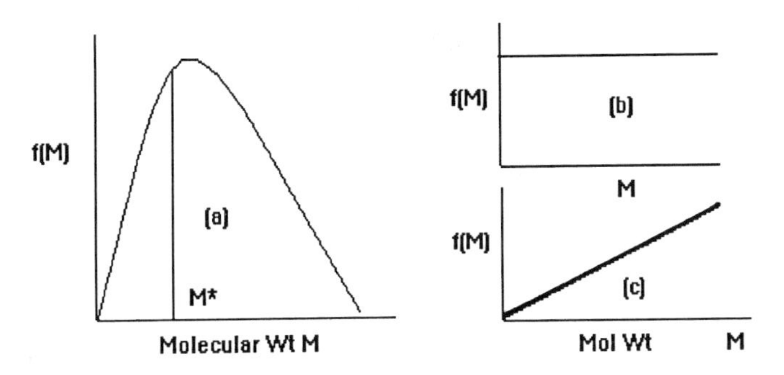

Figure 2.21 Molecular weight distributions, $f(M)$. *(a)* Typical polydisperse distribution where M^* separates the diffused equilibrium chains from the diffusing chains; *(b)* flat profile; *(c)* ramp profile.

Figure 2.21 shows typical normalized molecular weight distributions, where $f(M)$ represents the *number* fraction of chains with molecular weight M. We assume that we have no segregation of the individual molecular weight species. We obtain a good approximate solution to the complex polydispersity problem if we realize that at a given welding time t, the distribution is divided at M^* into equilibrium and dynamic contributions: those chains with $M < M^*$ have already reached their equilibrium state $H_\infty = M^{(3r-s)/4}$, but those chains with $M > M^*$ continue to diffuse according to $H(t) = H_\infty\,(t/T_r)^{r/4}$. The relation between t and M^* is

$$t = \tau_e (M^*/M_e)^3 \tag{2.8.3}$$

where τ_e is the entanglement relaxation time. The dynamic contribution for chains with $M > M^*$ can be expressed in terms of M^* via $t/T_r \sim (M^*/M)^3$ as

$$H(t) = M^{*(3r/4)} M^{(-s/4)} \tag{2.8.4}$$

Thus, contributions from $f(M)$ to $H(t)$ accumulate as M^* increases according to $M^* \sim t^{1/3}$ up to the maximum molecular weight M_∞. In Figure 2.21, this corresponds to the position of M^* moving from left to right with increasing time.

The average molecular property $H(t)$ contributed by all molecular weight species is therefore obtained from the sum of their individual equilibrium and dynamic contributions,

$$H(t) = \underbrace{\int_0^{M^*} f(M)\, M^{(3r-s)/4}\, dM}_{\text{(equilibrium)}} + \underbrace{\int_{M^*}^{M_\infty} f(M)\, M^{*(3r/4)}\, M^{-s/4}\, dM}_{\text{(dynamic)}} \tag{2.8.5}$$

in which the time dependence occurs through $M^* \sim t^{1/3}$ (Eq 2.8.3). One uses this relation by first choosing the molecular weight distribution $f(M)$, integrating to obtain $H(t)$, and then solving for individual molecular properties [$l(t), p, n, X$, etc.] using their respective r and s values from Table 2.1. The following example illustrates this.

Example: Flat Molecular Weight Distribution

We will determine the average contour length $l(t)$ for welding with a flat molecular weight distribution, as shown in Figure 2.21. This distribution approximates the broad and flat profile obtained with condensation polymerization at a very high extent of reaction, but its use here is primarily pedagogical. Substituting for $f(M) = 1/M_\infty$ in Eq 2.8.5, we obtain the general solution

$$H(t) = \frac{4}{[(4-s)M_\infty]} \left\{ M^{*(3r-s+4)/4} \left[\frac{(4-s)}{(3r-s+4)} - 1 \right] + M^{*(3r/4)} M_\infty^{(1-s/4)} \right\} \tag{2.8.6}$$

For the contour length $l(t)$, Table 2.1 gives $r = s = 2$; the solution in terms of M^* is

$$l(M^*) = \frac{2M^{*3/2}}{\sqrt{M_\infty}} - \frac{3/2\, M^{*2}}{M_\infty} \tag{2.8.7}$$

When $M^* = M_\infty$, the average diffused contour length is $L_\infty = M_\infty/2$. Substituting for $M^* \sim t^{1/3}$, $L_\infty \sim M_\infty/2$, and $\tau_\infty \sim M_\infty{}^3$, we obtain the scaling law for the flat distribution,

$$l(t) = 4L_\infty (t/\tau_\infty)^{1/2} [1 - \tfrac{3}{4}(t/\tau_\infty)^{1/6}] \tag{2.8.8}$$

This function behaves approximately as $l(t) \sim t^{1/2}$ for values of $l(t)/L_\infty$ up to 0.4 at $t/\tau_\infty \approx 0.04$, and when t/τ_∞ is greater than 0.04, the apparent time exponent decreases. Thus, welding is nearly complete at a small fraction of the longest relaxation time τ_∞.

Exercise

Consider the molecular weight distribution f(M) given by the linear ramp function

$f(M) = 2M/M_\infty{}^2$

as shown in Figure 2.21. Show that the average diffused contour length is given by

$l(t)/L_\infty = 2\,(t/\tau_\infty)^{1/2} - (t/\tau_\infty)$

where $\tau_\infty \sim M_\infty{}^3$ *and* $L_\infty = \tfrac{2}{3} M_\infty$. *This function initially behaves as* $l(t) \sim t^{1/2}$ *(approximately) up to* $l(t)/L_\infty \approx \tfrac{1}{2}$, *after which the exponent decreases.*

The reader may wish to analyze interdiffusion using the theoretical molecular weight distributions for linear step reaction (condensation) polymerization and chain reaction (free radical, ionic) polymerization. For condensation reactions, we have Flory's relations expressed in terms of the degree of polymerization, M, and extent of reaction, p,

$$f(M) = p^{M-1}(1-p) \tag{2.8.9}$$

where $p \leq 1$. With chain reaction polymerization,

$$f(M) = (1-p)^2 (M+1)\, p^M \tag{2.8.10}$$

With anionically polymerized polymers ("living polymerization"), $f(M)$ is essentially a Dirac delta function (single-valued spike at M), which, when used in Equation 2.8.5, generates the monodisperse result for $H(t)$.

2.8.3 Molecular Interrelations at Welding Interfaces

We present here the interrelationship between the molecular properties $H(t)$ for interfaces formed with (a) random-coil chains with initial random chain-end distribution in the bulk, (b) random-coil chains with initial surface segregation of chain ends, (c) rigid rods with random chain-end distribution, and (d) rigid rods with surface segregation of chain ends. These examples apply to a wide range of interfaces with different molecular architectures, compositions, and structures. The resulting properties and interrelationships can be used to analyze liquid crystal (smectic, nematic) interfaces, incompatible A/B interfaces, diblock A–B compatibilizers at A/B interfaces, coupling agents at polymer–metal interfaces, nails at wood–wood interfaces, short fibers at cement interfaces, tire cord pullout from rubber, fiber-reinforced composite interfaces, and related problems in material interfaces. We find these relations useful as we try to formulate a universal solution for evaluating the strength of a wide range of diverse interfaces.

To use Tables 2.2(a–d), relate a property H_1 in the vertical column to H_2 in the horizontal column, via $H_1 \sim H_2^a M^b$, where a and b are exponents: for example, the monomer depth X (H_1) is related to the number of chains n (H_2) by $X \sim nM$. The time dependence of these properties may be obtained from Table 2.1, or through $l \sim t^{1/2}M^{-1/2}$.

Interface with Random-Coil Chains and Random Chain-End Distribution

This interface is the one we have analyzed most intensively in this chapter and which is most commonly encountered in practical applications. There are no constraints among the $H(t)$ variables other than molecular weight M, and the interrelations are shown in Table 2.2(a).

Table 2.2(a) Interface with Random-Coil Chains and Random Chain-End Distribution

$H_1 \backslash H_2$	l	n	p	X	N	p_c	H_∞
Contour, l	l	n^2M^2	pM	X^2	$N^{2/3}M^{2/3}$	p_c^2	M
Chains, n	$l^{1/2}/M$	n	$p^{1/2}M^{-1/2}$	X/M	$N^{1/3}M^{2/3}$	p_c/M	$M^{-1/2}$
Bridges, p	l/M	n^2M	p	X^2/M	$N^{2/3}M^{-1/3}$	p_c^2/M	M^0
Depth, X	$l^{1/2}$	nM	$p^{1/2}M^{1/2}$	X	$N^{1/3}M^{1/3}$	p_c	$M^{1/2}$
Monomers, N	$l^{3/2}/M$	n^3M^2	$p^{3/2}M^{1/2}$	X^3/M	N	p_c^3/M	$M^{1/2}$
Bridges/chain, p_c	$l^{1/2}$	nM	$p^{1/2}M^{1/2}$	X	$N^{1/3}M^{1/3}$	p_c	$M^{1/2}$

Interface with Random-Coil Chains and Chain-End Segregation

In this case, the chain ends are segregated to the surface before welding commences and the number of chains, n, remains constant during interdiffusion. The interrelations between properties are given in Table 2.2(b).

Rigid-Rod Interface with Random Chain-End Distribution

This case describes liquid crystalline polymers, diblocks with M less than M_c, unentangled polymers, and brushes. The constraint in this case is that the number of bridges per chain, p_c, be equal to unity. The properties are given in Table 2.2(c).

Rigid-Rod Interface with Chain-End Segregation

In this system we have three constraints, $p_c = 1$, $p = n$, and $n =$ constant. The resulting properties are given in Table 2.2(d).

Table 2.2(b) Interface with Random-Coil Chains and Chain-End Segregation; n = constant

$H_1 \backslash H_2$	l	n	p	X	N	p_c	H_∞
Contour, l	l	0	p^2/n^2	X^2	N/n	p_c^2	M
Chains, n	0	n	0	0	0	0	$M^{-1/2}$
Bridges, p	$l^{1/2}n$	0	p	Xn	$N^{1/2}n^{1/2}$	$p_c n$	M^0
Depth, X	$l^{1/2}$	0	p/n	X	$N^{1/2}n^{-1/2}$	p_c	$M^{1/2}$
Monomers, N	ln	0	p^2n	X^2n	N	p_c^2n	$M^{1/2}$
Bridges/chain, p_c	$l^{1/2}$	0	p/n	X	$N^{1/2}/n^{1/2}$	p_c	$M^{1/2}$

Table 2.2(c) Rigid-Rod Interface with Random Chain-End Distribution; $p_c = 1$

$H_1 \backslash H_2$	l	n	p	X	N	p_c	H_∞
Contour, l	l	nM	pM	X	$N^{1/2}M^{1/2}$	p_c	M
Chains, n	l/M	n	p	X/M	$N^{1/2}M^{-1/2}$	0	M^0
Bridges, p	l/M	n	p	X/M	$N^{1/2}M^{-1/2}$	0	M^0
Depth, X	l	nM	pM	X	$N^{1/2}M^{1/2}$	p_c	M
Monomers, N	l^2/M	n^2M	p^2M	X/M	N	0	M
Bridges/chain, p_c	0	0	0	0	0	p_c	M^0

Table 2.2(d) Rigid-Rod Interface with Chain-End Segregation; $p_c = 1$; $p = n$ = constant

$H_1 \backslash H_2$	l	n	p	X	N	p_c	H_∞
Contour, l	l	0	0	X	N/n	0	M
Chains, n	0	n	p	0	0	0	M^0
Bridges, p	0	n	p	0	0	0	M^0
Depth, X	l	0	0	X	N/n	0	M
Monomers, N	ln	0	0	Xn	N	0	M
Bridges/chain, p_c	0	0	0	0	0	p_c	M^0

2.8.4 Relating Structure to Strength

We now take up the question of relating interface structure to strength. Table 2.1 and the appropriate Table 2.2 can be used to determine other properties not considered here and evaluate models relating interface structure to mechanical properties. If a mechanical property such as the fracture energy, $G_{1c}(t)$, is controlled by one of the properties $H(t)$, or a product or ratio of these properties, then the time dependence of welding is given by

$$G_{1c}(t) \sim t^{r/4} M^{-s/4} \tag{2.8.11}$$

This proportionality suggests experiments that can be done to determine the values of r and s and to probe relations between structure and strength. Furthermore, if we identify a relation between G_{1c} and one of the molecular properties, for example, $G_{1c} \sim l(t)$, where $r = s = 2$ (as developed in Chapter 7), then the interrelationship with all other molecular properties is

$$G_{1c} \sim H^{2/r} \tag{2.8.12}$$

We briefly discuss the interrelated interface properties presented in Table 2.2(a–d).

Contour Length, l(t)

For random coils and rigid rods, we expect

$$G_{1c} = G_{1c\infty}\, l/l_\infty \tag{2.8.13}$$

where $G_{1c\infty}$ is the virgin-state property. In Chapter 7 we will see that the molecular weight dependence is given by

$$G_{1c\infty} \sim M/M_c[1 - \sqrt{M_c/M}]^2 \tag{2.8.14}$$

for the range $M_c \leq M \leq 8M_c$, where M_c is the critical entanglement molecular weight ($M_c \approx 2M_e$). Eq 2.8.12 has relevance to polymer welding, with $G_{1c} \sim t^{1/2}$ (Chapter 8, 12), tack of uncured elastomers with $\sigma \sim t^{1/4}$ (Chapter 7, 8), healing of damage in polymer composites (Chapter 11), and crack healing (Chapter 12). The upper molecular weight limit, $M^* \approx 8M_c$ in Eq 2.8.14, applies to random coils and from it we conclude that polymers cannot be disentangled without breaking of bonds when M is greater than M^* [69]; this is discussed in Chapter 7. For rigid rods and fibers, M^* (or a critical fiber length) depends on the bond strength of the chain or fiber to the surrounding matrix, compared to the strength of the chain or the fiber fracture strength.

Bridges, p

Table 2.2(a) gives

$$G_{1c} = G_{1c\infty}(p/p_\infty) \tag{2.8.15}$$

and therefore

$$G_{1c} \sim pM \tag{2.8.16}$$

which is similar to the solution for rigid rods in Table 2.2(c).

This last solution is very interesting, and resembles the "nail solution" (Chapter 7). If we nail two pieces of wood together with p nails, the energy to separate the pieces is proportional to p. This is also true for fiber pullout from rubber: the stress required to pull p fibers from a rubber base is proportional to $p^{1/2}$, so $G_{1c} \sim p$. Similarly, for the critical stress intensity factor K_{1c} of short-fiber-reinforced cement interfaces, $K_{1c} \sim p^{1/2}$. The nail solution applies to rigid-rod polymers, polymers with $M < M_c$, extended chain A–B diblocks at A/B incompatible interfaces, and brushes (see Chapter 9).

Number of Chains, n

From Table 2.2(a), with $G_{1c} \sim l$, we have the dependence on the number of chains as

$$G_{1c} = G_{1c\infty} (n/n_\infty)^2 \tag{2.8.17}$$

so that $G_{1c} \sim n^2M^2$. The relation $G_{1c} \sim n^2$ is demonstrated in Chapter 9 and has relevance to diblocks with random-coil structure at incompatible interfaces with M greater than M_e, and coupling agents at polymer–metal interfaces. For unentangled systems and rigid rods, the relation changes to $G_{1c} \sim n$ [Table 2.2(c)].

Monomer Diffusion Depth, X

For random-coil chains, Tables 2.2(a) and 2.2(b) give

$$G_{1c} = G_{1c\infty} (X/X_\infty)^2 \tag{2.8.18}$$

This solution can be used to analyze A/B incompatible interfaces of thickness d_∞ (Chapter 9, 10), where $G_{1c} \sim d_\infty^2/R_g^2$. For rigid rods, we expect $G_{1c} \sim X/X_\infty$.

Thus, the molecular structure correlations in Table 2.1 and Table 2.2 can be used to examine molecular models of strength. For example, we know that weld strength increases with time as $G_{1c} \sim t^{1/2}$. We also know that the nail solution for simple chain pullout gives $G_{1c} \sim nL^2$. It has been suggested that the latter solution can describe the strength of interfaces with entangled random-coil chains; the validity of this idea can be examined as follows: if $G_{1c} \sim nL^2$, we find in Table 2.2(a) that $G_{1c} \sim n^5M^4$, and in Table 2.1 that the time dependence of welding should be $G_{1c} \sim n^{5/4}M^{-9/4}$. Neither of these is consistent with experimental data and so we find the suggestion nonpersuasive.

In summary, it may be an overly simplified argument to suggest that a single molecular entity, existing typically over an interface a few hundred angstroms thick, is responsible for the energy dissipation and behavior of a much larger deformation zone at the crack tip. At high fracture energies, other damage mechanisms absorb energy in the vicinity of the interface and the situation becomes more complex. For example, in peeling experiments, the viscoelastic contribution plays a dominant role in energy dissipation, as shown in Chapter 8. At low welding times and low fracture energies, we anticipate best agreement with the approach described above. Therefore, we suggest that the scaling laws presented in Tables 2.1 and 2.2 offer a useful framework for the evaluation of molecular models in terms of the static and dynamic properties of the polymer chains.

2.9 References

1. S. S. Voyutskii, *Autohesion and Adhesion of High Polymers*; Vol. 4 in series *Polymer Reviews*, H. F. Mark and E. H. Immergut, Eds.; Wiley–Interscience, New York; 1963.
2. S. S. Voyutskii, *Rubber Chem. Technol.* ***33***, 748 (1960).
3. S. S. Voyutskii, A. I. Shapalova, and A. P. Pisarenko, *Colloid J. USSR (Engl. Transl.)* ***19***, 279 (1957).
4. S. S. Voyutskii and V. L. Vakula, *J. Appl. Polym. Sci.* ***7***, 475 (1963).
5. S. S. Voyutskii, S. M. Yagnyatinskaya, L. Ya. Kaplunova and N. L. Garetovskaya, *Rubber Age* ***105***(2), 37 (1973).
6. H. H. Kausch and K. Jud, *Plast. Rubber Process. Appl.* ***2***, 265 (1982).
7. H. H. Kausch and K. Jud, *Macro-79 Proceedings of IUPAC*, Mainz, Germany, 1979.
8. K. Jud and H. H. Kausch, *Polym. Bull.* ***1***, 697 (1979); K. Jud, H. H. Kausch, and J. G. Williams, *J. Mater. Sci.* ***16***, 204 (1981).
9. H. H. Kausch, D. Petrovska, and R. F. Landel, in press.
10. H. H. Kausch, *Polymer Fracture*; Springer-Verlag, Berlin; 2nd ed., 1987.
11. H. H. Kausch, *Pure Appl. Chem.* ***55***, 833 (1983).
12. H. H. Kausch and M. Tirrell, *Annu. Rev. Mater. Sci.* ***19***, 341 (1989).
13. S. Prager and M. Tirrell, *J. Chem. Phys.* ***75***, 5194 (1981).
14. R. P. Wool and K. M. O'Connor, *J. Appl. Phys.* ***52***, 5953 (1981).
15. R. P. Wool and K. M. O'Connor, *J. Polym. Sci., Polym. Lett. Ed.* ***20***, 7 (1982).
16. R. P. Wool, B.-L. Yuan, and O. J. McGarel, *Polym. Eng. Sci.* ***29***, 1340 (1989).
17. P.-G. de Gennes, *J. Chem. Phys.* ***55***, 572 (1971).
18. M. Doi and S. F. Edwards, *J. Chem. Soc., Faraday Trans. II* ***74***, 1789 (1978).
19. M. Doi and S. F. Edwards, *J. Chem. Soc., Faraday Trans. II* ***74***, 1802 (1978).
20. M. Doi and S. F. Edwards, *J. Chem. Soc., Faraday Trans. II* ***74***, 1818 (1978).
21. M. Doi and S. F. Edwards, *J. Chem. Soc., Faraday Trans. II* ***75***, 38 (1979).
22. S. F. Edwards, *Proc. Phys. Soc.* ***92***, 9 (1967).
23. S. F. Edwards and K. E. Evans, *J. Chem. Soc., Faraday Trans. II,* ***77***, 1913 (1981).
24. M. Doi and S. F. Edwards, *The Theory of Polymer Dynamics*; Clarendon Press, Oxford; 1986.
25. P.-G. de Gennes, *C. R. Seances Acad. Sci., Ser. 2,* ***292***, 1505 (1981).
26. P.-G. de Gennes, *J. Chem. Phys.* ***72***, 4756 (1980).
27. P.-G. de Gennes, *C. R. Acad. Sci., Ser. 2,* ***307***, 1841 (1988).
28. R. P. Wool, *Rubber Chem. Technol.* ***57*** (2), 307 (1984).
29. Y.-H. Kim and R. P. Wool, *Macromolecules* ***16***, 1115 (1983).
30. A. Y. Lee and R. P. Wool, *Macromolecules* ***19***, 1063 (1986).
31. A. Y. Lee and R. P. Wool, *Macromolecules* ***20***, 1924 (1987).
32. A. Y. Lee and R. P. Wool, "Matrix Effects on the Orientation Relaxation of Linear Polymer Melts", *Polym. Prepr. (Am. Chem. Soc., Div. Polym. Chem.)* ***28*** (1), 334 (1987).
33. W. J. Walczak and R. P. Wool, *Macromolecules* ***24***, 4657 (1991).
34. J. F. Tassin and L. Monnerie, *Macromolecules* ***21***, 1846 (1988).
35. M. Doi, D. Pearson, J. Kornfeld, and G. Fuller, *Macromolecules* ***22***, 1488 (1989).
36. B. D. Lawrey, R. K. Prud'homme, and J. T. Koberstein, *J. Polym. Sci., Polym. Phys. Ed.* ***21***, 2402, (1988).
37. J. T. Koberstein, private communication of experimental results obtained by B. D. Lawrey, *A Study of Polymer Chain Segment Orientation and Relaxation in Isotopically Labelled*

Block Copolymers Using a Modulated Infrared Dichroism Technique; Ph.D. Thesis, Princeton University, Princeton, NJ; 1989.
38. F. Boue, M. Nierlich, G. Jannink, and R. C. Ball, *J. Phys. (Paris)* ***43***, 137 (1982).
39. K. Osaki, E. Takatori, M. Ueda, M. Kurata, T. Kotaka, and H. Ohnuma, *Macromolecules* ***22***, 2457 (1989).
40. K. Kremer, G. S. Grest, and I. Carmesin, *Phys. Rev. Lett.* ***61***, 566 (1988); K. Kremer and G. Grest, *J. Chem. Phys.* ***92***, 5057 (1990).
41. J. Skolnick, R. Yaris, and A. Kolinski, *J. Chem. Phys.* ***88***, 1407 (1988).
42. J. Skolnick and R. Yaris, *J. Chem. Phys.* ***88***, 1418 (1988).
43. W. Paul, D. W. Binder, and K. Kremer, *J. Chem Phys.* ***95***, 7726 (1991).
44. J. Klein and B. J. Briscoe, *Proc. R. Soc. London, A* ***365***, 53 (1979).
45. J. Klein, D. Fletcher, and L. J. Fetters, *Nature (London)* ***304***, 526 (1983).
46. E. J. Kramer, P. F. Green, and C. J. Palmstrom, *Polymer* ***25***, 473 (1983).
47. P. F. Green and E. J. Kramer, *J.Mater. Res.* ***1***, 202 (1986).
48. P. F. Green and E. J. Kramer, *Macromolecules* ***19***, 1108 (1986).
49. U. Steiner, G. Krausch, G. Schatz, and J. Klein, *Phys. Rev. Lett.* ***64***, 1119 (1990).
50. S. J. Whitlow and R. P. Wool, *Macromolecules* ***22***, 2648 (1989).
51. S. J. Whitlow and R. P. Wool, *Macromolecules* ***24***, 5926 (1991).
52. A. Karim, A. Mansour, G. P. Felcher, and T. P. Russell, *Phys. Rev. B: Condens. Matter* ***42***, 6846 (1990).
53. A. Karim, *Neutron Reflection Studies of Interdiffusion in Polymers*; Ph.D. Thesis, Northwestern University, Evanston, IL; 1990.
54. M. Doi, *J. Polym. Sci., Polym. Lett. Ed.* ***19***, 265 (1981).
55. K. S. Schweizer, *J. Chem Phys.* ***91***, 5802 (1989).
56. J. des Cloizeaux, *Macromolecules* ***23***, 3992 (1990).
57. W. M. Merrill, A. V. Pocius, B. V. Thakker, and M. Tirrell, *Langmuir* ***7***, 1975 (1991).
58. T. P. Lodge, N. A. Rotstein, and S. Prager, *Adv. Chem. Phys.* ***79***, 1 (1990).
59. T. P. Russell, V. R. Deline, W. D. Dozier, G. P. Felcher, G. Agrawal, R. P. Wool, and J. W. Mays, *Nature (London)* ***365***, 235 (1993).
60. G. Agrawal, R. P. Wool, W. D. Dozier, G. P. Felcher, T. P. Russell, and J. W. Mays, *Macromolecules* ***27***(15), 4407 (1994).
61. H. Zhang and R. P. Wool, *Macromolecules* ***22***, 3018 (1989).
62. H. Zhang and R. P. Wool, "Concentration Profiles at Amorphous Symmetric Polymer/Polymer Interfaces"; *Polym. Prepr. (Am. Chem. Soc., Div. Polym. Chem.)* ***31*** (2), 511 (1990).
63. M. Tirrell, D. Adolf, and S. Prager, "Orientation and Motion at a Polymer–Polymer Interface: Interdiffusion of Fluorescent-Labelled Macromolecules"; chapter in *Orienting Polymers: Proceedings of a Workshop Held at the IMA, University of Minnesota, Minneapolis, March 21–26, 1983*, J. L. Ericksen, Ed.; No. 1063 in series *Lecture Notes in Mathematics*, A. Dold and B. Eckmann, Eds.; Springer-Verlag, Berlin; 1984; p 37.
64. M. Tirrell, *Rubber Chem. Technol.* ***57***, 523 (1984).
65. P.-G. de Gennes, private communication.
66. R. P. Wool, "Dynamics and Fractal Structure of Polymer Interfaces"; chapter in *New Trends in Physics and Physical Chemistry of Polymers*, L.-H. Lee, Ed.; Plenum Press, New York; 1989; p 129.
67. R. P. Wool and J. M. Long, *Macromolecules* ***26***, 5227 (1993); R. P. Wool and J. M. Long, "Structure of Diffuse Polymer–Metal Interfaces", *Polym. Prepr. (Am. Chem. Soc., Div. Polym. Chem.)* ***31*** (2) 558, (1990).

68. R. P. Wool, *J. Elastomers Plast.* ***17*** (Apr), 107 (1985).
69. R. P. Wool and A. T. Rockhill, *J. Macromol. Sci. Phys. B* ***20***, 85 (1981).
70. R. P. Wool, *Macromolecules* ***26***, 1564 (1993).
71. D. Adolf, M. Tirrell, and S. Prager, *J. Polym. Sci., Polym. Phys. Ed.* ***23***, 413 (1985).
72. A. G. Mikos and N. A. Peppas, *J. Chem. Phys.* ***88***, 1337 (1988).
73. K. E. Evans and S. F. Edwards, *J. Chem. Soc., Faraday Trans. II* ***77***, 1891 (1981).
74. G. Agrawal and R. P. Wool, *Macromolecules*, in press.
75. R. P. Wool and H. Zhang, to be published.
76. M. Stamm, S. Huttenbach, G. Reiter, and T. Springer, *Europhys. Lett.* ***14***, 451 (1991).
77. P.-G. de Gennes, *Europhys. Lett.* ***15***, 191 (1991).
78. P.-G. de Gennes, *J. Phys. (Paris)* ***50***, 2551 (1989).
79. P.-G. de Gennes, *C. R. Acad. Sci., Ser. 2,* ***308***, 13 (1989).
80. F. Brochard-Wyart and P. Pincus, *C. R. Acad. Sci.* ***314***, 131 (1992).

3 Rouse Dynamics and Diffusion at Polymer Interfaces*

3.1 Introduction to Segmental Dynamics

In Chapter 2, the concentration profiles $C(x,t)$ for interfaces of two chemically identical monodisperse polymers were calculated using the reptation model [1]. These profiles were derived by considering only the primitive path reptation contribution, and had the unique feature of being discontinuous at the interface plane, as shown in Figure 2.13. The gap shrinks with the square root of the healing time, and disappears when $t = T_r$. However, Rouse motion is important before there is any significant contribution from reptation; in Rouse motion, chain segments diffuse across the interface and effectively eliminate the gap. Therefore, the complete concentration profile includes contributions from two mechanisms, reptation and Rouse–like motion of the chain and entangled sections of the chain.

3.1.1 Dynamics of Molecules, Small and Large

Consider the small molecule shown in Figure 3.1 consisting of three atoms of mass m_i ($i = 1, 2, 3$) at spatial positions X_i. The masses are connected by two bonds, which act as Hookean springs with force constants k_{ij} ($k_{1,2}$ and $k_{2,3}$). Each mass has three degrees of freedom, and the molecule as a whole has nine modes of motion ($3N = 9$). Three modes result in translation in space of the entire molecule, and three others involve rotation about the three directions. This leaves three internal modes of motion, which describe the bending and stretching modes of the bonds and are referred as the *normal modes* of motion or vibration. The normal modes involve specific displacements (*eigenvectors*) of the bonds and atoms, and occur at unique frequencies ν_i (*eigenvalues*), which have periods or relaxation times τ_i [$\tau_i = 1/(\nu_i c)$], where c is the speed of light (3×10^{10} cm/s). For example, the three vibrations of the C–C–C molecule shown in Figure 3.1 involve symmetric bond stretching (ν_s), asymmetric bond stretching (ν_a), and angle bending (ν_b), and have frequencies of about 1000 cm^{-1}.

The normal modes are calculated from Newton's second law of motion ($f = ma$, force = mass × acceleration), used in the form of Lagrange's equation of motion,

---

* Dedicated to R. H. Boyd

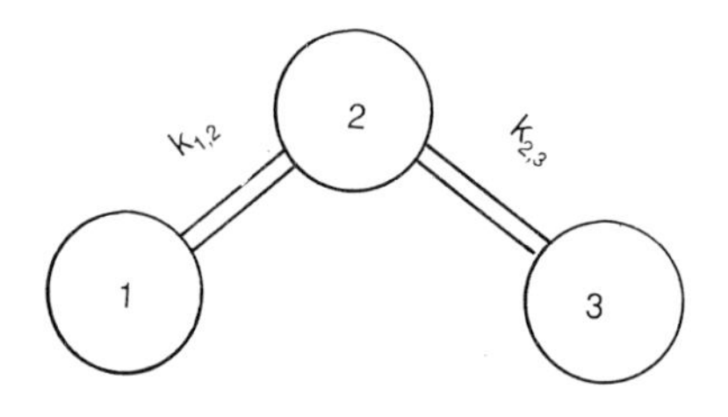

Figure 3.1 A triatomic oscillator with three masses and two springs, with spring constants $k_{1,2}$ and $k_{2,3}$.

$$dU/dx - d/dt\,(\delta K/\delta \dot{x}) = 0 \tag{3.1.1}$$

The potential energy, U, involves the sum of terms, $U_{ij} = \frac{1}{2} k_{ij}(x_i - x_j)^2$, due to stored energy in the bonds and angles, and the kinetic energy, K, is the sum of terms, $K_i = \frac{1}{2} m_i (dx_i/dt)^2$, due to the motion of the masses. The first term in Eq 3.1.1 (dU/dx) corresponds to force, and the second term [$d/dt\,(\delta K/\delta \dot{x})$, in which $\dot{x} = dx/dt$] is mass × acceleration. To obtain the frequencies and relaxation times from Eq 3.1.1, one constructs N simultaneous equations using the potential and kinetic energy relations, and solves the secular equation for the eigenvalues α_i and eigenvectors. The eigenvalues are introduced in the secular equation through the simple harmonic motion relation, $d^2x_i/dt^2 = \alpha_i x_i$, and are related to the vibrational frequencies by $\alpha_i = 4\pi^2 c^2 v_i^2$.

Larger molecules exhibit modes of motion similar to those of smaller molecules. For example, the Rouse relaxation time of unentangled chains is associated with a symmetric stretching or "breathing" motion similar to that for the symmetric stretching mode of the C–C–C group in Figure 3.1.

Exercise

Given a linear triatomic oscillator with masses $m_1 = m_2 = m_3$, and force constants $k = k_{1,2} = k_{2,3}$, use Lagrange's equation of motion to determine the relative magnitude of the symmetric and asymmetric bond stretching frequencies.

3.1.2 Rouse Dynamics of Polymer Chains

P. G. Rouse in 1953 considered the dynamics of unentangled linear polymer chains [2]. The model applies to chains in dilute solutions and to polymer melts where M is less than M_c. He took a random-walk chain and transformed it into a spring-bead chain (as shown in Figure 3.2). The Rouse chain is immersed in a continuum of fluid of viscosity η. The fluid opposes the motion of each bead with a friction μ. The springs represent Gaussian subchains with end-to-end vector b, and have a force constant k_R, derived from rubber elasticity theory, as

$$k_R = 3\,kT/b^2 \tag{3.1.2}$$

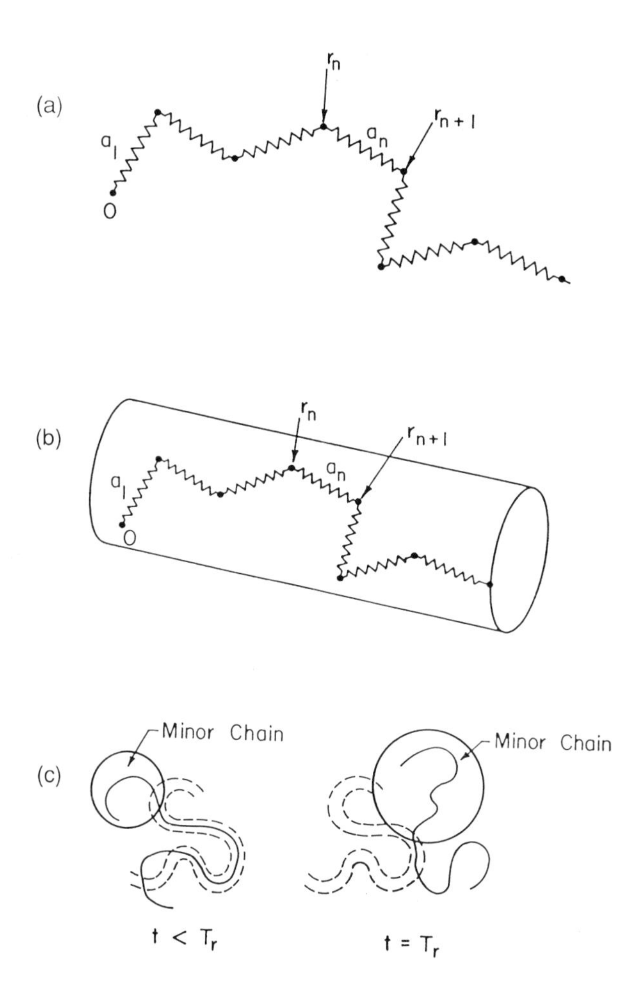

Figure 3.2 Schematic diagram representing the chain dynamics in different time regimes: (a) free Rouse relaxation of the chain segments between two entanglements ($\tau_0 < t < \tau_e$); (b) restricted Rouse relaxation inside a tube ($\tau_e < t < \tau_{RO}$); (c) the growth of minor chains by reptation ($\tau_{RO} < t < T_r$). Definitions: τ_0 is the relaxation time of the statistical segment, τ_e is the Rouse relaxation time of an entanglement segment, τ_{RO} is the Rouse relaxation time of the whole chain, and T_r is the reptation time.

where T is the fluid temperature and k is Boltzmann's constant. The N springs store energy in a manner identical to that in the small-molecule problem discussed above. The dynamics of the molecule is now controlled by three factors, (a) Brownian interactions with the surrounding fluid, which cause individual beads to move (white noise), (b) the hydrodynamic or drag forces of the beads in the solvent, and (c) the stored energy in the bonds, U. The drag forces, f, are of the form

$$f = \mu \, \mathrm{d}x/\mathrm{d}t \tag{3.1.3}$$

where μ is a friction factor related to solvent viscosity.

Rouse obtained the following important results. The model gives relaxation times, τ_i, associated with the normal modes of motion

$$\tau_i = \mu b^2 / [24\, kT \sin^2(i\pi/2N)] \tag{3.1.4}$$

where $i = 1, 2, \dots, N$. The longest relaxation time occurs at $i = 1$, which is the Rouse relaxation time, τ_{RO}. With the approximation that $\sin^2 x \approx x^2$ if $0 < x << 1$ (here, $x = \pi/2N$), Eq 3.1.4 gives the Rouse relaxation time as

$$\tau_{RO} \approx \mu b^2 N^2/(6\pi^2 kT) \tag{3.1.5}$$

Since the molecular weight, M, is directly proportional to N, the molecular weight dependence of the Rouse relaxation time at constant temperature is obtained from Eq 3.1.5 as

$$\tau_{RO} \sim M^2 \tag{3.1.6}$$

The decay of the end-to-end vector R correlation function is determined by

$$<R(t) - R(0)>^2 = R^2 \exp(-t/\tau_{RO}) \tag{3.1.7}$$

where $R^2 = Nb^2$. Thus, at $t = \tau_{RO}$ the molecule has diffused a distance approximately equal to R. Using the Fickian diffusion equation for the center of mass motion, $<X^2> = 2D_{RO}t$, where D_{RO} is the Rouse diffusion coefficient, the molecular weight dependence of the diffusion coefficient at $<X^2> = R^2$ and $t = \tau_{RO}$ is obtained as

$$D_{RO} \sim M^{-1} \tag{3.1.8}$$

Also, for low-molecular-weight species, the product of viscosity and diffusion coefficient is constant, or

$$\eta \sim M \tag{3.1.9}$$

This result is important for chain disentanglement involving chain pullout (Chapters 7–12).

The molecular weight dependence of the Rouse relaxation time, diffusion coefficient, and viscosity are considered classical and have received considerable experimental support for polymer melts with M less than M_c.

The monomer displacement function $r_n(t)$, which measures the displacement of individual monomers in the chain, manifests the correlations due to bond connectivity.

The monomer displacement function obeys the relation

$$<r_n(t) - r_n(0)>^2 \sim t^{1/2} \qquad \text{for } t < \tau_{RO} \tag{3.1.10}$$

and

$$<r_n(t) - r_n(0)>^2 \sim t \qquad \text{for } t > \tau_{RO} \tag{3.1.11}$$

The transition from $t^{1/2}$ to t^1 for normal Fickian diffusion is similar to that for the reptation model discussed in Chapter 2.

In this chapter, we use the theories proposed by Rouse, Doi and Edwards, de Gennes, Oono, and Zhang and Wool to examine chain dynamics with emphasis on the Rouse segmental motion. We consider the concentration profile contributed by segmental motion at times less than the relaxation time of the entanglement chain length, τ_e, and the Rouse relaxation time of the whole chain, τ_{RO}. The total depth profile, $C(x,t)$, contributed by both reptation and segmental motion is calculated for the symmetric monodisperse interface.

3.2 Chain Dynamics at a Polymer–Polymer Interface

In this section the chain dynamics is examined, with emphasis on the segmental motion. The concentration profiles $C(x,t)$ are calculated at $t < \tau_{RO}$ for the symmetric A(M)/A(M) interface, and the profiles are used to examine the molecular aspects.

3.2.1 Rouse Dynamics at the Interface

When two amorphous polymers are brought into good contact above T_g, the chain conformations at the interface tend to relax toward those in the bulk because of Brownian motion. Five different time regions can be identified (three are diagrammed in Figure 3.2), and are described as follows.

Statistical Segments

At $t < \tau_0$, motion takes place on the order of a single statistical segment, with relaxation time τ_0. Since the lateral displacements are quite small, the statistical segments feel neither the topological constraints [2–7] nor the chain connectivity [2]. Therefore the motion of the chain during this period can be treated as the free diffusion of the center of mass of statistical segments. The segmental diffusion coefficient is associated with the local “jump” process and is independent of the molecular weight [2]. By the end of this time period, the conformations of the statistical segments are relaxed.

Entanglement Segments

At $\tau_0 < t < \tau_e$, where τ_e is the Rouse relaxation time of the entanglement chain length, the lateral displacements become larger, and any disturbance can propagate farther along the chain backbone. As a result, the chain connectivity comes into play, and segments up to lengths approximating the entanglement length undergo Rouse relaxation [5, 7]. By this time period, the monomer diffusion distance is of the order of the radius of gyration, R_{ge}, of the entanglement chain with molecular weight M_e.

Rouse Relaxation of the Chain

At $\tau_e < t < \tau_{RO}$, a relaxation event on the chain can propagate as far as the chain ends in a time $\tau_{RO} \approx \tau_e(M/M_e)^2$. Attempted lateral displacements may be larger than R_{ge}. Consequently, the motion of the chain is determined not only by the chain connectivity but also by the topological constraints. The motion can be broken down into two components: along the tube contour, the chain undergoes free Rouse relaxation, while in the direction perpendicular to the tube contour, the chain undergoes hindered motion [5, 7]. If $<x_S^2>$ represents the mean square displacement of the monomers, then the larger the ratio $<x_S^2>/R_{ge}^2$, the more restricted is the lateral motion. This simply means that the monomers cannot diffuse farther than the tube wall at $t < \tau_{RO}$, unless other processes are important, such as "hairpins", which allow the chain to elbow its way into the other chains.

Reptation

At $\tau_{RO} < t < T_r$, the chain motion at the interface is described by the minor chain reptation model [1, 8, 9]. This time period is the most important for the processes of adhesion and welding, since optimal diffusion to depths of the order $X \approx R_g$ are achieved (Chapter 2).

Fickian Diffusion

At $t > T_r$, the chain loses its memory of its initial conformation and becomes a random coil. As a result, its motion can be described in terms of Fickian diffusion with $D \sim M^{-2}$ [5, 7].

3.3 Concentration Profile with Segmental Motion

We now examine the contribution from segmental motion to the concentration profile $C(x,t)$ for $t < T_r$.

3.3.1 Concentration Profile $C_S(x,t)$ Due to Free Segmental Motion

At $t < \tau_0$, the motion can be treated as the Fickian diffusion of the statistical segments, with a microscopic diffusion coefficient, D_0. If we assume a uniform initial distribution of segments, we obtain the following boundary conditions

$$C_S(x, t) = 0 \qquad \text{for } x > 0$$
$$C_S(x, t) = \rho / M_S \qquad \text{for } x < 0 \tag{3.3.1}$$

where M_S is the molecular weight of the statistical segment. The concentration profile is given by the Fickian result

$$C_S(x, t) = 0.5\,(\rho / M_S)\ \text{erfc}\,(x/2\sqrt{D_0\, t}) \qquad \text{for } t < \tau_0 \tag{3.3.2}$$

The characteristic penetration depth is of the order of the radius of gyration of the statistical segment, about 10 Å.

3.3.2 Concentration Profile $C_0(x,t)$ Due to Entanglement Relaxation

At $\tau_0 < t < \tau_e$, the chain connectivity governs the dynamics on length scales on the order of the entanglement length. We use the Langevin equation to describe the interdiffusion behavior of a Rouse chain of length M_e [5, 10, 11, 12], as

$$\partial \vec{R}(\tau, t)/\partial t = 3\,kT/\xi_0\, b^2 + \partial^2 \vec{R}(\tau, t)/\partial \tau^2 + \Theta\,(\tau, t) \tag{3.3.3}$$

where k is Boltzmann's constant, T is the absolute temperature, ξ_0 is the friction constant for the statistical segment of size b, and $R(\tau,t)$ is the position vector of the τ-th chain unit at time t. The first term on the right of Eq 3.3.3 represents the elastic force due to chain connectivity and the second term represents the random force due to thermal noise. $[\vec{R}(\tau,t)]^N_{\tau=0}$ specifies the instantaneous conformations of a chain of N statistical segments at time t.

The term $\Theta(\tau,t)$ in Eq 3.3.3 is a Gaussian white noise with zero mean and covariance given by

$$<\Theta(\tau, t)\ \Theta(\sigma, s)> = 2\,kTdI\ \partial(\tau - \sigma)\ \partial(t - s)/\xi_0 \tag{3.3.4}$$

where d is the spatial dimensionality and I is the unit tensor [16].

The origin of the coordinates is chosen so that the interface plane is at $x = 0$ and the x axis is perpendicular to the interface plane. The total number, $C_0(x,t)$, of statistical segments per unit volume at a distance x from the interface plane is calculated as [10]

$$C_0(x,t) = \frac{\rho}{M_e} \int_0^{\infty} \mathrm{d}x \int_0^{N} \mathrm{d}\tau \frac{i}{\sqrt{2\pi}\, R_{2\Theta}(\tau,t)} \exp\left\{-\frac{[x - R_{1x}(\tau,t)]^2}{2R_{2\Theta}(\tau,t)}\right\} R_x(\sigma,0) \tag{3.3.5}$$

with

$$R_{2\Theta}(\tau,t) = \frac{Nb^2}{3\pi^2}\{t' + \sum_{p=1}^{\infty} \hat{p}^{-2} \cos^2(\hat{p}\,\pi)\ [-\exp(-2\,\hat{p}^2\, t')]\} \tag{3.3.6}$$

$$R_{1x}(\tau, t) = \int_0^N \mathrm{d}\sigma\ G(\tau,\sigma|t)\ R_x(\sigma, 0) \tag{3.3.7}$$

The Green's function $G(\tau,\sigma\,|\,t)$ used above is given by [10, 13, 14, 15]

$$G(\tau,\sigma|t-s) = \frac{1}{N}\{1 + 2\sum_{p=1}^{\infty}\cos(\hat{p}\tau\cos(\hat{p}\sigma)\exp[-3\mu\, kT\hat{p}\,(t-s)/b^2]\} \tag{3.3.8}$$

where $\mu = 1/\xi_0$ is the mobility of the chain unit, p is a summation variable, $\hat{p}$ is defined as $p\pi/N$, and t' is the reduced time t/τ_e.

Although there is at this time no analytical solution for $R_x(\tau,0)$ because of the presence of free surfaces, we can obtain a numerical solution by executing a random walk in the presence of a reflecting wall at $x = 0$, as shown in Figure 3.3. Using a computer, we calculated the mean square displacement of the monomer position vector $\phi_\tau(t)$, defined by

$$\phi_\tau(t) \equiv <[R_x(\tau,t) - R_x(\tau,0)]^2> \tag{3.3.9}$$

and the mean square one-dimensional end-to-end distance, $<r_x^2>$. We did this for chains of N statistical segments both in the bulk (without surfaces) and near the surface, using Eq 3.3.5. We found that in the bulk

$$<r_x^2> = 0.34\,N \qquad \text{for } N = 10, 20, ..., 100 \tag{3.3.10}$$

which is consistent with Gaussian statistics, $<r_x^2> = N/3$. In the presence of a surface, we obtained $<r_x^2> = 0.24\ N,\ 0.30\ N,\ 0.34\ N$, when the chain ends were initially at

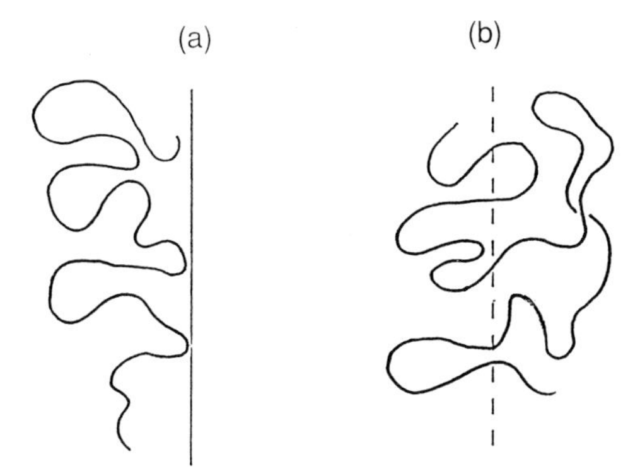

Figure 3.3 Schematic diagram describing the effect of the free surface on the chain conformations at (a) $t = 0$; (b) $t = \tau_{RO}$.

distances $5b$, $10b$, and $15b$, respectively, from the free surface. The change in the numerical factor is due to the fact that chains are not Gaussian near the free surface. They are in a compressed state in the direction perpendicular to the free surface, and are further compressed as the chain approaches the surface.

According to our calculations, the monomer displacement function behaved as $\phi(t) \sim t^{\beta}$; $\beta \approx ½$ at $t << \tau_e$, but increases with time. The correlated motion effects in the Rouse model are relatively weak, and require very high molecular weights at short times to be observable. For a given value of N, the exponent averaged over the entire Rouse relaxation time was between ½ and 1, depending on the magnitude of N. This loss of correlation is due to the high mobility of the chain ends, with the result that monomers near the center of the chain exhibit more correlated motion than those near the end. This may create difficulty in attempts to observe the double-correlated motion of Rouse chains in a tube where the asymptotic behavior is $\phi_{\tau}(t) \sim t^{1/4}$.

3.3.3 Concentration Profile $C_R(x,t)$ Due to Restricted Rouse Relaxation

At $\tau_e < t < \tau_{RO}$, a chain can be modeled as a Rouse chain confined in a tube. Here the dynamics are exceedingly complex due to the simultaneous actions of the chain connectivity and the topological constraints. The topological constraints are often modeled as a potential [17–19]. If we could model the potential in terms of the position vector, the method used for the free Rouse relaxation would be applicable for the restricted Rouse relaxation. However, since, according to the tube model, the modeling potential does not depend on the position vector but on how much a segment deviates from the center line of the tube, it is very difficult to analyze the diffusion behavior using the equation of motion.

Realizing that there we have no analytic solution at $\tau_e < t < \tau_{RO}$, we seek an approximate concentration profile contribution from segmental motion. We can find solutions by examining the tube model more closely. The model indicates that when

$t = \tau_{RO}$, the distribution of segments inside the tube is equilibrated. This implies that the concentration profile due to segmental motion is independent of the time at $t > \tau_{RO}$.

At $\tau_e < t < \tau_{RO}$, three forces control the segmental motion, namely, the random force arising from thermal noise, the elastic force arising from chain connectivity, and the repulsive force arising from the uncrossability of chains. The larger the displacement from the tube center line, the more hindered is the motion, and the slower is the lateral segmental diffusion. The maximum excursion of the segments is X_m, which is related to the radius of gyration of the entanglement molecular weight, R_{ge}. We show later that $X_m = 1.8\ R_{ge} N^{1/4}$, where $N = M/M_e$. We assume that the lateral segmental diffusion coefficient, $D_S(t)$, exponentially decreases with the ratio $<x_S^2>/X_m^2$, as

$$D_S(t) = D_0 \exp\left[-\frac{<x_S^2>}{X_m^2}\right] \qquad \text{for } \tau_e < t < \tau_{RO} \tag{3.3.11}$$

where $<x_S^2> = 2 D_0 t$ is the mean square displacement of a free segment in the absence of constraints.

Using Eq 3.3.11 we can now calculate $C_R(x,t)$. If we assume that the initial segmental distribution is random, we can describe the motion perpendicular to the interface plane by

$$\partial C_R(x,t)/\partial t = D_S(t)\, \partial^2 C_R(x,t)/\partial x^2 \qquad \text{for } t < \tau_{RO} \tag{3.3.12}$$

The solution with a constant diffusion coefficient can be used for Eq 3.3.12, but Dt must be replaced by the averaged product, $<D(t)t>$ [20, 21]. The averaged product is calculated, using Eq 3.3.11, as

$$<D(t)\, t> = 0.5\, X_m^2 [1 - \exp(-2\, D_0\, t/X_m^2)] \tag{3.3.13}$$

The solution to Eq 3.3.12 is obtained with respect to the initial conditions

$$C_R(x,t) = 0 \qquad \text{for } x > 0$$
$$C_R(x,t) = \rho/M_S \qquad \text{for } x < 0 \tag{3.3.14}$$

where ρ is the mass density. The solution is given by

$$C_R(x,t) = 0.5\,(\rho/M_S)\, \mathrm{erfc}\left(\frac{x}{\sqrt{2}\, X_m\, [1 - \exp(-2\, D_0\, t/X_m^2)]^{1/2}}\right)$$
$$\text{for } \tau_e < t < \tau_{RO} \tag{3.3.15}$$

At $t >> \tau_{RO}$, Eq 3.3.15 reduces to

$$C_R(x,t) = 0.5\,(\rho/M_S)\,\mathrm{erfc}\,[x/(\sqrt{2}\,X_m)] \tag{3.3.16}$$

This gives the maximum Rouse contribution to the concentration profile as a constant factor of width $8/\pi R_{ge} N^{1/4}$.

3.3.4 Concentration Profile $C(x,t)$ Due to Reptation and Rouse Motion

For the symmetric interface, the concentration profile $C_r(x,t)$ due to reptation is given from Chapter 2 by [1]

$$C_r(x,t)/(\rho/M_S) = [l(t)/L + x^2/(aL)]\,\mathrm{erfc}\,\{x/[2l(t)\,a]^{1/2}\} - [2l(t)/\pi a]^{1/2}\,(x/L)\,\exp\{-x^2/[2l(t)\,a]\} \quad \text{for } t < T_r \tag{3.3.17}$$

where a is the segment size of the minor chain, L is the chain contour length, and $l(t) = 2(D_1 t/\pi)^{1/2}$ where $D_1 \sim M^{-1}$.

At $t > \tau_{RO}$, the chain middle section of length $[L - 2l(t)]$ contributes to $C(x,t)$ through the segmental motion mechanism, while the two minor chains contribute to $C(x,t)$ through reptation. Consequently, $C(x,t)$ is expected to be the addition of the two weighted individual contributions, $C_r(x,t)$ (Eq 3.3.17) and $C_R(x,t)$ (Eq 3.3.15, with the weighting factor) as

$$C(x,t)/(\rho/M_S) = 0.5\,[1 - 2\,l(t)/L]\,\mathrm{erfc}\left(\frac{x}{\sqrt{2}\,X_m\,[1 - \exp(-2\,D_0\,t/X_m^2)]^{1/2}}\right) + [l(t)/L + x^2/(aL)]\,\mathrm{erfc}\,\{x/[2l(t)\,a]^{1/2}\} - [2l(t)/\pi a]^{1/2}\,(x/L)\,\exp\{-x^2/[2l(t)\,a]\} \quad \text{for } \tau_{RO} < t < T_r \tag{3.3.18}$$

It should be noted that $C(x,t)$ is also valid for $t < \tau_0$ because $D_S(t)$ reduces to D_0 for this time period. Indeed, at $t < \tau_0$, since $l(t)/L$ goes to zero and $2D_0 t/X_m^2$ is much smaller than unity, Eq 3.3.18 reduces to the solution to the Fickian diffusion equation. For $t >> \tau_{RO}$, $2D_0 t/X_m^2$ becomes much larger, and then Eq 3.3.18 can be further simplified to

$$C(x,t)/(\rho/M_S) = 0.5\,[1 - 2\,l(t)/L]\,\mathrm{erfc}[x/(\sqrt{2}\,X_m)] + [l(t)/L + x^2/(aL)]\,\mathrm{erfc}\,\{x/[2\,l(t)\,a]^{1/2}\} - [2\,l(t)/\pi a]^{1/2}\,(x/L)\,\exp\left[1 - \frac{x^2}{2\,l(t)\,a}\right]$$

$$\text{for } \tau_{RO} << t < T_r \qquad (3.3.19)$$

which is independent of D_0.

According to Eq 3.3.19, as the distance from the interface plane or the healing time increases, the contribution to the profile from the lateral segmental motion decreases rapidly, but the contribution from reptation increases. At $t \geq T_r$, $2l(t) = L$, and $C(x,t)$ is determined only by Fickian diffusion.

Figure 3.4 shows $C(x,t)$ computed at different healing times at 140 °C for polystyrene, using the self-diffusion coefficient D^* measured by Kramer *et al.* [22] for M = 500,000, and $D_0 = 15 \times 10^{-16}$ cm^2. The curves were plotted at time intervals of $T_r/15$ starting at τ_{RO} (curve 1) and ending at T_r (curve 15). The interdiffusion distance, x, was normalized by R_g to highlight this important region. At first glance, these profiles (Figure 3.4) appear normal, since the sharp gap at $x = 0$ due to reptation has been broadened by the Rouse contribution. However, the essential features of reptation become obvious again at high molecular weights and times in the range $\tau_{RO} < t < T_r$, as shown in Figure 3.5, where M = 20,000,000 [10].

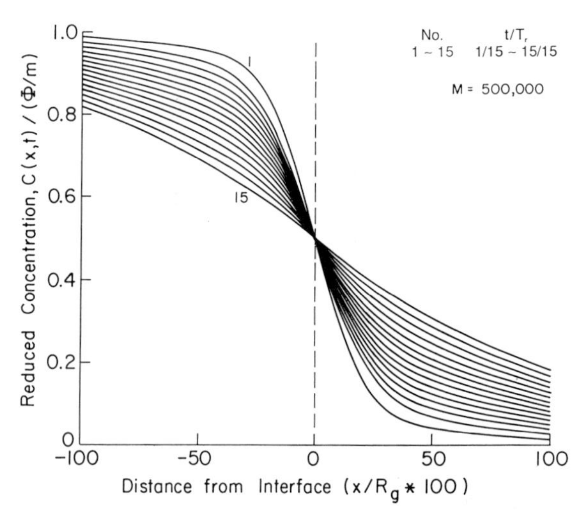

Figure 3.4 Concentration profiles for the symmetric monodisperse interface, A(M)/A(M), made of polystyrene of M = 500,000, at 140 °C for $(t/\tau_{RO}) \times 15 = 1, 2, 3, \ldots, 15$ (Zhang and Wool).

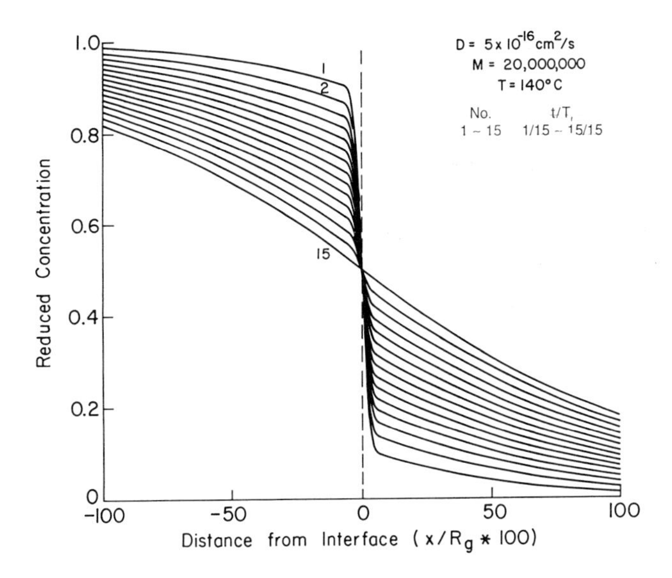

Figure 3.5 Concentration profiles for the symmetric PS/PS interface with $M = 20 \times 10^6$ at 140 °C and time intervals of $T_r/15$. Note the appearance of the gap at the interface ($x = 0$) plane (Zhang and Wool).

In the next section we examine how the Rouse and reptation contributions combine to affect the molecular properties derived as $H(t) = H_\infty(t/T_r)^{r/4}$ in Chapter 2.

3.4 Molecular Properties with Rouse and Reptation Dynamics

The effects of Rouse motion on the molecular aspects $H(t)$ of the symmetric interface, calculated in Chapter 2, are reevaluated in this section using the concentration profile given by Eqs 3.3.18 and 3.3.19.

3.4.1 Continuity at the Interface

In contrast with the discontinuity (gap) predicted from the concentration profile due to reptation, the concentration profile contributed by both reptation and segmental motion is continuous at the interface plane. In fact, as can be seen from Eq 3.3.19, $C(0,t) \equiv 0.5\ (\rho/M_S)$ at time $t > 0$. However, with very high molecular weight polymers, the discontinuity should be more readily observable when the long-range diffusion contribution from reptation far exceeds that of Rouse broadening around $x = 0$. We see the gap beginning to emerge in Figure 3.5 when $R_g/R_{ge} >> 1$.

3.4.2 Number of Monomers Crossing the Interface, $N(t)$

We calculate the number of monomers, $N(t)$, that cross a unit area of the interface plane by integrating the concentration profile with respect to x. $N(t)$ is given for the different relaxation regions as follows [10]:

Segmental Relaxation Effect on N(t)

$$N(t) = (\rho/M_S)(1/\sqrt{\pi})\, D_0^{1/2}\, t^{1/2} \qquad \text{for } 0 < t < \tau_0 \qquad (3.4.1)$$

Here $N(t) \sim t^{1/2}$, which is characteristic of Fickian diffusion.

Entanglement Relaxation Effect on N(t)

$$N(t) = \left(\frac{\rho}{M_0}\right)\left(\frac{X_m}{2\sqrt{\pi}}\right)\left[1 - \exp\left(\frac{-2\, D_0\, t}{X_m^2}\right)\right]^{1/2} \qquad \text{for } \tau_e << t < \tau_{RO} \qquad (3.4.2)$$

When $2D_0t << X_m$, then $N(t) \sim t^{1/2}$. At $t = \tau_{\mathrm{RO}}$, the contribution from Rouse motion is

$$N(\tau_{\mathrm{RO}}) = (\rho\, M_0)(0.5\, R_{ge}\, N^{1/2}) \qquad (3.4.3)$$

which depends on $M^{1/4}$.

Rouse and Reptation Effect on N(t)

$$N(t) = \left(\frac{\rho}{M_0}\right)\left(\frac{16\sqrt{2}}{3\pi^{5/4}}\right)\frac{\sqrt{a}(Dt)^{3/4}}{L} + \left(\frac{X_m}{\sqrt{4\pi}}\right)\left[1 - \frac{2\,l(t)}{L}\right] \qquad \text{for } \tau_{RO} < t < T_r \qquad (3.4.4)$$

At longer times

$$N(t) = (\rho/M_0)\,(1/\sqrt{\pi})\, D^{*\,1/2}\, t^{1/2} \qquad \text{for } t > T_r \qquad (3.4.5)$$

The second term in Eq 3.4.4 denotes the contribution from Rouse motion. When this contribution is small, $N(t)$ behaves as $t^{3/4}M^{-7/4}$, and then the crossover of the time dependence of $N(t)$ from $t^{3/4}$ to $t^{1/2}$ is observable at $t = T_r$. When the contribution from segmental motion is significant, so that the second term has to be taken into account, the crossover behavior still occurs but the slope of log $N(t)$ versus log t at $t < T_r$

becomes smaller than ¾. This behavior is confirmed by SIMS measurements (Chapter 5). Equation 3.4.5 can also be used to determine the self-diffusion coefficient.

3.4.3 Average Monomer Interpenetration Depth, *X*(*t*)

The average interpenetration depth, $X(t)$, at time t is defined as

$$X(t) = \int_0^\infty x\, C_i(x,t)\,\mathrm{d}x \Big/ \int_0^\infty C_i(x,t)\,\mathrm{d}x \qquad (3.4.6)$$

With $C_S(x,t)$, $C_R(x,t)$, and $C(x,t)$ from the last section, the following results were calculated for the different relaxation regions:

Segmental Relaxation Effect on X(t)

$$X(t) = (D_0 \pi/4)^{1/2}\, t^{1/2} \qquad \text{for } t < \tau_0 \qquad (3.4.7)$$

where we have $X(t) \sim t^{1/2}$, which is Fickian by construction.

Entanglement Relaxation Effect on X(t)

We did a simulation for $N = 17$, which is of the order of the number of statistical segments per entanglement length. We found that [10]

$$X(t) = (2/\pi)^{1/2} R_{ge} (t/\tau_e)^{0.36} \qquad \text{for } \tau_0 < t < \tau_e \qquad (3.4.8)$$

where the exponent of 0.36 is greater than the asymptotic Rouse slope of ¼. The difference is due to finite size effects where the chain ends have enhanced motion compared to the rest of the monomers. At the entanglement relaxation time,

$$X(\tau_e) = 0.8\, R_{ge} \qquad (3.4.9)$$

For example, with polystyrene, $M_e = 18{,}000$, and $R_{ge} = 37$ Å, so that $X(\tau_e) \approx 30$ Å.

Rouse Relaxation Effects on X(t)

$$X(t) = (\sqrt{\pi}/4)\, X_m [1 - \exp(-2 D_0 t / X_m^2)] \qquad \text{for } \tau_e < t < \tau_{RO} \qquad (3.4.10)$$

At the Rouse relaxation time,

$$X(\tau_{RO}) = 0.8\, R_{ge} (M/M_e)^{1/4} \qquad (3.4.11)$$

For polystyrene with $M = 10^6$ and $M_e = 18{,}000$, $X(\tau_{RO}) \approx 80$ Å.

Rouse and Reptation Effect on X(t)

$$X(t) = \frac{a l^2(t)/(4L) + (X_m^2/8)[1-2l(t)/L]}{[8l(t)/a\pi]^{1/2}[l(t)/(3L)] + (X_m/\sqrt{4\pi})\{[1-2l(t)]/L\}}$$

$$\text{for } \tau_{RO} < t < T_r \tag{3.4.12}$$

where $X_m = 1.8R_{ge}(M/M_e)^{1/4}$. Note that the second terms in both the denominator and the numerator of Eq 3.4.12 are due to segmental motion. If the second terms are small, Eq 3.4.12 becomes the reptation expression, $X(t) = 0.81R_g(t/T_r)^{1/4}$. The effect of the Rouse contribution on a log plot of $X(t)$ versus t is to reduce the slope below that for pure reptation, tending to flatten the response, depending on molecular weight. As t approaches T_r and the Rouse term vanishes, the slope becomes ¼. This is consistent with SIMS and neutron reflection experiments discussed in Chapters 5 and 6 [23–26].

When $t = T_r$, $l(T_r) = L/2$ and Eq 3.4.12 reduces to

$$X(T_r) = \tfrac{3}{4}(3\pi/8)^{1/2} R_g \tag{3.4.13}$$

or $X_\infty = 0.81R_g$. When $t > T_r$, the behavior in the Fickian region is

$$X(t) = (\pi D^*/4)^{1/2} t^{1/2} \qquad \text{for } t > T_r \tag{3.4.14}$$

The relation between $X(t)$ and $[\langle X^2\rangle^{1/2} = (2Dt)^{1/2}]$ is $X(t) = (\pi/8)^{1/2} \langle X^2\rangle^{1/2}$.

3.4.4 Relation Between Bulk and Interdiffusion Dynamic Exponents

The preceding analysis suggests that there is a difference between $X(t)$ for interdiffusion and $\phi_\tau(t)^{1/2} \equiv \langle[R_x(\tau,t) - R_x(\tau,0)]^2\rangle^{1/2}$ for the bulk. In the bulk, the Doi–Edwards theory predicts that $\phi_\tau(t)^{1/2} \sim t^{1/2}, t^{1/4}, t^{1/8}, t^{1/4}$, and $t^{1/2}$, respectively, in the five dynamic regions. However, by virtue of the interface, $X(t)$ is split into contributions from Rouse and reptation dynamics, and the above slopes are realized only for the first and last Fickian diffusion regions with $X(t) \sim t^{1/2}$. In both the Rouse relaxation regions and the reptation region, $X(t)$ contains mixed contributions that depend on molecular weight. There is no contradiction here with the Doi–Edwards theory; it is simply a matter of how an interface partitions and mixes the component dynamics compared to the bulk without an interface, so that the functions $\phi_\tau(t) \equiv \langle[R_x(\tau,t) - R_x(\tau,0)]^2\rangle$ and $X(t)^2$ are not identical in all relaxation regions. This difference presents interesting opportunities to explore molecular dynamics at interfaces, which we pursue in Chapters 5 and 6.

3.5 References

1. H. Zhang and R. P. Wool, *Macromolecules* ***22***, 3018 (1989).
2. P. E. Rouse, *J. Chem. Phys.* ***21***, 1272 (1953).
3. S. F. Edwards, *Proc. Phys. Soc.* ***92***, 9 (1967).
4. M. Doi and S. F. Edwards, *J. Chem. Soc., Faraday Trans. II* ***74***, 1789 (1978).
5. M. Doi and S. F. Edwards, *The Theory of Polymer Dynamics*; Clarendon Press, Oxford; 1986.
6. P.-G. de Gennes, *J. Chem. Phys.* ***55***, 572 (1971).
7. P.-G. de Gennes, *Scaling Concepts in Polymer Physics*; Cornell University Press, Ithaca, NY; 1979.
8. Y.-H. Kim and R. P. Wool, *Macromolecules* ***16***, 1115 (1983).
9. R. P. Wool, *Rubber Chem. Technol.* ***57*** (2), 307 (1984); R. P. Wool, *J. Elastomers Plast.* ***17*** (Apr), 107 (1985).
10. H. Zhang and R. P. Wool, in preparation.
11. G. Fleischer, *Polym. Bull. (Berlin, Germany)* ***9***, 152 (1983); *ibid.* ***11***, 75 (1984).
12. M. Antonietti, J. Coutandin, R. Grutter, and H. Scillescu, *Macromolecules* ***17***, 798 (1984).
13. S. Chandrasekhar, *Rev. Mod. Phys.* ***15***, 1 (1943).
14. Y. Oono, "Dynamics in Polymer Solutions. A Renormalization–Group Approach"; chapter in *Polymer–Flow Interaction*, Y. Rabin, Ed.; No. 137 in series *AIP Conference Proceedings*, R. G. Lerner, Ed.; American Institute of Physics, New York, 1985; p 187.
15. T. Myint-U and L. Debnath, *Partial Differential Equations for Scientists and Engineers*; North–Holland, New York; 3rd ed., 1987.
16. H. Zhang, R. P. Wool, and Y. Oono, in preparation.
17. G. Marrucci, *Macromolecules* ***14***, 434 (1984).
18. D. S. Pearson and E. Helfand, *Macromolecules* ***17***, 888 (1985).
19. L. G. Curro, D. S. Pearson, and E. Helfand, *Macromolecules* ***18***, 1157 (1985).
20. P. G. Shewmon, *Diffusion in Solids*; McGraw–Hill, New York; 1963.
21. J. Crank, *The Mathematics of Diffusion*; Clarendon Press, Oxford; 1975.
22. P. F. Green, C. J. Palmstrom, J. W. Mayer, and E. J. Kramer, *Macromolecules* ***18***, 501 (1985).
23. S. J. Whitlow and R. P. Wool, *Macromolecules* ***24***, 5926 (1991).
24. U. Steiner, G. Krausch, G. Schatz, and J. Klein, *Phys. Rev. Lett.* ***64***, 1119 (1990).
25. A. Karim, *Neutron Reflection Studies of Interdiffusion in Polymers*; Ph.D. Thesis, Northwestern University, Evanston, IL; 1990.
26. G. Agrawal, R. P. Wool, W. D. Dozier, and T. P. Russell, "Interdiffusion of Polymers: Comparison of Experiment, Simulation, and Theory for Chain-End Versus Chain Center Motion", *Bull. Am. Phys. Soc.* ***37*** (1), 728 (1992).

4 The Fractal Structure of Interfaces*

4.1 Introduction

When diffusion occurs at an interface, the concentration depth profile $C(x,t)$ varies smoothly as a function of the one-dimensional depth variable x, as demonstrated in the last two chapters. However, when the diffusion process is viewed in two or three dimensions, the interface is not smooth; indeed, it can be very rough. For example, Figure 4.1 shows two-dimensional views of a monomer–monomer interface, a polymer–polymer interface, and a polymer–metal interface [1, 2, 3]. The random nature of diffusion permits the formation of complex structures with fractal characteristics. The diffusion field in each case is divided into two parts: the part nearest the origin (dark in the figures) is connected to itself through a percolation criterion, and the nonconnected part involves small clusters of diffusing molecules that are not connected to the source of diffusion. The frontier separating the connected from the nonconnected regions is called the *diffusion front* by Sapoval, Rosso, Gouyet, and co-workers [4–8]. The diffusion front has fractal character, which provides an elegant method of describing the naturally rough structure of interfaces. The roughness and position of the diffusion front vary as functions of the average monomer interdiffusion depth, $X(t)$. Consequently, properties that are sensitive to connectivity, such as adhesion strength, mechanical modulus, thermal expansion, and electrical conductivity have a strong dependence on this type of structure. The concepts presented in this chapter apply to polymer interfaces but are of general importance to any diffuse interface.

In this chapter, we utilize the approach of Sapoval and co-workers in describing the fractal structure of interfaces. The concentration profile and structures shown in Figure 4.1 can be understood in terms of gradient percolation concepts, which provide a natural basis for describing the ramified (rough) and fractal characteristics of polymer–polymer interfaces. We begin with a discussion of *scalar percolation*, a concept introduced by Flory [9]. We then use fractal concepts to describe the natural roughness of percolation clusters in a concentration gradient. Finally, we present several examples of the fractal nature of polymer–polymer and polymer–metal interfaces.

---

* Dedicated to B. Sapoval

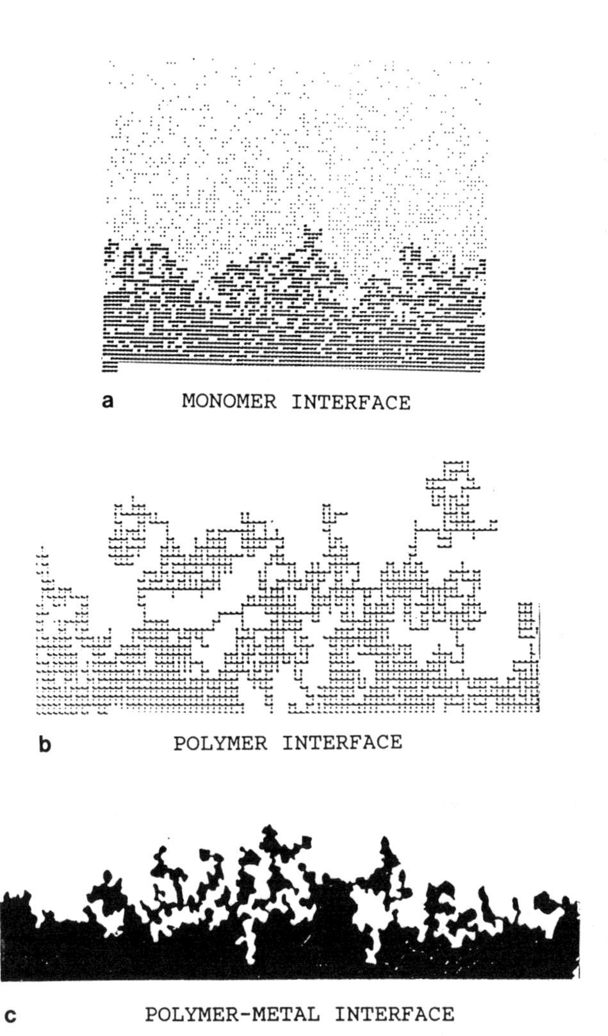

Figure 4.1 The ramified fractal nature of diffuse interfaces for (a, top) a computer-simulated 2-d monomer–monomer interface, where the heavy region represents the connected monomers on one side; (b, middle) a simulated 2-d random-coil polymer interface at the reptation time; and (c, bottom) electrochemically deposited silver diffusing in polyimide, with the unconnected metal atoms removed to show the fractal diffusion front of the connected metal atoms.

4.2 Percolation

Conduction percolation, involving connectivity between dispersed particles in a nonconducting matrix, is the basis for evaluating the structure of interfaces with concentration gradients. *Vector percolation* involves the transmission of forces through a lattice (2-d or 3-d) where a certain fraction of the bonds are broken or missing. To distinguish between the two forms of percolation, the conduction form has been referred to as *scalar percolation*; we will see that it is the basis for describing the fractal structure of interfaces in general. Vector percolation has many potential applications in molecular fracture, craze fibril breakdown, and network relaxation (Chapter 7). The next two sections provide the reader with a brief description of percolation.

4.2.1 Scalar Percolation

Scalar percolation theory concerns the connectivity of one component randomly dispersed in another [9–12]. A common example of this is gelation during the polymerization of monomers with multifunctional linkages. Percolation in two dimensions can be well described by the following example [13, 14]. Figure 4.2(a) shows a square lattice randomly populated with gray squares, with occupation probability p. *Connectivity* is defined as the progressive sum of contacts via first nearest neighbors of the gray squares with the lower border of black squares; second nearest neighbors can also be used. In Figure 4.2(b), all of the gray squares connected to the black border have been transformed into black squares. The percent connected is defined as the number of transformed (black) squares divided by the original number of gray squares.

An example of a larger 2-d percolation system (512×512) is presented in Figure 4.3. In the lattice in Figure 4.3(a), 58% of the lattice sites are occupied, but only 16.6% of the squares are connected to the bottom or top. The connected regions near the surface represent the connected clusters, so that the average connected depth can be determined by the average cluster size.

Figure 4.3(b) has a 1% higher occupation, 59%, with 63.6% connected, and Figure 4.3(c) has 60% occupation, with 78% of the squares connected. In Figure 4.3(b), the black squares span the entire lattice (that is, a connected pathway exists between the top and bottom). The minimum concentration at which this could happen on an infinite lattice is known as the *percolation threshold*, p_c. This is directly analogous to the gel point in nonlinear polymerizations, and the point at which electrical conductivity is obtained with metal particles randomly dispersed in a nonconducting matrix. Table 4.1 shows p_c values for different types of lattices. Because 3-d lattices (simple cubic and body-centered cubic) have a higher degree of freedom, the thresholds occur at lower concentration than in 2-d lattices (square and triangular).

Since percolation is an example of critical phenomena, many properties, $H(p)$, of the system follow scaling laws near p_c of the form

$$H(p) \sim (p - p_c)^m \tag{4.2.1}$$

where m is a *critical exponent*. The critical exponents of some selected scaling laws and their related properties in 2-d are given in Table 4.2, using the notation of Stauffer

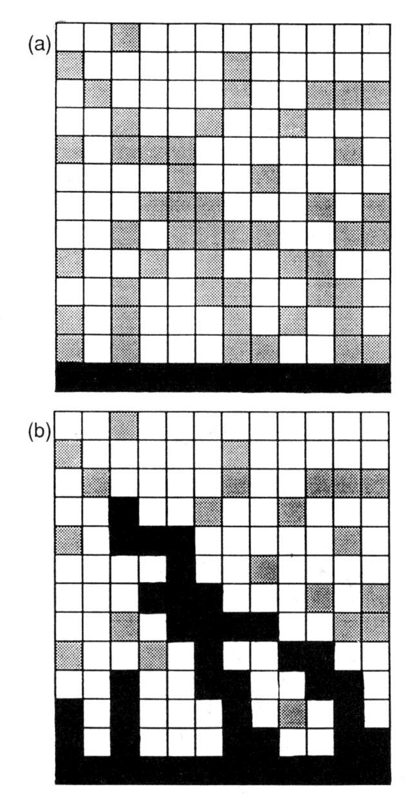

Figure 4.2 An illustration of connectivity via first nearest neighbors (Peanasky, Long, and Wool).

Table 4.1 Site Percolation Thresholds

Lattice type	p_c
Square	0.59275
Triangular	0.50000
Simple cubic	0.3117
Body-centered cubic	0.245

Table 4.2 Critical Exponents for Three Dimensions

Exponent-related property	Theoretical	Experimental
α, number of clusters	−0.60	−0.64
β, strength of infinite network	0.40	0.41
γ, divergence of second moment	1.8	1.9
σ, average cluster size	0.45	0.45
τ, cluster number at $p = p_c$	2.2	2.1

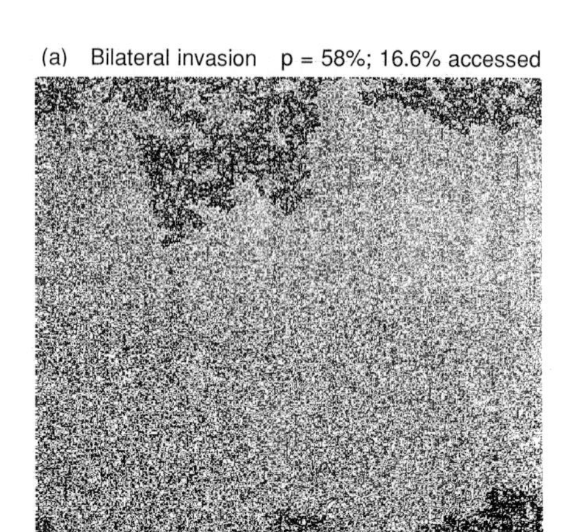

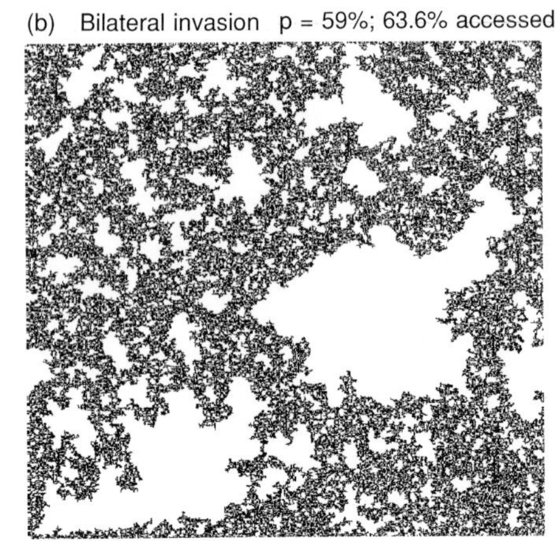

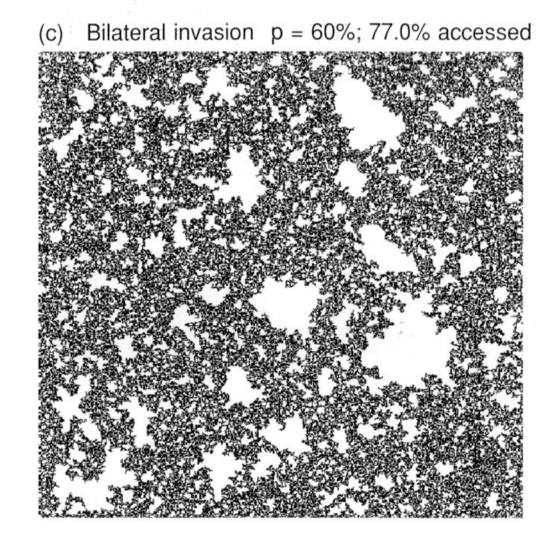

Figure 4.3 Effect of concentration on accessibility modeled by bilateral invasion from both the top and bottom sides. The square 512 × 512 has occupation probabilities of (a) 58%, (b) 59%, and (c) 60%. Accessibilities are displayed on each figure.

[10]. Note that these exponents are universal quantities: their values are independent of lattice type, at constant lattice dimensionality.

The concept of clusters is used for describing the interface structure. The average cluster size or correlation length, ξ, is

$$\xi \sim (p - p_c)^{-\nu} \tag{4.2.2}$$

where the critical exponent, ν, is 4/3 in two dimensions and 0.88 in three dimensions. When p is less than p_c, the lattice has aggregates of connected particles called *clusters*. As the percolation threshold is approached, the clusters aggregate to form a large cluster that spans the lattice.

4.2.2 Vector Percolation

The transmission of forces through a lattice as a function of the fraction p of bonds in the lattice has been analyzed by Kantor and Webman [15], Feng, Sen, and co-workers [16–19], Thorpe and co-workers [20–25] and our group [26]. The normalized elastic modulus, E/E_0, of the lattice as a function of p was found to obey relations similar to scalar percolation

$$E/E_0 = (p - p_c)^{m_v} \tag{4.2.3}$$

but with a different critical exponent m_v. For example, in two dimensions, the exponent for scalar percolation is about 1.3, compared with exponents in the range of 3 to 5 calculated from vector percolation theory. Vector percolation is therefore considered to belong to a different universality class than scalar percolation. This problem has application to the microscopic aspects of the strength of interfaces, where, for example, breakdown and fracture of entanglement networks in deformation zones at crack tips is an issue. Vector percolation has relevance in many other areas of polymer fracture, such as the fracture of cross-linked polymers, entanglement relaxation (Chapter 7) and disentanglement, random bond scission, craze fibril fracture, composite fatigue, and fracture of nets.

We examined the role of vector percolation in the fracture of model nets at constant strain and subjected to random bond scission. Figure 4.4 shows an experiment where a net with $N_0 = 10^4$ bonds is step strained (about 2%). As the bonds are randomly fractured, the elastic modulus ratio, E/E_0, decays, as shown in Figure 4.5. In the initial stages of relaxation, the stress relaxation is well represented by the *effective medium theory* (*EMT*): the modulus ratio as a function of the number of severed bonds, N, decreases as

$$E/E_0 = 1 - 3N/N_0 \tag{4.2.4}$$

The fraction of bonds, p, remaining in the lattice is related to N, the number of bonds severed, by

$$p = 1 - N/N_0 \tag{4.2.5}$$

The EMT analysis indicates that the stress relaxes in proportion to the number of bonds removed. The linear decrease of E/E_0 with N makes intuitive sense and is the current basis for many constitutive theories of polymers. An example is the Doi–Edwards theory of viscoelasticity, in which the stress relaxation modulus is proportional to the number of entanglements remaining in the melt, which is represented by an entanglement network. In our studies of model nets, we find that the

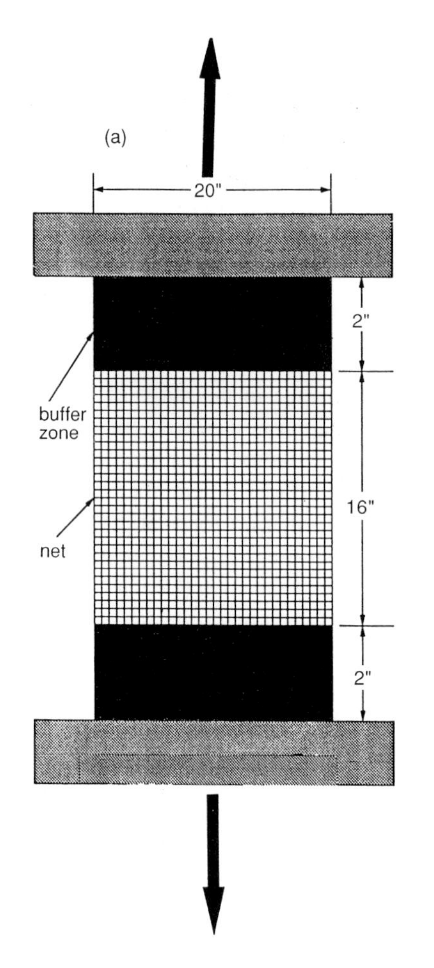

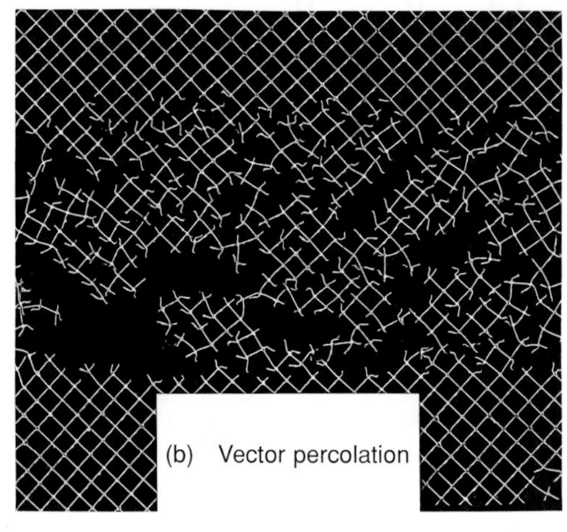

Figure 4.4 Vector percolation experiments: (a) A 2-d net is shown in preparation for a vector percolation test. The net is subjected to a step strain (about 1%), N bonds are fractured by random computer selection, and the stress is measured as a function of N. (b) An example of the tortuous connectivity in a step-strained net near p_c with random bond fracture (Wool, Bernaert, and Daley). This net is for illustration purposes only and is too small to conduct proper vector percolation studies. Nets must have more than 10^4 bonds if we are to observe vector percolation, as used in Figure 4.5.

EMT analysis is excellent for lattices of finite size containing fewer than about 2,000 bonds: the modulus ratio relaxes, with a slope given by $3/N_0$ and an intercept on the N axis of $N_0/3$, corresponding to an apparent percolation threshold, p_c', of ⅔. However, the bond conductivity percolation threshold, p_c, is ½, which suggests that mechanical relaxation occurs while the lattice remains conducting. This is an accidental agreement with the Feng and Sen theory of vector percolation, which correctly predicts that when vector percolation occurs, the lattice becomes very "floppy" and, though it could still conduct electricity ($p > ½$), it cannot transmit elastic forces.

For larger lattices, as shown in Figure 4.5, percolation develops after about 80% relaxation occurs; such a lattice near percolation is shown in Figure 4.4(b). For vector percolation to be observable, the lattice size must be large enough (about 10^4) so that, as percolation is approached, finite-sized clusters of fractured bonds do not completely relax the lattice by bridging the mechanical grips holding the lattice. In percolation

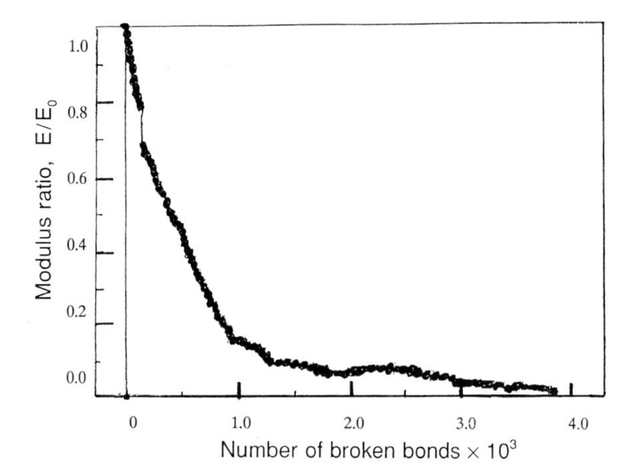

Figure 4.5 Modulus ratio relaxation in a 2-d polymer net (10^4 bonds) as a function of the number of broken bonds, N. The sudden drop near $E/E_0 \approx 0.8$ is due to fracture of a "hot bond". The stress decays towards zero as the percolation threshold is approached (about 5×10^3 broken bonds (Daley and Wool).

language, this means that the lattice length L must obey the relationship

$$L > (p - p_c)^{-\nu} \tag{4.2.6}$$

where $(p - p_c) = 0.05$ (within 5% of percolation) and ν is the critical exponent for the correlation length. Lobb and Forester demonstrated this relation for torsional rigidity relaxation of cylindrical nets with random bond scission. For most molecular systems, especially entanglement networks, the lattice size is effectively infinite compared to the above macroscopic experiment, and we can expect vector percolation to play an important role in the final stages of fracture. For these larger lattices, we have

$$E/E_0 \sim (N_c - N)^{\beta} \tag{4.2.7}$$

where $\beta \approx 3$, in the range of theoretical predictions of 3 to 5, and $N_c \approx N_0/2$.

Several interesting physical points are made by vector percolation analysis. As deduced from Figure 4.4, the stress distribution in the lattice bonds becomes highly nonuniform, so that some bonds are highly stressed and others bear little stress. The existence of highly stressed bonds is a prelude to molecular fracture, and parallels the "hot bonds" in conductivity percolation. In electrical circuits, hot bonds arise from high current density in single bonds in a tortuously connected network near the percolation threshold, so that large clusters of bonds are connected by a single bond. The concept of mechanical hot bonds is relevant to fracture of polymers in general. When polymers such as polypropylene and polyethylene are subjected to uniform tensile stresses, we and others (using infrared and Raman spectroscopy) have shown that the molecular stress distribution can be quite broad, even though the applied stress is well below the macroscopic fracture stress. The existence of overstressed bonds results in molecular fracture, microvoiding, eventual coalescence of voids, and macroscopic failure under moderate loads at long times. The relation between the spectroscopically observed molecular stress distribution and that predicted from vector percolation has not been elucidated, and would be an interesting topic to pursue.

Another point of interest is that only a fraction of the bonds must be fractured ($p > p_c$) before complete failure occurs. Thus, in a deformation zone at a crack tip, the crack advances through the zone by breaking a fraction ($1 - p_c$) of bonds or fibrils in a craze network (see Chapter 8). Similarly, for fracture controlled by chain pullout, it is not necessary to pull out every chain, or each chain all the way, from the eventual fracture surface.

4.2.3 Fractal Nature of Percolation Clusters

The fractal nature of percolation clusters has been described by several authors [27–29]. The percolation cluster is not dense and is highly ramified, as shown in Figure 4.4(b). The cluster is also self-similar, which means that it appears the same at different length scales of observation. Clouds in the sky can have this property, with the result that an observer in an airplane is never sure how far away they are. They appear the same close up or far away, because small pieces of the cloud are similar to much larger pieces.

Percolation clusters, like clouds and other fractal objects, do not fill space in a dense manner but tend to fill a fraction of space, so that the mass, M, versus distance, R, is given by

$$M \sim R^{D_c} \tag{4.2.8}$$

where D_c is the fractal dimension, typically less than the embedding dimension, d. For example, the fractal dimension of the percolation cluster shown in two dimensions (the embedding dimension) in Figure 4.3 is $D_c \approx 1.89$, and the fractal dimension of a cloud is about 2.5 in a 3-d sky. The fractal dimension is closer to the embedding dimension when the object is more dense. More ramified objects also tend to have a higher fractal dimension. A branched polymer chain is more dense than its linear counterpart of the same mass in the amorphous state.

Polymer chains in the melt are random walks and form fractal objects. In three dimensions, the radius of a random-walk chain depends on its mass, M, as $R \sim M^{1/2}$, so that

$$M \sim R^2 \tag{4.2.9}$$

where the fractal dimension $D_{RW} = 2$.

For self-avoiding chains (SAWs), R is related to M via

$$R \sim M^{3/(d+2)} \tag{4.2.10}$$

and the fractal dimension is related to the embedding dimension d by

$$D_{SAW} = (d+2)/3 \tag{4.2.11}$$

in which $d \leq 4$. Thus, a polymer chain in a good solvent (d = 3) has a fractal dimension D_{SAW} of 5/3. When it is compressed between two plates with monomer separation distance (d = 2), then D_{SAW} is 4/3. If d is 1, the SAW is a straight line with D_{SAW} of 1 but is nonfractal. When the solvent is changed from a "good" to a Θ solvent (Chapter 1, Section 1.2.3), the chains again become simple random walks for which D_{RW} is 2 in dimensions 2 and 3. Since a SAW is more expanded in space and is therefore less dense than the simple random walk, it has a lower fractal dimension.

In a poor solvent (d = 3), the chains tend to collapse into themselves and precipitate with

$$R \sim M^{1/3} \tag{4.2.12}$$

so that $D = d = 3$. In poor solvents, the chains lose their fractal properties, as expected. Similar events happen at interfaces when noninteracting particles become interacting and destroy the fractal structure by precipitation.

Some physical properties of polymers can be related to their fractal dimension. For example, the intrinsic viscosity, $[\eta]$, of polymers in dilute solution with hydrodynamic radius R behaves as

$$[\eta] \sim R^3/M \tag{4.2.13}$$

Substituting in the mass-to-radius relation, we obtain

$$[\eta] \sim M^{(3/D)-1} \tag{4.2.14}$$

For random-coil chains in Θ solvents, D is 2 and $[\eta] \sim M^{1/2}$; for good solvents, D is 5/3 and hence $[\eta] \sim M^{4/5}$, in agreement with the Mark–Houwink–Sakurada relation [30].

The fractal dimension, D_f, of the percolation cluster perimeter in Figure 4.3(b) is related to the correlation length critical exponent, v, via

$$D_f = (1+v)/v \tag{4.2.15}$$

One can deduce that the leading edge of the connected region of the diffusion field at interfaces has a similar fractal dimension and roughness characteristics determined by percolation theory.

Percolation theory could be used to improve adhesion at incompatible interfaces, as discussed in the following solved problem.

Problem

How can percolation be used to promote adhesion at polymer interfaces?

Potential Solution

One could use the surface roughness of percolation clusters to promote adhesion at interfaces. Consider the case of bonding a liquid-hardening epoxy to a smooth polyethylene surface at room temperature. The interface strength is fairly weak because of chemical incompatibility and the inability of the reacting epoxy to form primary bonds with the PE. However, the PE surface can be roughened if the polymer has been premixed with a volume fraction p of particles of diameter a, and the particles are removed by chemical or solvent techniques prior to the application of the epoxy liquid. The surface of a polymer containing randomly dispersed particles can be visualized as the surface one obtains by drawing a plane through the body of the material [or a line through Figure 4.3(a)]. The plane effectively samples the cluster size distribution function of aggregating particles. Subsequent removal of these clusters with fractal surfaces produces a very rough surface that can be mechanically interlocked by the epoxy.

When p is greater than p_c, the volume fraction of particles that can be removed, f, is determined as

$$f = (p - p_c)^{0.41}$$

where $p_c = 0.31$ is the percolation threshold. However, only a small number of particles near the surface must be removed to promote mechanical interlocking. When p is less than p_c, the surface fraction accessed is determined by the average cluster size distribution via

$$f = p_c^{\sigma}\, a / h |(p - p_c)^{\sigma}|$$

where $\sigma = 0.45$ and h is the PE film thickness. This equation is limited to values of p such that the percolation correlation length

$$\xi = |p - p_c|^{-\sigma}$$

is less than the film thickness. Thus, as p increases towards p_c, the depth of the surface roughening increases rapidly and eventually penetrates through the sample thickness at $p \leq p_c$. The value of p can be optimized to adjust the roughness depth and promote maximum bond strength at the interface, which has now become fractal.

4.2.4 Gradient Percolation

Consider the two-dimensional lattice diffusion of A atoms (shown in Figure 4.6) into B atoms (not shown), as discussed by Sapoval and co-workers [4–8]. The concentration profile of A atoms is given by the Fickian result

$$C(x) = \operatorname{erfc}(x/L_d) \tag{4.2.16}$$

where $L_d = 2(D_0 t)^{1/2}$ is the Einstein diffusion length at time t with diffusion coefficient D_0. The diffusion front (heavy line in Figure 4.6) is defined by those A atoms that are connected to the diffusion source at $x = 0$ via other A atoms and have a first or second B neighbor that is itself connected to the source of B atoms. The interface with a diffusion front is seen to consist of solid regions, with lakes or holes leading down to the seashore and unconnected islands of A atoms in the B sea. The presence of holes and the connectedness between atoms in the interface is important in the determination of the electrical and mechanical properties of the interface.

Figures 4.7 to 4.9 show larger scale (512 × 512 lattice) computer simulations of the diffusion front at diffusion lengths $L_d = 256$, 512, and 10,240 lattice units, respectively

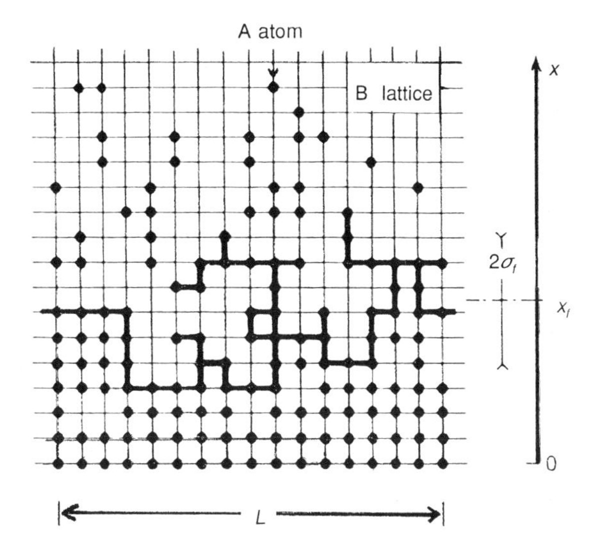

Figure 4.6 Gradient percolation is shown schematically for A atoms (bold) diffusing into B atoms (not shown) on a 2-d lattice. The concentration gradient decreases from bottom to top. The bold line represents the diffusion front that separates the nonconnected A atoms from the A atoms connected to the source of A atoms (bottom) (courtesy of Sapoval *et al.*).

Figure 4.7 Gradient percolation simulation of atoms diffusing on a 512^2 lattice with a diffusion length L_d of 256. The grey region represents the diffusion front between the connected (black) and nonconnected (white) atoms (Wool and Long).

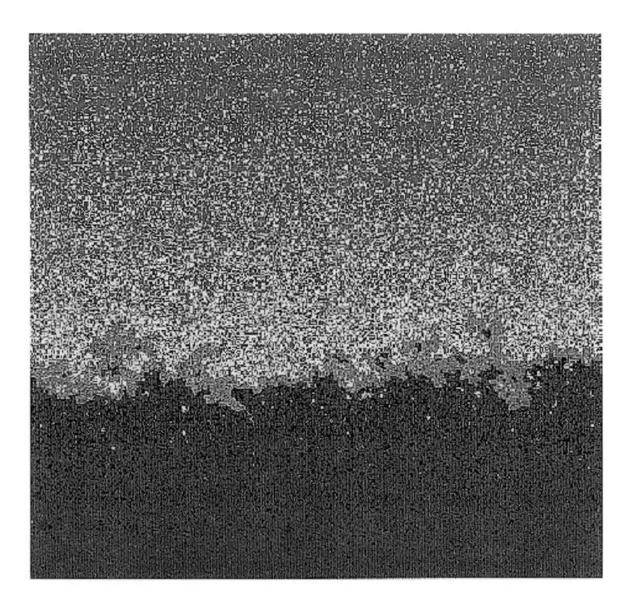

Figure 4.8 Gradient percolation of atoms with a diffusion length L_d of 512 on a 512^2 lattice.

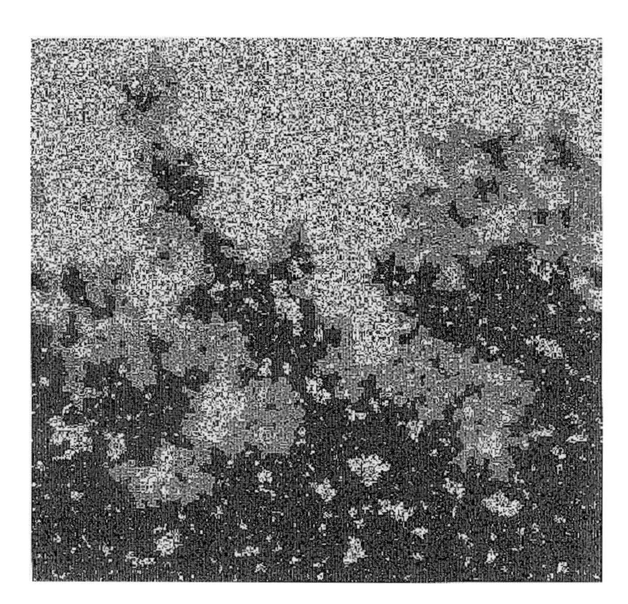

Figure 4.9 Gradient percolation of atoms with a diffusion length L_d of 10,240 on a 512^2 lattice.

[2]. The diffusion profiles were generated using the Fickian profile (Eq 4.2.16) with $p = 1$ (A atoms) at $x = 0$, at the bottom of each figure (black). Towards the top of the figure, the concentration of A atoms decreases (grey) at $x >> 0$. At any interdiffusion depth x, a site has probability p of containing an A atom, and the site is given a color depending on its occupation and connectivity. The diffusion front (white) becomes more extensive and ramified with increasing diffusion depth, L_d. The black to grey shades represent the connected A atoms (black), A-atom clusters (grey), and diffusion front (white). The A atoms that are connected to the $x = 0$ plane are shown in black. The connectivity criterion was identical to that used for the percolation algorithm in Figure 4.2. If the A atoms were metal, the black region represents those atoms that could conduct electricity introduced at $x = 0$. One might argue for polymers that the black region represents the area connected by entanglements to the A side of the interface. The black region extends up to the percolation threshold concentration, $C(x) = p_c$, at position $x(p_c)$. At lower concentrations, or diffusion distances x greater than $x(p_c)$, we are below the percolation threshold and the connectivity is lost.

The grey regions represent the clusters of A atoms that are not connected to the source at $x = 0$. The clusters are small at the top of the figure but increase in size in the vicinity of the percolating black region. Note that many grey islands exist in the black region. These are A atoms surrounded by B atoms and therefore not connected to the A source at $x = 0$. If the B atoms were nonconducting, then the grey A islands would not contribute to the black conduction zone.

The connected B zone extends down to the connected A zone, and the region where they meet is the diffusion front (white). The diffusion front can also be understood as the frontier that separates the connected and nonconnected A atoms. The roughness and width of the diffusion front increase with diffusion length, and the fractal character becomes more apparent, for example, in a comparison of Figure 4.7 ($L_d = 256$) with Figure 4.9 ($L_d = 10{,}240$). Note that in Figure 4.9, the white diffusion front contains many black islands; these represent conducting A atoms that are not in contact with B atoms but connect through the front. In the next section, the methods of characterizing the fractal nature of diffusion fronts are examined.

4.3 Characterization of Diffusion Fronts

The diffusion front of each of the interfaces shown in Figures 4.7–4.9 can be characterized by three macroscopic quantities: its position, x_f, in the concentration gradient; its width, σ_f; and the total number of particles on the front, N_f [4–8].

4.3.1 Position of the Diffusion Front, x_f

The concentration of particles at the mean front position, x_f, is identical to the percolation threshold, so that $p(x_f) = p_c$, where, for site percolation, $p_c = 0.593$. The simulation results for p_c are shown in Table 4.3. (In Table 4.3, each front parameter Y is analyzed in terms of the general scaling law $Y = AL_d^{\alpha}$, in which A and α are constants, and compared with our simulations for different lattice sizes.) The position of the front is proportional to the diffusion length L_d [4], and for a 2-d lattice with a Fickian error-function profile, we have from Eq 4.2.16 at concentration p_c

$$x_f = 0.378\, L_d \qquad (4.3.1)$$

or

$$L_d = 2.645\, x_f \qquad (4.3.2)$$

4.3.2 Fractal Dimension of the Diffusion Front, D

The fractal dimension, D, of the front in Figure 4.9 ($L_d = 10{,}240$) was determined from the slope of the log plot of mass versus radius (Figure 4.10). We obtained $D = 1.76$, which is in excellent agreement with the prediction of Sapoval and coworkers [4]. The mass of the front was measured as follows. Circles of radius R with their centers on the $x(p_c)$ line were drawn (by computer). The number of front particles (white) inside each circle was determined as a function of R, and the procedure was

Table 4.3 Analysis of Diffusion Fronts

Parameter	$Y = A\, L_d^{\alpha}$	256^2	512^2	1024^2	Theory	Sapoval
Front Width σ_f	$A(\sigma_f)$	0.859	0.608	0.597	~1	0.46
	$\alpha(\sigma_f)$	0.479	0.521	0.544	0.571	0.57±0.01
Width Noise $\delta\sigma_f^2$	A	1.56×10^{-3}	10^{-4}	2.46×10^{-5}	—	—
	α	1.469	1.621	1.764	—	—
Length of Front N_f	$A(N_f)$	1.074	0.949	0.918	0.97	0.96
	$\alpha(N_f)$	0.409	0.427	0.430	0.43	0.425
Length Noise δN_f^2	A	0.404	0.240	0.200	~1	~2
	α	1.333	1.400	1.437	1.428	1.5±0.05
Front Position x_f	$A(x_f)$	0.361	0.371	0.376	0.378	0.378
	$\alpha(x_f)$	0	0	0	0	0
Position Noise δx_f^2	$A\times10^3$	6.63	2.02	2.44	—	—
	α	1.462	1.400	1.472	—	—
Fractal Dimension	D	1.734	1.741	1.753	1.75	1.76
	δD	0.089	0.079	0.079	0.000	0.02
Frontier Breadth B	$A(B)$	6.962	5.98	5.366	—	—
	$\alpha(B)$	0.404	0.449	0.479	—	—
Breadth Noise δB^2	A	0.146	0.055	0.012	—	—
	α	1.126	1.254	1.432	—	—
Threshold p_c	p_c	0.598	0.596	0.594	0.593	0.593
	δp_c	0.015	0.010	0.007	0.000	0.000

averaged over many locations on the diffusion front x_f line. The procedure was repeated at least 50 times for different fronts generated at the same diffusion length. The analysis is applicable to radii in the range $R < \sigma_f$. When R exceeds the width of the front, the mass is directly proportional to R, and $D = 1$, as shown in Figure 4.10 with $L_d = 128$ and $R > 20$ lattice units.

We examined the fractal dimension in 256^2, 512^2, and 1024^2 lattices, with diffusion lengths up to 10^4 lattice units. Periodic boundary conditions were used on the lateral sides. We found that D was independent of diffusion length for $R < \sigma_f$ (shown in Figure 4.11 for the 1024^2 lattice). In this case, $D = 1.754 \pm 0.079$; the 0.079 standard deviation was from an average of 50 runs at each diffusion length. The scatter in the data shown in Figure 4.11 was found to be constant with increasing diffusion length.

The fractal dimension appeared to be bounded by the average range $1.675 \le D \le 1.833$. The value of $D = 1.833$ might correspond to portions of the front that have internal cluster character ($D_c = 1.89$) and the lower value of $D = 1.675$ may correspond to more open structures similar to diffusion-limited aggregation and viscous fingering with $D \approx 1.64$. However, we should be cautious about interpreting the scatter

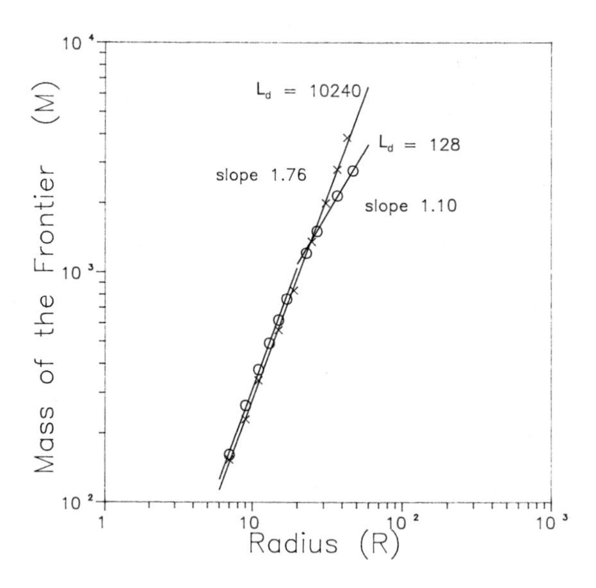

Figure 4.10 Mass of diffusion front versus radius. The slope on a log–log plot determines the fractal dimension (here, D = 1.76). Key: (–×–) L_d = 10,240 from Fig 4.9; (–○–) L_d = 128 (Wool and Long).

Figure 4.11 Fractal dimension, D, versus diffusion length, L_d, for a 1024^2 lattice.

of the diffusion front dimension in terms of other fractal structures. We found that D had a small dependence on lattice size: slightly lower D values (about 1.734) were obtained with the 256^2 rather than the 1024^2 lattice.

We conclude that the value of $D = 7/4$ predicted from percolation theory is an excellent convergence limit at large lattice size.

4.3.3 Width of the Diffusion Front, σ_f

As is evident from Figures 4.7–4.9, the width of the front, σ_f, increases with diffusion length, L_d. Sapoval *et al.* [4] analyze this problem as follows (reviewed in [17]). As $p(x)$ approaches p_c, the cluster size of A atoms (grey in Figure 4.9) begins to approach the width of the front. Thus the main length scale is related to the correlation length evaluated at $(x_f \pm \sigma_f)$. The concentration gradient $\mathrm{d}p/\mathrm{d}x$ prevents σ_f from becoming infinitely large at p_c. However, as the diffusion depth increases and the concentration gradient decreases, the width of the front increases and converges on the normal percolation case with uniform average concentrations.

Using Eq 4.2.2 for the cluster size as a function of $p(x)$, and expanding $p(x)$ around x_f, they obtain [4]

$$\sigma_f \approx [p(x_f \pm \sigma_f) - p_c]^{-\nu} \tag{4.3.3}$$

$$\approx [\sigma_f \, (\mathrm{d}p/\mathrm{d}x)_{x_f}]^{-\nu} \tag{4.3.4}$$

in which the proportionality constants are of the order of unity.

The concentration gradient $\mathrm{d}p/\mathrm{d}x$ is inversely proportional to the diffusion length L_d. From Eq 4.3.1, we have $(\mathrm{d}p/\mathrm{d}x)_{x_f} = 0.997/L_d$. Therefore, σ_f depends on L_d as

$$\sigma_f \approx L_d^{\alpha(\sigma)} \tag{4.3.5}$$

where the exponent $\alpha(\sigma)$ is related to the percolation exponent ν by

$$\alpha(\sigma) = \nu/(\nu + 1) \tag{4.3.6}$$

When $\nu = 4/3$ in two dimensions, then $\alpha(\sigma) = 1/D = 4/7$ and

$$\sigma_f \approx L_d^{0.57} \tag{4.3.7}$$

Computer simulations by Sapoval *et al.* were in excellent agreement with Eq 4.3.7 [4]. Results of our computer simulations are shown in Figure 4.12 (and Table 4.3) for the 1024^2 lattice. We obtained

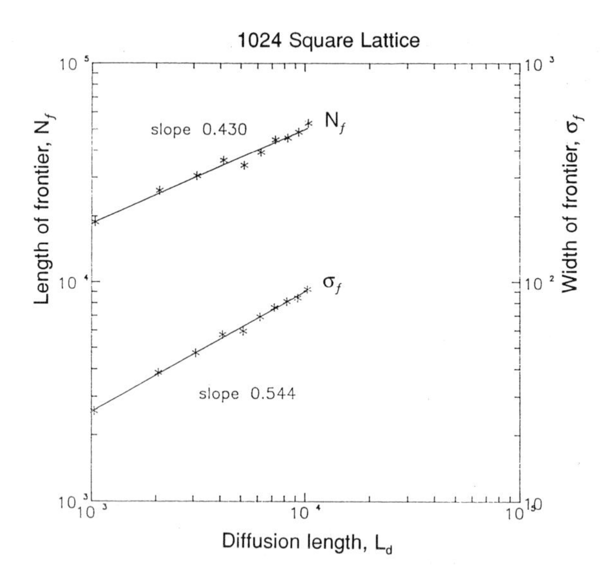

Figure 4.12 Number of particles on diffusion front, N_f, and front width, σ_f, versus diffusion length, L_d, for the 1024^2 lattice.

$$\sigma_f = 0.60\, L_d^{0.54} \tag{4.3.8}$$

which is in close agreement with the theoretical prediction. In this case, the standard deviation $\delta\sigma_f$ increases with L_d from 2.2 at $L_d = 1024$ to 16.75 at $L_d = 10{,}240$. This is to be expected, since the fluctuations in the position of the diffusion front are also increasing with L_d. Results for other lattices are listed in Table 4.3.

4.3.4 Span of the Diffusion Front, *B*

The span of the front, $B = 2|x_m - x_f|$, was determined, where x_m is the largest value of the diffusion depth that still contains the front. B is essentially the full width of the diffusion front in the depth direction, while σ_f describes the standard deviation. The span has significance for the design of electronic devices that encapsulate the front without leakage.

If $p_f(x)$ is the Gaussian concentration profile of the front, then the profile centered about x_f is determined in terms of N_f and B as

$$p_f(x) = [N_f / (\sigma_f L \sqrt{2\pi})]\ \exp - (x - x_f)^2 / 2\sigma_f^2 \tag{4.3.9}$$

Since $B = 2\,(x_m - x_f)$, the relation between B and σ_f is obtained at

$$p_f\ (x_m) = q/L \tag{4.3.10}$$

as the ratio

$$B/\sigma_f = 2\sqrt{2}\,[\ln N_f\ /(q\sigma_f\sqrt{2\pi})]^{1/2} \tag{4.3.11}$$

where q is the number of atoms that can be detected on the smallest cluster on the outer regions of the diffusion front, on the order of (but greater than) unity.

Substituting for the diffusion length dependence of N_f and σ_f on the right side of Eq 4.3.11, we obtain

$$B/\sigma_f = 2\sqrt{2}\{\ln[0.83\,L/(q L_d^{1/7})]\}^{1/2} \tag{4.3.12}$$

Thus, the span B is not directly proportional to the width σ_f, as might have been expected, but is modified by the $L_d^{1/7}$ factor on the right.

The computer simulation gives $B/\sigma_f = 4.82$, 5.97, and 5.72 for the 256, 512, and 1024 lattices, respectively, with $L_d = 1024$, and $B/\sigma_f = 4.05$, 5.05, and 4.93, respectively, with $L_d = 10{,}240$. Analysis of these results suggests that a value of $q \approx 6$ is required to obtain agreement between simulation and Eq 4.3.12. The ratio B/σ_f determined from simulations is expected to depend on the lattice size, since larger lattices have a higher probability of producing larger B values. With increasing diffusion length L_d, the ratio B/σ_f converges slowly to a value of $2\sqrt{2}$.

The noise in the width of the diffusion front can be expressed as the square of the standard deviation $\delta\sigma^2$, and was found to behave for the 1024^2 lattice as

$$\delta\sigma_f^2 = 2.5 \times 10^{-5}\,L_d^{1.76} \tag{4.3.13}$$

The exponent was found to be lower with the other lattices (Table 4.3), and the front factor is of the order of $1/L^2$. This result will be discussed with other features of the noise in the diffusion front in a later section.

4.3.5 Number of Atoms on Diffusion Front, N_f

The number of atoms on the front, N_f, is a very important parameter and readily quantifies the roughness of the interface in many practical situations. For example, in Figures 4.7–4.9, we would like to know how the length of the diffusion front (white line) changes with diffusion length L_d. Upon examining Figures 4.7–4.9, we see that N_f increases with diffusion length. The exact dependence can be derived as follows [4].

Consider a square box on the diffusion front, of length σ_f. Because the front is fractal, the box contains σ_f^D atoms, according to Eq 4.2.8. If the length of the interface is the lattice length L, then the interface contains L/σ_f fractal boxes, and the total number of atoms on the front is

$$N_f \sim L\sigma_f^{D-1} \tag{4.3.14}$$

where the proportionality constant is about 2. Substituting for σ_f from Eq 4.3.7, then

$$N_f \sim L\, L_d^{\alpha(N)} \tag{4.3.15}$$

where the exponent $\alpha(N)$ is

$$\alpha(N) = (D-1)\,(v/\,v+1) \tag{4.3.16}$$

In two dimensions, $D = 7/4$ and $v = 4/3$, so that $\alpha(N_f) = (1 - 1/D) = 3/7$ and

$$N_f \sim L\, L_d^{0.43} \tag{4.3.17}$$

Figure 4.13 shows our computer simulation analysis of this relation, where the length of the frontier, N_f, is plotted as a function of the diffusion length, L_d, for the 1024^2 lattice. The solid line in Figure 4.13 is given by the power law

$$N_f = 940.5\; L_d^{0.43} \tag{4.3.18}$$

which is in excellent agreement with theory. The front factor of 940 in Eq 4.3.18 is approximately equal to the lattice dimension (1024), as predicted by Sapoval *et al.* [4]. The front factor can be normalized by the lattice dimension as 940/1024 = 0.92. The standard deviation, δN_f, is also strongly dependent on L_d, and changes from 2163 at L_d = 1,024 to 10,704 at L_d = 10,240.

The interface roughness can be described by the number $S = N_f/L$. If we take a box on the front of length σ_f containing σ_f^D particles, then we have the simple relation (see Table 4.3) that

$$\sigma_f^D = 0.4\; L_d \tag{4.3.19}$$

Since $S = \sigma_f^D/\sigma_f$, and $B \approx 6\sigma_f$, the interface roughness number is conveniently obtained in terms of diffusion observables as

$$S = 2.4\; L_d/\,B \tag{4.3.20}$$

This number has been referred as the *Sapoval number* and is based on the peculiarity that the number of particles in a box of width σ_f is proportional to the diffusion length, as expressed by Eq 4.3.19. The actual number of particles on the front is related to S and the lattice length L by $N_f = SL$. The relation of interface roughness to fracture energy is discussed in Chapter 9.

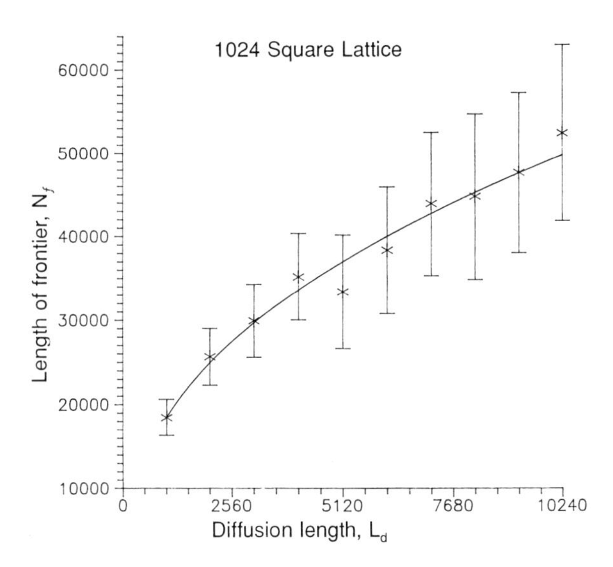

Figure 4.13 Length of frontier, N_f, versus diffusion length, L_d. The error bar heights represent the standard deviation δN_f for a minimum of 50 simulations on a 1024^2 lattice. The solid line is given by $N_f = 940\ L_d^{0.43}$ (Wool and Long).

4.3.6 Dynamics and Noise of the Diffusion Front

If we examine the diffusion front (white line) in Figure 4.9, it becomes apparent that small motions of single atoms on the front can disconnect or connect large clusters, and dramatically alter the position of the front. For example, a cluster containing σ_f^D atoms can be connected or disconnected from the front by the motion of a single atom.

The dynamic properties of the diffusion front were examined by Sapoval, Rosso, Gouyet, and Colonna [31]. They note that the most remarkable feature about front fluctuations is their very high frequency. If θ is the average jump time for a particle in the lattice, it is not necessary to wait this time for a jump to occur somewhere on the front, but rather a much shorter time, θ/N_f.

They define $\delta N_f(t)^2$ as the square of the change of the number of points on the frontier in the time interval t,

$$\delta N_f(t)^2 = \langle N_f(t) - N_f(0) \rangle^2 \tag{4.3.21}$$

At short times (Figure 4.14), the fluctuations are described by

$$\delta N_f(t)^2 \sim t^{2H} \tag{4.3.22}$$

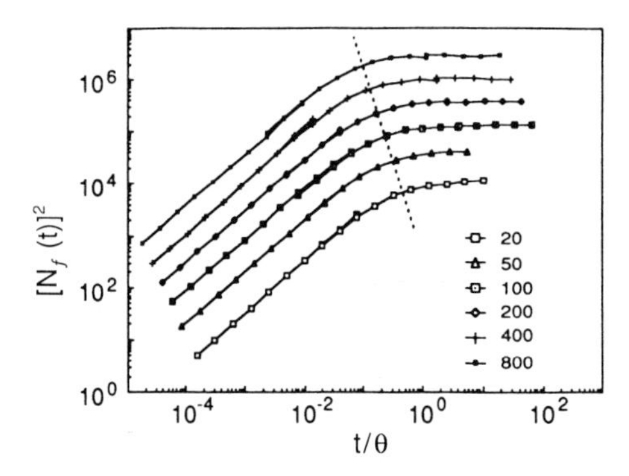

Figure 4.14 Behavior of the mean square fluctuation δN_f^2 in a time interval t (Eq 4.3.21) as a function of t/θ for several values of diffusion length (inverse gradient). At small time intervals, the fluctuations increase with an exponent of unity, corresponding to a $1/f^2$ power spectrum. Above a crossover time t_c, the exponent is zero, corresponding to a white noise spectrum. The dashed line represents the expected scaling behavior of t_c with L_d (courtesy of B. Sapoval, M. Rosso, J.-F. Gouyet, and J.-F. Colonna).

where $H = ½$ is the Hurst exponent corresponding to $1/f^2$ noise at high frequency. They find that at longer times and lower frequencies, the fluctuations stabilize. Above a crossover time t_c, H is equal to 0, corresponding to a white spectrum.

The variation of $\delta N_f(t)^2$ in the low frequency limit at $t > \Theta$ can be understood as follows [31]. The frontier is treated as L/σ_f independent percolating boxes of side σ_f, where each box represents the average cluster size that can be added to or subtracted from the frontier. Since each box is fractal, it contains σ_f^D particles, and the total fluctuation is

$$\delta N_f(t)^2 = (L/\sigma_f)\,\sigma_f^{2D} \tag{4.3.23}$$

Substituting for σ_f as a function of L_d, they obtain

$$\delta N_f(t)^2 = L\,L_d^{10/7} \tag{4.3.24}$$

which is independent of time at $t > t_c$. Their dynamic simulation gave an exponent of 1.5 ± 0.05, in good agreement with the theoretical value $10/7 = 1.428$.

We examined the predicted dynamics in Eq 4.3.24 using our static simulations by assuming that an observer in a box of length σ_f, when moved along the frontier to the next box, sees fluctuations corresponding to the standard deviation of N_f. In Figure 4.13, the standard deviation δN_f (average of 50 runs per diffusion length), represented by the error bars, increases with diffusion length for the $L = 1024$ lattice. A plot of δN_f^2 versus L_d in Figure 4.15 gives

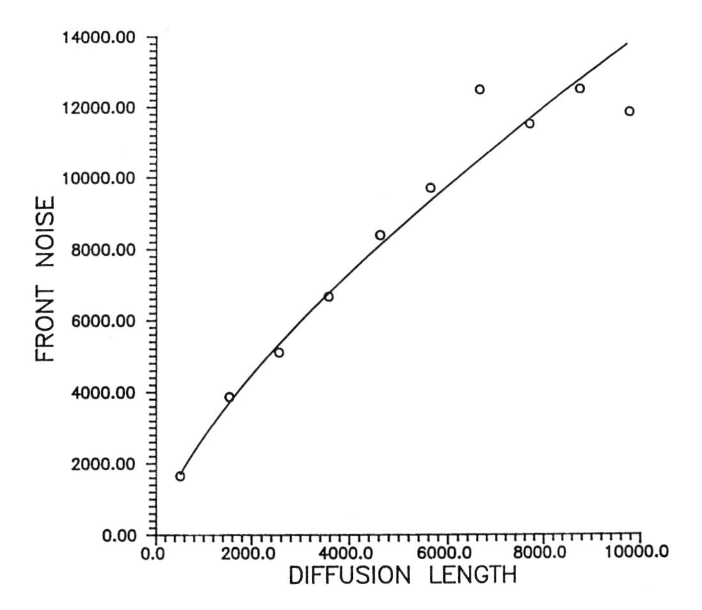

Figure 4.15 Frontier length noise, δN_f (error bars in Figure 4.13), versus diffusion length, L_d. The solid line through the data points was determined by the power law fit, $\delta N_f = 14.4\ L_d^{0.718}$. For a lattice length of $L = 1024$, this corresponds to $\delta N_f^2 = 0.2\ L\ L_d^{1.437}$.

$$\delta N_f^2\,(1024) = 0.2\,L\,L_d^{1.437} \tag{4.3.25}$$

For the 512^2 lattice, a similar analysis gives an exponent of 1.416 (Table 4.3). The exponents of 1.437 and 1.416 are in very good agreement with the theoretical exponent of 10/7 = 1.428. The front factor in this analysis is on the order of unity, as predicted by Sapoval *et al.* [31].

The fluctuations in the front width, $\delta\sigma_f^2$, can be obtained by taking the derivative $\delta N_f/\delta\sigma_f$ of the relation $N_f \approx 2L\sigma_f^{D-1}$, and solving for $\delta\sigma_f^2$

$$\delta\sigma_f^2 = \delta N_f^2\,/\,[4(D-1)^2\,L^2\,\sigma_f^{2(D-2)}] \tag{4.3.26}$$

Substituting both for δN_f^2 from Eq 4.3.24 and for the diffusion length dependence of σ_f from Eq 4.3.7, we obtain the diffusion length dependence of the width noise as

$$\delta\sigma_f^2 = 0.06\,/\,L\,L_d^{12/7} \tag{4.3.27}$$

The exponent of 12/7 is in good agreement with the computer simulation with $L = 1024$ (Table 4.3). The front factors, which are proportional to $1/L$, are of the correct order of magnitude and are predicted by Eq 4.3.27 to be 2.3×10^{-4}, 1.17×10^{-4}, and 5.8×10^{-5}, for the 256, 512, and 1024 lattices, respectively.

The "geometrical" nature of the noise in the diffusion front is surprisingly large, considering the smooth advance of the average position of the front. The fluctuations δN_f may have interesting physical consequences; for example, large impedance fluctuations can occur in electrical contacts. The ability to control the diffusion front roughness, N_f, of a polymer–metal interface and its molecular connectivity has interesting implications for development of the mechanical and electrical properties of composites and electronic material interfaces. Roughness promotes mechanical interlocking, but fractal characteristics with holes could reduce electrical conductance and affect signal speed in the metal layer. The thickness of the conducting strip is determined by the position of the fractal diffusion front. In superconductor/metal interfaces, the chemical potential difference seen by the Cooper pair of electrons causes a splitting so that one electron travels across the planar interface into the normal metal conductor and the other is reflected back into the superconductor. This phenomenon, known as *Andreev reflection*, is expected to be significantly altered by the presence of a fractal rough interface where the metal has diffused into the superconductor.

We find the fractal analysis of interfaces useful for examining the development of structure at reacting polymer/cement interfaces [40] and for investigating the biodegradation of composites containing degradable and nondegradable components, for example, starch and polyethylene blends. For materials in which the starch molecules are uniformly distributed in the PE matrix, biodegradation involves a percolation invasion process [13, 14]. However, when flow-induced concentration profiles develop, the starch near the surface is removed by microbes while the remaining material is encapsulated within the PE matrix. The surface separating the accessed material from the encapsulated material is similar to the fractal diffusion front shown in Figure 4.9.

4.4 Polymer–Metal Interfaces

The structure of polymer–metal interfaces was explored using computer simulation experiments and image analysis of transmission electron micrographs. As a starting point, a computer simulation was developed using gradient percolation. The simulation was then used to study the effects of lattice size and diffusion length on the behavior of the fractal dimension of diffusion fronts and other associated parameters. Once we were satisfied that the simulation was behaving appropriately we used it to examine the fractal dimension of a diffuse polyimide–silver (PI–Ag) interface.

4.4.1 Fractal Polymer–Metal Interfaces

Ordinarily, one forms a featureless polymer–metal interface by depositing the polymer on a flat metal surface. However, in vapor metallization processes, chemical vapor deposition, plasma-assisted vapor chemical deposition, and diffusion-controlled metallization, the metal atoms can diffuse considerable distances into the polymer substrate. Such processes are commonly used in the manufacture of Very Large Systems Integrated (VLSI), that is, high-performance electronic materials and integrated circuits. For example, Figure 4.16 shows a 2,500-Å titanium strip that was vapor deposited on a polyimide (MDA–ODA) substrate by Robertson and Birnbaum [32]. The PI–Ti interface is diffuse, with the Ti diffusing up to 1000 Å in the PI. The structure of the diffuse interface depends on the deposition conditions and annealing treatments. Metal atoms that bond strongly to the PI substrate are not expected to form deep diffuse layers [33]. Similar observations of diffuse interfaces were made using copper, silver, tungsten, and nickel [32, 33]. Diffuse metal interfaces occur under conditions of low deposition flux, high polymer substrate temperature, and low bonding potential between the metal and polymer.

Figure 4.17 shows a 3-d computer simulation of the process of metal atom deposition on a PI substrate, by Silverman [34]. When one-quarter of a monolayer of metal atoms has been deposited, a highly diffuse interface has formed. The 2-d view of this interface can be compared with those in Figures 4.7–4.9. Silverman also simulates the annealing process and finds that the aggregation of metal atoms with strong interactions into clusters considerably reduces the ramified nature of the interface. From a plot of log mass versus log radius, he finds that the clusters have a fractal dimension of 2.98, which reflects their globular compact structure. On the other hand, the aggregation of clusters with weak interactions gives a fractal dimension of 1.5 ± 0.10.

Several methods can be used to "image" the diffusion field in two dimensions. Figure 4.18 shows the technique used by Hashimoto *et al.* [35], where mutual

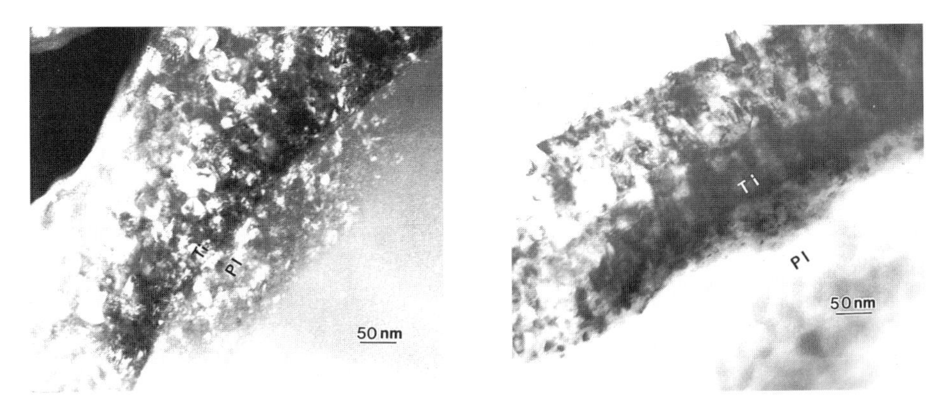

Figure 4.16 Transmission electron micrographs of a polyimide–titanium interface. The interface was formed by metallization of the PI substrate at 175 °C and was annealed at 320 °C in deuterium (760 torr). Diffusion of the Ti up to about 1000 Å in the PI phase is indicated (courtesy of Robertson and Birnbaum).

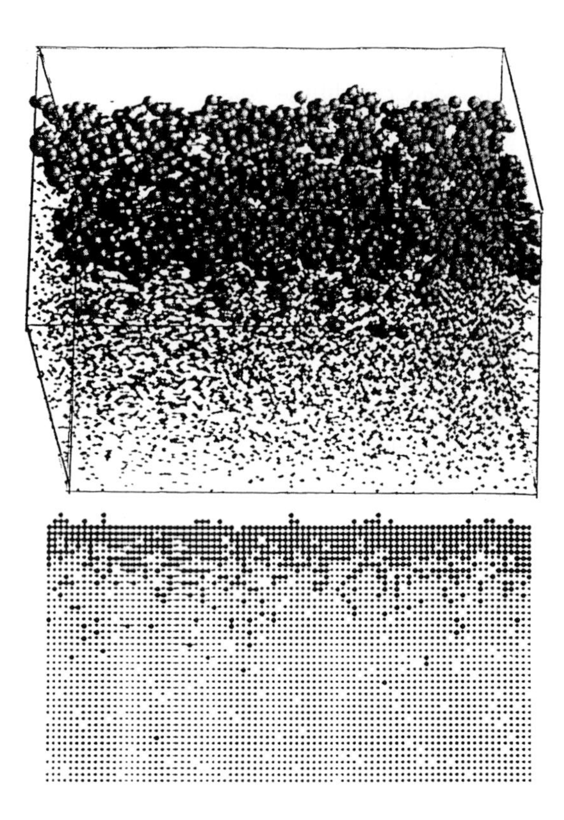

Figure 4.17 (Top) Computer-simulated three-dimensional view, at an angle to the top film surface, after deposition of one-quarter monolayer of metal atoms. The metal atoms are shown as larger spheres. (Bottom) Metal atom penetration profile into the film just after initial metallization (courtesy of Silverman).

interdiffusion of polystyrene and styrene–isoprene diblocks is marked by the preferential staining of the segregated isoprene phase. As diffusion proceeds, the highly ramified nature of the initially planar interface becomes apparent.

Mazur and co-workers [36, 37] and Manring [38] formed diffuse polyimide–silver interfaces by electrochemical deposition. A silver ion solution was permitted to diffuse into a polyimide film, and the ions were reduced from Ag^+ to Ag^0 on the cathode side of the film. The results are shown in Figure 4.19. A transmission electron micrograph (TEM) of an 800-Å slice of the film is shown (the white bar represents 1000 Å). This figure presents a near-2-d view of a 3-d diffusion field. The particles form a conducting black strip, with a highly diffuse region of silver particles. The closeup view of the particles on the right in Figure 4.19 shows a size distribution from about 50 Å to 500 Å.

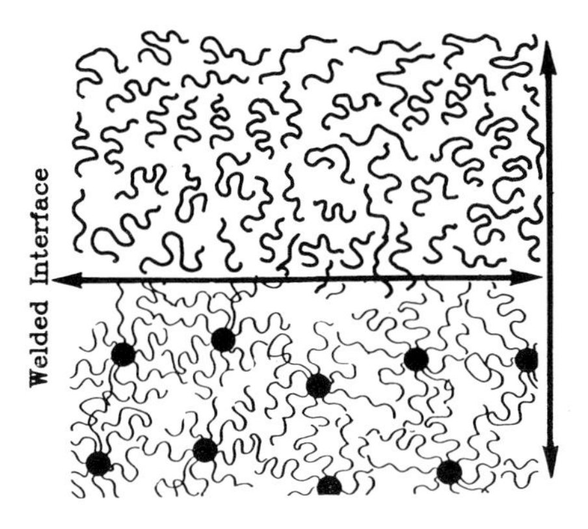

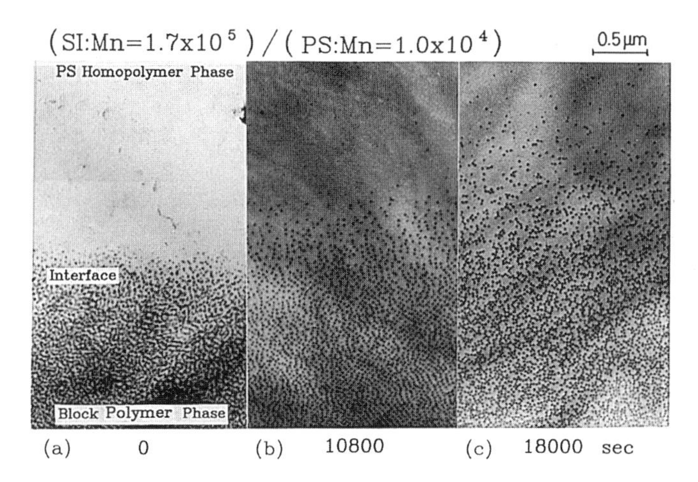

Figure 4.18 (Top) Schematic illustration of the interface formed by a styrene–isoprene (SI) diblock copolymer (segregated isoprene spheres) and a polystyrene (PS) homopolymer. (Bottom) Electron micrographs of the cross sections of SI–PS interfaces after welding at 150 °C for (a) 0 s, (b) 10,800 s, and (c) 18,000 s. The dark spheres (stained isoprene) show the progress of the mutual diffusion field in two dimensions (courtesy of S. Koizumi, H. Hasegawa, and T. Hashimoto, Kyoto Univ.).

Mazur found that the metal layer could not be delaminated from the polymer substrate and could be removed only by abrasion. This indicated excellent adhesion at the polymer–metal interface. However, for such diffuse interfaces one may ask, where is the interface? The fractal interface analysis suggests that the fractal diffusion front represents the outer frontier of connected silver. Adhesion should be enhanced by the very rough diffusion front, which would provide considerable mechanical interlocking in addition to high surface contact area for bonding.

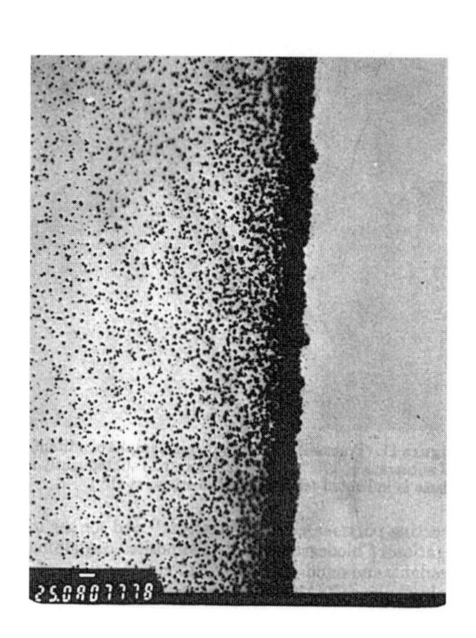

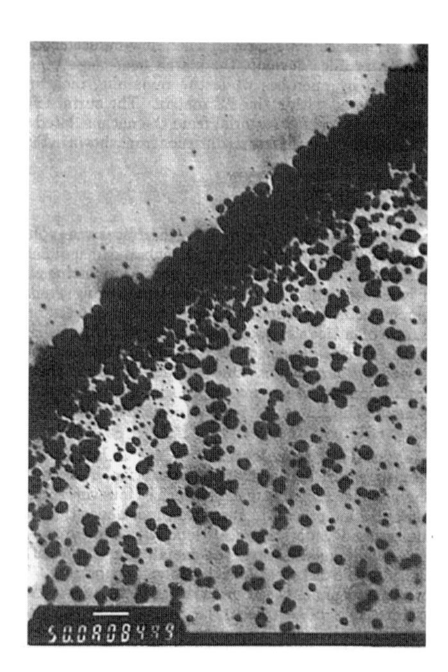

Figure 4.19 (Left) Transmission electron micrograph of a polyimide film containing a concentration gradient of electrochemically deposited silver particles. The TEM analysis was obtained by taking a 1000-Å slice through the interface. The bar marker is 1000 Å. (Right) A higher magnification of a similar PI–Ag interface shows a distribution of silver particle sizes ranging from about 50 Å to 500 Å (courtesy of S. Mazur *et al.*).

Comparing Figure 4.7 with Figure 4.19, we can propose that the black area represents the percolating conducting region and the grey region represents a polymer–metal interphase region, adjacent to the conducting metal strip, which has unique mechanical and electrical properties. For example, one can expect to have a gradient of mechanical moduli and thermal expansion coefficients in this interphase region.

However, there is a potentially serious problem for electronic materials with such diffuse interfaces. The technology used to produce the interfaces shown in Figures 4.16 and 4.19 is often used to make high-speed circuits, such as for computers. For such a system, the delay time for the electrical signals is determined by

$$s \sim h\sqrt{K'} \tag{4.4.1}$$

where h is the length of the strip and K' is the dielectric constant of the polymer substrate in contact with the conducting strip. Polymer dielectrics with large K' values slow the electrical signal because the electric field of the pulse induces a field of opposite sign in the dielectric, which slows it down. The magnitude of the opposing field is determined by the polarizability of the polymer molecules. The dielectric constant, in the low frequency limit, is proportional to the refractive index of the polymer. When air is used with suspended wire or aerogel technology, the ideal value of $K' = 1$ can only be approximated; typical K' values of polymer dielectrics in electronic materials are in the range of 2–4. Air can be used as the dielectric in suspended wire technology and in aerogels, but each has its unique application problems. Designers of high-speed computers take Eq 4.4.1 into account by making the distance h between contact points as small as possible using submicron VLSI technology, and by using dielectric materials with very low K' values, such as PI and polytetrafluoroethylene (Teflon™). However, with polymer–metal interfaces like those shown in Figure 4.17 and 4.19, the polymer material immediately adjacent to the conducting metal strip becomes highly modified by the nonconducting dissolved metal particles and the K' value may be substantially increased above its design value.

Figure 4.20 shows the effect of a uniform concentration of aluminum powder on the dielectric constant of a polyester resin. We made the plot from dielectric data obtained by Berger and McCullough [39], using composite samples at 20 °C and 100 kHz. The K' value increases from 3.2 at 0% metal to about 30 at 45% metal. The conductivity percolation threshold for this 3-d system is in the vicinity of 31% by volume of metal. At this concentration, $K' \approx 20$, or about 6 times its initial value. We cannot apply these results *in toto* to the polymer–metal interface with a metal concentration gradient, but we can expect similar changes to occur locally. Thus, the fractal nature of the interface may provide an advantage in mechanical properties, but could have adverse effects on some electrical properties and improve others at certain operating frequencies and current densities.

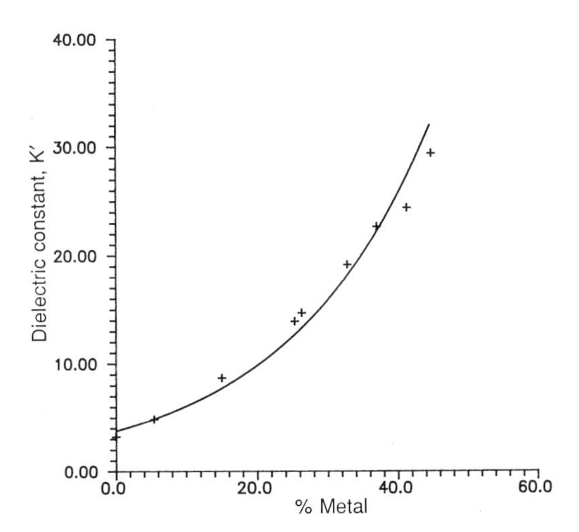

Figure 4.20 Dielectric constant versus volume fraction of aluminum powder in a polyester composite. Data were obtained at 20 °C and 100 kHz (data of Berger and McCullough).

The thermal properties of fractal polymer–metal interfaces have not been analyzed. One of the major factors in the failure of polymer–metal interfaces in electronic and composite materials is the difference in thermal expansion coefficients (about a factor of 10). With large temperature changes, often encountered during processing or use, significant interfacial stresses develop, which can result in failure of the material. The sharper the interface, the greater the mismatch and interfacial stress. Conversely, the more diffuse the interface, the lower the stress. We expect that the interfacial stresses, s_{ij}, are inversely proportional to the fractal diffusion front width, σ_f, and proportional to the temperature difference, $(T_2 - T_1)$, and thermal expansion coefficient difference between polymer and metal, $(\alpha_p - \alpha_m)$, so that

$$s_{ij} \approx (T_2 - T_1)(\alpha_p - \alpha_m) L_d^{-\nu/(\nu+1)} \tag{4.4.2}$$

where L_d is the diffusion length and ν is the critical exponent for the percolation correlation length. Thus, fractal interfaces should be more stable with respect to temperature changes with increasing L_d values.

4.4.2 Fractal Dimension of a Polyimide–Silver Interface

We analyzed the polyimide–silver interface TEM micrograph shown in Figure 4.19. Since the TEM image was not on a digital data file, we first had to develop a procedure to render the information machine readable, and had to analyze the image further before submitting the data to the fractal dimensional analysis (FDA) software [2]. The FDA software used in Figures 4.7–4.9 works with a two-dimensional binary lattice; the captured data are initially encoded with lattice site values ranging from 0 to 255.

The first step in the process was to use a light table and a charge-coupled-device video camera to convert the TEM image to a computer file. The camera was connected to a Truevision Vista board in an IBM PC/AT. Utilizing Truevision's Vista-Tips software, we obtained a grey-scale image at a resolution of 756 × 486 pixels. Next, we rescaled the captured image via color imaging and histogram analysis from the original set of values (0 to 255) to a resolved range of 0 to 7, using an IBM 7350 image processor with Hlips image analysis software. The rescaled image then became an input file to the FDA software, which was modified to allow for selective dichotomization of the cell values. This modification allowed us to reexamine the image with different levels of occupation. Presented in Figure 4.21 are three analyses of the PI–Ag image, which vary with respect to how stringent the criterion for cell occupation was set. The color code is the same as shown in Figures 4.7–4.9, where the connected metal region (black) is separated from the polyimide phase (grey), which is interspersed with nonconnected metal particles (grey), separated by the frontier (white).

The fractal dimension of the frontier was found to be sensitive to the image resolution, and ranged in Figure 4.21 from 1.492 to 1.619. By digitizing the image and using the FDA software, we obtained the diffusion front and front density profile shown in Figure 4.22. The breadth or span of the profile, B, was 42 digitized units,

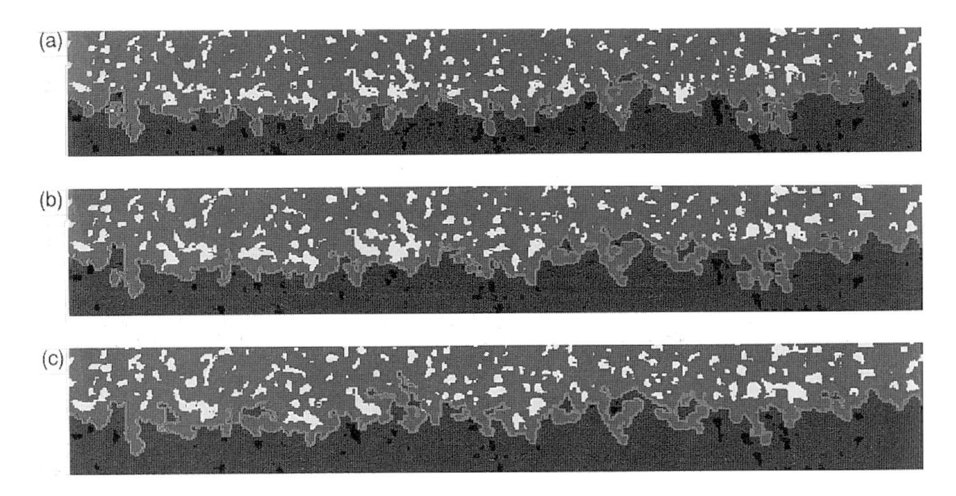

Figure 4.21 Image analysis of the PI–Ag interface (from Figure 4.19). The gradient percolation analysis of the images (a–c) obtained under different resolution conditions is given in Table 4.4 (Wool and Long).

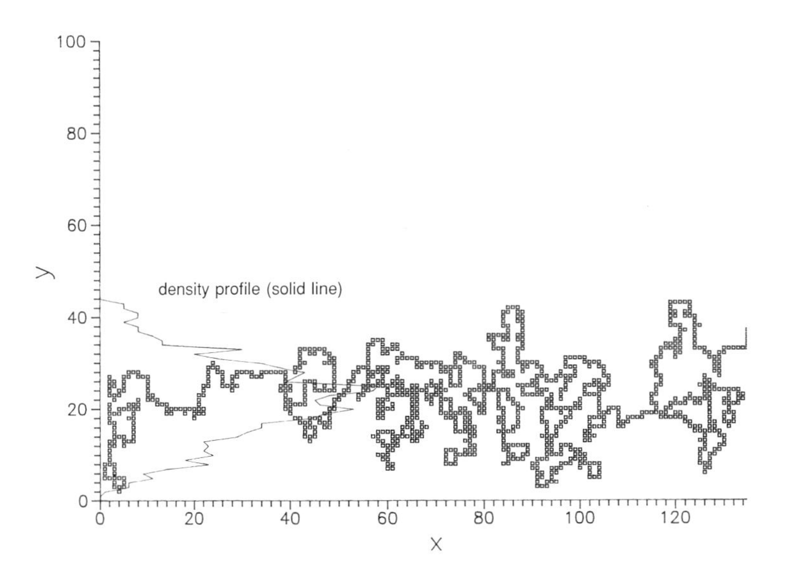

Figure 4.22 Diffusion front (squares) and density profile (solid line) of the PI–Ag interface (from Figure 4.19) [analysis in Table 4.4 Image (d)].

which corresponds to about 3,000 Å on the real silver interface (Figure 4.19). The mass of the frontier, N_f, was 1056 in Figure 4.22. The width of the front was determined from a Gaussian analysis using

Table 4.4 Analysis of PI–Ag Interface

Property Source	Image (a) Fig 4.21(a)	Image (b) Fig 4.21(b)	Image (c) Fig 4.21(c)	Image (d) Fig 4.22
Front position, x_f	26	32	37	37
Diffusion length, L_d	68.77	84.64	97.86	98
Lattice length, L	486	486	486	136
Fractal Dimension, D	1.492	1.619	1.572	1.73
$1/D$	0.670	0.618	0.636	0.578
$(1 - 1/D)$	0.330	0.382	0.364	0.422
Front span, B	43	43	52	44
Front width, σ_f	8.453	8.563	8.664	7.33
σ_f, theory	7.83	7.14	8.488	6.511
B/σ_f	5.1	5.0	6.0	6.0
B/σ_f, theory	6	6	6	6
Front length, N_f	2015	2013	2147	1056
N_f, theory	1806	2436	2371	866
$N_f\,\sigma_f/L\,L_d$	0.51	0.42	0.39	0.58
$N_f\,\sigma_f/L\,L_d$, theory	0.44	0.44	0.44	0.44
Sapoval no., S	3.84	4.72	4.52	5.34

$$\sigma_f = N_f/(M_{max}\sqrt{2\pi}) \tag{4.4.3}$$

where N_{max} is the number of particles at the peak. From Figure 4.22, we have $N_{max} = 58$ and $N_f = 1056$, so that $\sigma_f = 7.26$. The ratio B/σ_f was 6.06, in excellent agreement with theory and simulation for this lattice length. The position of the front occurs at $x_f = 37$ (Figure 4.22 has a 15-point offset on x_f) resulting in a diffusion length, L_d, of 98.

A plot of the diffusion front mass (in Figure 4.22) as a function of radius gave a fractal dimension, D, of 1.73 (shown in Figure 4.23), which is in very good agreement with the Sapoval *et al.* theory and computer simulations [4]. The fractal dimension is quite sensitive to the resolution used to analyze the image. If we take a true frontier with $D = 1.75$ and analyze it with decreasing resolution, then the fractal dimension decreases to a lower bound determined by a self-avoiding walk (SAW) such that $D = 1.33$ [41]. As the resolution decreases, the harbors and small bays close up and are not detected, and the front loses its ramification and eventually converges to an SAW with a lower fractal dimension. Figure 4.23 also shows the result of an image analysis of a TV image of a computer-simulated diffusion field that had a true fractal dimension of 7/4. The image analysis method slightly reduces the fractal dimension, but very little difference is seen between the two data sets in Figure 4.23.

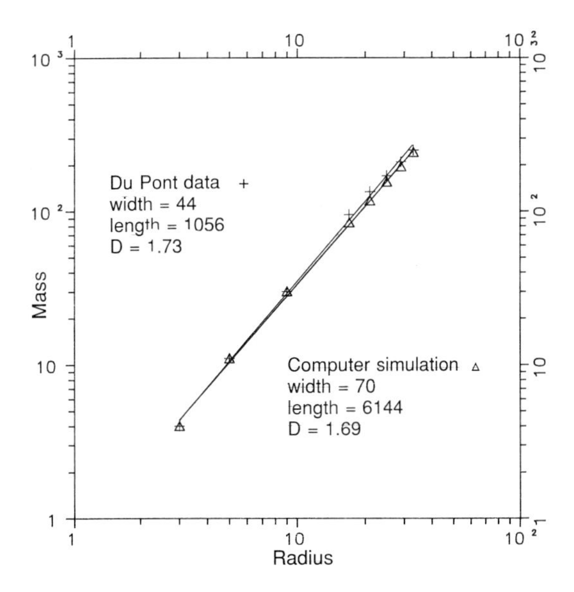

Figure 4.23 Fractal analysis of the diffusion front mass versus radius of the PI–Ag diffusion front. Key: (–+–) Analysis of Mazur's Du Pont data from Figure 4.22 gave $D = 1.73$. (–△–) Image analysis of computer-simulated frontier (Figure 4.8) with a true value of D of 1.75 gave $D = 1.69$.

From the image analysis of the PI–Ag interfaces shown in Figures 4.21 and 4.22, we can evaluate several quantitative aspects of the frontier theory. The frontier equations will be used in the following forms:

$$N_f = 0.92\, L_d^{1-1/D} \tag{4.4.4}$$

$$\sigma_f = 0.46\, L_d^{1-D} \tag{4.4.5}$$

$$B/\sigma_f \approx 6 \tag{4.4.6}$$

$$L_d = 2.645\, x_f \tag{4.4.7}$$

$$N_f\, \sigma_f/(L\, L_d) = 0.44 \tag{4.4.8}$$

$$S = 2.4\, L_d/B \tag{4.4.9}$$

Equation 4.4.8 states the invariant ratio of the front factors in the previous equations, since the exponents cancel. This equation also suggests that the number of front particles per box of width σ_f is proportional to the diffusion length, L_d. We proceed by taking the experimental values of N_f, σ_f, B, and D, and using x_f to determine L_d, and we compare them with the values calculated from the above equations. The experimental value of D was used in each equation to obtain the prediction. The results are given in Table 4.4.

The results show considerable support for the general theory. In particular, it should be noted that the exact fractal dimension of 1.75 is not necessary to describe the quantitative aspects of the interface. The exponents for N_f, σ_f, etc., can be expressed in terms of the observed fractal dimension.

4.5 Three-Dimensional Diffusion Fronts

Most of the structural features of the 2-d lattice carry over to the 3-d case, with a few notable exceptions, as observed by Rosso, Gouyet, and Sapoval [42]. However, the basic picture of the front remains the same in terms of gradient percolation producing a highly ramified interface and a diffusion front with fractal characteristics. In the next section, we compare 2-d and 3-d diffusion.

4.5.1 A Comparison of 2-d and 3-d Diffusion Fronts

Figure 4.24 summarizes the major features of the 2-d diffusion behavior for an A/B interface on a square lattice with a linear concentration gradient of A in B. The position of the diffusion front for the A molecules (occupied sites) is defined with respect to the first-neighbor connectivity, which coincides with the position of the front for the B molecules (or vacant sites) defined with respect to first- and second-neighbor connectivity. For the 2-d case, the A and B percolation thresholds form a matching pair related by

$$p_{CA} + p_{CB} = 1 \tag{4.5.1}$$

However, for 3-d diffusion, the positions of the diffusion fronts occur at different concentrations, so that the two fronts interpenetrate each other, rather than being contiguous (2-d). Figure 4.25 shows the 3-d concentration profile with positions of the diffusion fronts. It is found that the 3-d front is highly extensive and essentially involves all molecules with concentration $p \geq p_c$. This feature is particularly true of polymers where the monomer connectivity is intrinsic at x less than R_g. The front extends over a concentration range $p_{CA} \leq p \leq (1 - p_{CB})$ and the density of sites occupied by the front is very close to the overall density of occupied sites.

The results of a 3-d simulation by Rosso *et al.* [42] are shown in Figure 4.26. Only sites connected to the bottom plane at $p = 1$ are shown; this produces the 3-d diffusion front. The spongy structure is highly porous and is not unlike the connected gel structure one obtains in normal 3-d percolation with a constant concentration near $p_c = 0.3117$ (cubic lattice). In the vicinity of p_c, the front has a fractal dimension, D,

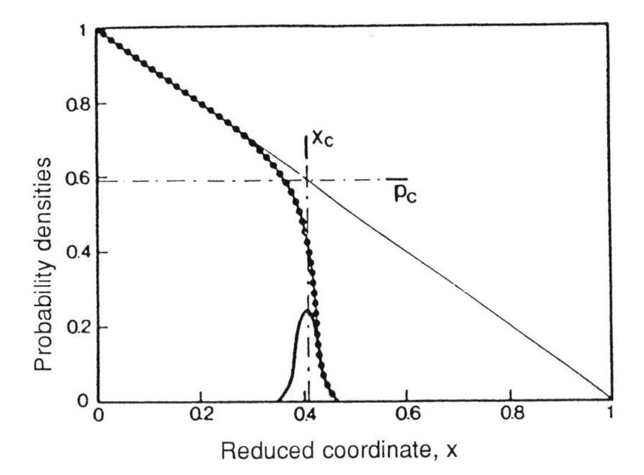

Figure 4.24 Summary of 2-d (square lattice) scalar percolation profiles: The variation with diffusion depth, x, is shown for the following: probability density p for occupied sites (A sites, thin straight line); probability density $p_{\infty A}$ for the occupied sites connected to the $p = 1$ line (A infinite cluster, dotted line); probability density p_{fA} of the front (thick Gaussian line). The probability density of A sites is equal to the percolation threshold p_c in the square lattice at the position x_c. (Courtesy of Rosso, Gouyet, and Sapoval).

of about 2.5. However, the front behaves as a normal solid with D equal to 3, over the entire range. Since the B molecules highly interpenetrate with the A front, the potential exists for increased adhesive strength via mechanical interlocking and the very large number of new sites available for A–B chemical bonds.

For a linear concentration gradient, the enhancement factor for the number of contacts between A and B is determined approximately as a function of diffusion length by

$$N_f \approx 1.3\ L_d \tag{4.5.2}$$

Eq 4.5.2 is derived from the integral between 0 and x_f in the concentration profile, $p(x) = 1 - x/3L_d$, with $x_f \approx 2\ L_d$. Since L_d can be very large, the enhancement factor can be substantial. The exponent for L_d is unity in Eq 4.5.2, which contrasts with the 2-d case where $N_f \sim L_d^{0.43}$.

Thus, in summary, the 3-d diffusion field can be divided into a connected region and nonconnected regions in terms of gradient percolation concepts; the position of the front occurs at the percolation threshold. The number of particles on the front, N_f, is essentially equal to the total number of diffused particles with $p \geq p_c$. The shape of the front is no longer Gaussian, as in the 2-d case (perhaps on one side of x_f) because it coincides with most of the concentration profile at $p \geq p_c$. The position of the front increases with L_d and the forward width (at $x > x_f$) of the front σ_f behaves as $\sigma_f \sim L_d^{\alpha(\sigma)}$, in which $\alpha(\sigma) = \nu/(1 + \nu)$.

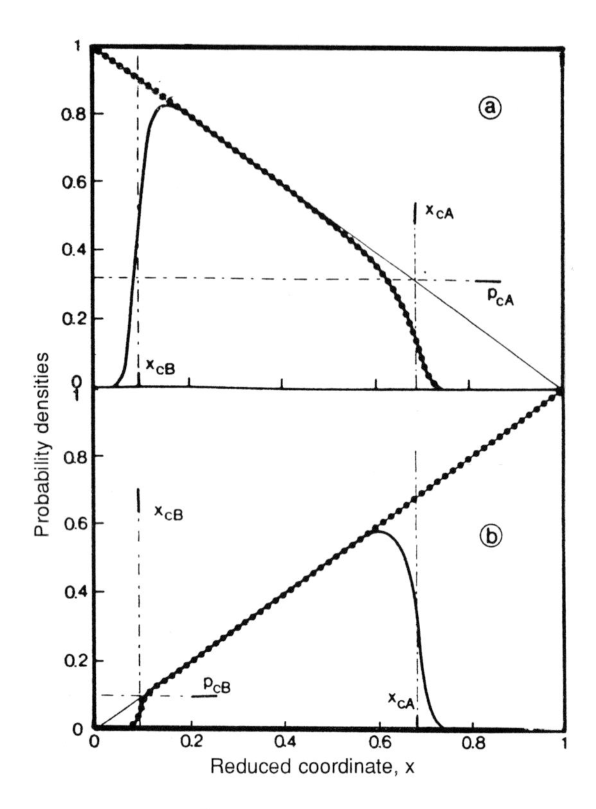

Figure 4.25 (a) Same as Figure 4.24 but for the cubic lattice. Key: x_{cA} is where the probability density of A sites is equal to the percolation threshold p_{cA} in the cubic lattice (first neighbors); x_{cB} is where the probability density of B sites is equal to the percolation threshold p_{cB} in the cubic lattice (first, second, and third neighbors). (b) Same as (a), but for empty sites (B sites, thin straight line) and empty sites connected to the $p = 0$ plane (infinite B cluster, dotted line). The front is shown by the thick line. (Courtesy of Rosso, Gouyet, and Sapoval).

4.6 Fractal Polymer–Polymer Diffusion Fronts

In this section, we consider the nature of the interface formed by the joining of two identical pieces of amorphous polymer. Such is the case in the welding of polymers in the melt, the lamination of composites, the coalescence of powder and pellet resin, the drying of latex paints, the internal weld lines formed during injection and compression molding, the tack of uncured elastomers, and a multitude of processing operations where interdiffusion is required for the development of full strength at an interface [1]. The analysis of the fractal nature of amorphous polymer diffusion fronts is complicated by the interpenetrated nature of the random-coil chains that allows

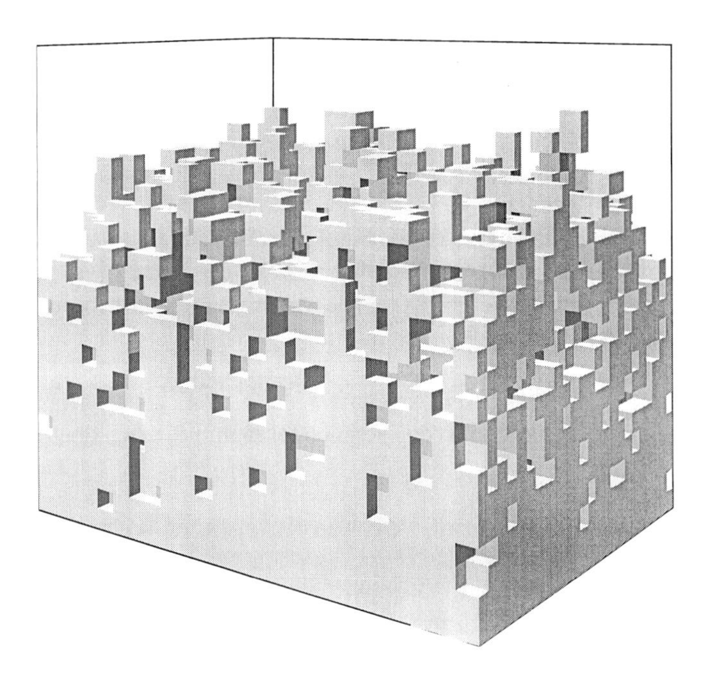

Figure 4.26 Three-dimensional view of diffusion front (courtesy of Rosso, Gouyet, and Sapoval).

several statistical segments or monomers to occupy a single lattice site. The computer simulation of polymer diffusion on a 2-d and 3-d lattice was done in a manner that kept track of both the occupied number of lattice sites, N_L, and the total number of monomers in the multiple occupied lattice sites, N_m.

Using the same algorithm as for monomer diffusion fronts, we examined the connectivity in polymers under gradient percolation conditions in terms of connected lattice sites, and identified the diffusion front as the boundary between nonconnected and connected sites. We could quantify the diffusion front in terms of either the occupied lattice sites, as in the monomer case, or of the total number of monomers, where multiple occupancy per lattice site was permitted for random-walk chains. Individual monomers on polymer chains do not occupy the same space, of course, but the statistical segments, or equivalent random-walk steps of the real chain, representing up to about 20 monomers per step, highly interpenetrate each other. The multiple occupation concept is further broadened if one considers mechanical connectivity in terms of entanglements, where the average network strand representing the amorphous polymer network is determined by the critical entanglement molecular weight, M_c.

The interdiffusion of polymer chains can be divided into four dynamic regions, which can be described in terms of characteristic length scales at characteristic relaxation times. These are (see Chapter 3 for details)

1. Rouse relaxation of entanglement lengths, resulting in diffusion lengths $L_d \approx R_{ge}$, where R_{ge} is the radius of gyration of the entanglement segment;
2. Rouse relaxation of the whole chain in a tube of topological constraints, so that $L_d \approx R_{ge} (M/M_e)^{1/4}$, where M_e is the entanglement molecular weight;
3. reptation of the whole chain, so that $L_d \approx R_g$, where R_g is the radius of gyration of the whole chain; and
4. long-range Fickian diffusion of the whole chain, with $L_d >> R_g$ and the chains essentially behaving as Brownian particles exhibiting center-of-mass motion.

Our principal interest here is to examine the largest region of importance in the welding of interfaces, the reptation region, with $L_d < R_g$ and times $t < T_r$, the reptation time, followed by the transition to normal Fickian diffusion at $L_d > R_g$ and $t > T_r$. In this study, the diffusion length, L_d, is determined in terms of the average monomer interdiffusion distance, $<X>$, with consideration for multiple occupation of lattice sites.

4.6.1 Diffusion Fronts of Polymer Interfaces with Reptation

We simulated interdiffusion of polymer chains using random-coil chains with a reptation algorithm (Chapter 2). In this algorithm, a random-walk chain of constant length is moved by a process that randomly selects one end monomer, moves the monomer one step in a random direction on a 2-d or 3-d lattice, and subtracts a monomer from the other end. This process is repeated many times and, as the chain ends explore new positions in space, the whole chain (center of mass) eventually moves a distance $X \approx R_g$ at $t = T_r$. This mechanism results in Fickian diffusion of the center of mass of the chain at t greater than T_r. However, the monomer displacement function is correlated and behaves as $X \sim t^{1/4}$ at times t less than T_r. Figure 4.27 shows a typical ramified diffusion field for the polymer interface in two dimensions. We obtained this diffusion front using the same technique as we employed for monomer diffusion. The front is potentially more ramified initially due to the connectivity of monomers within a given chain. At t less than T_r, most of the chains contribute to the front and very few chains have escaped to longer diffusion distances.

Figure 4.28 shows N_m (number of monomers on the diffusion front), as a function of $<X>$ (average monomer interdiffusion distance), for random-coil chains of molecular weights $N = M/M_e$, in the range 60–100. Two regions of behavior are observed.

For t less than T_r, the front is discontinuous, since the concentration profile for reptating chains has a gap or discontinuity at $C(0,t)$. Thus, the time dependence of N_m can be well approximated by a summation of the average contributions from single chains as

$$N_m \approx n(t) <X>^{d-1} \tag{4.6.1}$$

where $n(t)$ is the number of chains crossing unit area (3-d) or unit length (2-d) of the initial interface plane and $<X>$ is the average monomer interpenetration distance. In

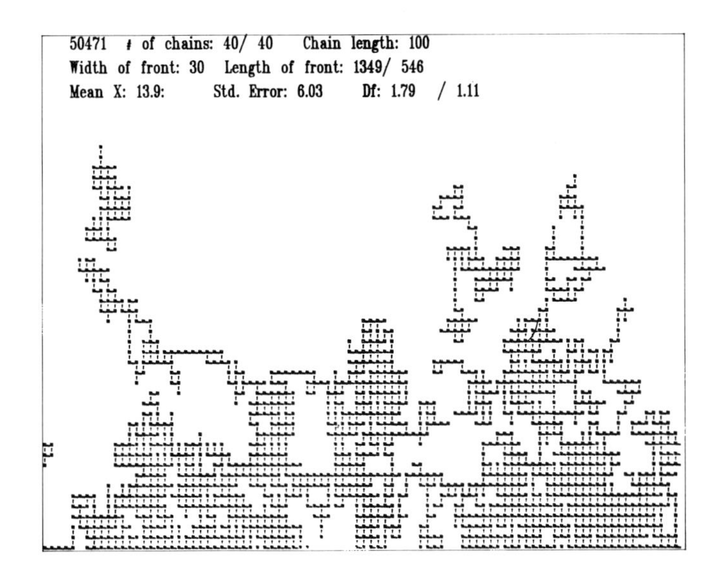

Figure 4.27 Polymer–polymer interface (one side) simulation of reptating chains on a 2-d lattice. The phantom chains of length 100 steps moved independently and were highly interpenetrated. The light grey chains at top of figure are not connected to the interface plane (lower edge of figure). The irregular heavy line marks the diffusion front (Wool and Long).

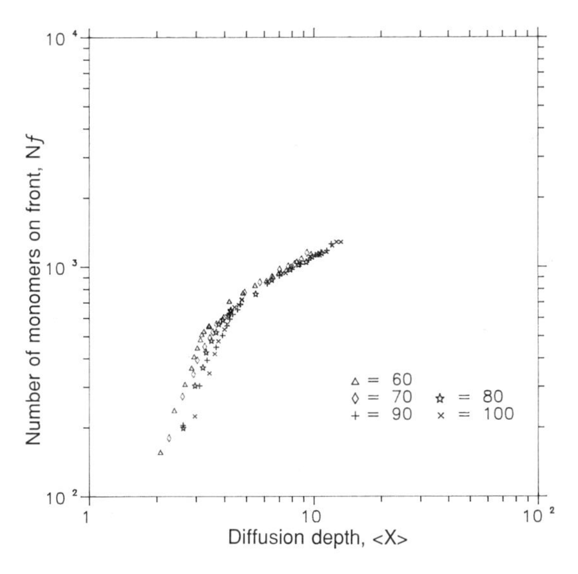

Figure 4.28 Number of monomers on the polymer diffusion front, N_m, versus the average monomer diffusion depth <X> for chains of length 60–100 steps (Wool and Long).

two dimensions, the contribution to the front from a single random-walk chain is the span of the walk, which has the same scaling law as $<X>$. In three dimensions, the individual chain contribution to the front is determined by the average interpenetration contour length $l(t)$, which is proportional to $<X>^2$.

To evaluate Eq 4.6.1 in terms of the fractal diffusion front theory, we need to convert from a time dependence to a diffusion length dependence of N_m. The average monomer interdiffusion length, $<X>$, is determined by (Chapter 2)

$$<X> \approx R_g (t/T_r)^{1/4} \qquad \text{for } t < T_r \tag{4.6.2}$$

and

$$<X> \approx R_g (t/T_r)^{1/2} \qquad \text{for } t > T_r \tag{4.6.3}$$

The number of chains crossing the interface is determined by the initial distribution of chain ends. If the chain ends are uniformly distributed in space, we have shown that the number of chains crossing is given by

$$n(t) \approx n_\infty \, (t/T_r)^{1/4} \tag{4.6.4}$$

where $n_\infty \sim M^{-1/2}$. In terms of the monomer diffusion depth, the number of chains crossing the interface is

$$n(t) \sim <X>/M \tag{4.6.5}$$

Thus, the number of monomers on the front for dimension d is

$$N_m \sim <X>^d/M \qquad \text{for } t < T_r \tag{4.6.6}$$

In the special case where the chain ends are segregated at the interface at $t = 0$,

$$n_s(t) \sim t^0 M^{-1/2} \qquad \text{for } t < T_r \tag{4.6.7}$$

and the number of monomers on the front is

$$N_{ms} \sim <X>^{d-1}/M^{3/2} \qquad \text{for } t < T_r \tag{4.6.8}$$

When t is much larger than T_r and $<X>$ is much larger than R, all correlated motion effects are lost and the polymer diffusion front should attain characteristics similar to the monomer diffusion front with

$$N_m \sim <X>^{1-1/D} \tag{4.6.9}$$

where D is the fractal dimension in two dimensions.

Figure 4.28 plots log N_m versus log $<X>$ for simulated interdiffusion of reptating random walks in two dimensions with molecular weight ratios, M/M_e, of 50–100 statistical units and with a uniform initial distribution of chain ends. At t less than T_r, the plot on a log–log scale has a slope of about 2 for all molecular weights, and at constant $<X>$, N_m varies approximately as the inverse of M, in agreement with Eq 4.6.6. The latter relation contrasts with the monomer diffusion front with $N_f \sim <X>^{0.43}$.

At times less than the reptation time, the diffusion front, though very rough, is not considered to be fractal since the criterion of self-similarity is not obeyed at length scales r greater than $<X>$ due to discontinuities in the interface plane. This concept may need to be modified when consideration is given to short-range segmental motion. Rouse-like motion of short chain segments was not considered in this simulation and the results reflect only the concentration profile for monomers on the primitive path of the reptating chain. The time-dependent scaling law for N_m at t less than T_r is identical to that for bridges, $p(t) \sim t^{1/2} M^{-3/2}$.

In three dimensions, at t less than T_r, the number of monomers on the diffusion front is well approximated by the total number of monomers that have diffused across the interface, $N(t)$. The number of chains crossing the interface, $n(t)$, is the same in 3-d as in 2-d, but the single-chain contribution to the front is proportional to $<X>^2$, so

$$N(t) \approx 1/2\, R_g (t/T_r)^{3/4} \tag{4.6.10}$$

Substituting for $<X>$ from Eq 4.6.2, we have the 3-d result

$$N_m \sim <X>^3 / M \tag{4.6.11}$$

Eq 4.6.11 is supported by SIMS studies of an HPS/DPS interface (Chapter 5). Thus in three dimensions, the polymer interface is extremely rough compared to the monomer diffusion case where $N_m \sim <X>$.

4.6.2 Polymer Diffusion Fronts at $<X>$ Much Greater Than R_g

At $t = T_r$, since $<X> \sim M^{1/2}$, $N_m(T_r)$ becomes independent of molecular weight, as indicated for the 2-d case. This effect is seen in Figure 4.28 where N_m for all molecular weights converges on a common line at long times. In this region, the fine structure of the chain disappears in the dynamic sense and the motion of the chains is represented by simple center-of-mass motion. Thus, when polymer chains behave as monomers at long times, we expect to recover the simple result for the number of monomers on the diffusion front at t much greater than T_r, as

$$N_m \sim <X>^{0.43} \tag{4.6.12}$$

This is shown in Figure 4.28 where the slope converges to 0.43 for all molecular weights at $<X>$ much greater than R_g of each chain.

Table 4.5 shows the results of the mass-radius fractal analysis of the polymer diffusion fronts as a function of molecular weight for the two cases,

$$N_m \sim R^{D(m)} \qquad \text{(mass fractal)} \qquad (4.6.13)$$

$$N_L \sim R^{D(L)} \qquad \text{(lattice-site fractal)} \qquad (4.6.14)$$

where N_m is the total number of monomers on the front with consideration for multiple occupancy of lattice sites due to chain interpenetration, N_L is the number of occupied 2-d lattice sites on the diffusion front, independent of multiple occupation, and $D(m)$ and $D(L)$ are the corresponding fractal dimensions. The analysis was done in the time region $t \approx 10\ T_r$ and length scales $R >> R_g$ for each molecular weight, so that all correlated motion effects were lost and the diffusion profile was Fickian. It can be seen (Table 4.3) that the fractal dimension for the total mass of monomers on the diffusion front, $D(m)$ is about 1.75 for all molecular weights, and is in good agreement with the monomer diffusion case where $N_f \sim R^{7/4}$ in two dimensions. Interestingly, the fractal dimension of the diffusion front evaluated with respect to the occupied lattice sites is consistently lower, with $D(L)$ of about 1.57. This may be due to a polymer connectivity effect between lattice sites causing a smoothing of the fractal structure and resulting in a lower fractal dimension.

Table 4.5 Polymer Fractal Dimensions

Molecular weight	Front mass fractal $D(m)$	Lattice-sites fractal $D(L)$
60	1.77	1.53
70	1.78	1.56
80	1.63	1.48
90	1.76	1.60
100	1.72	1.62

In three dimensions, the fractal dimension of the diffusion front is about 2.5 at long diffusion times, similar to the monomer case.

4.7 Summary of Fractal Analysis of Polymer Interfaces

When diffusion occurs at an interface, the concentration profile $C(x,t)$ varies smoothly as a function of the one-dimensional depth, x. However, when the diffusion process is viewed in two or three dimensions, the interface is not smooth—in fact, it is very rough. The random nature of diffusion permits the formation of complex structures with fractal characteristics. In this chapter, we used gradient percolation theory, developed by Sapoval and co-workers, to examine the structure and properties of diffuse interfaces formed by metallization of polymer substrates and welding of symmetric amorphous polymer interfaces. Gradient percolation separates the connected from the nonconnected region of the diffusion field. The edge of the connected region is the (fractal) diffusion front. We examined the ramified diffuse interface structure in terms of the diffusion front's width, σ_f, length, N_f, position, x_f, breadth, B_f, fractal dimension, D, and noise in these properties, $\delta\sigma_f^2$, δN_f^2, δx_f^2, and δB_f^2, respectively, as functions of the diffusion length, L_d. We obtained the following computer simulation and theoretical results: width, $\sigma_f \sim L_d^{1/D}$, $\delta\sigma_f^2 \sim L_d^{3/D}$; front length, $N_f \sim L_d^{(1-1/D)}$, $\delta N_f^2 \sim L_d^{(2-1/D)}$; position, $x_f \sim L_d$, $\delta x_f^2 \sim L_d^{(2-1/D)}$; and breadth $B_f \approx 6\,\sigma_f$, $\delta B_f^2 \sim L_d^{(2-1/D)}$, where $D = 7/4$. The simulation results were in very good agreement with an experimental analysis of diffuse polyimide–silver interfaces.

For welding of polymer–polymer interfaces, we examined the diffusion front for reptating chains of molecular weight M, and found that the interface became fractal at diffusion distances L_d greater than the radius of gyration $R_g \sim M^{1/2}$, and at times t longer than the reptation time T_r. At t less than T_r, and L_d less than R_g, self-similarity was lost due to the correlated motion of the chains creating gaps in the interface. However, the interface was very rough and the diffusion front was determined by $N_f \sim L_d^d/M$, where the superscript d is the dimensionality (2 or 3). When L_d was much greater than R_g, the polymer diffusion front behaved as the monomer case with $N_f \sim L_d^{(1-1/D)}$. The fractal nature of diffuse interfaces plays an important role in controlling the physical properties of polymer–polymer and polymer–metal interfaces.

4.8 References

1. R. P. Wool, B.-L. Yuan, and O. J. McGarel, *Polym. Eng. Sci.* ***29***, 1340 (1989).
2. R. P. Wool and J. M. Long, *Macromolecules* ***26***, 5227 (1993); R. P. Wool and J. M. Long, "Structure of Diffuse Polymer–Metal Interfaces", *Polym. Prepr. (Am. Chem. Soc., Div. Polym. Chem.)* ***31*** (2), 558, (1990).
3. R. P. Wool, "Dynamics and Fractal Structure of Polymer Interfaces"; chapter in *New Trends in Physics and Physical Chemistry of Polymers*, L.-H. Lee, Ed.; Plenum Press, New York; 1989; p 129.
4. B. Sapoval, M. Rosso, and J.-F. Gouyet, *J. Phys. Lett.* ***46***, L149 (1985).
5. R. M. Ziff and B. Sapoval, *J. Phys. A: Math. Gen.* ***19***, L1169 (1986).
6. M. Kolb, J.-F. Gouyet and B. Sapoval, *Europhys. Lett.* ***3***, 33 (1987).
7. M. Rosso, B. Sapoval, and J.-F. Gouyet, *Phys. Rev. Lett.* ***57***, 3195 (1986).
8. B. Sapoval, M. Rosso, J.-F. Gouyet, "Fractal Interfaces in Diffusion, Invasion and Corrosion"; Section 3.2.3 in *The Fractal Approach to Heterogeneous Chemistry*, D. Avnir, Ed.; John Wiley & Sons, Chichester; 1989; p 227.
9. P. J. Flory, *Principles of Polymer Chemistry*; Cornell University Press, Ithaca, NY; 1953.
10. D. Stauffer, *Introduction to Percolation Theory*; Taylor and Francis, London and Philadelphia; 1985.
11. D. Stauffer, A. Coniglio, and M. Adam, *Adv. Polym. Sci.* ***44***, 103 (1982).
12. S. R. Broadbent and J. M. Hammersley, *Proc. Cambridge Philos. Soc.* ***53***, 629 (1957).

13. J. S. Peanasky, J. M. Long, and R. P. Wool, *J. Polym Sci: Part B, Polym Phys.* ***29***, 565 (1991).
14. R. P. Wool, D. Raghavan, S. Billieux, and G. C. Wagner, "Statics and Dynamics in Biodegradable Polymer–Starch Blends"; In *Biodegradable Polymers and Plastics*, M. Vert, J. Feijen, A. Albertsson, G. Scott and E. Chiellini, Eds.; Royal Society of Chemistry, London; 1992; p 111.
15. Y. Kantor and I. Webman, *Phys. Rev Lett.* ***52***, 1891 (1984).
16. S. Feng, P. N. Sen, B. I. Halperin, and C. J. Lobb, *Phys. Rev. B: Condens. Matter* ***30***, 5386 (1984).
17. S. Feng, B. I. Halperin, and P. N. Sen, *Phys. Rev. B: Condens. Matter* ***35***, 197 (1987).
18. S. Feng and P. N. Sen, *Phys. Rev. Lett.* ***52***, 216 (1984).
19. S. Feng, M. F. Thorpe, and E. J. Garboczi, *Phys. Rev. B: Condens. Matter* ***31***, 276 (1985).
20. H. He and M. F. Thorpe, *Phys. Rev. Lett.* ***54***, 2107 (1985).
21. E. J. Garboczi and M. F. Thorpe, *Phys. Rev. B: Condens. Matter* ***31***, 7276 (1985).
22. M. F. Thorpe and E. J. Garboczi, *Phys. Rev. B: Condens. Matter* ***35***, 8579 (1987).
23. W. Tang and M. F. Thorpe, *Phys. Rev. B: Condens. Matter* ***36***, 3798 (1987).
24. W. Tang and M. F. Thorpe, *Phys. Rev. B: Condens. Matter* ***37***, 5539 (1988).
25. H. Yan, A. R. Day, and M. F. Thorpe, *Phys. Rev. B: Condens. Matter* ***38***, 6876 (1988).
26. R. P. Wool, Y. Bernaert, G. Agrawal, and M. A. Daley, "Vector Percolation Relaxation Analysis of 2d Networks with Random Bond Scission", *Bull. Am. Phys. Soc.* ***36***, 792 (1991).
27. B. B. Mandelbrot, *The Fractal Geometry of Nature*; W. H. Freeman, San Francisco; 1982.
28. B. B. Mandelbrot, *Fractals, Form, Chance and Dimension*; W. H. Freeman and Co., San Francisco; 1977.
29. F. Family and T. Vicsek, Eds., *Dynamics of Fractal Surfaces*; World Scientific, River Edge, NJ; 1991.
30. M. Kurata and Y. Tsunashima, "Viscosity–Molecular Weight Relationships and Unperturbed Dimensions of Linear Chain Molecules"; chapter in *Polymer Handbook*, J. Brandrup and E. H. Immergut, Eds.; Wiley-Interscience, New York; 3rd ed., 1989; p VII-2.
31. B. Sapoval, M. Rosso, J.-F. Gouyet, J.-F. Colonna, *Solid State Ionics* ***18/19***, 21 (1986).
32. I. M. Robertson and H. Birnbaum, unpublished results, Materials Research Laboratory, University of Illinois, Urbana, IL.
33. P. S. Ho, R. Haight, R. C. White, and B. D. Silverman, *J. Phys., Colloq. (C5, Interface Sci. Eng. '87)* ***C5***– 49 (1988).
34. B. D. Silverman, *Macromolecules* ***24***, 2467 (1991).
35. S. Koizumi, H. Hasegawa, and T. Hashimoto, *Macromolecules* ***23***, 2955 (1990).
36. S. Mazur and S. Reich, *J. Phys. Chem.* ***90***, 1365 (1986).
37. S. Mazur, P. S. Lugg, and C. Yarnitzky, *J. Electrochem. Soc.* ***134***, 346 (1987).
38. L. E. Manring, *Polym. Commun.* ***28***, 68 (1987).
39. M. A. Berger and R. L. McCullough, *Compos. Sci. Technol.* ***22***, 81 (1985).
40. P. G. Desai, J. F. Young, R. P. Wool, "Cross-Linking Reactions in Macro-Defect Free Cement"; chapter in *Advanced Cementitious Systems: Mechanisms and Properties*, F. P. Glasser, G. J. McCarthy, J. F. Young T. O. Mason, and P. L. Pratt, Eds.; Vol. 245 in series *Materials Research Society Proceedings*; Materials Research Society, Pittsburgh, PA; 1992; p 179.
41. B. Sapoval, private communication, 1991.
42. M. Rosso, J.-F. Gouyet, and B. Sapoval, *Phys. Rev. Lett.* ***57***, 3195 (1986).

5 SIMS Analysis of Polymer Interfaces*

5.1 Introduction

My primary objective in this chapter is to explore the structure of symmetric amorphous polymer interfaces using experimental techniques. In Chapters 2–4, we showed how the structure of the interface could be expressed in terms of a set of molecular properties $H(t) = H_\infty(t/T_r)^{r/4}$ (see Table 2.1), and how this set could be derived from the concentration depth profile $C(x,t)$. However, the interface properties $H(t)$ were based on the reptation model developed by de Gennes [1] and Edwards [2–4]. Therefore, we need to conduct experiments that explore the validity of the reptation model, evaluate $H(t)$ from the concentration profiles, and address ancillary questions, such as, Where are the chain ends initially at the surface: segregated or uniformly distributed? Investigation of the reptation model at interfaces requires an experimental technique with an important feature: the ability to measure depth profiles at distances x less than R_g and diffusion times t less than T_r. This region is important because it is where the unique correlated motion effects due to the snakelike motion of reptating chains is manifest.

Spectroscopic techniques for surface and interface analysis were reviewed by Tirrell and Parsonage [5], and are given in Table 5.1. Depth profiles can be measured by forward recoil spectroscopy (FRES) [6, 7], infrared (IR) spectroscopy [8, 9], small-angle neutron scattering (SANS) [10, 11], and specular neutron reflection (SNR) [12–14]. FRES has been used successfully to explore bulk properties via long-range interdiffusion, particularly on polystyrene interfaces. IR and attenuated total reflectance (ATR) IR are limited by a depth resolution of about 1000–10,000 Å, but provide useful information at large interdiffusion depths. SNR has been developed to examine short-range interdiffusion in polymers, with a resolution of about 5–10 Å. We discuss this technique in the next chapter. SANS has a resolution of about 10 Å but requires special sample preparation, for example, stacking of interfaces. Secondary ion mass spectroscopy (SIMS) is another technique for measuring polymer diffusion; we have used it recently to examine interdiffusion in polystyrene interfaces [15].

SIMS has a high sensitivity to hydrogen and deuterium, so it is useful for tracer studies, and has the ability to measure depth profiles directly. Furthermore, it can monitor elements to much greater depths than some of the above methods. While it

---

* Dedicated to G. Zerbi

Table 5.1 Techniques for Polymer Surface Analysis [5].

Technique	Sampling depth / depth resolution / lateral resolution	Information content / comments
scanning electron microscopy (SEM)	10 nm / 10 nm / 5 nm	direct image of surface topography / Auger and X-ray mode can give chemical composition
transmission electron microscopy (TEM)	∞ / – / 3 nm	2-d composition profile / special sample preparation required
scanning tunneling microscopy (STM)	– / 0.1 nm / 0.1 nm	molecular imaging and surface topography / conductive substrate required
atomic force microscopy (AFM)	– / 0.1 nm / 0.5 nm	molecular imaging and surface topography
X-ray reflectometry (XR)	∞ / 1 nm / –	composition profiles and roughness / interpretation requires model predictions
neutron reflectometry (NR)	∞ / 1 nm / –	composition profiles / interpretation requires model predictions
ellipsometry (ELLI) (VASE)	∞ / 1 nm / –	film thickness, refractive index profile / limited contrast
X-ray photoemission spectroscopy (XPS) (ESCA)	10 nm / 0.1 nm / 10 μm	surface composition, composition profiles / radiation damage and sample charging are problems
high resolution energy loss spectroscopy (HREELS)	1 nm / 1 nm / 1 μm	surface composition, vibrational spectra / radiation damage and charging are problems
static secondary ion mass spectroscopy (SSIMS)	1 nm / – / 1 μm	surface composition / complex spectra
dynamic secondary ion mass spectroscopy (DSIMS)	∞ / 13 nm / –	composition profile / etch rate is composition-dependent
forward recoil spectrometry (FRES)	μm / 80 nm / –	composition profile
Rutherford backscattering spectrometry (RBS)	μm / 30 nm / –	composition profile of marker / requires heavy elements
nuclear reaction analysis (NRA)	μm / 13 nm / –	hydrogen and deuterium composition profile / resolution is depth-dependent
infrared attenuated total reflection (IR-ATR)	μm / – / –	vibrational spectra of polymer surface adjacent to ATR crystal surface
infrared grazing incidence reflection (IR-GIR)	mm / – / –	vibrational spectra of thin films and adsorbed layers

provides the concentration profile directly, its depth resolution, which is about 50–100 Å for polymers, is not as good as that of SNR, which has a resolution of about 5 Å. However, useful information can be obtained at distances less than the random-coil size if high-molecular-weight polymers are used, as we show in this chapter.

I now describe the SIMS method of Whitlow and Wool [15] for analyzing polymer interfaces, and examine the short-range interdiffusion behavior of protonated/deuterated polystyrene chains. Our experiments represent a critical test of the reptation model, where a crossover in the time dependence of several scaling laws occurs at the reptation time. We also use selectively deuterated matching *triblocks*, chains with the center deuterated 50% (HDH), and chains with the ends deuterated 50% (DHD), to examine the snakelike anisotropic motion of reptating chains. The experiment using HDH/DHD interfaces is known as the *ripple experiment* [16, 17] and gives substantial support for the minor chain reptation model set forth in Chapter 2. The initial location of the chain ends, the temperature dependence of the self-diffusion coefficients, and the thermodynamics of interaction of HPS with DPS are also examined.

5.2 Theoretical Considerations

5.2.1 Reptation and Interdiffusion

We consider the interdiffusion of two sets of chains, shown in Figure 5.1. The first set consists of *fully deuterated polystyrene* (*DPS*) interdiffusing with *normal polystyrene* (*HPS*) with a range of molecular weights. The second set consists of matching triblocks, where the center of one chain is deuterated 50% (HDH) and the ends of the other chain are deuterated 50% (DHD); chains of this type are used for the ripple experiment. For both the HPS/DPS and HDH/DHD interfaces, the molecular weights are roughly equal on both sides of the interface. The evolution of the minor chains at a polymer–polymer interface provides a convenient method of determining molecular properties, $H(t)$, and concentration profiles, $C(x,t)$, as presented in Chapter 2.

The number of monomers, $N(t)$, crossing a symmetric amorphous interface as a function of time, t, and molecular weight, M, is given for the uniform initial chain-end distribution by [18–21]

$$N(t) \sim t^{3/4} M^{-7/4} \qquad \text{for } t < T_r \tag{5.2.1}$$

$$N(t) \sim t^{1/2} M^{-1} \qquad \text{for } t > T_r \tag{5.2.2}$$

where T_r is the reptation time. The case with surface segregation is treated later in this chapter and investigated by experiment. A second important minor chain property is the average monomer interpenetration depth, $X(t)$, given by

$$X(t) \sim t^{1/4} M^{-1/4} \qquad \text{for } t < T_r \tag{5.2.3}$$

$$X(t) \sim t^{1/2} M^{-1} \qquad \text{for } t > T_r \tag{5.2.4}$$

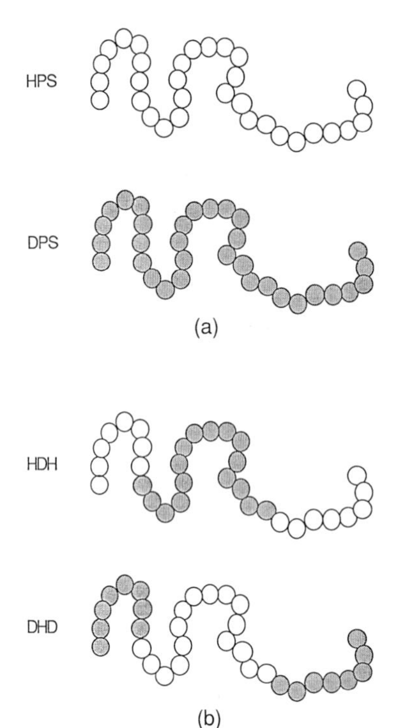

Figure 5.1 The selectively deuterated chains used in the SIMS analysis of interface structure: (a) HPS/DPS; (b) HDH/DHD.

We can derive all other minor chain properties $H(t)$ from $N(t)$ and $X(t)$ (see Table 2.1).

The number of monomers crossed, $N(t)$, and the average monomer interdiffusion distance, $X(t)$, can be measured experimentally from the depth profile, $C(x,t)$:

$$N(t) = \int_0^\infty C(x,t)\,\mathrm{d}x \tag{5.2.5}$$

$$X(t) = \frac{\int_0^\infty x\,C(x,t)\,\mathrm{d}x}{\int_0^\infty C(x,t)\,\mathrm{d}x} \tag{5.2.6}$$

Depth profiles of bilayer samples healed at times both greater than and less than T_r can be used to evaluate scaling laws for $N(t)$ and $X(t)$, as shown schematically in Figure 5.2. The simultaneous crossover in slopes from ¾ to ½ for $N(t)$ and ¼ to ½ for $X(t)$ at the reptation time is highly characteristic of the reptation theory.

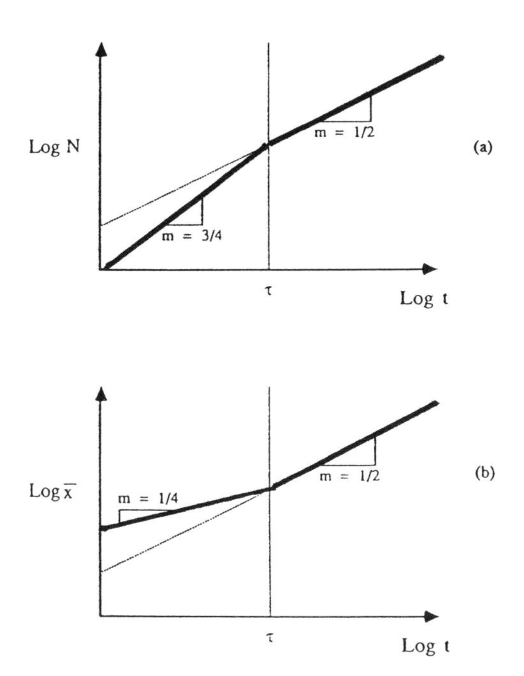

Figure 5.2 Qualitative prediction of the transition at the reptation time, T_r, for (a) the number of monomers crossing the interface, N, and (b) the average interpenetration distance, X.

5.2.2 Measurement of Diffusion

The fundamental equation for the study of isothermal diffusion is Fick's second law,

$$\frac{\partial c}{\partial x} = \frac{\partial}{\partial x}\left(D\frac{\partial c}{\partial x}\right) \tag{5.2.7}$$

where the diffusion coefficient, D, may be constant or depend on concentration. The Matano–Boltzmann method [22, 23] is the approach commonly used with a concentration-dependent diffusion coefficient, and the Grube method [22] is used when D is independent of concentration or is a weak function of composition. The solution to Fick's second law with D constant (Grube method) results in a closed-form error-function solution. The mathematical analysis is more complex when D is not constant. The Matano–Boltzmann method necessitates a solution to Eq 5.2.7 of the form

$$\frac{\partial c}{\partial t} = \frac{\partial}{\partial x}\left[D(c)\,\frac{\partial c}{\partial x}\right] \tag{5.2.8}$$

and the diffusion coefficient is

$$D = -\frac{1}{2t}\frac{\partial x}{\partial c}\int_{c_0}^{c} x \, \mathrm{d}c \tag{5.2.9}$$

which requires a numerical or graphical analysis. The *Matano interface* (Figure 5.3) must first be located before Eq 5.2.9 can be employed. The point where the diffusant areas A and B are equal identifies the Matano interface at $x = 0$. For symmetric polymer pairs, the maximum in the profile derivative locates the Matano interface. Polymer diffusion coefficients can be functions of composition, as demonstrated by Kramer *et al.* [24]; in this case the use of the Matano–Boltzmann approach over the Grube method is necessary. If deuterated polymers are used, additional problems can develop, such as immiscibility [28] and thermodynamic slowing down (TSD) due to the deuterium isotope effect [29, 30].

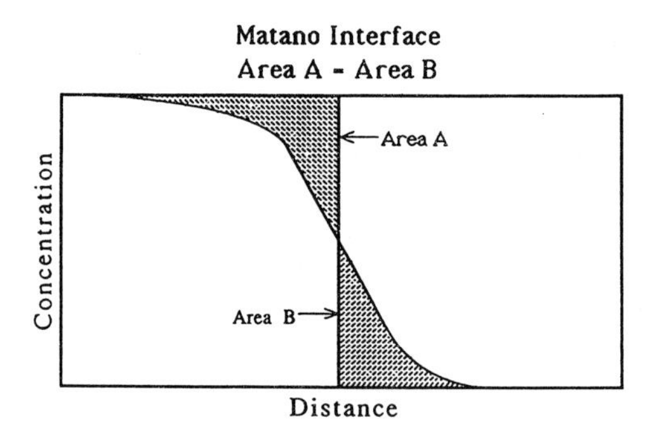

Figure 5.3 Position of the Matano interface.

5.2.3 Temperature Dependence of Diffusion

The temperature dependence of the diffusion coefficient, D, controls the rate of welding at interfaces and can be described by two approaches. The first is the Arrhenius activation energy process, given by

$$D(T) = D_0 \exp(-Q/RT) \tag{5.2.10}$$

where Q is the activation energy and D_0 is a pressure-dependent constant. The second approach is empirical, given by the Vogel relation

$$\log(D/T) = A - B/(T - T_\infty) \tag{5.2.11}$$

For polystyrene with a number average molecular weight M of 255,000, the values of the empirical constants in Eq 5.2.11 are: A = −9.49 (l/K^2)log(cm^2/s·K) [24, 30], B = 710 (l/K)log(cm^2/s·K), and T_∞ = 322 K = 49 °C. The molecular weight dependence of A is given by

$$A = -9.49 - 2\log\left(\frac{M}{255{,}000}\right) \tag{5.2.12}$$

The temperature range for Eq 5.2.11 with these values is 120 °C to 219 °C [30], while the Arrhenius law is most useful for high temperatures (T above 155 °C).

The temperature dependence of welding is related to the above analysis through $G_{1c} \sim H(T)^{2/r}$, where $H(T) = H_\infty[t/T_r(T)]^{r/4}$, in which the temperature dependence of the reptation time, $T_r(T)$, is obtained from $T_r(T) = R^2/[3\pi^2 D(T)]$. Thus, if $r = 2$ for fracture energy, the activation energy for the rate of welding, Q_w, is related to the activation energy for diffusion, Q, by $Q_w \approx ½Q$.

5.2.4 Thermodynamics of Interaction

Many analytical methods used in the study of polymer diffusion rely on isotopic labeling [6–15]. All of the approaches assume that polymers and their chemically identical deuterated analogs form ideal solutions. Results from SANS [28, 29] and FRES [30] measurements indicate that binary mixtures of normal and deuterated polystyrenes are characterized by a small, but positive, Flory–Huggins interaction parameter, χ. Phase equilibrium and mutual diffusion are affected by a positive χ. The mutual diffusion coefficient should undergo a *thermodynamic slowing down* (*TSD*) in the proximity of the critical composition, c_{cr}, where

$$c_{cr} = N_{\mathrm{H}}^{1/2}/(N_{\mathrm{H}}^{1/2} + N_{\mathrm{D}}^{1/2}) \tag{5.2.13}$$

N_{H} and N_{D} are the degrees of polymerization of the protonated and deuterated polymer, respectively. The diffusion coefficient for a blend is given by Kramer *et al.* [24],

$$D = \Omega(c)\,[\chi_s(c) - \chi] \tag{5.2.14}$$

where

$$\chi_s = ½\left[\frac{1}{N_{\mathrm{D}}\,c} + \frac{1}{N_{\mathrm{H}}(1-c)}\right] \tag{5.2.15}$$

and

$$\Omega(c) = 2c[D_{\mathrm{D}}^{*}N_{\mathrm{D}}(1-c) + D_{\mathrm{H}}^{*}N_{\mathrm{H}}\,c] \tag{5.2.16}$$

The χ term in Eq 5.2.14 corrects the χ_s term for the noncombinatorial entropy of mixing and (positive) enthalpic contributions to the thermodynamic driving force. Thermodynamic slowing down should be more pronounced as χ approaches $\chi_s(c)$ or, alternatively, as the temperature approaches the upper critical solution temperature (UCST) [29]. When χ nears zero, or at temperatures sufficiently far from the critical point, thermodynamic slowing down should not be observed [28, 29]. The distance scale ($x < 500$ Å) that we explore here may be too small to allow the larger scale concentration fluctuations to develop and promote the TSD effect.

5.3 Secondary Ion Mass Spectroscopy

The basis for SIMS is the ejection of charged atoms and molecules caused by an impinging ion beam (Figure 5.4). The primary ion beam is rastered over a selected region to erode the sample surface [31–33]. Charged species are filtered and then detected by a mass spectrometer tuned to an element of interest at that depth. The present discussion concentrates on a brief review of the SIMS process as it is used for depth profiling. References are available with details on SIMS [31–33], other surface characterization techniques [33–35], and the fundamentals of sputtering [36, 37].

5.3.1 Sputtering

Impingement of energetic ions affects the sample surface in two ways. The first is a loss of surface material, known as *sputtering*. The second effect involves changes in the target structure, known as *atomic mixing*. In depth profiling, sputtering is desirable

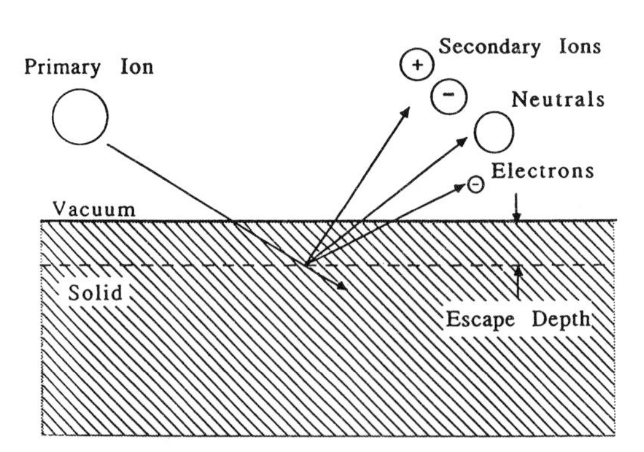

Figure 5.4 Schematic of the sputtering process caused by an impinging primary ion, resulting in secondary ions, neutral atoms, and electrons.

but atomic mixing results in a loss of depth resolution. The sputtering process is shown schematically in Figure 5.4. The incoming ion dissipates a large portion of its energy in the vicinity of the sample surface, transferring energy and momentum to a region around the point of impact. Target atoms within a certain (escape) depth have enough kinetic energy to overcome the surface potential and escape from the sample. The escape depth is usually quite small, typically only a few monolayers. Most sputtered species are neutral, although a small number undergo charge exchange with their local environment to produce positive and negative secondary ions (secondary current). The primary ion beam current is held constant and the secondary current is measured as a function of time.

Several approaches are available for the quantitative analysis of raw profiles [32, 38, 39]. The secondary ion current is usually converted to an absolute concentration by the use of standards. The present study employs diffusion couples where the concentration of deuterium varies from 100% to essentially zero. Since only relative differences in concentration are important, there is no need to convert the secondary ion current to an absolute concentration. Sputtered particles may be atoms or molecules. The number and type increase with decreasing primary ion current density. This fact has been used to characterize polymer surfaces in the static (low energy) SIMS mode [40]. Depth profiling requires energies in the dynamic SIMS mode (ion flux, $j_p \geq 10$ nA/cm^2). Dynamic and static SIMS have long been important tools for the analysis of semiconductors. However, relatively little work has been done on depth profiling of polymers [12, 41, 42].

5.3.2 Atomic Mixing

Two principal contributions to atomic mixing are *recoil implantation* and *cascade mixing*. For a given primary ion energy, the energy transferred from the primary ion to a target atom depends on the distance of closest approach, known as the *impact parameter*, P. For atoms of comparable mass, recoil implantation occurs when the impact parameter is small. The collision is similar to that between two billiard balls: the primary ion elastically transfers a significant portion of its energy and momentum, and the target atom recoils deep into the sample.

If the impact parameter is large, collisions produce cascade mixing. Less energy is transferred to the target atom because the interaction between the ion and the target atom now involves a screened Coulombic potential in place of the hard-sphere potential that is active for recoil implantation. Instead of recoiling into the sample, the target atom displaces its neighboring atoms, producing secondary cascades. There is a general homogenization of all the near-surface atoms affected by the cascade.

The probability of transferring a given amount of energy is usually expressed in terms of the area through which an ion trajectory must pass if the energy transfer is to occur. This area is called the *differential stopping cross section* and is given in terms of the impact parameter, P, as

$$d\sigma = 2\pi P \, dP \tag{5.3.1}$$

Obviously, the probability of energy transfer increases with increasing impact parameter. Consequently, cascade mixing is more probable than recoil implantation and is the dominant contribution to atomic mixing.

The thickness of the region affected by atomic mixing is usually on the order of the primary ion range [32]. This parameter characterizes the distribution of depths where primary ions lose all their energy. For a given target, it is influenced by the primary ion acceleration voltage and the masses of the interacting species. Monte Carlo (TRIM 86) and molecular dynamic [38] programs are used to calculate primary ion ranges as well as other parameters of interest in ion beam interactions, and can be used to evaluate the effects of different primary ions and experimental conditions.

5.3.3 Distortion of the Depth Profile

Atomic mixing has the adverse effect of broadening the measured profile, $I(z)$, of an initially sharp interface (Figure 5.5(a)). The ideally sharp interface is given by the step function $c(z)$. The measured profile is given by the secondary ion current $I(z)$. The abscissa in Figure 5.5(a) is shown both as sputter depth z and as time t to denote the need for depth scale calibration.

Broadening of the measured profile relative to the ideal profile is quantified by the depth resolution, Δz [Figure 5.5(a)]. There are several definitions of Δz, but the most common is the depth range where the signal drops from 84% to 16% of the maximum secondary ion current [32, 33]. This is an arbitrary definition that corresponds to twice the standard deviation of the resolution or response function (Figure 5.5(b)). The resolution function accounts for all effects that distort a measured profile compared to its true profile [33]; instrumental effects as well as sample–instrument interaction are included. The response function is assumed to be Gaussian and independent of sputtering depth. When the true profile is a step function, the response function is found as the derivative of the measured profile [33].

The optimal depth resolution is determined by the atomic mixing range, hence by the primary ion range [32]. Atomic mixing, dominated by cascade mixing, depends on the primary ion energy, the atomic numbers of the interacting species, and the angle between the primary beam and sample surface. Low-energy, high-atomic-number ions at small ion-beam-to-sample-surface angles can be used to minimize atomic mixing [33, 43–47].

A number of other effects can severely degrade the optimal depth resolution for a given sample–instrument system. These artifacts can be divided into instrument factors and ion matrix effects [33, 48]. Instrument factors depend on the particular SIMS being utilized. Ion matrix effects include selective sputtering, initial surface roughness, and sample charging. Selective sputtering is a problem only for multi-elemental targets

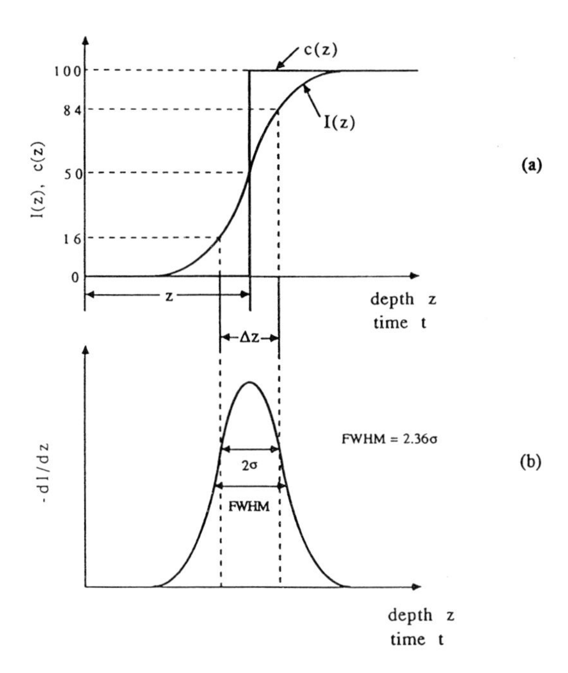

Figure 5.5 Broadening (a) of the measured profile, $I(z)$, relative to the initial step profile, $c(z)$, quantified by the depth resolution, $v(z)$. The measured profile is a convolution of the true profile with the response function, (b). In most cases, the response function is the (negative) derivative of $I(z)$ measured for a sharp interface.

whose components have significantly different sputter yields. Fortunately, selective sputtering is not an issue in the present work; the sputtering rates for hydrogen, deuterium, and carbon are slightly different, but the differences are small. SEM micrographs of bilayers that were not completely sputtered through to the silicon substrate did not exhibit cones or other selective sputtering artifacts. Exceptional care must be taken to make sure the sample surface is flat, especially with procedures employed for making polymer diffusion couples. On a rough surface, a uniform sputtering rate produces signals simultaneously from different depths, consequently degrading the depth resolution.

Polymers are insulators and they become charged when bombarded with ions. Secondary electrons lost during sputtering leave a net positive charge on the specimen [32], and interactions between the secondary ions and the matrix are adversely affected. Several approaches have been used to solve this problem [31], including gold coating, flooding the sputtered area with electrons, and using conductive masks and grids. Another approach, which can be quite difficult in practice, is to change the sample surface potential during data acquisition.

5.4 SIMS Analysis Methods

5.4.1 SIMS Testing

A Cameca IMS-5f Secondary Ion Mass Spectrometer (Figure 5.6) was used to do the depth profile of the HPS/DPS bilayers. The incoming Cs^+ primary ion beam, with a diameter of 60 μm, was rastered over a 500 μm × 500 μm area. Secondary ions (D^+, H^+, and C^+) were produced over the entire rastered area but a mechanical aperture was used to monitor only a 60-μm circular area in the center of the crater [15], as shown in Figure 5.7. Rastering promotes uniform sputtering within the analyzed area and promotes a flat crater bottom. Mechanical gating helps to prevent artifacts like memory effect and sputter redeposition from the crater edges [44]. Without rastering and gating, these artifacts would adversely affect the depth profile.

Observation of positive secondary ions requires a positive sample bias. The accelerating voltage of a positively charged primary ion is reduced as it travels towards the sample. For the conditions used, the relative accelerating voltage is +8.16 kV, the difference between the primary ion accelerating voltage (12.66 kV) and the sample bias (4.5 kV).

In contrast, the observation of negatively charged ions requires a negative sample bias. For the same voltage magnitudes, the relative accelerating voltage would be +17.16 kV, double that for positive secondary ions. The primary ion range and atomic mixing effects would consequently extend over a greater distance, degrading the depth resolution.

Several types of primary beams were available on the Cameca IMS-5f: Cs^+, O_2^+, O^-, and Ar^+. Cesium has the greatest stopping power, by virtue of its mass. Cesium also offers greater beam stability than the other choices. TRIM 86 (Transport of Ions in Matter) calculations show that Cs^+ at +8.16 kV has a much smaller penetration depth in PS (around 150 Å) than O^- or Ar^+ at the same energy (over 275 Å). Extraction of cesium ions from the source in the primary beam column requires a minimum of +12.66 kV.

5.4.2 SIMS Data

The positive carbon secondary current was measured in the HPS/DPS bilayer to monitor stability of the profile, as shown in Figure 5.8. After steady-state sputtering is achieved, a constant C^+ current indicates that the testing is proceeding smoothly. Secondary currents for H^+ and D^+ are the measured quantities. The current intensities in Figure 5.8 are plotted in real time on a log intensity versus channel or cycle number (time unit), to facilitate quantitative analysis of the depth profiles.

An important ion matrix effect in SIMS is mass interference, where signals from different atoms (or molecules) with the same ratio of mass to charge interfere with one another. For example, a doubly charged Si^{++} atom (mass 28) would interfere with

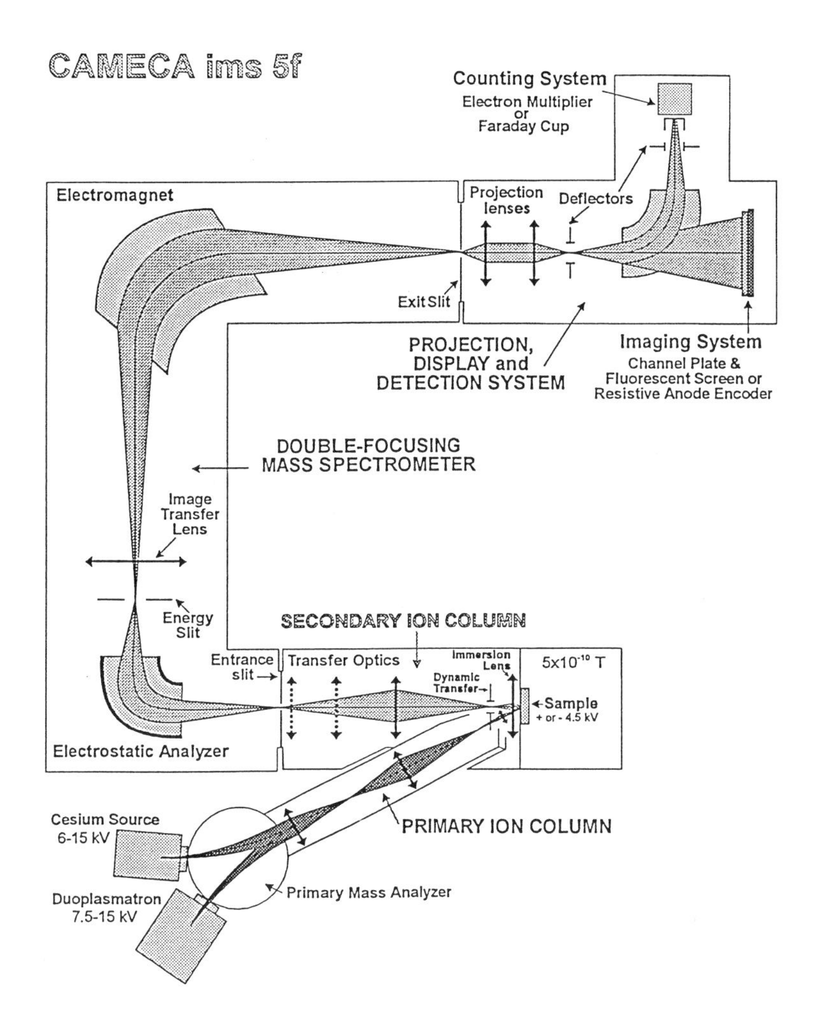

Figure 5.6 Instrumentation for SIMS Cameca IMS-5f.

measurements of Al^+ (mass 14). Control samples of hydrogenated, deuterated, and centrally deuterated triblock copolymer (HDH) were analyzed to assess H_2^+ interference with the D^+ signal. We found that H_2^+ interference was negligible, which is most likely attributable to its very low isotope abundance (0.015%). The large dynamic range of the deuterium signal, usually several decades (Figure 5.8), makes the D^+ signal less susceptible to H_2^+ interference.

Introduction of a small amount of water vapor is unavoidable during sample changes in the SIMS sample chamber. Hydrogen from the water interferes with the H^+ signal. The vacuum system (10^{-6} torr) decreases the partial pressure of water vapor as the test proceeds, resulting in a hydrogen signal that is a function of time. This effect

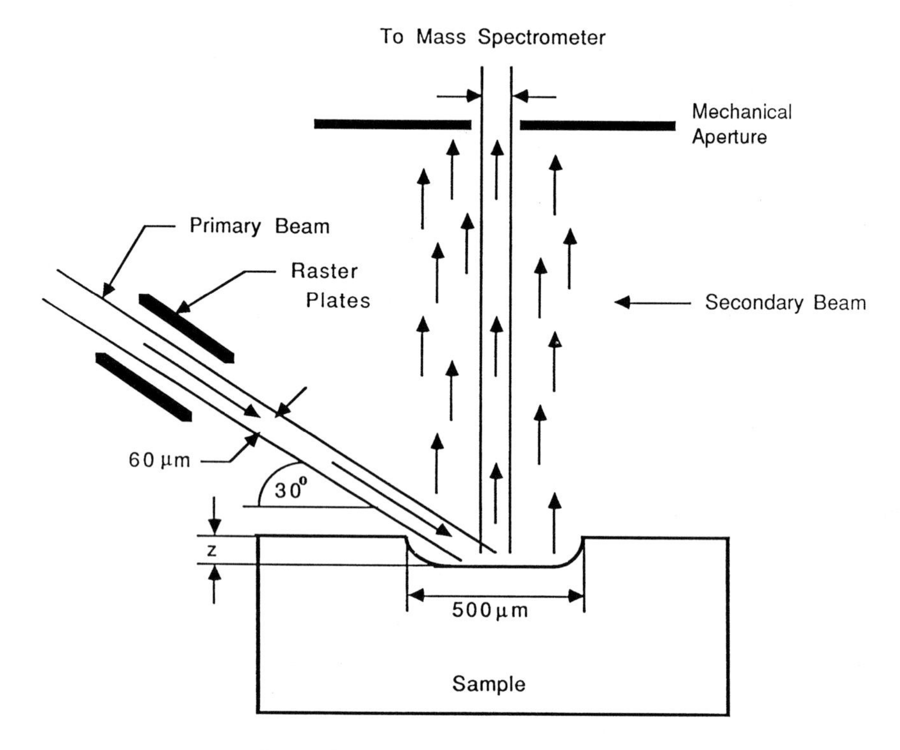

Figure 5.7 Sample–ion beam interaction.

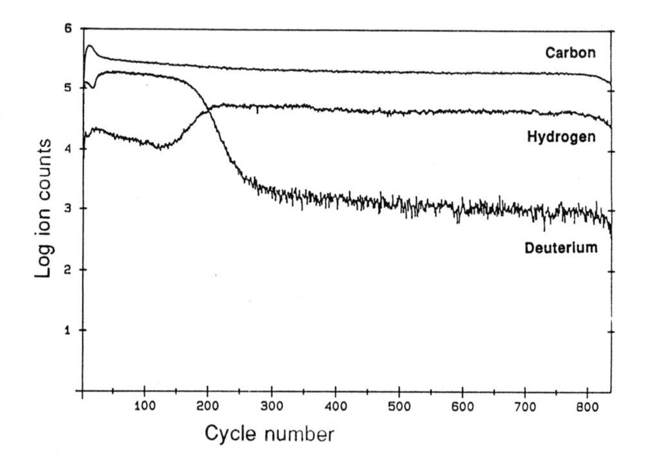

Figure 5.8 Raw profile from SIMS testing. This plot shows the log of the secondary ion signal versus cycle number (time) (Whitlow and Wool [15]).

is most clearly seen at the beginning of the depth profile on a linear scale, as shown later. After sputtering through the initial surface layer is complete (about 50 channels or 250 Å) the H^+ signal continues to decrease. There is no steady-state sputtering before the transition zone is entered. Uncertainty in the location of the lower H^+ plateau and its time-dependent nature render the hydrogen signal problematic for analysis. In this work, the D^+ signal was used to analyze the concentration profiles.

5.5 SIMS Data Analysis Methods

5.5.1 Depth Scale Calibration

There are several methods for calibrating the depth scale [32, 33]. The initial work by Whitlow and Wool (1989) utilized a stylus technique to measure crater depth after testing. We found that this approach introduces errors for polystyrene because the diamond-tipped stylus does not penetrate the polymer to the same depth during each measurement. A second approach, if the sample is homogeneous, is to assume a constant erosion rate, z'. The crater depth is then $z = z't$. Several techniques can be used to determine z [32, 33]. One is to measure the time to sputter through a film of known thickness. Slight variation in the primary ion current during testing introduces great uncertainty to depths found by this method.

The best approach for depth calibration of polymer samples is as follows [15]. Before the Au coating step [15], the gradient in thickness of each sample due to film casting is carefully measured by ellipsometry. The clip end of the sample is the datum for distance along the sample length. Thicknesses at several positions are measured to obtain the sample thickness gradient; thicknesses elsewhere are determined by linear interpolation. After SIMS testing, the distance from the clip end to the center of each crater is measured. The gradient analysis is used to determine the precise bilayer thickness of each crater. Although the thickness gradient may be significant over the length of the sample, it becomes negligible over the 60-μm analyzed region. For example, one of the largest gradients we measured was about 1100 Å from sample end to end; the thickness change over the 60-μm analyzed area was then about 4 Å. Therefore, the gradient is small enough that the depth resolution is unaltered, but it must be measured in order to determine the crater thickness accurately.

The final step in depth calibration is to find the time needed to sputter a particular crater to a known depth. A raw depth profile is shown in Figure 5.8. Each channel (or cycle number) on the abscissa is the time it takes to scan through the masses being measured. For example, if C^+, D^+, and H^+ are each measured for 1 second, then each cycle number is 3 seconds. Since the mass spectrometer requires a short time to switch between masses, each cycle is actually slightly longer than 3 seconds. All secondary signals decrease when the cascade mixing region reaches the silicon substrate. We found that we can accurately determine the point of contact by observing the spot where a secondary ion current departs from its level plateau; in Figure 5.8 this occurs near cycle 800, corresponding to the bilayer thickness (for this crater) minus the cascade mixing region [51]. The time scale is converted to one of depth by subtraction of the thickness of the cascade mixing region (calculated with TRIM 86) from each thickness. Figure 5.9 shows the data from Figure 5.8 converted to a linear count versus depth scale. Depth calibration in this manner avoids problems associated with assuming a constant sputtering rate and has higher accuracy than the stylus technique.

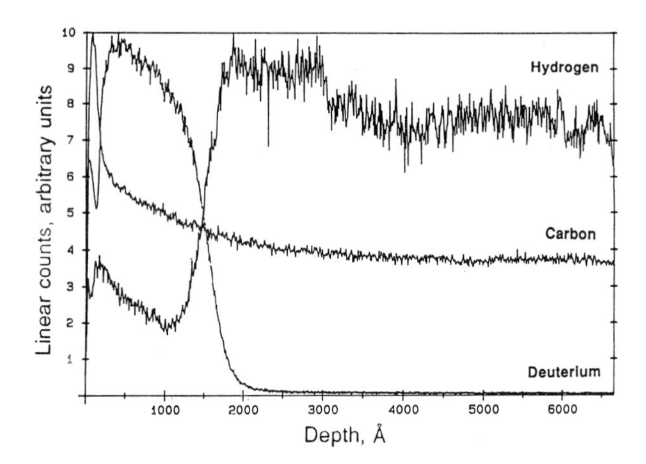

Figure 5.9 Figure 5.8 after depth scale calibration. The distance scale is in terms of total sputter depth, z, through the HPS/DPS bilayer.

5.5.2 Deconvolution of SIMS Profiles

The true $c(z')$ and measured $I(z)$ concentration profiles are related by the convolution integral

$$I(z) = \int_{-\infty}^{\infty} c(z')g(z-z')\,\mathrm{d}z' \tag{5.5.1}$$

The resolution function $g(z)$ can be obtained from the derivative of the measured profile for a sharp interface [52]. The depth resolution is twice the standard deviation of the response function [33].

The solution to Eq 5.5.1 is readily obtained if $I(z)$, $c(z)$, and $g(z)$ are all Gaussian functions. The true broadening, Δz_0, is given in terms of the measured broadening, Δz_m, and the instrument Gaussian broadening, Δz_g, by [52]

$$\Delta z_0 = (\Delta z_m^2 - \Delta z_g^2)^{1/2} \tag{5.5.2}$$

where

$$I(z) = \operatorname{erf}\left(\frac{z}{\Delta z_m}\right) \tag{5.5.3}$$

$$c(z') = \operatorname{erf}\left(\frac{z}{\Delta z_0}\right) \tag{5.5.4}$$

$$g(z-z') = \frac{1}{2\pi \Delta z_g^2} \exp\left[-2\left(\frac{z-z'}{\Delta z_g}\right)\right] \tag{5.5.5}$$

The resolution function, Δz_g, is assumed to be independent of depth. The true broadening, Δz_0, is obtained by correcting the measured broadening for the effect of the system response.

Other deconvolution methods must be used when the measured profile has a more general shape, or when the resolution function cannot be approximated by a Gaussian probability function. Two approaches to obtain the general solution to Eq 5.5.1 are the Fourier transform and van Crittert methods [54–56]. Comparison of the two procedures shows that convergence of the van Crittert method produces a solution equivalent to the best Fourier transform [56]. The van Crittert method, unlike the Fourier transform, produces unique results with physical significance.

The van Crittert method is an iteration procedure that gives the solution

$$c^i(z) = c^{i-1}(z) + I(z) - \int_{-\infty}^{\infty} c^{i-1}(z')\, g(z-z')\, \mathrm{d}z' \tag{5.5.6}$$

$$c^0(z) \equiv I(z) \tag{5.5.7}$$

As n goes to infinity, $c^n(z)$ goes to $c(z)$. The applicability and convergence of this method have been examined in depth [53, 54, 57, 58]. This is an excellent approach to recovering the true profile as long as $\Delta z_m \geq \sqrt{2}\Delta z_g$. When the depth resolution is greater than 70% of the measured profile, Eq 5.5.6 may not converge to $c(z)$ in a realistic number of iterations.

Profiles from interdiffused bilayers were deconvoluted using Eq 5.5.1. We calculated minor chain properties $N(t)$ and $X(t)$ from the deconvoluted profiles using Eqs 5.2.5 and 5.2.6. We computed diffusion coefficients at various concentrations by the Matano–Boltzmann method using Eq 5.2.9.

5.6 Depth Profile Results of HPS/DPS Interfaces

5.6.1 Interface Design

At T_r, the average monomer interpenetration distance $X(T_r)$ for a symmetric polymer interface is given by (Chapter 2)

$$X(T_r) = 0.81\, R_g \tag{5.6.1}$$

R_g is given by

$$R_g = (C_\infty jM/6M_0)^{1/2} b_0 \tag{5.6.2}$$

where C_∞, j, M_0, and b_0 are the characteristic ratio, number of bonds per monomer, monomer molecular weight, and bond length, respectively. To observe small-scale motion at t less than T_r and $X(t)$ less than $X(T_r)$, $2X(t)$ must exceed the depth resolution Δz of about 100 Å. For the 929K(HPS)/1082K(DPS) bilayer system, we calculate $X(T_r)$ as follows: $M = 1{,}082{,}000$; M_0(deuterated) = 112; $C_\infty = 10$; $b_0 = 1.54$; and $j = 2$; then $R_g = 276$ Å and $X(T_r) = 223$ Å. Since the width of the interface is $2X(T_r) = 446$ Å, a resolution of 100 Å should provide useful information in a considerable monomer interpenetration range at t less than T_r.

Figure 5.10 shows typical profiles obtained from the 929K(HPS)/1030K(DPS) bilayers at 146 °C and 720 minutes diffusion time. The symmetry of the profiles permits a diffusion analysis from several sections. However, at shallow depths ($X \approx 500$ Å), the protonated profile may be contaminated with H from water vapor, so the deuterated profile was used to extract the D values.

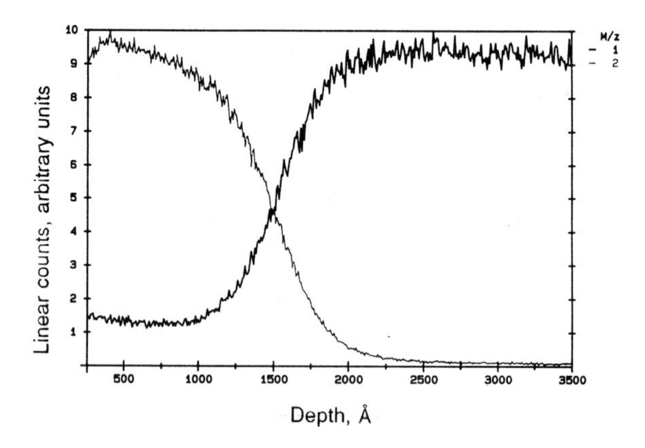

Figure 5.10 The protonated and deuterated depth profiles for a 929K(HPS)/1030K(DPS) interface healed at 146 °C for 720 min (Whitlow and Wool).

5.6.2 Temperature Dependence of Interdiffusion

The diffusion coefficients, D, were first evaluated at long times (t greater than T_r) as functions of temperature and molecular weight. The value of D was calculated by the Matano method for concentrations c, in the range $0.1 \leq c \leq 0.9$. The data (average D values) are summarized in Table 5.2. The temperature dependence of D was analyzed using both the Arrhenius and Vogel methods.

Table 5.2 Diffusion Coefficients in HPS/DPS

Anneal temp, °C	Anneal time, min	Monomers, counts·Å	Interpenetration distance, Å	First moment, counts·Å·°	Diffusion coefficient, cm²/s
		(a) 93K(HPS)/111K(DPS) results			
123.2	362	89.03	312	2.779×10^4	1.993×10^{-16}
128.2	90	51.29	258	1.326×10^4	4.657×10^{-16}
130.4	90	138.7	395	5.475×10^4	1.345×10^{-16}
132.4	90	217.5	550	1.196×10^5	2.664×10^{-15}
134.8	90	280.8	715	2.009×10^5	4.415×10^{-15}
140.0	30	256.9	700	1.798×10^5	1.245×10^{-14}
		(b) 169K(HPS)/199K(DPS) results			
125.4	362	45.13	218	9.824×10^3	1.403×10^{-16}
133.0	90	56.51	269	1.520×10^4	9.013×10^{-16}
135.8	90	90.17	351	3.168×10^4	1.719×10^{-15}
138.2	90	103.0	381	3.920×10^4	2.045×10^{-15}
146.6	30	137.8	443	6.102×10^4	8.611×10^{-15}
		(c) 591K(HPS)/693K(DPS) results			
138.2	362	34.35	140	4.794×10^3	2.026×10^{-16}
145.0	90	22.91	93	2.142×10^3	3.679×10^{-16}
151.5	90	90.18	252	2.272×10^3	3.099×10^{-15}
159.2	90	166.9	504	8.407×10^3	1.205×10^{-14}
166.2	90	203.6	628	1.279×10^3	1.777×10^{-14}
173.1	30	167.3	595	9.960×10^3	4.547×10^{-14}
		(d) 929K(HPS)/1082K(DPS) results			
142.0	362	38.77	127	4.932×10^3	1.522×10^{-16}
149.6	90	25.39	106	2.681×10^3	3.930×10^{-16}
153.2	90	45.83	171	7.853×10^3	1.117×10^{-15}
156.2	90	98.13	295	2.896×10^4	3.500×10^{-15}
160.2	90	82.37	312	2.575×10^4	3.866×10^{-15}
169.8	30	100.6	326	3.277×10^4	1.246×10^{-15}

Figure 5.11 shows a Vogel plot of log (D/T) versus $B/(T - T_\infty)$. The constants A and B from the Vogel analysis (Eq 5.2.11) of our data are given in Table 5.3 for each

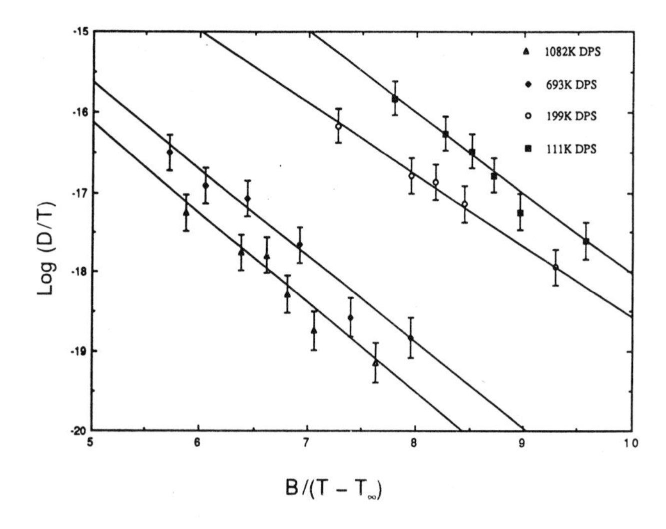

Figure 5.11 Diffusion coefficients for all molecular weights fitted with the Vogel relation, log (D/T) versus $B/(T - T_\infty)$.

Table 5.3 Vogel Constants for HPS/DPS Bilayers

mol. wt. of DPS	constant A		constant B	
	Green & Kramer	Expt.	Green & Kramer	Expt.
111K	−8.768	−7.531	710	753
199K	−9.2748	−9.8292	710	617
693K	−10.358	−10.194	710	778
1082K	−10.745	−10.490	710	808

bilayer and compared with FRES-derived data reported by Green and Kramer [60, 61]. The FRES method measured $D = 8.8 \times 10^{-15}$ cm^2/s for 900K(HPS)/915K(DPS) at 174 °C. The value becomes D_{FRES}(1082K, 174 °C) = 6.29×10^{-15} cm^2/s when scaled to 1082K DPS according to $D \sim M^2$. If Green and Kramer's A and B constants (Table 5.3) are used, the Vogel analysis predicts D_{FRES}(1082K, 174 °C) = 1.72×10^{-14} cm^2/s, which is about 2.7 times larger than that obtained by scaling with molecular weight. The Vogel analysis for the 1082K sample with our values in Table 5.3 predicts D_{SIMS}(1082, 174 °C) = 5.1×10^{-15} cm^2/s, which is about 3.3 times smaller than Green and Kramer's value. At 170 °C, Green and Kramer's constants are in better agreement with our experimental data.

5.6.3 Molecular Weight Dependence of Diffusion at t Greater Than T_r

The self-diffusion coefficients, D, obtained at t greater than T_r are shown as a function of molecular weight at 175 °C and compared with those obtained by other techniques (from [49] and [50]) in Figure 5.12. The data were shifted to a reference temperature of 175 °C. For the Vogel and Arrhenius methods of shifting the data, the scaling laws at 175 °C are given by

$$D \sim M^{-2.1+0.2} \quad \text{(Vogel)} \tag{5.6.3}$$

$$D \sim M^{-2.1+0.2} \quad \text{(Arrhenius)} \tag{5.6.4}$$

with correlation coefficients of 0.976 and 0.952, respectively.

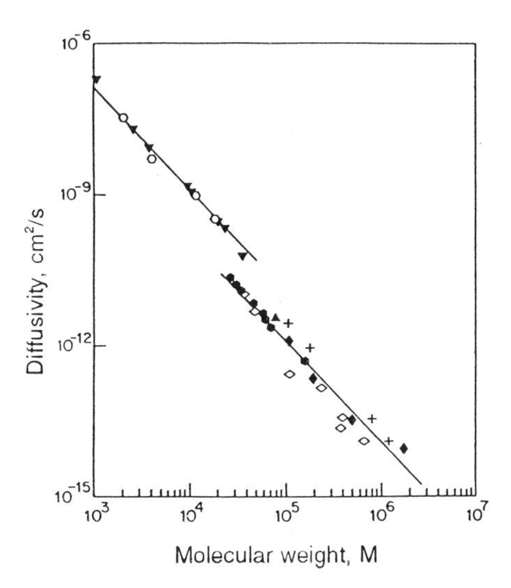

Figure 5.12 Comparison of diffusion coefficients measured by SIMS with those obtained by other techniques [49]. Diffusion coefficients are shifted to 175 °C by an Arrhenius relation with Q = 75 kcal/mol. SIMS values are denoted by the + symbol.

In Figure 5.12, we extrapolated the data to a reference temperature of 175 °C using an Arrhenius method with an activation energy of Q = 75 kcal/mol, as used by Tirrell [49]. This value is higher than our Q values obtained at this temperature, but similar errors may have resulted when we shifted the other data. However, the order of magnitude of the self-diffusion coefficients is correct, and the agreement between SIMS and other methods in the same molecular weight range (10^5–10^6) is considered to be satisfactory.

5.6.4 Diffusion Coefficients at t Less Than T_r

At times less than the reptation time, and diffusion distances less than the radius of gyration, the chain configurations remain affected by the reflecting boundary condition imposed by the original surface. Interdiffusion at t less than T_r involves the relaxation of the nonequilibrium chain configurations [18]. Zhang and Wool analyzed the short-time dynamics and its contribution to both Rouse and reptation relaxation processes in Chapter 3 [63]. The compressed chain configurations have a rapid relaxation response in the region where t is less than τ_e, where τ_e is the Rouse relaxation time for the entanglement length of molecular weight M_e. Segmental motion dominates the interdiffusion process for times less than τ_{RO}, where τ_{RO} is the Rouse relaxation time of the whole chain. For times between τ_{RO} and T_r, the reptation mechanism is predicted to dominate the interdiffusion process. Since the monomer interdiffusion contains significant contributions from segmental motion of chains in nonequilibrium conformations, the diffusion coefficient D' should be considered apparent since it is not a measure of true center-of-mass motion.

The apparent diffusion coefficients, D', were measured for the 929K(HPS)/1082K(DPS) bilayer at 146 °C and t less than T_r. The reptation time at these conditions is about 317 min. Figure 5.13 shows D' versus healing time. The D' values decay from high values at t less than T_r to a constant value D_∞, at t greater than T_r. The high D' values at t less than T_r are largely due to segmental motion. Substitution of Green and Kramer's A and B constants (Table 5.3) in Eq 5.2.11 gives a value of D_∞ of 3.7×10^{-16} cm^2/s, which is in close agreement with our data.

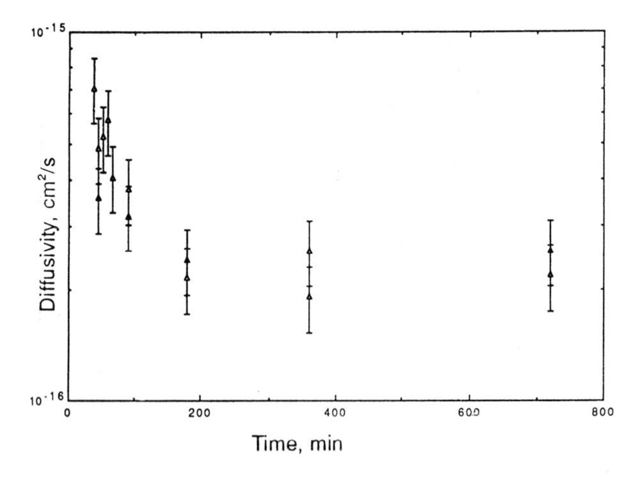

Figure 5.13 Diffusion coefficients as a function of time for 929K(HPS)/1082K(DPS) healed at 146 °C (Whitlow and Wool).

The relation between D' and D_∞ can be approximated as follows. When t is equal to T_r, the diffusion distance is of the order of the chain's end-to-end vector, R, and $R^2 \sim D_\infty T_r$. At times t less than T_r, the experimental interdiffusion diffusion distance, $X(t)^2$, measured from the concentration profile, is related to the apparent diffusion coefficient by $X(t)^2 \sim D't$. Thus, the ratio of the apparent to the long-time diffusion coefficients is

$$D'/D_\infty = [X(t)^2/R^2]/(T_r/t) \tag{5.6.5}$$

We can readily show from Eq 5.6.5 that the ratio D'/D can be much greater than unity. Consider the time range $\tau_{RO} < t < T_r$, where τ_{RO} is the Rouse relaxation time of the chain and is related to the reptation time via $T_r = 3\ M/M_e\ \tau_{RO}$. The monomer diffusion function (from Chapter 2) is $X(t)^2/R^2 = (t/T_r)^{1/2}$. We substitute this in Eq 5.6.5, and find the ratio of diffusion coefficients is

$$D'/D_\infty \simeq (T_r/t)^{1/2} \tag{5.6.6}$$

For example, from Figure 5.13 and Table 5.4, T_r is about 360 minutes. At t of about 37 minutes, $\sqrt{(T_r/t)} \approx 3$ and $D'/D \approx 7/2.2 \approx 3$. Thus, in this time range, $D' \sim t^{-1/2}$ and a plot of log D' versus log t of the data in Table 5.4 gives an approximate slope of $-\frac{1}{2}$, in agreement with the reptation theory. At the Rouse time, $D'/D = \sqrt{(3M/M_e)}$; when $M = 10^6$ and $M_e = 18{,}000$, then $D'/D \approx 13$. At t less than τ_{RO}, the ratio D'/D can be much larger. Similar results were obtained by SNR (Chapter 6) by Karim *et al.* [65, 66].

The apparent self-diffusion coefficients shown in Figure 5.13 do not represent the actual center-of-mass motion. In fact, the opposite behavior occurs: the real center-of-mass diffusion coefficient, D, is less than the equilibrium value at distances $X(t)$ less than R_g. The relaxation of the compressed configurations results in very little displacement of the chain's center of mass, $X(t)_{cm}$ [18]. This behavior is largely dependent on the static rather than the dynamic properties of the chains, and should be observed for Rouse- as well as reptation-controlled diffusion.

5.6.5 Number of Monomers, $N(t)$, Diffused at t Less Than T_r

The number of monomers, $N(t)$, crossing the interface plane is determined from the integral of the concentration profile, Eq 5.2.5. Figure 5.14 shows a plot of log $N(t)$ versus log t for the 929K/1082K bilayer at 146.6 °C. The reptation time, T_r, was calculated using [1]

$$T_r = R^2/(3\pi^2 D) \tag{5.6.7}$$

where D is measured at t much greater than T_r.

Table 5.4 Time Series Results of Diffusion of 929K(HPS)/1082K(DPS) Healed at 146.6 °C

Anneal time, min	Monomers, counts·Å	Interpenetration distance, Å	First moment, counts·Å°	Diffusion coefficient, 10^{-16} cm^2/s
37.5	10.52	161	3499	7.047
45.0	15.99	114	3682	3.572
45.0	16.26	135	4609	4.865
52.5	14.66	164	3359	5.225
60.0	17.34	159	5130	5.785
67.5	10.72	146	6229	4.077
90.0	26.19	153	6651	3.195
90.0	43.50	166	7227	3.785
180.0	47.46	197	9329	2.430
180.0	47.22	180	8491	2.166
360.0	63.61	243	15490	1.925
360.0	75.49	277	20870	2.567
720.0	155.70	384	59750	2.565
720.0	107.10	364	38980	2.201

The self-diffusion coefficient, D, was obtained by the method suggested by Zhang and Wool [21] using measured $N(t)$ values under Fickian diffusion conditions at long times, via

$$D = \pi N^2(t)/t \qquad (5.6.8)$$

for t greater than T_r. The proportionality constant in the relation $N(t) \sim t^{1/2}$ at t greater than T_r is $(D/\pi)^{1/2}$. Using Eq 5.6.8 and the $N(t)$ experimental data in Table 5.4, we obtain a value of D of about 0.83×10^{-16} cm^2/s for $t \geq 360$ minutes. Substituting for this value of D and $R = 677$ Å in Eq 5.6.7, we find T_r is about 317 min.

In Figure 5.14, at t less than T_r, the data in the range $t \leq 360$ min are described by

$$N(t) \approx 0.66\, t^{0.8} \qquad (5.6.9)$$

The slope of 0.8 is in good agreement with the predicted slope of 0.75. The exponent of 0.75 applies to polymers without chain-end segregation. With segregation, we would have an exponent of 0.5 at t less than T_r. At t much greater than T_r, $N \sim t^{1/2}$ and the expected slope of 0.5 is shown for comparison. The number of monomers diffused at the reptation time, $N(T_r) = N_\infty$, is obtained from Eqs 5.6.7 and 5.6.8 as

$$N_\infty = \sqrt{(2/\pi^3)}\, R_g \qquad (5.6.10)$$

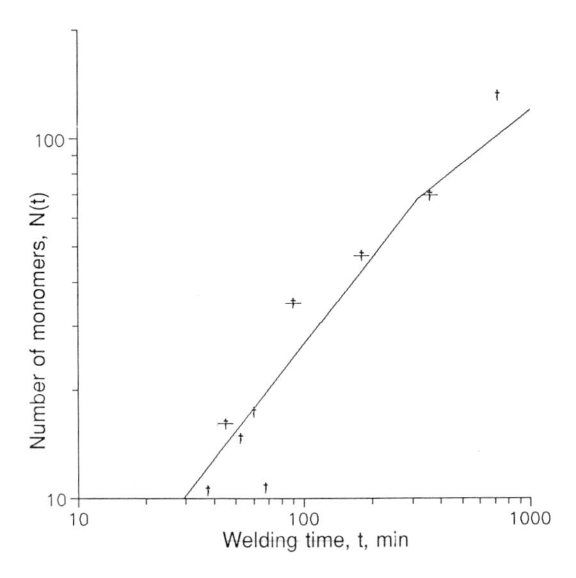

Figure 5.14 Number of monomers, N, crossing the interface as a function of time for 929K(HPS)/1082K(DPS) healed at 146 °C. The lines are drawn according to the reptation prediction, for which $N \sim t^{3/4}$ at $t < T_r$ and $N \sim t^{1/2}$ at $t > T_r$, where $T_r = 317$ min (Whitlow and Wool).

or roughly, $N_\infty \approx R_g/4 \approx 70$ Å. The front factor (0.66) in Eq 5.6.9 can be compared with $N_\infty/T_r^{3/4} \approx 0.93$, which is of a similar order of magnitude. Considering the complexity of this experiment, the overall agreement with the reptation prediction $N(t) = N_\infty(t/T_r)^{3/4}$ is quite good. While the data in Figure 5.14 exhibit scatter, especially at the earliest times where the SIMS resolution factor works against us, we can say with certainty that the exponent at t less than T_r is greater than 0.5. This favors a situation where the chain ends are not segregated at $t = 0$, which is discussed later in Section 5.6.7.

5.6.6 Monomer Interdiffusion Depth, $X(t)$, at t Less Than T_r

The average monomer interpenetration distance, $X(t)$, was measured from the normalized first moment of the concentration profile, Eq 5.2.6. To compare with the scaling law predictions, Eqs 5.2.3 and 5.2.4, we plotted log $X(t)$ against log t, as shown in Figure 5.15. Again, at t greater than T_r, $X(t) \sim t^{1/2}$, as expected for normal Fickian diffusion. At t less than T_r, where T_r is about 317 minutes, the data deviate from the extrapolated $t^{1/2}$ slope and lie closer to the theoretical $t^{1/4}$ slope. A least squares fit of the data for $0 < t \leq 360$ minutes gives

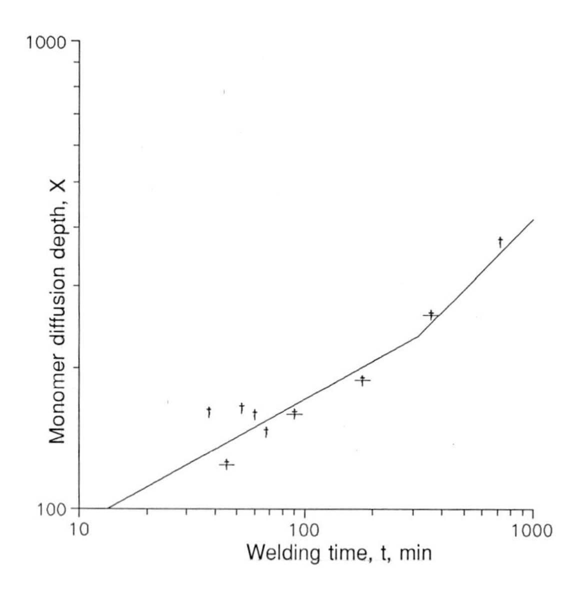

Figure 5.15 Average monomer interdiffusion distance, X, as a function of time for 929K(HPS)/1082K(DPS) healed at 146 °C. The transition in slope from ¼ to ½ at $t = T_r$ and $X \approx R_g$ is drawn in accordance with the reptation prediction (Whitlow and Wool).

$$X(t) = 49.9\, t^{0.27} \qquad (\text{Å}/\text{min}^{0.27}) \qquad (5.6.11)$$

Reptation theory predicts $X(t) = (0.81R_g/T_r^{1/4})\,t^{1/4}$. With $R_g = 276$ Å and $T_r = 317$ min,

$$X(\text{theory}) = 53\, t^{1/4} \qquad (\text{Å}/\text{min}^{1/4}) \qquad (5.6.12)$$

and $X_\infty = 0.81R_g = 223$ Å. The close agreement between Eqs 5.6.11 and 5.6.12, in both front factor (53) and exponent (¼), is highly supportive of the reptation model.

Over the entire healing range, $0 < t \le 720$ minutes, we obtain $X(t) = 38.5\, t^{0.33}$. In the range $180 \le t \le 720$ minutes, we have $X(t) = 14.3\, t^{0.50}$. Thus, at short times the data are well represented by $X(t) \sim t^{1/4}$, which changes to $X(t) \sim t^{1/2}$ at long times. The predicted crossover from $t^{1/4}$ to $t^{1/2}$ at $X_\infty \approx 223$ Å is consistent with Figure 5.15. This agreement is necessary but not sufficient to prove the reptation model. We must still demonstrate that chains actually move like snakes, which we do later in this chapter.

The interdiffusion distance at the Rouse time $\tau_{RO} = 2$ min is obtained as $X(\tau_{RO}) = 60$ Å from Eq 5.6.11, and is comparable to the tube diameter R_{tube}, as discussed by Graessley [59]. R_{tube} should be related to the end-to-end vector, R_e, of the entanglement molecular weight, M_e. With $R_{tube} = 1.31\, R_e$ (the factor 1.31 describes the diameter of the sphere that completely contains the entire chain of end-to-end vector R), $M_e = 18{,}000$, $R_e = 90$ Å, and $R_{tube} = 118$ Å. If $X(\tau_{RO}) = R_{tube}/2$ because of symmetry at the interface, then $R_{tube}/2 = 59$ Å, in agreement with the interdiffusion distance for the Rouse time. From Chapter 3, we calculated $X(\tau_e) \approx 30$ Å and $X(\tau_{RO}) \approx 80$ Å.

The above SIMS results are consistent with recent studies of interdiffusion in symmetric HPS/DPS interfaces using specular neutron reflection (SNR) methods by Karim, Felcher, and co-workers [65, 66] and by Zhang and Wool [63] (discussed in Chapter 6). Recently, Steiner *et al.* [62] examined the interface formation of a HPS/DPS interface using forward scattering (FRES) from nuclear reactions produced by a ^{3}He beam from a Van de Graaff accelerator. The molecular weights of the DPS and HPS polymers were 1.03×10^6 and 2.89×10^6, respectively. They observed that the width, $w(t)$, of the interface increases approximately as $w(t) \sim t^{0.34}$ at 160 °C, which they interpret in terms of the diffusion of partially miscible polymers with thermodynamic slowing down (TSD). At 160 °C, the interface broadened up to 1000 Å and remained constant at times greater than 50,000 seconds (833 minutes), which is consistent with the TSD effect. These times are much greater than the reptation time. At 140 °C, they obtained $w(t) \sim t^{0.27}$, up to times of 10^6 seconds. At this temperature an upper depth plateau indicating restricted growth of the interface was not observed in the depth range studied. The data are not sufficiently detailed in the region of the reptation time and depths less than the radius of gyration to determine if the transition in slope from ¼ to ½ occurs before the onset of TSD at much longer times.

5.6.7 Location of Chain Ends at Time Zero

We enquire from the HPS/DPS experiment if the chain ends were initially segregated at the surface of each polymer at $t = 0$. This possibility was suggested by de Gennes [64], and affects the interpretation of adhesion data. Chain-end segregation could be favored as a means of minimizing the loss of conformational entropy of the chain with reflecting boundary conditions at a surface. If the chain ends are uniformly distributed in the bulk, the number of monomers crossing the interface behaves as $N = N_\infty (t/T_r)^{3/4}$ and the average depth as $X = X_\infty (t/T_r)^{1/4}$, with N and X related at $t \leq T_r$ by

$$N = N_\infty X^3 / X_\infty^3 \tag{5.6.13}$$

So we expect $N \sim X^3$, with a factor $N_\infty / X_\infty{}^3$; for $t > T_r$, $N \sim X$ (Fickian diffusion).

For the surface-segregated case, the chain ends ($n_\infty \sim M^{-1/2}$ in number) within a radius of gyration of the surface, are at the surface at $t = 0$. We have $N_s \sim n_\infty \langle l \rangle$, so $N_s = N_\infty (t/T_r)^{1/2}$, and since $X(t)$ stays the same, the number of monomers crossing is

$$N_s = N_\infty X^2 / X_\infty^2 \tag{5.6.14}$$

With segregation (subscript s), we have $N_s \sim X^2$, with a front factor of $N_\infty / X_\infty{}^2$. Again, as in the previous case, when t is greater than T_r, the segregation effect is forgotten by the diffusing chains and $N \sim X$.

From Eqs 5.6.9 and 5.6.11, the quantitative relation between N and X should be

$$N = 5.3 \times 10^{-6} X^3 \tag{5.6.15}$$

where the exponent of 3.0 is derived from the ratio of exponents 0.804/0.268. The slope transition from 3 to 1 at X_∞ is indicated in Figure 5.16. The data cross-plot between N and X in Figure 5.16 tends to exaggerate the data scatter already apparent for $N(t)$ and $X(t)$ in Figures 5.14 and 5.15, respectively. However, the correlation between the data averages resulting in Eq 5.6.15 is better. While the exponent of 3.0 in Eq 5.6.15 agrees very well with the nonsegregated case (Eq. 5.6.13), the front factor comparison provides additional support. Comparing the front factors using $N_\infty \approx R_g/4$ and $X_\infty \approx 0.8R_g = 223$, we obtain $N_\infty/X_\infty^3 \approx ½R_g^2 \approx 7 \times 10^{-6}$ for the nonsegregated case, which is of the same order of magnitude as the experimental data (5.3×10^{-6}). For the segregated case we obtain $N_\infty/X_\infty^2 \approx 0.4/R_g \approx 1.4 \times 10^{-3}$, which is about three orders of magnitude different from the nonsegregated case. We conclude that chain-end segregation has very little effect and that interdiffusion experiments can be well described by a uniform chain-end distribution. This point is reinforced by other SIMS and neutron reflection experiments (Chapter 6).

Let us examine the argument that a chain end on the surface lowers the energy of the chain below that for making a loop by reflection of the end from the surface. For highly entangled polymers, this effect is expected to be felt only by the chain ends

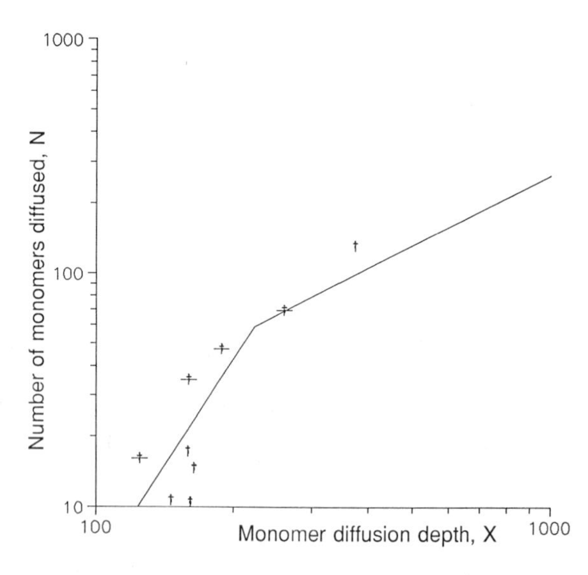

Figure 5.16 Number of monomers, N, versus monomer depth, X, for the 929K(HPS)/1082K(DPS) interface. The line has a slope of 3.0 at $X < 200$ Å, consistent with a uniform initial chain-end distribution.

existing within a distance equal to the tube diameter from the surface. The entanglement structure screens out the effect for chains that are more than one entanglement length, R_{ge}, from the surface. Let us assume that at M_c, a fraction β of all the chain ends are segregated to the surface; we derive the surface enrichment factor, f_R, as a function of molecular weight

$$f_R = \beta R_{ge} / R_g \tag{5.6.16}$$

and

$$f_R = \beta (M_c / M)^{1/2} \tag{5.6.17}$$

Note that $f_R = \beta$ at M_c, and at $M/M_c = 33$ (such as for PS with M = 1,000,000), $f_R = 0.17\,\beta$. The β factor should depend on the loop energy, Q_{loop}. In the extreme case when Q_{loop} is very high, $\beta = 1$, and surface segregation becomes important, particularly when M is near M_c.

The result, $N \sim X^3$, is also important for evaluating the number of monomers on the fractal diffusion front, N_f, discussed in Chapter 4. The roughness of the fractal-like diffusion front of a polymer interface can be characterized in terms of the number of monomers per unit area existing on the connected diffusion front. We note that at t less than T_r, the interface roughness, $N_f(t)$, in three-dimensional space (3-d) is very well described by $N_f(t) = N_\infty(t/T_r)^{3/4}$. When t is much greater than T_r, $N_f \approx 1.3\,X$, so that $N_f(t\text{,3-d}) \sim R_g(t/T_r)^{1/2}$. In two dimensions, the roughness of the front evolves as $N_f(t\text{,2-d}) \sim [R_g(t/T_r)^{1/2}]^{1-1/D_f}$, where the fractal dimension, D_f, is 1.75. The roughness could be important, for example, in diffusion-controlled chemical reactions, where the new sites for reaction are determined by the availability of exposed sites on the diffusing polymer front.

5.7 Do Molecules Move like Snakes?

5.7.1 The Ripple Experiment

We present here a simple experiment, conceived to test the reptation model and the minor chain model used to derive the interface structure function, $H(t)$, in a very direct manner [16, 17, 69]. Consider two layers of polystyrene with chain architectures shown in Figure 5.17 and described previously (Figure 5.1). Each layer is composed of selectively deuterated polystyrene chains. In one of the layers, the central 50% of the monomers are deuterated. This constitutes a triblock copolymer of labeled and normal polystyrene that is denoted HDH (¼H–½D–¼H). In the second layer, the labeling has been reversed so that the two end fractions (25% each) of the chain are deuterated. This is denoted DHD (¼D–½H–¼D). At temperatures above the glass transition temperature of the polystyrene (~ 100 °C), the polymer chains begin to

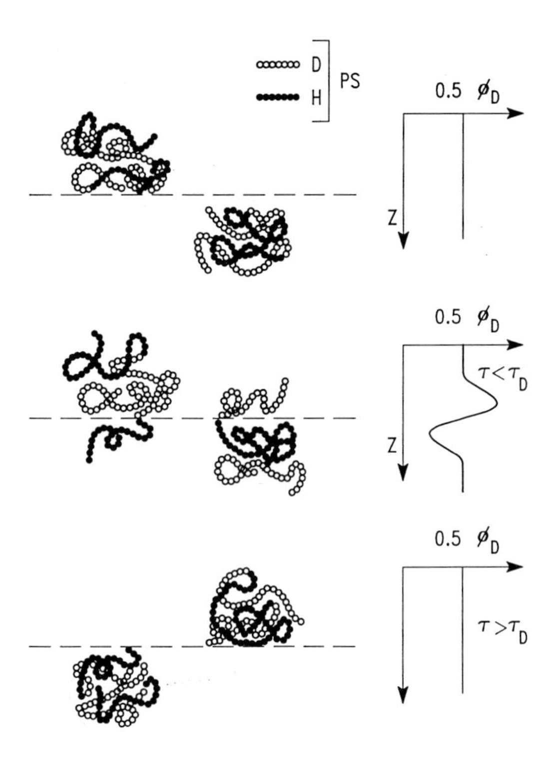

Figure 5.17 The ripple experiment at the interface between a bilayer of HDH and DHD labeled polystyrene, showing the interdiffusion behavior of matching chains. The protonated sections of the chain are marked by filled circles. The D concentration profiles are shown on the right. (a) The initial interface at $t = 0$. The D concentration profile is flat, since there is 50% deuteration on each side of the interface. (b) The interface after the chain ends have diffused across ($x < R_g$). The deuterated chains from one side enrich the deuterated centers on the other side, *vice versa* for the protonated sections, and the ripple in the depth profile of D results. A ripple of opposite sign occurs for the H profile. (c) The interface when the molecules have fully diffused across. The D profile becomes flat [16, 69].

interdiffuse across the interface. If the motion of the polymer were indifferent for each portion of the molecule, the concentration of deuterium across the interface would remain constant. However, if the motion were biased towards reptation, a ripple in the concentration profile, as described below, would be displayed.

The motion of Rouse chains is also anisotropic in the bulk. However, at an interface, central segments of a Rouse chain can diffuse across the interface, and the difference in center and end segment displacements is very small compared to a reptating chain. For a reptating chain, lateral motion of the central segment of the chain is permitted up to depths approximating the tube diameter, beyond which the central segments must follow the chain ends in a snakelike fashion.

In Figure 5.17, HDH and DHD are depicted by open and filled circles; the open circles represent the deuterium-labeled portions of the molecule, and the filled circles are the normal (protonated) portions of the chains. Initially, the average concentration of the labeled portions of the molecules is 0.5, as seen along the normal to the interface. If the chains reptate, the chain ends diffuse across the interface before the chain centers. This leads to a "ripple" or excess of deuterium on the HDH side, and a depletion on the DHD side of the interface, as indicated in the concentration profile shown at the right in Figure 5.17. However. when the molecules have diffused distances comparable to R_g, the ripple vanishes and a constant concentration profile at 0.5 is again found.

The ripple profile $C(x,t)$ shown in Figure 5.17 represents the enrichment (chain ends minus chain center) of deuterium on one side and protons on the other side of the interface, and is characterized as follows: $C(x,t)$ is the profile shape; $C(0,t)$ is the peak height; $N(t)$ is the area of the peak; δX is the peak-to-peak separation; and τ_m is the time for the peak to reach maximum peak height. These entities can be characterized in terms of the diffused contour length $\langle l \rangle$. The average chain length that has diffused across the interface is given (Chapter 2) by $\langle l \rangle = L/2\ (t/T_r)^{1/2}$, where $L \sim M$ is the contour length. Note that when t is equal to T_r, $\langle l \rangle$ is $L/2$, since each chain has two minor chains. This length relation can be used to determine the approximate extent of the HDH/DHD ripple evolution, provided that $\langle l \rangle \leq L/4$.

The fraction of chain, F, remaining in its tube and which has not crossed the interface is given by

$$F = 1 - 4/\pi^{3/2}\,(t/T_r)^{1/2} \tag{5.7.1}$$

For the symmetric HDH/DHD triblock with ¼–½–¼ construction, we expect the maximum effect in the ripple to occur near $F \approx$ ½ at the time $\tau_m \approx T_r/2$, or $T_d/4$, where $T_d \approx 2T_r$ is the complete escape time. The ripple peak height behaves approximately as $C(0, t) = ½\ (t/T_d)^{1/2}$, for $t \leq \tau_m$, so that at $\tau_m \approx T_d/4$, the maximum peak height $C(0, \tau_m) \approx 0.25$. The approximation here relies on the assumption that the early stages of the ripple are identical to the symmetric interdiffusion case. This assumption is good when F is much less than ¼ but it breaks down near $F \approx$ ¼. Agrawal and Wool showed by computer simulations that the maximum enrichment occurs at values of $C(0, \tau_m)$ less than 0.25 (closer to 0.13) and times less than $T_r/2$ (closer to $T_r/4$) because of the Gaussian broadening of the contributions from chain centers as they diffuse across the interface [67–69]. The ripple peak evolution has the same scaling law as do bridges, $p(t)$ (Chapter 2), and hence its maximum is independent of molecular weight. However, the area under the ripple behaves as $N(t)$, and its lateral width as $X(t)$, so that the maximum ripple area and width both depend on molecular weight.

The integrated ripple signal gives the number of monomers crossing the interface, $N(t)$, as $N(t) \sim R_g(t/T_r)^{3/4}$, when $t \leq \tau_m$. At $\tau_m \approx ¼\ T_r$, we obtain $N(\tau_m) \approx 0.01\ R_g$ (with units of Å) [67]. The peak-to-peak separation, δX is approximately 0, and is independent of time, which is supported by both theory and simulation. The SIMS experiment then gives δX as the instrumental resolution. At times greater than τ_m, the contribution from the chain centers increases, thereby leading to a decrease in the ripple intensity.

We undertook computer simulations to calculate the concentration profiles due interdiffusion between HDH and DHD as a function of chain length and deuteration fraction [17, 69]. Sample profiles are shown in Figure 5.18. These profiles were obtained by simulating interdiffusion between random-walk chains at the HDH/DHD interface, as described in Chapter 2. The chains were initially constrained by the surfaces, since we used reflecting boundary conditions. The chains interdiffused by reptating on a two-dimensional lattice, and the diffusion was studied up to times greater than T_r. The simulations show that the ripple profile increases up to a maximum amplitude of about 0.13 at $T_r/4$ before it begins decreasing.

5.7.2 SIMS Analysis of the Ripple Experiment

To perform the ripple experiment, J. Mays (University of Alabama, Birmingham) synthesized suitably labeled polystyrene chains with matching HDH/DHD architecture using anionic polymerization [16]. With size exclusion chromatography we found the weight average molecular weight of the DHD polymer, M_w, was 2.5×10^5, with a polydispersity, M_w/M_n of 1.04. An identical reaction scheme was used to generate the HDH polystyrene with, of course, the reverse labeling, and M_w of 2.25×10^5. The fractions of deuterium labeling in each chain were 0.57 (HDH) and 0.5 (DHD). Complete details of the sample preparation are given in reference [16].

Figure 5.19 shows the excess of D as a function of depth for the different annealing times at 118 °C. The rise and decay of the maxima and minima in the profiles follow the expected behavior (shown by computer simulation) once the instrumental resolution function is taken into account. The time over which the density ripple is visible corresponds to the expected reptation time, T_r, of about 2,200 minutes, and the maximum in the ripple peak occurs near τ_m of about ½ T_r. More directly, the experiment shows that the distances over which the maximum and minimum extend are of the same order of magnitude as the radius of gyration of the polystyrene (indicated in Figure 5.19). These results were checked for possible artifacts: the layer thicknesses are large enough to avoid effects from the air and substrate interfaces; the results were independent of the ordering of the layers; more importantly, the experiment was designed to have the same chemical potential for all segments in the vicinity of the internal interface.

Our results demonstrate the reptative motion of the polymer chains and, coupled with the previous analysis of HPS/DPS interfaces, provide convincing support for the minor chain reptation model and the interface structure function, $H(t)$, developed in Chapter 2. We could not observe isotropic fluctuations of chain segments across the interface in this experiment because of the nature of the labeling. In fact, nearly 120 minutes (slightly less than the Rouse relaxation time of the whole chain) were required before any excess was observable. The lack of a ripple signal is expected when Rouse

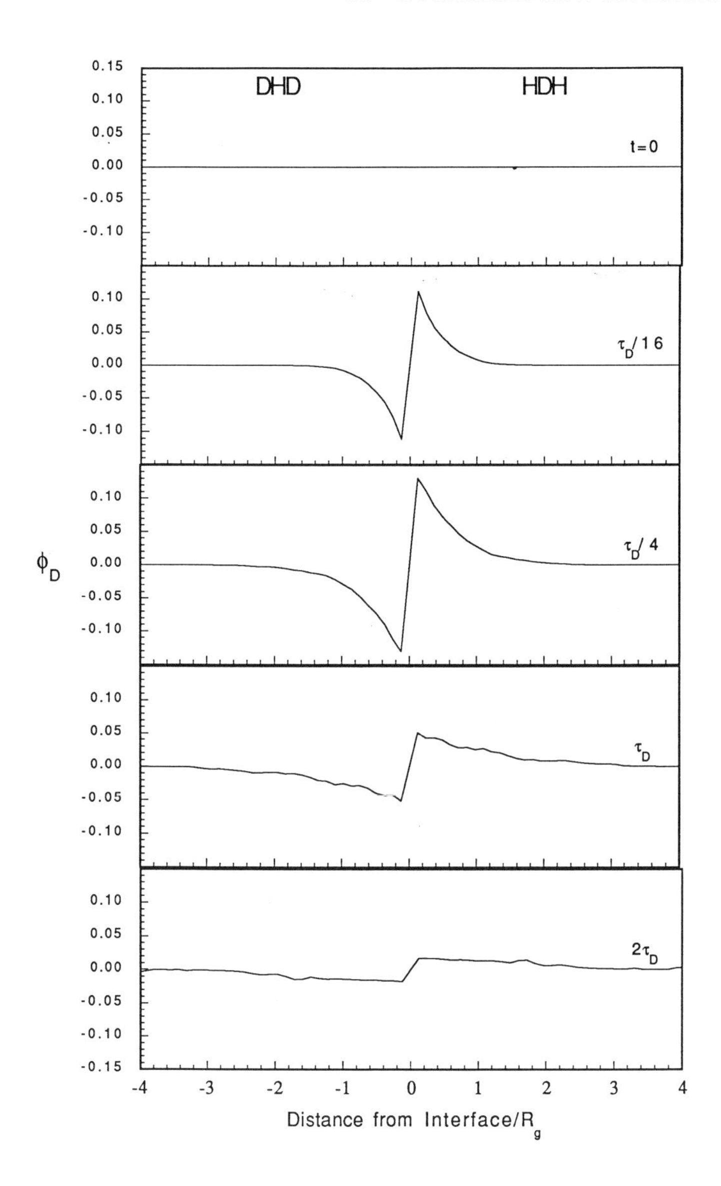

Figure 5.18 Simulated concentration profiles for the HDH/DHD ripple experiment as a function of time. The interdiffusion distance and time are normalized with respect to the radius of gyration, R_g, and the reptation time, τ_d, respectively (Agrawal *et al.* [17, 69]).

motion of all the segments occurs, resulting in initial diffusion to depths on the order of the tube diameter. The spatial separation of the maximum and the minimum in the profile of excess D was time invariant, but the breadth of the signal increased and later decreased with time. The amplitude of the excess reached a maximum at τ_m of about

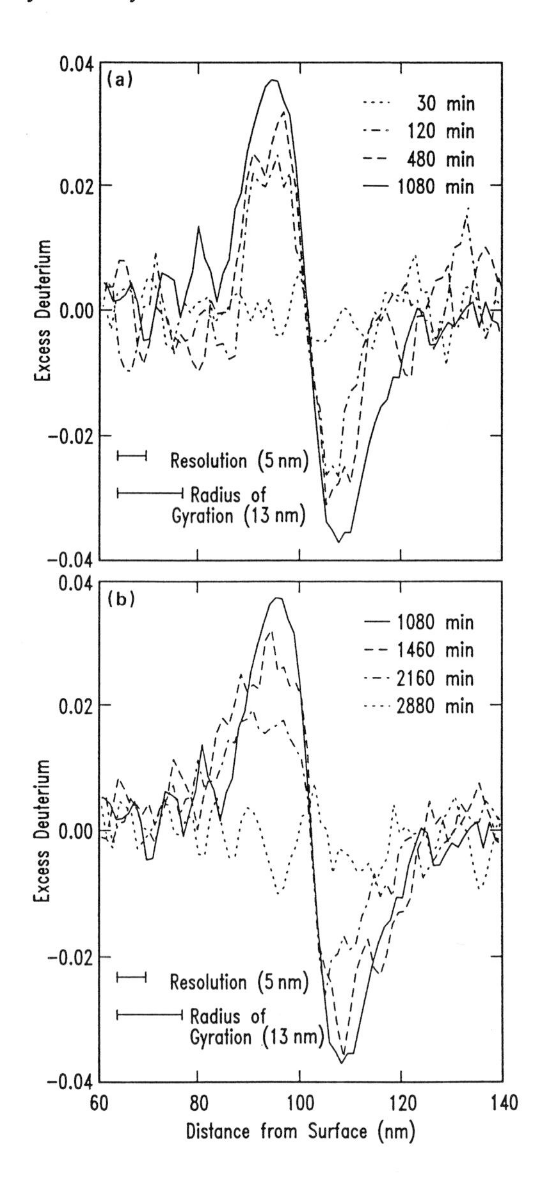

Figure 5.19 SIMS analysis of the ripple experiment. Excess deuterium depth profiles for the HDH/DHD bilayer annealed at 118 °C; (a) shows the ripple increasing from 30 min to 1080 min; (b) shows the ripple amplitude decreasing with increasing time from 1080 min to 2880 min [16, 69].

1,080 minutes and had vanished by 2,880 minutes, which corresponds well to the calculated reptation time, T_r, of about 2,200 minutes.

To establish reptation as the sole reason for the observed ripple, we needed to show that the ripple effect does not occur for unentangled chains at M less than M_c, where Rouse motion is expected to dominate. We found this to be the case, with both SIMS and neutron reflection analysis of HDH/DHD interfaces [17]. In fact, if the reptation hypothesis is correct, there exists a "visibility" molecular weight limit of order $4M_e$, below which we do not see the ripple in HDH/DHD experiments. The labeled chain ends must have a radius of gyration greater than that of the entanglement molecular weight M_e, in order to be distinguishable from the chain centers. For PS, $4M_e \approx 70{,}000$ and it is significant that we did not see a ripple with samples of molecular weight 50,000. We also need to demonstrate that other dynamics models do not produce ripples of a similar magnitude. Rouse dynamics produces very low amplitude propagating ripples [69] that are readily distinguishable from reptation ripples. Neutron reflection, with a resolution of about 5–10 Å (about an order of magnitude better than SIMS) was used to quantify the ripple profiles (Chapter 6) and provided a better comparison between theory, computer simulation, and experiment.

5.7.3 Comment on Reptation Dynamics and Rheology

The SIMS experiments above and the neutron reflection experiments in the next chapter provide compelling evidence for the validity of the reptation dynamics model. The experiments show that the chains tend to move in a snakelike manner and manifest the correlated motion dynamics that are concomitant with the constrained behavior of a chain relaxing in a tube. For our purposes of elucidating the development of structure at interfaces, the reptation model is sufficient. However, these experiments, while supporting reptation, do not automatically extend support to all other theories built on reptation, which may contain additional assumptions.

The Doi–Edwards theory of polymer rheology requires additional assumptions that relate polymer dynamics to mechanical properties via constitutive relations. The derivation of the stress tensor and constitutive laws has been quite controversial. According to Curtiss and Bird [70–72], shortcomings of the Doi–Edwards constitutive model persist and involve (a) an inadequate description of viscosity versus shear rate, (b) an inability to describe the "rod climbing" effect, (c) a poor description of the molecular weight dependence of the zero-shear compliance for polydisperse fluids, (d) lack of a description of recoil after cessation of steady-state shear flows, and (e) other issues, such as temperature profiles in nonisothermal flow. They propose that these rheological problems are better addressed by the Curtiss–Bird *phase-space theory*. This kinetic theory was developed using Kirkwood–Irving statistical mechanics of irreversible processes and provides constitutive laws that describe a wide range of rheological phenomena. Curtiss and Bird's major criticism is that the Doi–Edwards theory has not taken into account the Stokes-law type of expression in describing polymer dynamics; furthermore, Doi and Edwards have not given an appropriate description of the stress tensor starting from the phase-space description of the polymer fluid. The latter approach would have been more internally consistent in terms of the kinetic theory and would permit one to determine in a more fundamental manner where assumptions were being made.

Curtiss and Bird claim that modifications to the original Doi–Edwards theory (involving chain-end fluctuation, nematic orientation coupling, constraint release, etc.) may provide "cut and paste" remedies but do not address fundamental problems and may even deviate from the original reptation model. This criticism, based on rheological data, does not invalidate the reptation model, as many had maintained, but perhaps points to the need for a better constitutive formulation. We would expect the new theory to encompass reptation and also explain the many rheological phenomena not addressed by the Doi–Edwards theory that cause it to be currently considered deficient. In fact, R. J. J. Jongschaap [72] (University Twente, Netherlands) reexamined the Doi–Edwards theory with reptation, and, paying particular attention to the stress tensor formulation, ended up with the Curtiss–Bird-type constitutive equation (epsilon = ½), rather than that proposed by Doi and Edwards. Thus, it seems that reptation can coexist within the Curtiss–Bird framework of polymer rheology when the constitutive equation is derived appropriately. While the controversy about constitutive laws may be mitigated to some extent by these developments, there may still exist a gap in our fundamental understanding relating dynamics of complex structures to mechanical properties, possibly involving connectivity.

5.8 Miscibility of DPS in HPS

With high-molecular-weight HPS/DPS systems, Bates *et al.* demonstrated that phase separation can occur if the Flory–Huggins interaction parameter $\chi(T)$ is greater than χ_{cr} [28]. Since our molecular weights are high, the potential exists for both phase separation and thermodynamic slowing down of diffusing chains. With equal molecular weights (both 1,000,000), $\chi_{cr} = 2/N$, and since $N = 10^6/104 \approx 10^4$, $\chi_{cr} = 2.0 \times 10^{-4}$. The temperature dependence of the Flory–Huggins interaction parameters is given by Bates *et al.* [28] as

$$\chi(T) = 0.2/T - 2.9 \times 10^{-4} \qquad (5.8.1)$$

At $T = 419.2$ K (146 °C), $\chi(146) = 1.87 \times 10^{-4}$. Since χ(T) is less than χ_{cr}, we expect that DPS is miscible in HPS under these conditions. This expectation is supported by the experimental results where diffusion at distances greater than the radius of gyration of the polymers was observed by SIMS. However, we should be concerned with thermodynamic slowing down (TSD), since Eq 5.8.1 considers only the equilibrium miscibility criterion. The molecular weights used by Steiner *et al.* [62] were such that $\chi(T)$ was greater than χ_{cr} at several temperatures, including 140 °C, and the TSD effect was observed at diffusion distances of the order of 1000 Å.

The experimental design used by Whitlow and Wool was different from that used by Green and Doyle [30], who made their measurements using FRES and reported the

TSD effect. Green and Doyle's approach utilized bilayer samples with slightly different compositions (10%), and they assumed that D was independent of composition for their analysis. Their results are shown in Figure 5.20; it can be seen that the diffusion coefficient decreases at the critical concentration, c_{cr}, of 0.5, which is indicative of the TSD effect. The decrease in D is more pronounced with decreasing temperature. The mutual diffusion coefficient measured at c_{cr}, $D(c_{cr})$, is deduced from Eqs 5.2.13–5.2.16 as

$$D(c_{cr}) = D^* [1 - \chi(T)/\chi_c] \tag{5.8.2}$$

where D^* is the self-diffusion coefficient in the absence of TSD. For our SIMS experiments with M of 10^6 and T of 146 °C, $D/D^* = 0.065$ and we could expect a large TSD effect.

Our results are shown in Figure 5.21 for M of 10^6 and welding temperatures of 142–170 °C at times t greater than T_r. These plots are representative of those at the other temperatures at which we conducted tests. The large TSD effect that was

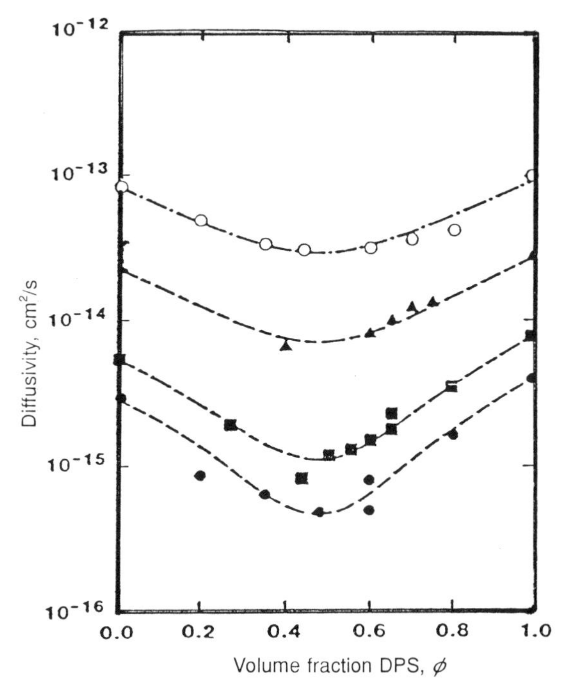

Figure 5.20 FRES results for diffusion coefficients as a function of composition (data of Green and Doyle). Data were taken at • 166 °C, ▪ 174 °C, ▲ 190 °C, and ○ 205 °C. Thermodynamic slowing down becomes more significant as the temperature decreases towards the upper critical solution temperature (UCST) at critical concentrations of 0.5 [30].

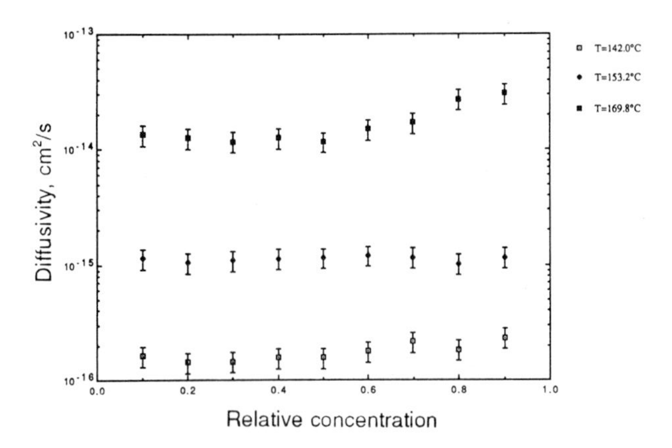

Figure 5.21 Diffusion coefficients from 929K(HPS)/1082K(DPS) as a function of relative concentration at three temperatures, for the investigation of thermodynamic slowing down (TSD) at the critical concentration, *c*, of 0.49.

anticipated in the vicinity of c_{cr} (Eq 5.8.13) was not observed, but is expected to occur at longer diffusion distances; the distances explored by the SIMS method in this work were too short for the composition fluctuations to develop and slow the interdiffusion. The enhanced diffusion due to segmental motion (Figure 5.13) at t less than T_r further supports our conclusions. Neutron reflection analysis of HPS/DPS blend interfaces by Karim, Felcher, and co-workers also failed to detect the TSD effect for deuterated/protonated blends of PS with varying compositions [65, 66]. Again, the depth range (x less than 500 Å) in their study was comparable to that used in our SIMS study. The absence of a TSD effect in both the SIMS and neutron reflection studies is most fortunate for the designing and interpreting of short-range interdiffusion data.

The data of Steiner *et al.* [62] (discussed in Section 5.6.6) show a clear TSD at long times and at interface widths on the order of 1000 Å in the range 140–160 °C. Since the depth resolution was about 300 Å, it is difficult to be certain of results at depths much less than $R_g \approx 300$ Å. However, they report deconvoluted interface widths of about 300 Å after annealing times at 140 °C of about 300 minutes, which agree with our data. The upper plateau of the interface width was at about 800, 1000, and 1,200 Å at 150, 160, and 170 °C, respectively. They did not attain a plateau at 140 °C.

5.9 Summary of SIMS Analysis

In this chapter, we outlined the details of the SIMS method to measure interdiffusion at symmetric polystyrene HPS/DPS interfaces. We make the following points in summary:

1. The SIMS method yields concentration profiles with a depth resolution of about 100 Å.
2. The SIMS method permitted relatively easy measurement of the self-diffusion coefficient, D, with minimal data processing. At long times, the raw SIMS data can be directly analyzed to obtain D values. At short times and distances (x less than R_g), deconvolution procedures improved the quality of the data.
3. We find that the molecular weight dependence of the self-diffusion coefficient scales as $D \sim M^{-2}$, in agreement with other methods, particularly FRES. The temperature dependence of D is best described by the Vogel analysis.
4. At short times, the apparent self-diffusion coefficient is greater than the long-time equilibrium value. This result can be interpreted as the segmental motion contribution to monomer interdiffusion, since little center-of-mass motion occurs at distances less than the radius of gyration.
5. The short-time results for both the number of monomers crossed, $N(t) \sim t^{3/4}$, and the average monomer interpenetration distance, $X(t) \sim t^{1/4}$, are in good agreement with the reptation scaling law predictions for the static and dynamic properties of polymer melts.
6. We investigated the surface segregation of chain ends through the relationship between N and X. We found that $N \sim X^3$ for high-molecular-weight polymers, indicating little or no surface segregation effects on interdiffusion.
7. With matching HDH/DHD interfaces we could observe the snakelike motion of reptating chains directly. The ripple experiment provided convincing proof of the minor chain reptation model of interdiffusion at symmetric polymer interfaces.
8. We did not observe the thermodynamic slowing down due to the deuterium isotope effect in the SIMS experiment in the depth range explored, x less than 500 Å.

5.10 References

1. P.-G. de Gennes, *J. Chem. Phys.* ***55***, 572 (1971).
2. M. Doi and S. F. Edwards, *J. Chem. Soc., Faraday Trans. II* ***74***, 1789 (1978).
3. M. Doi and S. F. Edwards, *J. Chem. Soc., Faraday Trans. II* ***74***, 1802 (1978).
4. M. Doi and S. F. Edwards, *J. Chem. Soc., Faraday Trans. II* ***74***, 1818 (1978).
5. M. V. Tirrell, and E. E. Parsonage, "Polymer Surfaces and Interfaces with Other Materials"; Chapter 14 in *Structure and Properties of Polymers*, E. L. Thomas, Ed.; Vol. 12 in series *Materials Science and Technology*, R. Cahn, P. Haasen, and E. J. Kramer, Eds.; VCH Publisher, Weinheim, Germany; 1993.
6. P. F. Green, P. J. Mills, C. J. Palmstrom, J. W. Mayer, and E. J. Kramer, *Phys. Rev. Lett.* ***53***, 2145 (1984).
7. P. J. Mills, P. F. Green, C. J. Palmstrom, J. W. Mayer, and E. J. Kramer, *Appl. Phys. Lett.* ***45***, 957 (1984).
8. J. Klein and B. J. Briscoe, *Proc. R. Soc. London, A* ***365***, 53 (1979).
9. J. Klein, D. Fletcher, and L. J. Fetters, *Nature (London)* ***304***, 526 (1983).
10. C. R. Bartels, W. W. Graessley, and B. J. Crist, *J. Polym. Sci., Polym. Lett. Ed.* ***21***, 495 (1983).

11. G. C. Summerfield and R. Ullman, *Macromolecules* ***20***, 401 (1987).
12. T. P. Russell, A. Karim, A. Mansour, and G. P. Felcher, *Macromolecules* ***21***, 1890 (1988); T. P. Russell, private communication: Symposium on Light, X-Ray and Neutron Scattering and Reflection from Polymers, 200th Am. Chem. Soc. Meeting, Washington, DC, Aug. 1990; T. P. Russell, *Mater. Sci. Rep.* ***5***, 171 (1990).
13. R. A. L. Jones, E. J. Kramer, M. H. Rafailovich, J. Sokolov, and S. A. Schwarz, *Phys. Rev. Lett.* ***62***, 280 (1989).
14. R. J. Composto, R. S. Stein, E. J. Kramer, R. A. L. Jones, A. Mansour, A. Karim, G. P. Felcher, *Physica B (Amsterdam)* ***156/157***, 434 (1989).
15. S. J. Whitlow and R. P. Wool, *Macromolecules* ***22***, 2648 (1989); *ibid.* ***24***, 5926 (1991).
16. T. P. Russell, V. R. Deline, W. D. Dozier, G. P. Felcher, G. Agrawal, R. P. Wool, and J. W. Mays, *Nature (London)* ***365***, 235 (1993).
17. G. Agrawal, R. P. Wool, W. D. Dozier, G. P. Felcher, T. P. Russell, and J. W. Mays, *Macromolecules* ***27*** (15), 4407 (1994).
18. Y.-H. Kim and R. P. Wool, *Macromolecules* ***16***, 1115 (1983).
19. R. P. Wool, B.-L. Yuan, and O. J. McGarel, *Polym. Eng. Sci.* ***29***, 1340 (1989).
20. R. P. Wool and J. M. Long, *Macromolecules* ***26***, 5227 (1993).
21. H. Zhang and R. P. Wool, *Macromolecules* ***22***, 3018 (1989).
22. R. E. Reed-Hill, *Physical Metallurgy Principles*; Van Nostrand, New York; 2nd ed., 1973.
23. C. Matano, *Jpn. J. Phys.* ***8***, 109 (1932).
24. E. J. Kramer, P. F. Green, and C. J. Palmstrom, *Polymer* ***25***, 473 (1983).
25. W. Wan and S. L. Whittenburg, *Macromolecules* ***19***, 925 (1986).
26. D. B. Kline and R. P. Wool, *Polym. Eng. Sci.* ***28***, 52 (1988).
27. P. F. Green and E. J. Kramer, *J. Mater. Res.* ***1***, 202 (1986).
28. F. S. Bates, G. D. Wignall, W. C. Koehler, *Phys. Rev. Lett.* ***55***, 2425 (1985).
29. F. S. Bates and G. D. Wignall, *Macromolecules* ***19***, 934 (1986).
30. P. F. Green and B. L. Doyle, *Phys. Rev. Lett.* ***57***, 2407 (1986); P. F. Green and B. L. Doyle, *Macromolecules* ***20***, 2471 (1986).
31. W. Katz and J. G. Newman, *MRS Bull.* ***12*** (6), 40 (1987).
32. C. W. Magee and R. E. Honig, *SIA, Surf. Interface Anal.* ***4*** (2), 35 (1982).
33. A. Benninghoven, F. G. Rüdenaur, H. W. Werner, Eds., *Secondary Ion Mass Spectrometry*; Vol. 86 in series *Chemical Analysis*, P. J. Elving and J. D. Winefordner, Eds., John Wiley & Sons, New York; 1987.
34. H.-M. Tong and L. T. Nguyen, Eds., *New Characterization Techniques for Thin Polymer Films*; Wiley–Interscience, New York; 1990.
35. P. F. Green and B. L. Doyle, "Ion Beam Analysis of Thin Polymer Films"; Chapter 6 in *New Characterization Techniques for Thin Polymer Films*, H.-M. Tong and L. T. Nguyen, Eds.; Wiley–Interscience, New York; 1990.
36. P. Sigmund, "Fundamentals of Sputtering"; chapter in *Secondary Ion Mass Spectrometry SIMS IV*, A. Benninghoven, J. Okano, R. Shimizu, H. W. Werner, Eds.; No. 36 in *Springer Series in Chemical Physics*; Springer-Verlag, Berlin; 1984; p 1.
37. P. Williams, "Secondary Ion Mass Spectrometry"; Chapter 7 in *Condensed Matter*, S. Datz, Ed.; Vol. 4 in series *Applied Atomic Collision Physics*, H. S. W. Massey, E. W. McDaniel, and B. Bederson, Eds.; Academic Press, Orlando, FL; 1983; p 327.
38. J. P. Biersack, "Ion Ranges and Energy Deposition in Insulators"; chapter in *Ion Beam Modification of Insulators*, P. Mazzoldi and G. W. Arnold, Eds.; Vol. 2 in series *Beam Modification of Materials*, Elsevier: Amsterdam; 1987; p 1.
39. S. Hofmann, *SIA, Surf. Interface Anal.* ***4*** (2), 148 (1982); *ibid.* ***2*** (2), 56 (1980).

40. D. Briggs and A. B. Wooton, *SIA, Surf. Interface Anal.* ***4***, 109 (1982); D. Briggs, *SIA, Surf. Interface Anal.* ***9***, 391 (1986).
41. R. Chujo, T. Nishi, Y. Sumi, T. Adachi, H. Naito, and H. Frentzel, *J. Polym. Sci., Polym. Lett. Ed.* ***21***, 487 (1983); R. Chujo *et al.*, *Bull. Inst. Chem. Res. Kyoto Univ.* ***66***, 312 (1988).
42. S. J. Valenty, J. J. Chera, D. R. Olson, K. K. Webb, G. A. Smith, and W. Katz, *J. Am. Chem. Soc.* ***106***, 6155 (1984).
43. H. H. Anderson, *Appl. Phys.* ***18***, 131 (1979).
44. S. Hofmann; chapter in *Secondary Ion Mass Spectrometry SIMS III*, A. Benninghoven, J. Giber, J. Lászl6, M. Riedel, and H. W. Werner, Eds.; No. 19 in Springer Series in Chemical Physics, Springer-Verlag, Berlin; 1982; p 186.
45. P. Sigmund and A. Gras-Marti, *Nucl. Instrum. Methods* ***168***, 389 (1980).
46. W. O. Hofer, W. O. and U. Littmark; chapter in *Secondary Ion Mass Spectrometry SIMS III*, A. Benninghoven, J. Giber, J. Lászl6, M. Riedel, and H. W. Werner, Eds.; No. 19 in Springer Series in Chemical Physics, Springer-Verlag, Berlin; 1982; p 201.
47. K. Wittmaack, *Nucl. Instrum. Methods* ***B7/8***, 750 (1985).
48. P. Williams and J. E. Baker, *Nucl. Instrum. Methods* ***182/183***, 15 (1981).
49. M. Tirrell, *Rubber Chem. Technol.* ***57***, 523 (1984).
50. H. H. Kausch and M. Tirrell, *Annu. Rev. Mater. Sci.* ***19***, 341 (1989).
51. K. Wittmaack, *Nucl. Instrum. Methods* ***168***, 343 (1980).
52. P. M. Hall, J. M. Morabito, and N. T. Panousis, *Thin Solid Films* ***41***, 341 (1977).
53. S. Hofmann and J. M. Sanz, "Depth Resolution and Quantitative Evaluation of AES Sputtering Profiles"; Chapter 7 in *Thin Film and Depth Profile Analysis*, H. Oechsner, Ed.; No. 37 in series *Topics in Current Physics*; Springer-Verlag, Berlin; 1984; p 141.
54. A. F. Carley and R. W. Joyner, *J. Electron Spectrosc. Relat. Phenom.* ***16***, 1 (1979).
55. P. H. van Crittert, *Z. Phys. (Engl. Transl.)* ***69***, 298 (1931).
56. G. K. Wertheim, *J. Electron Spectrosc. Relat. Phenom.* ***6***, 239 (1975).
57. H. H. Madden and J. E. Houston, *J. Appl. Phys.* ***47***, 3071 (1976).
58. P. S. Ho and J. E. Lewis, *Surf. Sci.* ***55***, 335 (1976).
59. W. W. Graessley, "Entangled Linear, Branched and Network Polymer Systems—Molecular Theories"; chapter in *Synthesis and Degradation; Rheology and Extrusion*; Vol. 47 in series *Advances in Polymer Science*, H.-J. Cantow *et al.*, Eds.; Springer-Verlag, Berlin; 1982; p 67.
60. P. F. Green and E. J. Kramer, *Macromolecules* ***19***, 1108 (1986).
61. P. F. Green, *Ion Beam Analysis of Diffusion in Polymer Melts*; Ph.D. Thesis, Cornell University, Ithaca, NY; 1985.
62. U. Steiner, G. Krausch, G. Schatz, and J. Klein, *Phys. Rev. Lett.* ***64***, 1119 (1990).
63. H. Zhang and R. P. Wool, "Concentration Profiles at Amorphous Symmetric Polymer/ Polymer Interfaces"; *Polym. Prepr. (Am. Chem. Soc., Div. Polym. Chem.)* ***31*** (2), 511 (1990).
64. P.-G. de Gennes, *C. R. Acad. Sci., Ser. 2*, ***307***, 1841 (1988).
65. A. Karim, A. Mansour, G. P. Felcher, and T. P. Russell, *Phys. Rev. B: Condens. Matter* ***42***, 6846 (1990).
66. A. Karim, *Neutron Reflection Studies of Interdiffusion in Polymers*; Ph.D. Thesis, Northwestern University, Evanston, IL; 1990.
67. G. Agrawal, R. P. Wool, W. D. Dozier, G. P. Felcher, T. P. Russell, and J. W. Mays, "Short Time Interdiffusion at Polymer Interfaces: Reptation?", *Bull. Am. Phys. Soc.*, ***38***, Mar, 659 (1993).

68. G. Agrawal, R. P. Wool, W. D. Dozier, G. P. Felcher, T. P. Russell, and J. W. Mays, results on HDH/DHD interfaces to be published.
69. G. Agrawal, *Short Time Interdiffusion at Polymer Interfaces: A Probe to Study Polymer Motion*; Ph.D. Thesis, University of Illinois, Urbana, IL; 1994.
70. C. F. Curtiss and R. B. Bird, *J. Chem. Phys.* ***74***, 2016 (1981).
71. R. B. Bird, R. C. Armstrong, and O. Hassager, *Dynamics of Polymer Liquids*, John Wiley & Sons, New York; 2nd ed., 1987.
72. R. B. Bird, private communication, 1993.

6 Neutron Reflection Analysis of Polymer Interfaces*

6.1 Introduction

Neutron reflection is a surface analysis technique whose application to polymers has been quite recent. Because of the excellent reflection contrast of deuterated versus protonated polymers in low-grazing-angle reflection, this method is ideally suited to examining the fine structure of polymer interfaces. The excellent depth resolution (about 10 Å) allows us to quantitatively examine both Rouse and reptation contributions to the development of structure during short-range interdiffusion at amorphous polymer interfaces. In this chapter we examine the structure of symmetric (equal molecular weights) and asymmetric (different molecular weights) HPS/DPS polystyrene interfaces. Partially deuterated polymers labeled in the center (HDH) or ends (DHD) are used to examine the microscopic details of the segmental motion at HDH/HPS, DHD/HPS, and HDH/DHD interfaces. We include a neutron reflection analysis of the ripple experiment at HDH/DHD interfaces and explore further how the chains move in a snakelike manner at the interface.

6.1.1 Principles of Neutron Reflectivity

Specular neutron reflection (SNR) studies of polymer interfaces have been conducted in recent years by Karim, Felcher, and co-workers [1–4], Wool and co-workers [5–8], Stamm *et al.* [9], Steiner and co-workers [10, 11], Russell and co-workers [12–15], Kramer and co-workers [16–18], and others [19, 20]. SNR is a powerful technique for studying interdiffusion at amorphous polymer interfaces because it offers high resolution, of the order of 10 Å, and gives a strong optical contrast between hydrogen and deuterium. The method for investigating polymer interfaces using neutron reflection is illustrated in Figure 6.1 for an HDH/DHD interface. A highly collimated beam of pulsed neutrons is incident at low grazing angles (θ of about 1°) on an HDH/DHD film pair on a flat silicon (Si) substrate. At such low grazing angles, the neutron beam probes the entire surface. The incident beam is partially reflected from several surface and internal interfaces, namely, air/HDH, HDH/DHD, and DHD/Si.

* Dedicated to G. Felcher

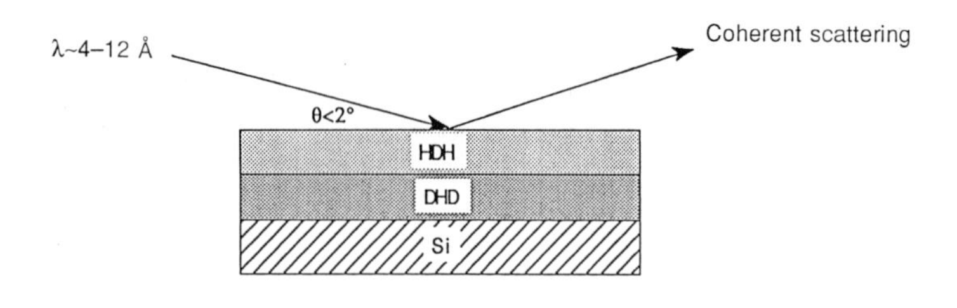

Figure 6.1 Schematic representation of specular reflectivity of neutrons from an HDH/DHD interface.

If the bottom polymer layer is made much thicker than the top layer, the reflections are primarily due to air/HDH and HDH/DHD. The methods and principles of neutron and X-ray reflection at polymer interfaces have been reviewed by T. P. Russell [12].

A few comments should give the reader a grasp of the method. Radiation incident on an interface composed of materials with different refractive index n is totally reflected if the incident angle is less than a critical angle θ_c. The critical angle is derived from Snell's law for a vacuum/material interface as $\cos \theta_c = n$, where n is the refractive index of the material. If the incident angle is greater than θ_c, the extent to which radiation is reflected from the surface or interface depends on the momentum transfer (in the depth or z direction) in the material on either side of the interface. This is where we gain most of the information in SNR experiments. The incident momentum, k_0, is reduced by an amount k_1 because of the interaction of the radiation with the molecules in the material below the surface. The *Fresnel reflectivity* (*reflectance, reflected intensity*), R, depends on the square of the difference, r. Thus, in general, if we know the relationship between the interaction loss k_1 and the material composition, we can measure the local material composition by measuring R.

The interaction of neutrons with an atomic nucleus can be described by the interaction potential $V_0(z)$,

$$V_0(z) = (2\pi h^2/m)\, N'(z)\, b(z) \tag{6.1.1}$$

where h is Planck's constant, m is the neutron mass, and $b(z)$ and $N'(z)$ are the average scattering amplitude and atom density at the depth z, respectively [20, 21]. For neutrons with kinetic energy E, the interaction potential is related to the material's refractive index $n(z)$ by [22–25]

$$n(z) = [1 - V_0(z)/E]^{1/2} \tag{6.1.2}$$

E can be expressed in terms of the de Broglie's wavelength α, as

$$E = h^2/(2\,m\alpha^2) \tag{6.1.3}$$

Although the nuclear force is attractive, $V_0(z)/E$ is always positive, resulting in $n(z)$ values smaller than unity. Consequently, the neutrons may be totally or partially reflected at a boundary between two media with different refractive indices, depending on the incident angle θ, as shown in Figure 6.1.

For the simplest case, in which there is only a single sharp interface between vacuum and a homogeneous material (1) at $z = 0$, the reflectivity R has been calculated as a function of the incident momentum $[k_0 = (2/\alpha)\sin\theta]$ and the scattering factor $(4\pi b_1 N_1)$ in medium 1 as [25]

$$k_1 = (k_0^2 - 4\pi\, b_1 N_1)^{1/2} \tag{6.1.4}$$

where b_1 is the scattering amplitude of medium 1, α is the wavelength, and N_1 is the atom density of the medium [11]. The reflectance coefficient r is defined as $(k_0 - k_1)/(k_0 + k_1)$, and the Fresnel reflectance R is r^2, which can be written out as

$$R = \left\{\frac{1 - [1 - (k_1/k_0)^2]^{1/2}}{1 + [1 - (k_1/k_0)^2]^{1/2}}\right\}^2 \tag{6.1.5}$$

When k_0 is much greater than k_1 (for sharp interfaces), $R \sim (k_1/k_0)^4$; when k_0 is large, Rk_0^4 is customarily plotted against neutron momentum.

The concentration profile at a polymer–polymer interface can be thought of as a collection of sharp slices of varying refractive index. For the histogram of layers of different thicknesses and different refractive indices, the reflectance coefficient at the boundary between the ith and $(i+1)^{\text{th}}$ layers has been given as [24]

$$r_{i,i+1} = (k_{1,i} - k_{1,i+1})/(k_{1,i} + k_{1,i+1}) \tag{6.1.6}$$

and the total reflectance of this system is a suitable combination of the products of the reflectivities at the single boundaries.

For the interface systems discussed in this chapter, shown in Figure 6.1, the reflectivity can be calculated as [26]

$$R \approx \frac{r_{0,1}^2 + r_{1,2}^2 + 2\,r_{0,1}\,r_{1,2}\cos(2\,k_1\,d_1)}{1 + r_{0,1}^2 + r_{1,2}^2 + 2\,r_{0,1}\,r_{1,2}\cos(2\,k_1\,d_1)} \tag{6.1.7}$$

where d_1 is the thickness of the upper (HDH) layer, and $r_{0,1}$ and $r_{1,2}$ are the reflectance at the boundary vacuum/HDH and the boundary HDH/DHD, respectively [26]. When interdiffusion takes place at an HDH/DHD interface, it is necessary to calculate the scattering length profile per unit volume, denoted by b/v, via the monomer concentration profile $C(z,t)$.

The scattering amplitude, b/v, for deuterium is about 4.5 times that for hydrogen, and provides excellent contrast at interfaces. For the mixture formed during interdiffusion at an HPS/DPS interface, b/v can be calculated as a function of depth z and welding time t by the rule of mixtures [11, 25], as

$$(b/v)_{\text{mixture}} = C_{\text{HPS}}(z,t)\,(b/v)_{\text{HPS}} + C_{\text{DPS}}(z,t)\,(b/v)_{\text{DPS}} \tag{6.1.8}$$

where $C_{\text{HPS}}(z,t)$ and $C_{\text{DPS}}(z,t)$ are the reduced monomer concentration for HPS and DPS, respectively. The concentrations are the same as the volume fractions, and $(b/v)_{\text{HPS}}$ and $(b/v)_{\text{DPS}}$ are their respective scattering lengths per unit volume. The values of $(b/v)_{\text{HPS}}$ of 1.43×10^{-10} cm^{-2} and $(b/v)_{\text{DPS}}$ of 6.5×10^{-10} cm^{-2} are used in the next section. For fractionally deuterated chains, such as HDH, b/v is determined from the averaged sum of the components.

6.2 Analysis of Symmetric Interfaces

6.2.1 Concentration Profiles for a Symmetric Interface

The concentration profile $C(z,t)$ for the symmetric polymer interface has two components, $C_{\text{seg}}(z,t)$ contributed by segmental motion from all the segments of a chain, and $C_{\text{rep}}(z,t)$ from reptation, as presented in Chapter 3. Taking the sum of the two components weighted with respect to the contour length gives a useful approximation for the complete profile,

$$C(z,t) = [1 - 2l(t)/L]\, C_{\text{seg}}(z,t) + C_{\text{rep}}(z,t) \tag{6.2.1}$$

where the prefactor $[1 - 2\,l(t)/L]$ is the fraction of the middle portion of a chain that is still trapped in its initial tube at time t. For times greater than τ_{RO}, the concentration profile can be approximately written as

$$\begin{aligned} C(z,t) = {} & 0.5[1 - 2l(t)/L]\,\text{erfc}[z/(X_m)] \qquad \text{for } t > \tau_{RO} \\ & + [l(t)/L + z_2/(aL)]\,\text{erfc}\{z/[2al(t)]^{1/2}\} \\ & - [2l(t)/\pi a]^{1/2}\,(x/L)\exp\{1 - z_2/[2al(t)]\} \end{aligned} \tag{6.2.2}$$

where $X_m \approx 1.8R_{ge}(M/M_e)^{1/4}$, and R_{ge} is the radius of gyration of the entanglement molecular weight M_e.

The reflectivity for the diffused system can be calculated as follows. The concentration profile is first sliced into n strips (on the computer) and the reduced monomer concentration as well as the refractive index for each strip is calculated under the condition of equal spacing of refractive index or, correspondingly, equal spacing of monomer content. From the indices calculated from the concentrations and scattering volumes, the reflectance is calculated first for the silicon/n^{th} slice and then for the nth slice/$(n-1)^{th}$ slice, and so forth. The last calculation is done for the first slice/vacuum. The combination of all the individual slices gives the calculated reflectivity, R(calc), which is then compared with its measured value, R(expt). If the predicted profile $C(z,t)$ accurately represents the actual profile, we expect that R(expt) and R(calc) are equal at all values of neutron momentum.

The neutron reflectivity measurements were performed using the reflectometer POSY-2, which is located at the Intense Pulsed Neutron Source at Argonne National Laboratory and has been discussed in detail by Felcher [25].

6.2.2 Welding and Neutron Reflection Analysis of HPS/DPS Interfaces

The bilayer HPS/DPS sample consisted of a bottom layer of protonated polystyrene (HPS) (M_w = 650,000, M_w/M_n = 1.04), and a top layer of deuterated polystyrene (DPS) (M_w = 708,000, M_w/M_n = 1.03) [5]. The bilayer on a silicon substrate was welded at 125 °C according to methods described by Zhang, Agrawal, and Wool [5–8, 30]. In order to design the welding experiment [27], we needed to estimate the three characteristic times, the Rouse time τ_{RO}, the tube equilibrium time τ_e, and the reptation time T_r. We therefore used the following relations:

$$\tau_e = 27\pi^2 \tau_{RO}^3 / T_r^2 \tag{6.2.3}$$

$$T_r = R^2 / 3\pi^2 D^* \tag{6.2.4}$$

$$\tau_{RO} = T_r / 3(M/M_e) \tag{6.2.5}$$

We used Green and Kramer's data (in Chapter 5) measured for M of 255,000 at 125 °C, and the relation $D \sim M^{-2}$, to obtain T_r = 1850 min, τ_{RO} = 17 min, and τ_e = 23 s for our polystyrene sample with M of 650,000.

For welding times less than the reptation time, we used the concentration profile in Eq 6.2.2 to calculate the reflectivity, Rk^4. Figure 6.2 shows the results for the as-cast sample, that is, without further heat treatment. The dashed line was calculated for t of 1 second and D_0 of 5 Å^2/s. The measured reflectivity (plotted dots) is compared with the calculated (dashed) line on the plot as a function of neutron momentum. Note that no adjustable parameters are needed once we fit the data at $t \approx 0$ in Figure 6.2. We then weld for different times to start the interdiffusion, and compare the measured reflectance profile with that calculated using Eq 6.2.4, as shown in Figure 6.3.

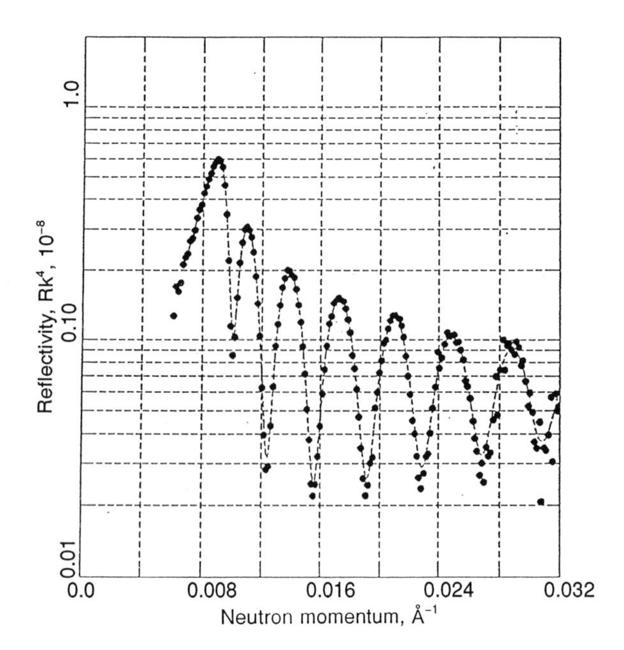

Figure 6.2 The measured reflectivity (plotted as dots) for the 650K(HPS)/708K(DPS) interface welded at 125 °C. The dashed line was calculated for t = 1 s (Zhang *et al.*).

At τ_e of 23 seconds, the interface thickness is $X(\tau_e) \approx 0.8\, R_{ge} \approx 30$ Å. Entanglement segments from all parts of the chain contribute to this diffusion process. Considering the short time span involved, this diffusion process occurs almost instantaneously in our isothermal welding experiment. However, we will see this diffusion depth more readily in the HDH/HPS interface study.

After 10 minutes, which is longer than τ_e but shorter than τ_{RO}, the average diffusion distance is determined by $X(t) \approx 0.8\, R_{ge} N^{1/4}(t/\tau_{RO})^{1/8}$, and since $R_{ge} = 37$ Å, and $N = 650/18 = 36.1$, then $X(10\text{ min}) \approx 68$ Å. Figure 6.3 shows the experimental data compared with the calculated reflectivity in the range $\tau_{RO} < t < T_r$, for welding times (a) 42 minutes, (b) 120 minutes, and (c) 242 minutes. The measured reflectivity data are in good agreement with the calculated reflectivity. As the interface broadens during welding, the reflectivity profiles decrease their amplitudes but maintain constant distances between the peaks. The peak-to-peak distance is related to the top layer thickness, which is typically about 780 Å in these experiments. Figure 6.4 shows representative concentration profiles during welding of the symmetric 650K interface at 480 minutes and 1149 minutes. The gap at the interface plane due to pure reptation has been eliminated by the segmental motion.

The average interdiffusion distance $X(t)$ with segmental motion is determined from Eq 6.2.2 in the range $\tau_{RO} < t < T_r$, by the normalized relation

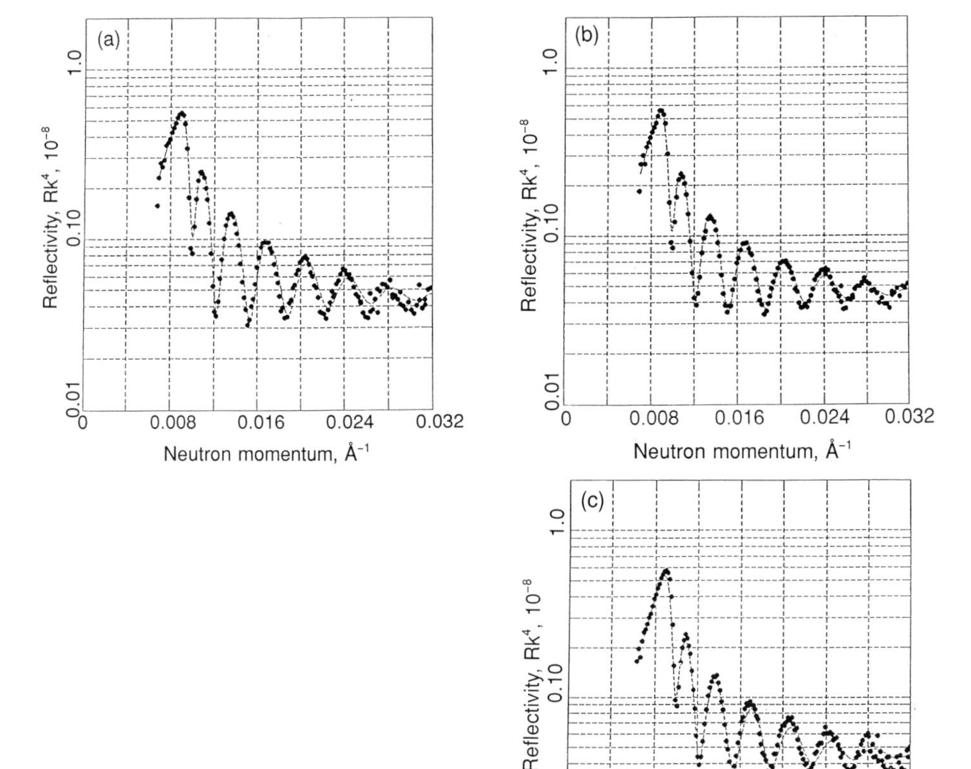

Figure 6.3 The measured reflectivity (plotted as dots) for the 650K(HPS)/708K(DPS) interface welded at 125 °C. The dashed lines were calculated for welding times of (a) 42 min; (b) 120 min; and (c) 242 min (Zhang *et al.*).

$$X(t)/X_{\infty} = \frac{\tau + N^{-1/2}(1-\tau^{1/2})}{\tau^{3/4} + N^{-1/4}(1-\tau^{1/2})} \tag{6.2.6}$$

where $\tau = t/T_r$, $X_{\infty} = 0.81\ R_g$, $N = M/M_e$ and $\tau_{RO} = \frac{1}{3}N$. Figure 6.5 shows the behavior of $X(t)/X_{\infty}$ versus t/T_r for M = 650,000 and M = 1,000,000. Note that because of segmental motion, the slope is very small at early times but it eventually approaches ¼ at longer times.

The above results are in agreement with the symmetric HPS/DPS studies of Karim, Felcher, and co-workers [1–4]. Figure 6.6 shows a plot of the interface thickness $X(t) = \langle z^2 \rangle^{1/2}$ as a function of time for HPS/DPS interfaces with $M \approx 200{,}000$. The data were obtained at different temperatures and shifted to a reference T of 120 °C using the William–Landel–Ferry (WLF) equation (Section 8.7.3) for the shift factor a_T

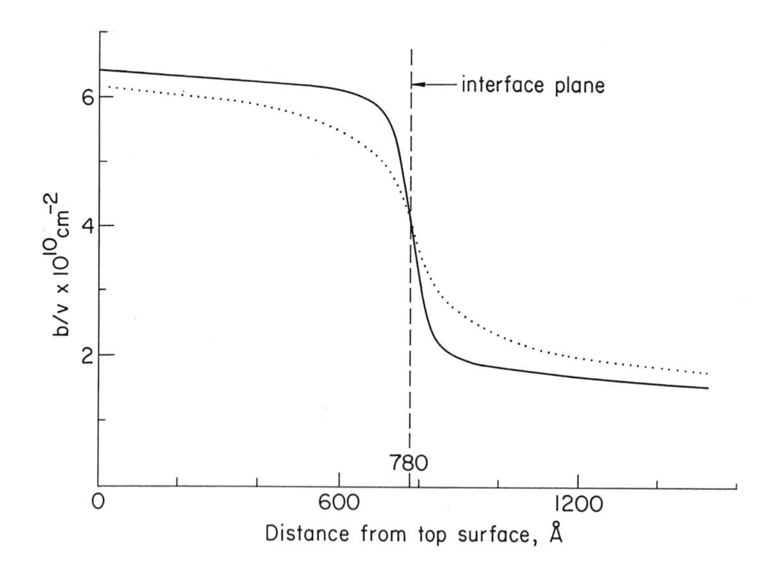

Figure 6.4 The calculated refractive index profile used to evaluate the reflectivity experiments for welding of the 650K polystyrene interface at t = 480 min (solid line) and 1149 min (dotted line) at 125 °C (Zhang *et al.*).

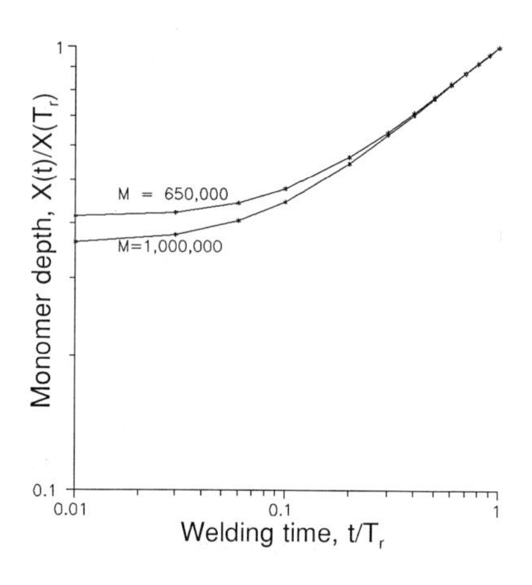

Figure 6.5 The normalized monomer diffusion depth $X(t)/X_\infty$ versus t/T_r for HPS/DPS interfaces with molecular weights of 650,000 and 1,000,000.

$$\log_{10} a_T = -9.06(T-120)/[69.81+(T-120)] \tag{6.2.7}$$

The data are superimposed and give excellent information on the initial diffusion regions at t much less than T_r. When M is 200K, Eq 6.2.6 predicts that $X(\tau_{RO}) = X_\infty/N^{1/4}$, and with N of 11.11 and X_∞ of about 100 Å, we calculate $X(\tau_{RO}) \approx 55$ Å. The predicted $X(\tau_{RO})$ and X_∞ agree well with Karim's data (Figure 6.6).

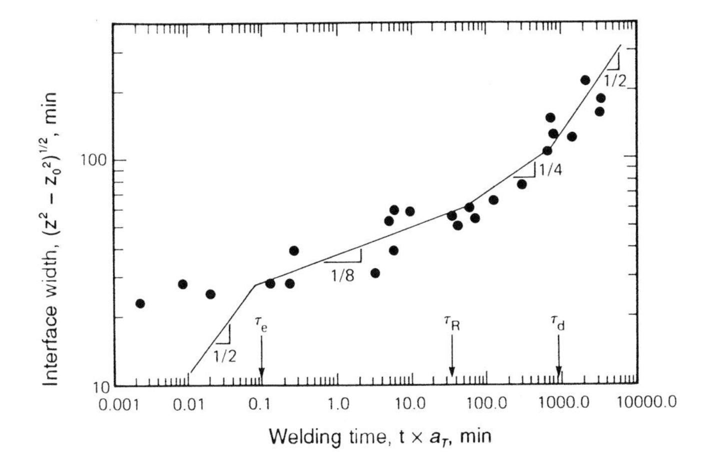

Figure 6.6 A master curve for interdiffusion depth versus time for the symmetric interface with M = 200,000, welded at 120 °C. The entanglement Rouse relaxation time, τ_e, the Rouse relaxation time for the whole molecule, τ_{RO}, and the reptation time, T_r, are indicated. The solid line is the prediction of the reptation theory (courtesy of Karim *et al.*).

The solid line indicates the characteristic slopes for the different dynamic regions according to the Doi–Edwards theory. The data are not sufficient to make direct comparisons with slope transitions from Fickian to Rouse (entanglement), to Rouse (chain), to reptation, and to Fickian as welding time increases. However, as with the SIMS data in Chapter 5, the data are in excellent qualitative agreement with theory and clearly indicate that simple Fickian diffusion cannot describe the data at t less than T_r. The reptation-to-Fickian transition at T_r appears as the strongest change in slope, from $t^{1/4}$ to $t^{1/2}$. However, as discussed in Chapter 3, the bulk dynamic slopes are not cleanly observed at interfaces because of the partition and mixing of Rouse and reptation contributions. Therefore, the $X(t)$ data increase smoothly and monotonically over most of the dynamic regions, as noted in Figure 6.6.

Reiter and Steiner used both neutron reflection and nuclear reaction analysis (NRA) to examine reptation arguments for the 752K(HPS) /660K(DPS) interface welded at 120 °C [10]. SNR was used to examine interdiffusion depths up to 200 Å, and NRA for depths greater than 100 Å. The interface width $w(t)$, corrected for initial broadening, behaved as $w(t) \sim t^{0.24}$ for t less than T_r (SNR) and $w(t) \sim t^{0.53}$ for t greater than T_r (NRA). The transition from $t^{1/4}$ (reptation) to $t^{1/2}$ (Fickian) behavior occurred near T_r and $w(t) \approx 0.8\ R_g$, which is consistent with the study by Karim, Felcher, and co-workers [1–4] and our work. That three different laboratories, using SIMS, SNR, and NRA, should be in such close agreement on their independent measurements of this dynamic transition is reassuring, if not remarkable.

Reiter and Steiner argue [10] (and in private communication G. Reiter has continued to maintain) that the gap at the interface plane decreased with an approximate $t^{1/4}$ dependence, suggesting that the chain ends had initially segregated to the surface. They arrived at this conclusion by subtracting a Rouse contribution from the combined Rouse plus reptation concentration profile. In the nonsegregated case, the gap decreases with a $t^{1/2}$ dependence. They described $C(x,t)$ empirically using the sum of two error functions, with appropriate decay functions for Rouse and reptation dynamics. Since the pure reptation function cannot be described in terms of a single error function, some difficulty may occur with the deconvolution methodology to reveal the gap decay function of $t^{1/4}$ at the interface plane. However, their data analysis supports arguments in favor of chain end segregation.

The apparent self-diffusion coefficient D' versus welding time ($\ln t/T_r$) is shown in Figure 6.7 [1, 2]. When t is less than T_r, we see a large increase in D'. These data are consistent with the SIMS data and can be approximately described by

$$D'/D_\infty = 0.32(T_r/t)^{1/2} \qquad \text{for } t < T_r \qquad (6.2.8)$$

where D_∞ is the time-independent diffusion coefficient at t much greater than T_r. I discuss further the behavior of the diffusion coefficient at t less than T_r in Chapter 5.

6.3 Interfaces with Different Molecular Weights

6.3.1 Introduction to M_1/M_2 Asymmetric Interfaces

Interfaces formed from chemically identical polymers but with different molecular weights (M_1 and M_2) are encountered in blends of different batches of homopolymer, in recycled plastics, and in reacting bilayers, where one layer has a different reaction time. When we have differences in molecular weight, we have unequal fluxes of molecules across the interface. This causes the interface plane to shift by an amount $u(t)$ into the low-molecular-weight side, as shown in Figure 6.8 [28]. The concentration profile $C(x,t)$ with Rouse and reptation dynamics was derived for the asymmetric interface, M_1/M_2, by Zhang and Wool [5, 6, 28] for HPS with M_1 = 4,434,000 and DPS with M_2 = 203,000. The 203K chains diffuse rapidly to the right while the 4,000K chains barely move to the left at t less than T_{rs} (reptation time of short chains).

The position of the interface $u(t)$, as determined by the inflection point of the profile, moves into the lower molecular weight side. In bilayer films in which the low-molecular-weight species is on top, we would observe a decrease in the film thickness, which can be readily measured by neutron reflection. The reptation model predicts that

$$u(t) \sim t^{3/4} \qquad (6.3.1)$$

This time dependence is similar to the monomer diffusion dependence for symmetric interfaces, where $N(t) \sim t^{3/4}M^{-7/4}$, and is subject to the initial location of the chain ends, so that with chain end segregation we expect that $u(t) \sim t^{1/2}$.

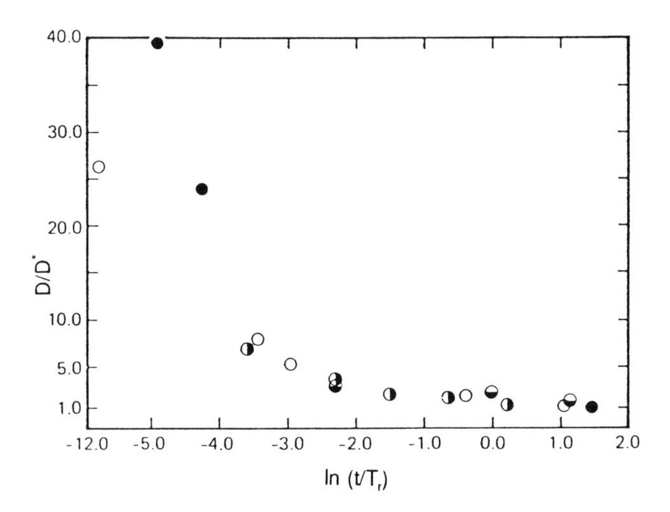

Figure 6.7 The time dependence of the diffusion coefficient D, measured from the reflectivity data, divided by self-diffusion coefficient D^*, of PS (M_w = 233,000) in the bulk at 120 °C as a function of the logarithm of t/T_r (courtesy of Karim *et al.*).

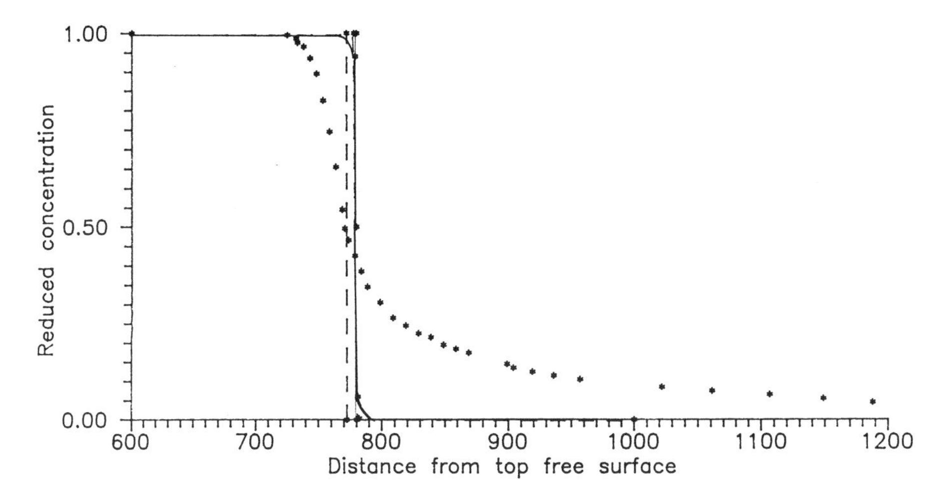

Figure 6.8 The concentration profile for the 4,000K(HPS)/203K(DPS) asymmetric interface welded at 120 °C. The interface position $u(t)$, evaluated at $C(x,t) = 0.5$, moves from right to left into the low-molecular-weight (203K) side, as the uneven flux moves from left to right into the high-molecular-weight (4,000K) side. As the interface moves, the top layer becomes thinner (Zhang *et al.*).

6.3.2 Displacements of Interface Plane, *u*(*t*)

The numbers of the monomers diffused, $N_{HPS}(t)$ and $N_{DPS}(t)$, can be deduced from the concentration profile, shown in Figure 6.8. It is clear that $N_{DPS}(t)$ is larger than $N_{HPS}(t)$, which means that there is a net transport of monomers from the low-molecular-weight side to the high-molecular-weight side. If the initial interface plane has a displacement $u(t)$ relative to the edge(s) of the bilayer film, then the volume of the region confined by the interface plane at $t = 0$ and that at time t is the product of the cross-sectional area, A, of the bilayer film and $u(t)$. Since the density of the film remains constant, there should be $[(\rho/M_0)u(t)A]$ moles of monomers in the region just defined. This number of monomers should be identified with the net transport of monomers, that is,

$$(\rho / M_0)\, u(t)\, A = (\rho / M_0)\, [N_{DPS}(t) - N_{HPS}(t)]\, A \quad (6.3.2)$$

Then we have

$$u(t) = N_{DPS}(t) - N_{HPS}(t) \quad (6.3.3)$$

Considering the reptation contribution only, we have

$$u(t) = [16(2a)^{1/2} / 3\pi^{5/4})\, (Dt)^{3/4} / L \quad (6.3.4)$$

Substituting for Dt, we have

$$u(t) = [(\tfrac{2}{3}\pi)^{1/2}]\ (t / T_r)^{3/4}\, R_g \quad (6.3.5)$$

and

$$u(T_r) = 0.46\, R_g \quad (6.3.6)$$

Considering both the reptation and segmental motion contribution, we have

$$u(t) = (\tfrac{2}{3}\pi)^{1/2}\, (t / T_r)^{3/4}\, R_g - 8^{-1/2}\, (t / T_r)^{1/2}\, R_{ge} \quad (6.3.7)$$

In the next section we explore the relevance of reptation and segmental motion on $u(t)$ with neutron reflection experiments on M_1/M_2 asymmetric bilayers.

6.3.3 Neutron Reflection Measurements on Asymmetric Interfaces

If these measurements are to be successful, the healing temperature and the molecular weights, film thicknesses, and healing times have to be selected properly. If we evaluate the monomer flux from $C(x,t)$ calculations, we see that the ratio of molecular weights $r = M_1/M_2$ must be larger than 10. In the work we report, r was 20; moreover, M_1 and M_2 were selected such that the free energy of mixing was negative and the reptation time of the shorter chain was large enough to make the healing procedure meaningful. For M_2 of 203,000, at 120 °C, T_r is 581 minutes and τ_{RO} is 18 minutes.

Welding was done for 10 minutes ($t < \tau_{RO}$), 73 minutes ($t \approx T_r/8$), 145 minutes ($t \approx T_r/4$), 291 minutes ($t \approx T_r/2$), and 436 minutes ($t \approx 3T_r/4$). The neutron reflection data were calculated from the concentration profiles with Rouse and reptation dynamics. In each SNR figure, we determined the thickness of the DPS film by matching the experimental periods, and the segmental diffusion coefficient D_0 by matching the amplitudes. We found that at t = 10 min, D_0 became a little bit smaller than the initial D_0 value at t = 1 s, but at t > 10 min, D_0 values were constant. This may be related to a resistance to interdiffusion arising from elastic forces produced when longer chains are swollen by shorter chains. We also found that the thickness at t = 10 min was about 10 Å greater than that of the as-cast film. This thickening effect is attributed to relaxation of the biaxial orientation in the DPS films caused by spinning [6].

The thickness of the top DPS layer decreases with welding time in proportion to the movement of the interface plane (Figure 6.8). Figure 6.9 shows log $u(t)$ versus log t plots, where the uppermost line, 1, is calculated using Eq 6.3.5 (reptation dynamics) and has a slope of ¾. Line 2 is calculated using Eq 6.3.7 (Rouse and reptation dynamics) and has a slope a little bit smaller than ¾. It can be seen that the experimental plot (line 3) agrees with the Rouse-plus-reptation prediction. The experimental slope of ¾ provides additional support for the initial uniform chain-end distribution function.

6.4 Selectively Deuterated HDH/HPS Interfaces

In this experiment, we deuterated the center of a PS molecule and observed its motion at an HDH/HPS interface by neutron reflection experiments [5, 29, 30]. The object of this experiment was to examine segmental motion at times of the order of the Rouse time τ_{RO}. This experiment also gave insight into the microscopic details of the chain motion and addressed such questions as the ability of chains to interdiffuse by hairpin formation or "looping" through the interface. The important issue of the snakelike motion of chains could be examined with HDH/HPS (center visible), HDH/DPS (ends visible), or with DHD/DPS (center visible) and DHD/HPS (ends visible). Most of these interfaces were examined by Agrawal and co-workers in collaboration with Felcher and co-workers [30, 31]. In Section 6.4.3, we report that the matching HDH/DHD interface with 50% deuteration gives convincing evidence for reptation.

6.4.1 Experimental Details for the HDH/HPS Interfaces

The top HPS layer (M_w = 233,000, M_w/M_n = 1.06) was obtained from Polymer Laboratories. The bottom HDH layer was made of a centrally deuterated triblock

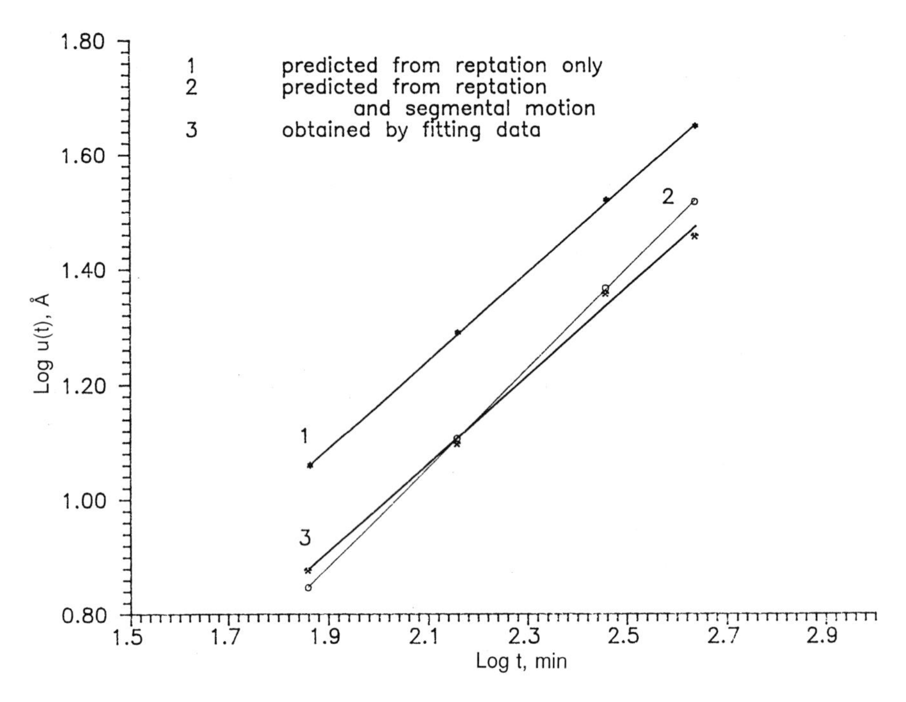

Figure 6.9 The motion of the interface $u(t)$ versus welding time for the asymmetric M_1(203K)/M_2(4,000K) interface. Line 1 represents the pure reptation prediction, with a slope of ¾; line 2 represents the prediction with Rouse plus reptation contributions, with slope of about ¾; line 3 represents the experimental data, with slope of about ¾ (Zhang *et al.*).

copolymer, with the degree of polymerization of the protonated segments N_H = 402 and N_D = 958. The bilayers were made by casting a thick layer of HDH on the Si substrate and placing a thin HPS layer on top. The samples were welded at 120 °C for t = 10, 90, 150, and 255 minutes. The reptation time, T_r, was about 425 minutes for the HDH chains.

6.4.2 Neutron Reflection Results of HDH/HPS Interfaces

Figure 6.10 shows the interface width $X(t)$ due to the centrally deuterated section as a function of time [30, 31]. The graph is interesting in that it shows a very rapid increase in the first 10 minutes and then levels off until at least 90 minutes, when it takes off slowly. This is completely consistent with the segmental motion concepts involving rapid monomer motion initially due to the Rouse relaxation of the entanglements, followed by hindered Rouse relaxation of the whole chain. This experiment does not support hairpin models of interdiffusion.

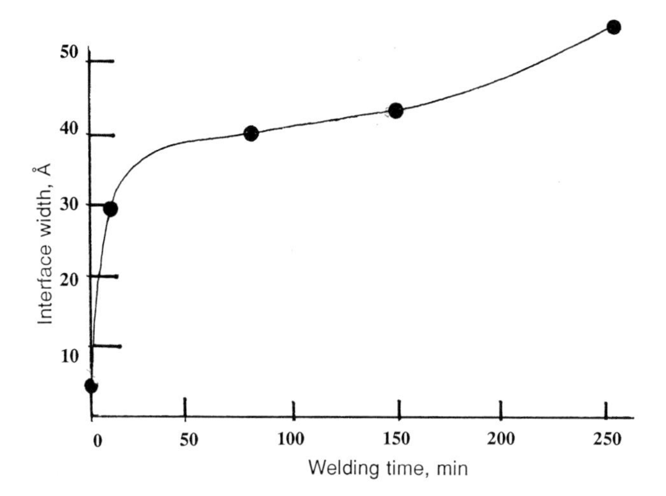

Figure 6.10 The interface width $X(t)$ versus welding time for the 203K(HPS)/183K(DPS) interface welded at 120 °C. The centrally deuterated fraction was 0.42 (Agrawal, Zhang *et al.*).

At the Rouse relaxation time of the entanglement segments, τ_e is about 5 seconds, and the interdiffusion distance should be $X(\tau_e) \approx 0.81\, R_{ge} \approx 30$ Å. In this experiment we observed $X \approx 30$ Å as the first welding data point at 10 minutes. At the Rouse relaxation time of the whole chain, when τ_{RO} is about 14 minutes, we can write that $X_D(\tau_{RO}) \approx 0.81\, (f_D\, M_{HDH}/M_e)^{1/4}\, R_{ge}$. With $M_{HDH} = 183{,}000$, $M_e = 18{,}000$, $f_D = 0.42$, and $R_{ge} = 37$ Å, we expect that $X_D(\tau_{RO}) \approx 43$ Å. This value corresponds to the value of the plateau in Figure 6.10.

Agrawal and co-workers [30, 31] also examined the diffusion of chain ends using DHD/HPS interfaces and found that the chain ends diffuse across the interface before the chain centers, consistent with reptation concepts.

6.4.3 Welding of HDH/DHD Interfaces

We report the results of interdiffusion between thin film layers of polystyrene triblocks (HDH/DHD) made by Agrawal and co-workers [7, 30] and discussed in Chapter 5. These polymers are symmetric triblocks composed of segments of polystyrene–d8 (dPS) and polystyrene–h8 (PS). In one material (DHD), the composition is 25–50–25 percent by weight dPS/PS/dPS. In the HDH polymer, the complementary structure is 25/50/25 percent by weight PS/dPS/PS. The molecular weight and the amount of deuteration were designed to be the same in both layers so that initially there was no variation of concentration across the HDH–DHD interface. Thus, with the HDH/DHD matching pairs, if the chain ends and the chain centers diffuse across the interface at the same rate, no variation in the concentration of the H or D segments across the interface is found. However, if anisotropic motion of the chain occurs, as in reptation, the chain ends diffuse across the interface first, thus enriching the HDH layer by dPS and the DHD layer by PS. Contributions from Rouse motion of the chains can be examined at times of order τ_{RO} for chains with M much larger than M_c and for matching pairs with M approximately equal to M_c.

Reflectivity profiles at several annealing times are shown in Figure 6.11. Reflectivity, R, is plotted as a function of the neutron momentum normal to the surface, $k_{z,0} = 2\pi \sin\theta/\lambda$, where θ is the grazing angle of incidence and λ is the wavelength. It is instructive to analyze the raw profiles. Two types of oscillation are apparent: those with the higher frequency relate to the interference of the neutrons reflected from the surface and the silicon substrate (for example, at 10 minutes), while those with the low frequency (for example, at 450 minutes) represent the interference between the surface and the HDH/DHD interface. The amplitude of the latter oscillations, particularly for 0.01 Å$^{-1} < k_{z,0} < 0.02$ Å^{-1}, increases until 450 minutes and then decreases with further annealing. This amplitude depends on the contrast at the polymer–polymer interface, that is, the difference in scattering length density of the two layers at the interface. Therefore, it is evident that the HDH layer becomes enriched with D and the DHD layer with H.

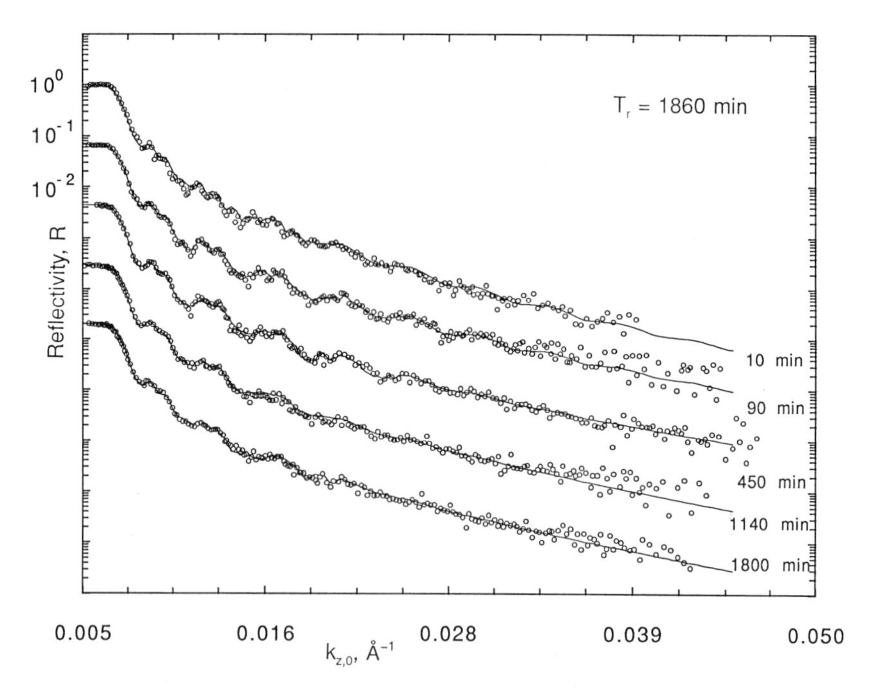

Figure 6.11 Reflectivity profiles for the ripple experiment using 225K(HDH)/250K(DHD) matching triblock interface welded for the indicated times at 118 °C (Agrawal *et al.*).

The best fits to the reflectivity profiles (shown as the solid lines in Figure 6.11) were obtained using the scattering length density profiles shown in Figure 6.12. The only independent parameters required to fit the reflectivity profiles were the amplitude and width of the ripple. In Figure 6.12 (a), the amplitude or the height of the ripple, which characterizes the enrichment at the interface, increases to a maximum value at

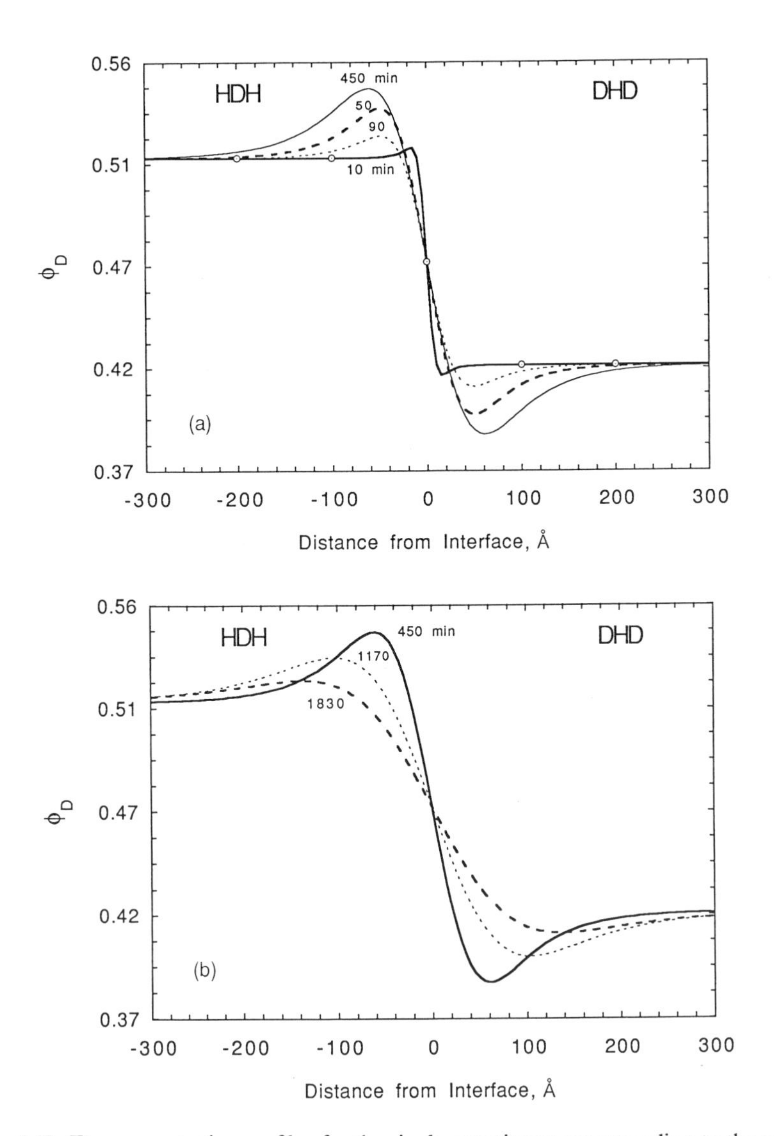

Figure 6.12 The concentration profiles for the ripple experiment corresponding to the neutron reflection profiles in Figure 6.11. (a) The ripple growth for times up to the peak maximum at 450 min. (b) The ripple decay for times greater than 450 min (Agrawal *et al.*).

450 minutes (which is close to $T_r/4$) and then decays [Figure 6.12 (b)]. A quantitative comparison of the experimental height of the ripple contribution with that predicted from the reptation model is given in Figure 6.13 as a function of annealing time. As can be seen, agreement is good between the predicted and experimental profiles. The time dependence of the ripple peak $C(0,t)$ is predicted to behave approximately as $C(0,t) \approx \frac{1}{4}\,(t/T_r)^{1/2}$, for $t \leq T_r/4$. Its maximum value is independent of molecular weight and occurs at $T_r/4$, such that $C(0,T_r) \approx 0.125$, as observed in Figure 6.13.

It must be emphasized that the ripple in the concentration profile, signifying the anisotropic motion of the chain segments at an interface, persists well beyond τ_{RO}, nearly to T_r. Significantly, experiments with matching pairs of molecular weights with $M \approx M_c$ do not show the ripple when the data are analyzed by SIMS or neutron reflection, and therefore preclude explanations based on Rouse dynamics. Note also in Figure 6.12 that at 10 minutes, the ripple contribution from Rouse motion is very small, as expected. The results of the ripple experiments with HDH/DHD pairs also agree with related work using HDH/HPS and DHD/HPS pairs, which show that the chain ends diffuse faster than the chain centers over distances of the order of R_g, and at annealing times in the range $\tau_{RO} < t < T_r$. When t is less than τ_{RO}, interdiffusion is dominated by segmental dynamics of entanglements and the chain ends appear to diffuse at about the same rate as the chain centers.

These results conclusively demonstrate that the motion of polymer molecules at interfaces is anisotropic. Given the correspondence of the growth and persistence of

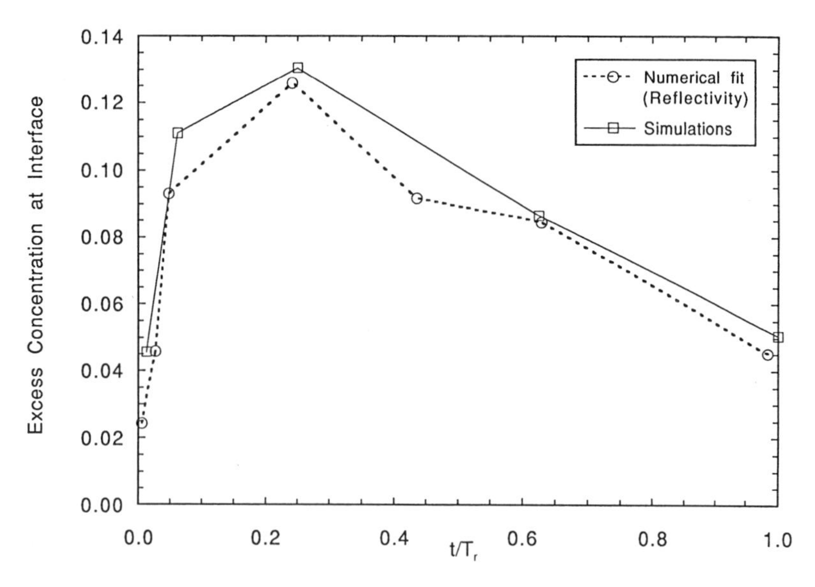

Figure 6.13 Comparison between a computer simulation of the ripple peak height and the neutron reflection experimental results for the HDH/DHD interface (Agrawal *et al.*).

the ripple at the interface with the times calculated from the reptation model, these results strongly favor reptation as the governing mechanism of polymer–polymer interdiffusion. The molecular weight dependence of the ripple evolution contains much additional information, and is being analyzed [7, 30, 31].

6.4.4 Final Comment on Chain-End Segregation

As a final comment in this chapter on chain-end segregation, we note that the neutron reflection experiments on HDH/DHD interfaces were able to sensitively determine whether chain-end segregation had occurred on either the HDH or the DHD surfaces: no chain-end segregation was observed, which agrees with studies by Zhao *et al.* [19].

6.5 References

1. A. Karim, A. Mansour, G. P. Felcher, and T. P. Russell, *Phys. Rev. B: Condens. Matter* ***42***, 6846 (1990).
2. A. Karim, *Neutron Reflection Studies of Interdiffusion in Polymers*; Ph.D. Thesis, Northwestern University, Evanston, IL; 1990.
3. A. Karim, A. Mansour, G. P. Felcher, and T. P. Russell, *Physica B (Amsterdam)* ***156/157***, 430 (1989).
4. G. P. Felcher, R. O. Hilleke, R. K. Crawford, J. Haumann, R. Kleb, and G. Ostrowski, *Rev. Sci. Instrum.* ***58***, 609 (1987).
5. H. Zhang and R. P. Wool, "Concentration Profiles at Amorphous Symmetric Polymer/Polymer Interfaces"; *Polym. Prepr. (Am. Chem. Soc., Div. Polym. Chem.)* ***31*** (2), 511 (1990).
6. H. Zhang, *Concentration Profiles at Amorphous Polymer–Polymer Interfaces*; Ph.D. Thesis, University of Illinois, Urbana, IL; 1990.
7. G. Agrawal, R. P. Wool, W. D. Dozier, G. P. Felcher, T. P. Russell, and J. W. Mays, *Macromolecules* ***27***, 4407 (1994).
8. H. Zhang and R. P. Wool, papers submitted to *Macromolecules.*
9. M. Stamm, S. Huttenbach, G. Reiter, and T. Springer, *Europhys. Lett.* ***14***, 451 (1991).
10. G. Reiter and U. Steiner, *J. Phys. II (Journal de Physique II)* ***1***, 659 (1991).
11. U. Steiner, G. Krausch, G. Schatz, and J. Klein, *Phys. Rev. Lett.* ***64***, 1119 (1990).
12. T. P. Russell, A. Karim, A. Mansour, and G. P. Felcher, *Macromolecules* ***21***, 1890 (1988).
13. T. P. Russell, *Mater. Sci. Rep.* ***5***, 171 (1990).
14. T. P. Russell, A. Menelle, W. A. Hamilton, G. S. Smith, S. K. Satija, and C. F. Majkrzak, *Macromolecules* ***24***, 5721 (1991).
15. T. P. Russell, V. R. Deline, W. D. Dozier, G. P. Felcher, G. Agrawal, R. P. Wool, and J. W. Mays, *Nature (London)* ***365***, 235 (1993).
16. R. A. L. Jones, L. J. Norton, E. J. Kramer, R. J. Composto, R. S. Stein, T. P. Russell, A. Mansour, A. Karim, G. P. Felcher, M. H. Rafailovich, J. Sokolov, and S. A. Schwarz, *Europhys. Lett.* ***12***, 41 (1990).
17. R. J. Composto, R. S. Stein, E. J. Kramer, R. A. L. Jones, A. Mansour, A. Karim, G. P. Felcher, *Physica B (Amsterdam)* ***156/157***, 434 (1989).

18. R. A. L. Jones, E. J. Kramer, M. H. Rafailovich, J. Sokolov, and S. A. Schwarz, *Phys Rev. Lett.* ***62***, 280 (1989).
19. W. Zhao, X. Zhao, M. H. Rafailovich, J. Sokolov, R. J. Composto, S. D. Smith, M. Satkowiski, T. P. Russell, W. D. Dozier, and T. Mansfield, *Macromolecules* ***26***, 561 (1993).
20. S. A. Werner and A. G. Klein, "Neutron Optics"; Chapter 4 in *Neutron Scattering*, S. K. Sköld and D. L. Price, Eds.; Vol. 23 in series *Methods in Experimental Physics*, R. Celotta and J. Levine, Eds.; Academic Press, Orlando, FL; 1986; p 259..
21. E. Fermi, *Phys. Rev.* ***71***, 666 (1947).
22. G. L. Squires, *Introduction to the Theory of Thermal Neutron Scattering*; Cambridge University Press, Cambridge; 1978.
23. L. Schiff, *Quantum Mechanics*; McGraw–Hill, New York; 1949.
24. J. Als-Nielsen, "Solid and Liquid Surfaces Studied by Synchrotron X-Ray Diffraction"; Chapter 5 in *Structure and Dynamics of Surfaces II*; W. Schommers and P. von Blanckenhagen, Eds.; Vol. 43 in series *Topics in Current Physics*; Springer-Verlag, Berlin; 1987; p 181.
25. G. P. Felcher, R. O. Hilleke, R. K. Crawford, J. Haumann, R. Kleb, and G. Ostrowski, *Rev. Sci. Instrum.* ***58***, 609 (1987).
26. O. S. Heavens, *Optical Properties of Thin Solid Films*; Dover, New York; 1965.
27. P.-G. de Gennes, *Scaling Concepts in Polymer Physics*; Cornell University Press, Ithaca, NY; 1979.
28. H. Zhang and R. P. Wool, work submitted for publication. .
29. R. P. Wool, H. Zhang, and G. Agrawal, "Structure and Concentration Profiles at Amorphous Polymer Interfaces: Theory and Experiment"; *PMSE Preprints; Proceedings of the American Chemical Society Division of Polymeric Materials: Science and Engineering* ***67***, 165 (1992).
30. G. Agrawal, *Short Time Interdiffusion at Polymer Interfaces: A Probe to Study Polymer Motion*; Ph.D. Thesis, University of Illinois, Urbana, IL; 1994.
31. G. Agrawal and R. P. Wool, paper submitted to *Macromolecules*.

7 Welding and Entanglements*

7.1 Introduction

Welding is defined as the process in which thermoplastics are united, fused, or brought into intimate contact. The materials are softened by heat or solvents, brought into contact, and held together under pressure until the weld cools or the solvent evaporates. Polymers can be welded by several techniques, including hot plate welding, hot air welding, vibrational welding, friction welding, solvent welding, dielectric welding, adhesive bonding, surface chemical modification, ion beam surface modification, resonance heating, and other more elaborate but less common techniques [1]. We are interested not only in the specific act of welding two polymers together using welding tools (ultrasonic horns, hot air jet, etc.), but also in strength development at generic polymer–polymer interfaces, for example, processing of powder and pellet resin, internal weld lines of polymer melts during extrusion and injection molding, lamination of composites, coextrusion, and autoadhesion of uncured linear elastomers.

In this chapter, we consider the problem of strength development at polymer–polymer interfaces in terms of the properties of random-coil chains discussed in Chapters 2–6 [2–5]. When two pieces of molten polymer are brought into contact, wetting or close molecular contact (van der Waals) first occurs, followed by interdiffusion of chain segments back and forth across the wetted interface. We enquire as to the mechanical energy, G, required to separate the two pieces after a contact time, t, as a function of time (t), temperature (T), contact pressure (P), and molecular weight (M), of the linear random-coil chains, that is,

$$G = W(t, T, P, M) \tag{7.1.1}$$

where W is the welding function to be determined.

In the previous chapters we used molecular dynamics models to describe interdiffusion at a symmetric interface and determine a molecular description of the interface structure as a function of the variables t, T, P, and M. To relate structure to strength, we first need a molecular connectivity relation that allows us to develop microscopic deformation models for chain disentanglement and fracture. The microscopic deformation processes in the deformation zone at the crack tip control the

* Dedicated to A. Gent

macroscopic fracture mechanics and fracture energy G_{1c} of the interface. We begin with a discussion of the fracture and peel adhesion tests used to measure interface strength; we go on to develop relations for mechanical connectivity between chains using a new entanglement model; and we use this model to relate structure to strength via disentanglement and bond rupture.

7.2 Fracture

7.2.1 Griffith Theory of Fracture

Configurations of fracture mechanics tests used to evaluate the strength of materials and interfaces are shown in Figure 7.1 [6]. The fracture mechanics specimens can be double cantilever beams (DCB), compact tension (CT), wedge cleavage (WC), single-edged notch (SEN), peel adhesion, blister tests, etc. Similar configurations have been designed by ASTM to determine adhesive strength [7–11]. Each test in Figure 7.1 involves an initial crack of length a. The sample is loaded in tension with a load P, the mechanical energy is stored in the deformed specimen at constant crack length, and finally fracture occurs as the crack suddenly propagates. The Griffith approach to brittle tensile fracture for such a material with a crack is given in terms of the stored elastic strain energy, U, and the energy, S, to create new surface area [12]. This elegant theory, the cornerstone of modern fracture mechanics and the basis for *linear elastic fracture mechanics* (*LEFM*), states that the incremental change in strain energy, dU, with crack length, da, exceeds the energy to create surface area, dS, or

$$(\partial U/\partial a) \geq (\partial S/\partial a) \tag{7.2.1}$$

If we have an elliptical crack in a flat specimen, where the major axis a is much greater than the minor axis b, the change in energy as a function of crack extension is given by

$$U = \pi\sigma^2 a^2 B/E \tag{7.2.2}$$

where σ is the applied stress (normal to the crack direction), B is the thickness of the specimen, and E is the tensile modulus. Differentiating, we have

$$\partial U/\partial a = 2\pi\sigma^2 aB/E \tag{7.2.3}$$

The incremental crack advance, ∂a, requires an increase in the surface area of $\partial S = 4BG\partial a$, and the surface energy term is

$$\partial S/\partial a = 4BG \tag{7.2.4}$$

where G is the energy required to create unit surface area. Substituting Eqs 7.2.3 and 7.2.4 in Eq 7.2.1 and solving for the critical stress at fracture σ_c, we obtain the important Griffith result

$$\sigma_c = [2\,GE/(\pi a)]^{1/2} \qquad \text{(plane stress)} \qquad (7.2.5)$$

The factor of 2 in the numerator accounts for a crack in the middle of a plate; if we cut the plate in two to make two equal fracture mechanics specimens with edge cracks as shown in Figure 7.1(i)–(iv), then the factor of 2 becomes unity, which we consider henceforth.

The condition of "plane stress" means that the only stresses operating on the crack surfaces are those in the x–y plane of the sample, that is, σ_x and σ_y (y is the direction of applied stress and x is the direction of crack propagation). Triaxial stress is absent ($\sigma_z = 0$) if specimens are very thin (B much smaller than a). With thicker specimens, the strains are constrained to the x–y plane and Eq 7.2.5 is modified by the $(1 - \nu^2)$ term

$$\sigma_c = [GE/\pi a(1 - \nu^2)]^{1/2} \qquad \text{(plane strain)} \qquad (7.2.6)$$

where ν is Poisson's ratio. For a perfectly elastic material undergoing a constant-volume tensile deformation, ν is 0.5, and is determined from the ratio of the lateral contraction to the longitudinal extension. The plane–strain correction means that the critical stress is 15.5% larger than that for a plane–stress fracture case. This results in the fracture "lips" often observed on the edges of the fracture surfaces of plastics where plane stress (with shear yielding) is operating on the outside edges, but plane strain (with dilational crazing) occurs in the interior of a thick specimen.

Design Problem

The Griffith theory is often used to examine reliability and material performance with known load applications. For example, a thin-walled plastic with modulus $E = 1$ GPa and fracture energy $G = 1000$ J/m^2 is designed for loads not exceeding 10 MPa. What is the largest surface crack a^* that the material can tolerate?

Letting σ_c be the maximum application load and solving for a^* using the Griffith equation,

$$a^* = GE/\pi\sigma_c^2 = \frac{(1000\ \text{N·m/m}^2)(10^{-9}\ \text{M/m}^2)}{(3.142)(10 \times 10^6\ \text{N/m}^2)^2} = 3.2\ \text{mm}$$

Under these application conditions, the critical-size crack would be visible to the eye, perhaps.

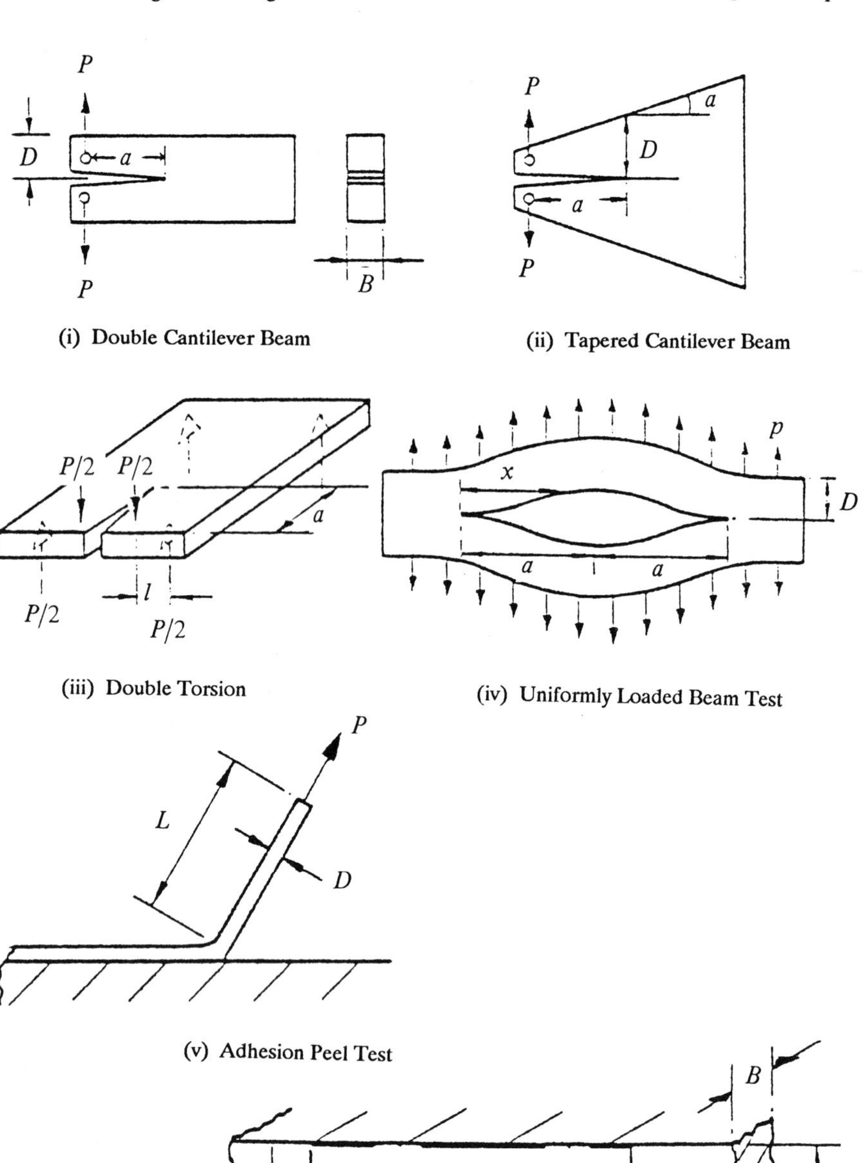

(i) Double Cantilever Beam

(ii) Tapered Cantilever Beam

(iii) Double Torsion

(iv) Uniformly Loaded Beam Test

(v) Adhesion Peel Test

(vi) Parallel Strip

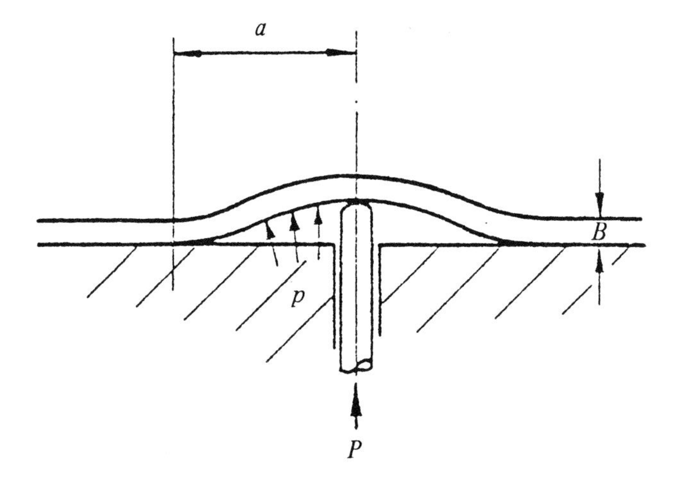

(vii) Blister Test

Figure 7.1 Configurations of fracture mechanics tests. [Note (i–vi) are on facing page]: (i) double cantilever beam; (ii) tapered cantilever beam; (iii) double torsion; (iv) uniformly loaded beam test; (v) parallel strip; (vi) adhesion peel test; (vii, above) blister test. (Courtesy of Gordon Williams).

7.2.2 Fracture Energy, G_{1c}, and Toughness, K_{1c}

The fracture energy, G, has been interpreted in LEFM as the sum of all energies required to create the surface, and is related through Eq 7.2.1 to the critical strain energy release rate per unit crack advance, G_{1c}, where the "c" means "critical" and the "1" means Mode I tensile crack opening. (Mode II involves in-plane shear of the crack surfaces, and Mode III involves out-of-plane torsion.) G_{1c} is commonly referred as the *fracture energy* of a plastic and has units of J/m^2. For a given material, the quantities G_{1c} and modulus, E, are constants, and the fracture variables, crack length a and critical stress σ_c, can be separated out in the form

$$\sigma_c^2 \pi a = EG_{1c} \tag{7.2.7}$$

Since the term EG_{1c} is constant, the left side must also be constant and its square root is given in terms of the stress intensity factor, K_{1c} (with typical units of MPa·m$^{1/2}$).

$$K_{1c} = \sigma_c (\pi a)^{1/2} \tag{7.2.8}$$

K_{1c} is also commonly called "fracture toughness", since it is a measure of the stress concentration factor at the crack tip when fracture occurs. G_{1c} and K_{1c} are related by

$$G_{1c} = K_{1c}^2 / E \qquad \text{(plane stress)} \tag{7.2.9}$$

and in plane strain, the right side is multiplied by $(1 - \nu^2)$.

For highly brittle materials, Griffith proposed that the critical strain energy release rate, G_{1c}, is related to the surface energy, Γ, of the solid by

$$G_{1c} = 2\Gamma \tag{7.2.10}$$

In principle, if one independently knows the surface energy, Γ, of a solid, for example, by contact angle measurements, one should be able to predict its fracture energy simply by using Eq 7.2.10. For example, how does the fracture energy, G_{1c}, of polystyrene compare with its surface tension, Γ, of 0.04 J/m^2 at room temperature? We find, for high molecular weights of about 200,000, that G_{1c} is about 1000 J/m^2, which is about 10^4 times greater than 2Γ (0.08). However, at very low molecular weights, below the critical entanglement molecular weight, the Griffith limit is approached, as discussed in Chapter 8. The difference in the high molecular weight result is due to the additional surface work required to create the craze zone and rough fracture surfaces compared to a smooth liquid surface.

7.2.3 Strain Energy Density and Fracture

Several important proportionalities can be deduced from the Griffith theory. The strain energy density, U_0, in the material is given by the uniaxial approximation

$$U_0 = \sigma^2 / 2E \tag{7.2.11}$$

Thus, at the critical stress, we have from the above relations

$$G_{1c} = 2\pi a U_{0c} \tag{7.2.12}$$

where U_{0c} is the critical strain energy density. The relation $G_{1c} \sim U_{0c}$ will be used in a later section when we introduce a microscopic mechanism by which we compute the stored strain energy required to pull a chain from the interface.

In the Dugdale model of fracture [6, 13] where a crack propagates through a thin deformation zone at the crack tip, the fracture energy is determined by

$$G_{1c} = \sigma^* \delta \tag{7.2.13}$$

where σ^* is the average stress creating the deformation zone at the crack tip and δ is the crack-opening displacement (COD) at fracture. From the Griffith theory we can argue that $\sigma_c \sim \sigma^*$ so that

$$\delta \sim \sigma^* / E \qquad (7.2.14)$$

For welding at interfaces, the modulus E is independent of both time and molecular weight and so we expect δ and σ to have similar scaling laws.

7.2.4 Fracture Measurements

ASTM D 5045 provides a convenient test method for determining the plane–strain fracture toughness, K_{1c}, and strain energy release rate, G_{1c}, of plastic materials [73]. A standard compact tension specimen is a single-edged notch specimen with proportions shown in Figure 7.2. The recommended sample width is $W = 2B$, and the crack length a should be selected so that $0.45 < a/W < 0.55$. A sharp notch is prepared by machining with a saw and then sharpening with the tap of a razor blade. (The latter action can be made reproducible by dropping a known weight from a given height on the blade). The length of the razor-sharpened portion of the crack must be at least twice the saw blade width (not to scale in Figure 7.2). Fatigue sharpening the notch is also highly desirable. A crosshead speed, s, of not more than 10 mm/min at 23 °C is recommended for the test.

A validity criterion for K_{1c} in this test is based on the size inequality

$$2.5\,(K_{1c} / \sigma_y)^2 < B, a, (W - a) \qquad (7.2.15)$$

where σ_y is the yield stress, B is the sample thickness, a is the crack length and $(W - a)$ is the ligament length (see Figure 7.2). The criteria are chosen such that B is large enough to ensure plane strain and $(W - a)$ is large enough to avoid excessive

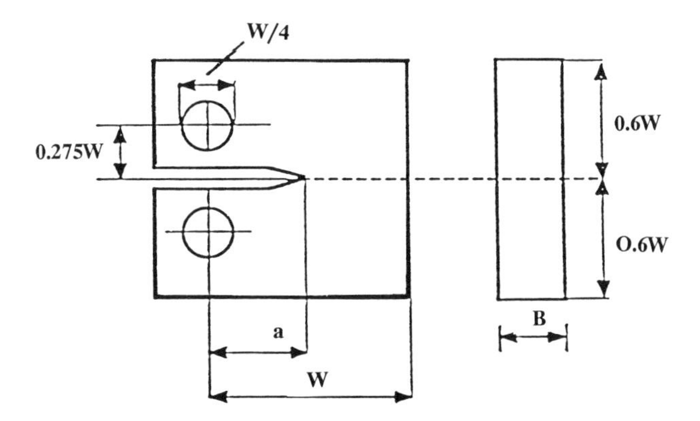

Figure 7.2 Compact tension test configuration.

plasticity in the ligament. For the CT specimen, the fracture toughness is determined by ASTM D 5045 as [73]

$$K_{1c} = (P/BW^{1/2})\, f(x) \tag{7.2.16}$$

where P is the critical load at fracture, $x = a/W$, and the function $f(x)$ is

$$f(x) = [(2+x)(0.886 + 4.64x - 13.32x^2 + 14.72x^3 - 5.6x^4)]/(1-x)^{3/2} \tag{7.2.17}$$

The function $f(x)$ has values ranging from $f(0.45) = 8.34$ to $f(0.55) = 11.36$.

Example of Toughness Measurement

Problem: *The following data were obtained from a compact tension fracture test of polystyrene:*

crack length	$a = 9.1$ *mm*
thickness	$B = 3.1$ *mm*
width	$W = 19.6$ *mm*
critical load	$P = 50$ *N*
modulus	$E = 3.0$ *GPa*
crosshead speed	$s = 2.5$ *mm/min*
yield stress	$\sigma_y = 28$ *MPa*

Determine the fracture toughness K_{1c} *and the critical strain energy release rate for polystyrene under these conditions. Was the test valid?*

Solution: *Using* $x = 9.1/19.6 = 0.464$, *and substituting all values in Eq 7.2.16, we obtain* $K_{1c} = 1.0$ *MPa* $m^{1/2}$. *The fracture energy in plane strain is determined by*

$$G_{1c} = (1 - \nu^2) K_{1c}^2 / E \tag{7.2.18}$$

where $\nu \approx 0.5$. *Thus,* $G_{1c} = 250$ J/m^2.

Exercise

Check the validity of this test using Eq 7.2.15.

7.3 Peel Adhesion

7.3.1 Introduction

Adhesive fracture and peeling of thin strips have been studied by Gardon [15], Gent [16–20], Hamed [21, 22], Williams [23], and others, either from rigid substrates or unconstrained T-test configurations, shown in Figure 7.3. Peel adhesion is a useful and

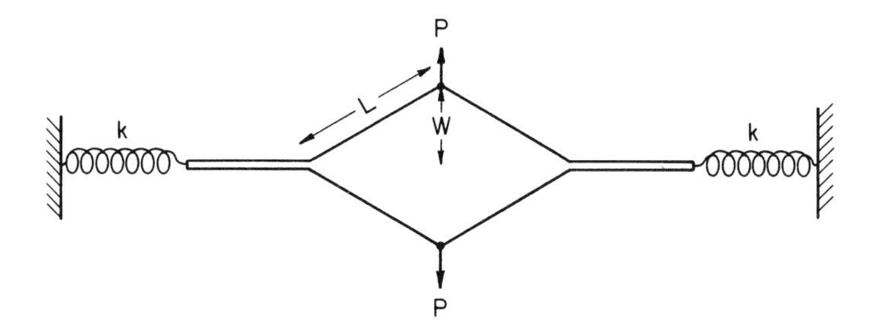

Figure 7.3 In-plane adhesive fracture between two thin strips. The adhesive strips are attached to lateral flexible constraints of modulus k. The strips, glued face to face, are debonded with a force P, normal to the constraint direction.

popular method of determining the strength of polymer interfaces. In this section, we examine fracture relations for a modified peel adhesion test configuration [24] for symmetric interfaces, where the stored strain energy, U, is well separated from the energy to create unit fracture surface area. The mechanics of centrally debonding two adhesive strips with lateral flexible elastic constraints is shown in Figure 7.3. The strips are glued together by autoadhesion, and the deformation direction is normal to the plane of the tapes and constraints. This analysis provides a quantitative solution for the peel force as a function of displacement, with the adhesive fracture energy per unit area of the interface and the constraint modulus as parameters. This study, while providing a solution to an interesting adhesive fracture configuration, is also of importance in the modeling of aspects of microstructural damage and healing in many materials (Chapters 11, 12), as will become apparent from the tensile characteristics of the peel experiment.

7.3.2 Peel Adhesion Energy Balance

The work input, $\mathrm{d}W = P\mathrm{d}w$, to the system shown in Figures 7.3 and 7.4 is balanced by changes in strain energy and the creation of new surface area due to fracture. A differential energy balance can be written as

$$P\,\mathrm{d}w = \mathrm{d}U + 2b\Gamma\,\mathrm{d}L \tag{7.3.1}$$

where P is the instantaneous force, w is the displacement of the centrally debonded point in a direction normal to the bonded tape surface, U is the strain energy stored in the springs of the lateral constraints, L is the length of debonded surface in one quadrant, Γ is the fracture energy per unit surface area of tape, and b is the width of the tape. Some of the parameters are shown in Figure 7.3 and 7.4.

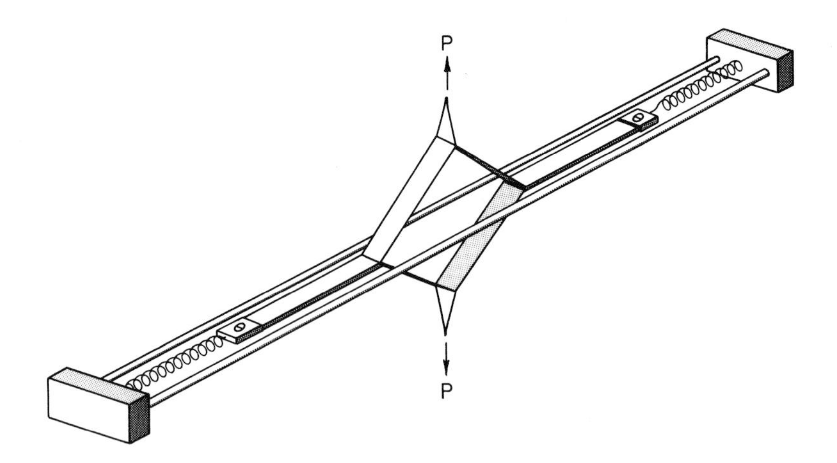

Figure 7.4 Experimental apparatus used for the peel testing of adhesive thin tapes. The tapes are glued face to face and constrained at their ends by metal springs maintained at a fixed separation distance.

The load P is obtained from Eq 7.3.1 as

$$P = dU/dw + 2b\Gamma\ dL/dw \tag{7.3.2}$$

The strain energy change, being a function of the independent variables L and w, can be written as

$$dU = (\partial U/\partial w)_L dw + (\partial U/\partial L)_w dL \tag{7.3.3}$$

Substituting Eq 7.3.3 in Eq 7.3.2, we obtain the governing equation for the load,

$$P = (\partial U/\partial w)_L + dL/dw[(\partial U/\partial L)_w + 2b\Gamma] \tag{7.3.4}$$

When P is less than the critical load to cause fracture, then

$$dL/dw = 0 \tag{7.3.5}$$

and Eq 7.3.4 is reduced to the elastic solution

$$P = (\partial U/\partial w)_L \tag{7.3.6}$$

At fracture, Eq 7.3.6 is still valid, and we infer from Eq 7.3.4 that

$$dL/dw[(\partial U/\partial L)_w + 2b\Gamma] = 0 \tag{7.3.7}$$

Also, at fracture

$$dL/dw \neq 0 \tag{7.3.8}$$

and Eq 7.3.7 becomes

$$(\partial U/\partial L)_w = -2b\Gamma \tag{7.3.9}$$

Eq 7.3.9 can be solved for L as a function of w, which when substituted in Eq 7.3.6 gives the solution for the peel force as a function of extension. Note that Eq 7.3.9 is similar to Eq 7.2.1, which is Griffith's criterion of fracture for a brittle material, and states that when sufficient energy is stored, the crack can advance at constant displacement and no further work input is required.

7.3.3 Force–Displacement Solution

The initial boundary conditions for the peel test shown in Figure 7.3 are $L = 0$, $U = 0$, $P = 0$ at $w = 0$. Other conditions on the system require the displacement to be non-negative and the lateral contraction to be zero. The strain energy U, peel force P, and geometric relations can be deduced from the equilibrium configuration of the peel test as

$$U = \tfrac{1}{2}k[2L^2 - 2L(L^2 - w^2)^{1/2} - w^2] \tag{7.3.10}$$

$$P = kw[L(L^2 - w^2)^{-1/2} - 1] \tag{7.3.11}$$

$$L^2 = w^2 + C^2 \tag{7.3.12}$$

In these equations, k is a linear elastic spring constant of the lateral constraint, and C is the quadrant crack length, or projection of L onto the plane of the constraint. Eq 7.3.11 can be derived from static equilibrium or by partial differentiation of U with respect to w at constant L as indicated in Eq 7.3.6.

We solve this system of equations as follows. From Eq 7.3.9 and Eq 7.3.10, we obtain

$$(\partial U/\partial L)_w = k[2L - (2L^2 - w^2)(L^2 - w^2)^{1/2}] = -2b\Gamma \tag{7.3.13}$$

Letting

$$\alpha = 2b\Gamma/k \tag{7.3.14}$$

and rearranging for w in Eq 7.3.13, we have

$$w^4 + w^2(4L\alpha^2) - (4L^3\alpha + \alpha^2L^2) = 0 \tag{7.3.15}$$

Solving this quadratic equation and using the boundary conditions to select the positive or negative roots, we obtain

$$w(L) = [\tfrac{1}{2}(\alpha + 2L)(4L\alpha + \alpha^2)^{1/2} - 2L\alpha - \alpha^2/2]^{1/2} \tag{7.3.16}$$

This equation can now be substituted in Eq 7.3.11 to evaluate the peel force as a function of crack surface length.

$$P = kw(L)\{L/[L^2 - w^2(L)]^{1/2} - 1\} \tag{7.3.17}$$

By inverting Eq 7.3.16, we obtain L as a function of w in the following form [24]

$$L = \tfrac{1}{4}\alpha\,[(2\alpha/3 + X(w)^2 - \alpha^2] \tag{7.3.18}$$

where

$$\begin{aligned} X(w) = {} & [2\alpha^2 w(w^2/\alpha^2 - 1/27)^{1/2} - \alpha^3/27 + 2\alpha w^2]^{1/3} \\ & + [2\alpha^2 w(w^2/\alpha^2 - 1/27)^{1/2} + \alpha^3/27 - 2\alpha w^2]^{1/3} \end{aligned} \tag{7.3.19}$$

This inversion was obtained with an algebraic root restriction that $w \geq \alpha/\sqrt{27}$; however, this restriction does not apply in Eq 7.3.16, and, for $w < \alpha/\sqrt{27}$, L can be evaluated numerically.

Substituting Eq 7.3.18 in Eq 7.3.11, we obtain the solution for $P(w)$ as

$$P/kw = [1 - (4\,w/\alpha\{[\tfrac{2}{3} + X(w)/\alpha]^2 - 1\})^2]^{-1/2} - 1 \tag{7.3.20}$$

Equation 7.3.20 represents the peel force with increasing displacement, w.

At any point during the peel fracture, we can reverse the displacement to obtain the unload curve. The unload curve can be described by

$$P = kw[L_0(L_0^2 - w^2)^{-1/2} - 1] \tag{7.3.21}$$

where L_0 is the surface crack length at maximum displacement, derived from Eq 7.3.18. Equation 7.3.21 is valid only if $dL/dw = 0$, that is, if further crack propagation or closure does not occur during unloading.

The peel equations contain only two constants, k and $2b\Gamma$ ($\alpha = 2b\Gamma/k$), which are usually known or can be evaluated. Thus, the theoretical derivation readily lends itself to experimental verification.

7.3.4 Experimental Test of Peel Adhesion

A model peel test was constructed to evaluate the solutions given by Eqs 7.3.16–7.3.21 above. The experimental apparatus is shown in Figure 7.4. Commercially available transparent adhesive Scotch™ (3M) tape ($8 \times 0.5 \times 0.0025$ in) was used. The tape strips were adhered face to face and supported in the rigid frame by light flat grips attached to metal springs 1.5 in long. The spring constants were determined independently as k = 2.925 lb/in. The Young's modulus of the tape was $E = 1.2 \times 10^5$ psi. This value is much greater than the modulus of the springs and justifies the inextensibility assumption.

The adhesive fracture energy, Γ, was determined from a peel experiment without spring constraints. Applying Eq 7.3.2, since $dU = 0$, we have

$$P \mathrm{d}w = 2 b \Gamma \mathrm{d}L \tag{7.3.22}$$

which, for a constant peel force over a finite deformation range, reduces to

$$\Gamma = P / 2b \tag{7.3.23}$$

Eq 7.3.23 is most commonly used in peel–adhesion experiments. With P = 1.516 lb, b = 0.5 in, we have Γ = 1.516 in·lb/in^2 (273 J/m^2), and thus α = 0.518 in.

The peel tests were conducted at room temperature and an extension rate of 20 in/min. The force–displacement behavior of the peel test is shown in Figure 7.5. The system was deformed to a maximum displacement of w = 1 in and then unloaded to w = 0 at the same rate. Equation 7.3.20 was used to describe the loading curve with values of k and α as determined above. The unloading curve was described by Eq 7.3.21 in which L_0 = 1.173 in was determined from Eq 7.3.18 with w = 1 in. Figure 7.5 indicates good agreement between theory and experiment.

7.3.5 Healing of the Adhesive Fracture Energy

If healing of the debonded sections does not occur in the rest state, the next reload path follows the last unload elastic curve until the peel curve is reached at the previous maximum displacement (say, w = 1 in). At that point, adhesive fracture commences with increasing w. However, if the debonded surfaces become wet and adhere in the rest state, the reload curve is intermediate between the peel curve and the unload curve. Thus, the degree of recovery or rehealing depends on the value of the time-dependent adhesive fracture energy at each point on the healing surfaces. We have used these concepts to evaluate healing of microscopic damage and crack closure phenomena in materials.

For symmetric interfaces, the following scaling law for healing of the adhesive fracture energy is suggested:

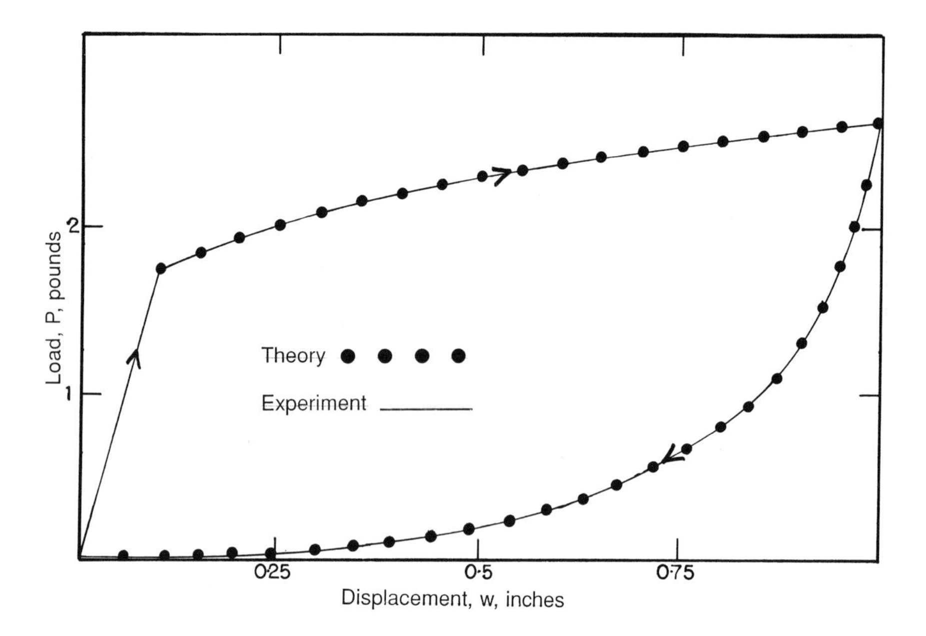

Figure 7.5 Experimental results for the adhesive fracture peel force of the tapes, compared with predictions for both loading and unloading displacements; α = 0.518 in, Γ = 1.516 lb, and k = 2.925 lb/in.

$$\Gamma(t) = \Gamma_\infty [t/\tau]^\beta \tag{7.3.24}$$

where Γ_∞ is the fully healed fracture energy, τ is a characteristic relaxation time, β is an exponent, t is the time, and $t \le \tau$. For adhesives made with linear polymers of molecular weight M, we speculate that $\Gamma(t)$ depends on the average interpenetration contour length $l(t)$, β = ½, and $\tau \sim M^3$.

The molecular weight dependence of the healing rate depends on the molecular weight dependence of Γ_∞ and τ. If we let $\Gamma_\infty \sim M^\epsilon$, then it follows from the last equation that

$$\Gamma(t,M) \sim t^\beta M^{\epsilon - 3\beta} \tag{7.3.25}$$

Experimental exponents of $M^{-1/2}$ are consistent with β = ½ and $\epsilon \approx 1$ (Chapter 8).

As shown in Figure 7.5, the mechanical behavior of the peel experiment is qualitatively similar to deformation characteristics of many materials such as filled elastomers, semicrystalline polymers with stacked lamellar morphologies [25], and other systems in which plastic deformation and elastic fracture mechanisms are important. For example, Figure 7.6 shows the mechanical behavior of hard elastic

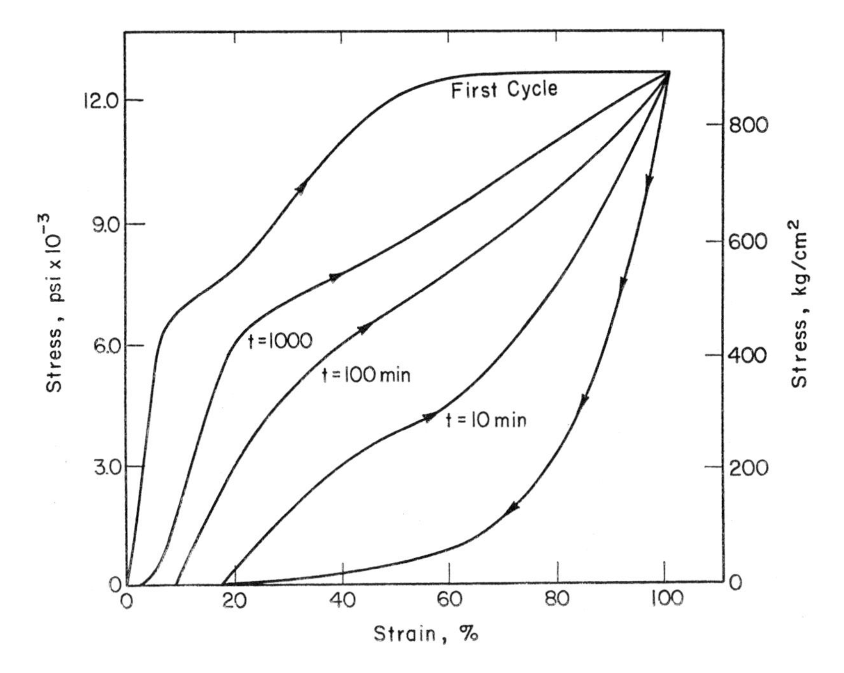

Figure 7.6 Mechanical healing behavior of hard elastic polypropylene fibers. The stress-strain response is shown, with healing times between the first and second loading.

polypropylene (HPP) fibers as a function of healing time between the first and second loading cycles [25]. The unusual stress–strain response for a semicrystalline fiber is due to reversible cavitation occurring between the crystalline lamellae, which are stacked normal to the stretching direction. Cavitation accommodates large strains, and healing occurs above the glass transition temperature (–5 °C) in the rest state between deformation cycles (Chapter 11).

Using the peel–adhesion analogy to examine microscopic fracture in materials, we can use the stress–strain curves to partition the total work, W, into stored elastic energy, U, and surface energy, S, so that

$$W = U + S \tag{7.3.26}$$

Thus, in Figure 7.6, W is determined from the total area under the load–displacement curve during peel fracture, or, analytically by integration of $P(w)\mathrm{d}w$, using Eq 7.3.20; the unused elastic energy, U, is determined from the area under the unload curve, or by integration of Eq 7.3.21; and the fracture energy is determined by subtraction, $W - U$. Typically, $U \approx \frac{1}{2} W$, first pointed out by Griffith, who showed generally that for the elastic plate with a propagating crack, the stored elastic energy is one-half the externally applied work.

The reader is encouraged to apply the latter concept and the peel adhesion analysis to attempt the following problems.

Problem 7.3.1

For the peel test with lateral constraints (Figure 7.3), determine the reload response (Eq 7.3.21) when healing occurs after unloading for times $t/\tau = 0, 0.2, 0.5$ *and* 1.0. *Assume* $\beta = ½$ *in Eq 7.28, and use the same experimental conditions and constants described above.*

Problem 7.3.2

Using the peel test (Figure 7.3) analogy and the HPP stress–strain data in Figure 7.6, determine the energy to debond polypropylene crystalline lamellae. Assume that the lamellar thickness is 100 angstroms. What are the values of β *and* τ *(Eq 7.3.25) for healing of HPP lamellae, and how can these values be related to the structure of the amorphous layer between lamellae?*

7.4 Polymer Entanglements

The Griffith concept of fracture has considerable value if one can determine how stored energy is consumed in forming the fracture surface. To address welding and healing problems, a stored strain energy approach to fracture is adopted, which considers both chain pullout via disentanglement and chain fracture mechanisms at the interface. We first develop a connectivity model to understand mechanical properties of interdiffusing entangled polymer chains at interfaces.

7.4.1 Entanglement Model for Random-Coil Polymers

We use an entanglement model that provides mechanical connectivity between chains and relates the interface structure to the breakdown process of the deformation zone at a crack tip. Entanglements develop from the interpenetration of random-coil chains, as discussed in Chapter 1, and are important in the determination of rheological properties. The subject has been extensively reviewed by Graessley [26] and by Aharoni [27], and more recently has been investigated by Kavassalis and Noolandi [28]. Of particular importance is the role of entanglements in controlling melt viscosity η, where the viscosity at low molecular weights behaves as a simple Rouse fluid with $\eta \sim M$, but at high molecular weight, $\eta \sim M^{3/4}$. The crossover to a high-viscosity entangled fluid occurs at the critical entanglement molecular weight, M_c. In this section, we propose a structural origin for M_c, which we use to predict the onset of an entanglement network and provide a basis for a disentanglement approach to fracture.

An entangled amorphous linear chain network is shown schematically in Figure 7.7. The bridge theory of connectivity in an amorphous network, which we proposed [29–31], requires that the number of chain segments, p, crossing any load-bearing plane exceed the number of chains, n, by $p > 3n$. When p is less than $3n$, a network cannot form and the chains readily slip apart by Rouse motion. When p is greater than $3n$, the chains are sufficiently interpenetrated to form an entanglement network, and relaxation occurs by diffusion in the presence of entanglement constraints. At $p \approx 3n$, the polymer chains are critically connected and the average bridge structure has three chain segments.

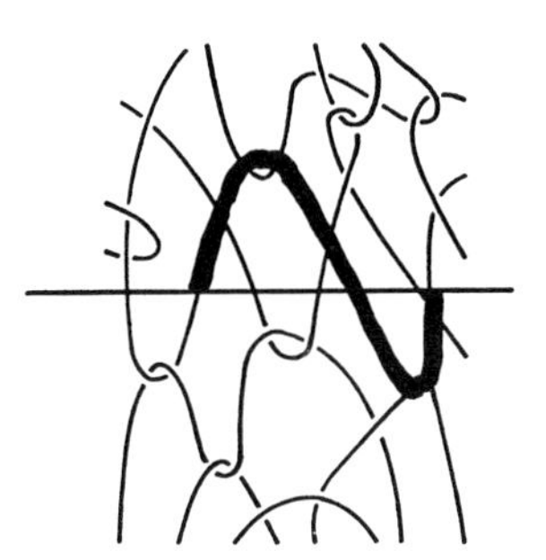

Figure 7.7 Entanglements in a polymer melt. The bold segment represents the critical length (M_c) to obtain a connected network.

We can explore the existence of an entanglement network in an isotropic concentrated melt by counting the number of bridges and chains intersecting an arbitrary plane, as shown in Figure 7.7. A *bridge* is a segment of chain that crosses the plane three times. It is sufficiently long to complete one circular loop through the plane. The bridge is capable of transmitting forces across the plane in the melt for a time that depends on the relaxation of this chain segment. The number of chain segment crossings per unit area, p, is independent of molecular weight in the virgin state, $p \sim M^0$. However, the number of chains intersecting the same plane decreases with increasing molecular weight as $n \sim M^{-1/2}$. Thus, by varying the molecular weight, we can reach a state where the number of bridges is comparable to the number of chains and an incipient network forms.

We define the number of bridges per chain, p_c, as

$$p_c = \tfrac{1}{2}(p/n - 1) \tag{7.4.1}$$

so that when $p_c = 1$, $p = 3n$, where the factor of three considers the three crossings per bridge. When $p_c = 1$, the entanglement network is critically connected; this value corresponds to a percolation threshold. The percolation threshold can be approached from above if p is decreased during stress relaxation (see section 7.4.6 on percolation corrections to viscosity), or if the number of chains is increased, say, by chain fracture; or $p_c = 1$ can be approached from below if M is increased, and hence n is decreased, at constant p. When p_c is less than 1, each chain contributes less than one bridge and the melt is not connected in a network. When p_c is greater than 1, an entanglement network exists and we can determine M_c from the condition that $p_c = 1$, or $p = 3n$. This argument can be readily tested because if the hypothesis is valid, M_c can be determined without any fitting parameters for all random-coil polymers, by a simple calculation of p and n, as follows.

The number of chains intersecting an arbitrary plane through random-coil chains is given in terms of the molecular weight and random-coil parameters as [31]

$$n = 1.31\,[C_\infty j/(M_0 M)]^{1/2}\, b\, \rho\, N_a \qquad (7.4.2)$$

where N_a, ρ, b, M_0, C_∞, and j are Avogadro's number, density, bond length, monomer molecular weight, characteristic ratio, and the number of bonds per monomer, respectively.

The number of chain segments crossing a unit area is given by

$$p = 1/a \qquad (7.4.3)$$

where a is the projection of the cross-sectional area of the chain segment on the plane, as shown in Figure 7.8. When $p = 3n$ at M_c, then Eq 7.4.3 provides that $3an = 1$.

The cross-sectional area of the chain segment, a, could be determined from the monomer volume, according to $a = (M_0/\rho N_a)^{2/3}$. However, this approach tends to ignore monomer aspect ratio, and conformational structure such as helical segments. A more useful approach is to determine the molecular area from unit-cell dimensions (when they exist) via the volume/length ratio, V/L, as

$$a = \sqrt{2}\, z M_0/(C \rho N_a) \qquad (7.4.4)$$

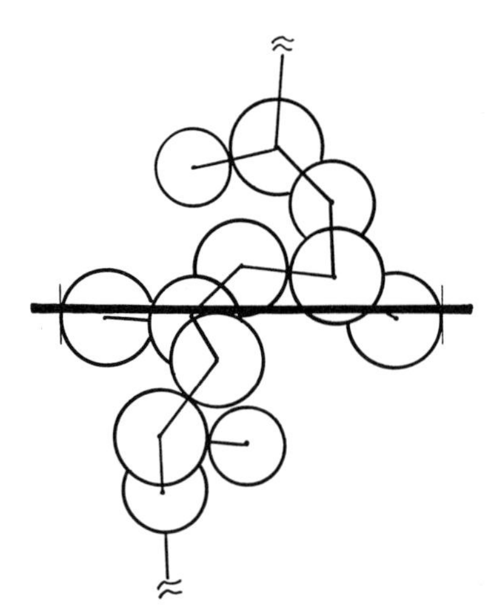

Figure 7.8 Chain segment with cross-sectional area a (note: $M_c \sim a^2$).

where the length $L = C$ is the c-axis length (backbone direction) of the unit cell, the volume is $V = zM_0/(\rho N_a)$, and z is the number of monomers per c-axis length. The factor of $\sqrt{2}$ accounts for the random orientation of chains crossing the plane, in contrast to their perfect alignment in a crystal. During a mechanical deformation, as the melt becomes anisotropic and the entanglement density changes, the change in the orientation factor of the chains must be considered.

Substituting for a, n, and p, and solving for M_c at $p_c = 1$, we obtain the expression for the critical entanglement molecular weight as

$$M_c = 30.89\,(zb/C)^2\,jM_0\,C_\infty \tag{7.4.5}$$

in which all quantities are known, or can be estimated. Also, we have $(zb/C)^2j \approx 1$ for many polymers, and a good approximation for M_c for vinyl ($j = 2$) polymers is

$$M_c \approx 30\,C_\infty\,M_0 \tag{7.4.6}$$

Note that $M_c \sim C_\infty$, which means that stiffer chains have generally higher M_c values.

Kavassalis and Noolandi [28] defined N_e as the length of chain to form one loop, and N_c as the number of steps to make one bridge. They devised a theory, called the KN theory, for N_e and N_c that is based on the number of other random-coil chains effectively constraining one particular chain in a tube as a function of molecular weight. They find that N_e depends on molecular weight and reaches a lower asymptotic value with increasing molecular weight. N_c is related to N_e via $N_c = 27/4\ N_e$, but because N_e is the length of chain to form one loop, our theory provides that

$$N_c = 9/4\,N_e \tag{7.4.7}$$

Generally, one finds from experiment that $N_c \approx 2\,N_e$, with minor deviations.

The KN theory determines M_c as

$$M_c \sim M_0^2/(\rho^2\,C_\infty^3\,b^6) \tag{7.4.8}$$

where the proportionality constant contains an unknown universal coordination number. Eq 7.4.8 suggests that $M_c \sim C_\infty^{-3}$, which differs strongly from our result, that $M_c \sim C_\infty$. Aharoni [27] analyzed a large body of experimental M_c and structural data, and found the empirical correlation $N_c \approx 10\ C_\infty^2$. There was enough scatter in the data to cast doubt on the exponent, 2, but the general trend of N_c increasing with C_∞ was convincing. Wu [32] finds from data analysis that $M_e \sim C_\infty^2$, in agreement with Aharoni. He also finds that the yield stress and the stress to induce crazing, σ_c, in glassy polymers are both proportional to C_∞, and that $\sigma_c \sim M_c^{-1/2}$.

In our model, N_c, the number of steps to make one bridge, is determined from Eq 7.4.5 as

$$N_c \approx 30 \tag{7.4.9}$$

This means that a random walker starting at a plane needs an average of about 30 steps to pass through the plane twice more. This problem is similar to the coin toss statistics problem where one bets a penny on heads or tails N times, and keeps track of the number of times one's winnings or debt returns to zero, $Z(N)$. The number of crossings from winning to losing, $I(N)$, is equal to half the number of returns to zero, plus one, or

$$I(N) = 1 + Z(N)/2 \tag{7.4.10}$$

where

$$Z(N) = 1 + \sum_{0}^{\infty} r\,W(N,r) / \sum_{0}^{\infty} W(N,r) \tag{7.4.11}$$

The function $W(N,r)$ is

$$W(N,r) = (N-r)!/[N!\,(N/2-r)!] \tag{7.4.12}$$

where r ranges from zero to infinity.

Equation 7.4.10 gives the number of crossings after N of 30 steps as $I(30) = 2.76$, which is in fair agreement with our random-coil analysis with $I(30) = 3$. This value also agrees with a random-walk computer simulation analysis with $I(30) = 2.74$ and with Bernoullian statistics, which gives $I(30) = 3.19$. The value of $N_c \approx 30$ appears ubiquitous for entanglement segments with three crossings.

The relation for the molecular weight dependence of the number of bridges per chain is $p_c \sim M^{1/2}$. Since $p_c = 1$ at M_c, it follows that the proportionality constant is $M_c^{-1/2}$, and

$$p_c = (M/M_c)^{1/2} \tag{7.4.13}$$

Thus, the total number of bridges per unit area in the virgin state, or at a fully healed interface, is given by $p_\infty = p_c\, n$, as

$$p_\infty = 1.31\,[C_\infty\, j/(M_0\, M_c)]^{1/2}\, b_0 N_a \tag{7.4.14}$$

Note that Eq 7.4.14 is independent of the chain molecular weight, M. This equation can also be used in the scaling law for bridges during welding of symmetric interfaces, via

$$p(t) = p_\infty\,(t/T_r)^{1/2} \tag{7.4.15}$$

The entanglement model presented above, unlike all others, contains no adjustable parameters, it contains only known constants, and therefore its utility can be tested by comparisons of theoretical and experimental M_c values.

7.4.2 The Critical Entanglement Molecular Weight, M_c

In this section we examine the ability of the entanglement model to predict values for M_c using Eq 7.4.5

$$M_c = 30.89\,(zb/C)^2\,jM_0\,C_\infty$$

and the short formula, $M_c \approx 30\ M_0\ C_\infty$, in Eq 7.4.6. In the calculations below, the crystal data were obtained from Tadokoro's book [33], and M_c data from review papers by Aharoni [27], Wu [32], Graessley [26], and Ferry [34]. Table 7.1 lists the molecular parameters used to investigate the entanglement model.

Table 7.1 Molecular Properties of Polymers

Polymer	Bond length b, Å	c-axis C, Å	Monomers /c-axis, z	Mol. wt., M_0	Char. ratio, C_∞	j
polyethylene	1.54	2.55[a]	1	28	6.7	2
polystyrene	1.54	6.5[b]	3	104	10	2
polypropylene	1.54	6.5[b]	3	42	5.8	2
poly(vinyl alcohol)	1.54	5.51[a]	2	44	8.3	2
poly(vinyl acetate)	1.54	6.5[b,c]	3	86	9.4	2
poly(vinyl chloride)	1.54	5.1[a]	2	62.5	7.6	2
poly(methyl methacrylate)	1.54	10.4[d]	4	100	8.2	2
poly(ethylene oxide)	1.51[e]	19.48[f]	7	44	4.2	3
polycarbonate	1.43[e]	20.8	2	254	2.4	12

a: planar zigzag; b: 3/1 helix; c: assumed; d: isotactic; e: average; f: 7/2 helix.

Polyethylene

With the parameters for polyethylene (PE) shown in Table 7.1, the entanglement model gives $M_c \approx 4{,}200$, which agrees with experimental values of $M_c \approx 4{,}000$. A planar zigzag (all *trans*) conformation of the PE chain in an orthorhombic unit cell was used to describe the cross-sectional area of the molecule. In the melt, we can expect many non-*trans* gauche conformers to be populated. However, the model is not very sensitive to the exact average conformational details because of compensating effects in the helix term, α, given by

$$\alpha = C/(zjb) \tag{7.4.16}$$

Here, large excursions of the internal rotation angles result in only small changes in the length C, and when z increases, C also increases. However, changes in α are amplified by the α^2 contribution, because $M_c \sim a^2$, where a is the cross-sectional area of a chain segment (Eq 7.4.4). The equation for M_c may be rewritten

$$M_c = 30.89\, C_\infty M_0 / (\alpha^2 j) \qquad (7.4.17)$$

The factor $\alpha^2 j$ is the difference between the exact and short formulas for M_c. Note that $\alpha^2 j \approx 1$ for a 3/1 helix with $j = 2$. For a 2/1 all-*trans* PE helix, $\alpha = 0.827$, but for a 3/1 *trans*–gauche PE helix, $\alpha = 0.703$. The latter conformation would give $M_c \approx 5{,}500$. The model predicts that M_c increases with increasing cross section of the chain, as $M_c \sim a^2$, corresponding to a helix with a greater number of monomers per turn.

The short formula (Eq 7.4.6) gives $M_c = 30 C_\infty M_0 = 5{,}600$. Thus, for molecules related to the PE family, such as low density polyethylene (LDPE), high density polyethylene (HDPE), and linear low density polyethylene (LLDPE), and polymers containing large $-CH_2$ sequences, such as polybutadiene, nylon 6, nylon 6–6, and nylon 6–10, we expect to have similar M_c values in the vicinity of $M_c \approx 4{,}000$–$5{,}000$, with some small variation due to differences in C_∞ and M_0. When the polymer molecule becomes highly polar or rigid, the model is not expected to work, since new mechanisms of transmitting forces are introduced.

Polystyrene

For PS, we obtain $M_c \approx 32{,}000$ from the data in Table 7.1. The experimental value is given as $M_c = 31{,}200$, although values as high as 36,000 and 38,000 have been reported [26, 34]. The cross-sectional area of a PS molecule was determined from the isotactic approximation for the helix factor, $\alpha = 0.703$. Syndiotactic and atactic configurations have similar cross sections and also give good agreement with theory. To obtain $M_c = 36{,}000$, the helix factor needs to change from $\alpha = 0.703$ to $\alpha = 0.654$, which represents a change from a 3/1 to about a 7/2 helix. The short formula gives $M_c = 31{,}200$.

Derivatives of PS, such as poly(α-methyl styrene) (PAMS), should have similar M_c values. For PAMS, the reported value [26] is $M_c = 28{,}000$ with $C_\infty = 10.5$ and $M_0 = 118$. Using Eq 7.4.17, we calculate that $\alpha^2 j = 1.37$ and $\alpha = 0.828$, which suggests that PAMS forms a tighter helical structure than PS ($\alpha \approx 0.703$). If PAMS had the same α value as PS, it would have a predicted M_c of 39,000. If the methyl group is on the phenyl ring in the *ortho* position, poly(*o*-methyl styrene) (POMS) forms a 4/1 helix with a periodicity of $C = 8.1$ Å and $\alpha^2 j = 0.66$. Thus, if $C_\infty \approx 10$ for POMS, we expect $M_c \approx 55{,}000$, which is much higher than that of PS. The experimental value has not been reported so comparison is not possible.

Polypropylene

For PP, we obtain (from Table 7.1) $M_c = 7{,}600$, which agrees well with the experimental value, $M_c \approx 7{,}000$. Although the isotactic configuration can adopt many local conformations in the melt, the average helix factor of $\alpha = 0.7$ and $\alpha^2 j = 0.99$ is not expected to be changed. The short formula gives $M_c = 7{,}300$.

Poly(vinyl alcohol)

For PVOH, we obtain (from Table 7.1) M_c = 7,000, in good agreement with experiment, M_c = 7,500. Even though PVOH is atactic, it adopts a planar zigzag conformation in the crystalline state, giving $\alpha = 0.89$ and $\alpha^2 j = 1.6$. Since the α value for a planar zigzag conformation cannot be increased further in the melt, we would have to increase C_∞ from 8.3 to 8.89 to obtain better agreement with experiment. The short formula gives M_c = 11,000, which is much higher than the experimental value because of the poor assumption in this case that $\alpha^2 j = 1$.

Poly(vinyl acetate)

For PVAc, we predict (Table 7.1) that M_c = 25,200, which agrees well with M_c = 24,500 from experiment. In this case, we did not have unit cell dimensions for this atactic amorphous polymer but could easily guess the helix factor. Bulky side groups in vinyl polymers tend to produce 3/1 helices with $\alpha = 0.703$. The short formula gives M_c = 24,300, which is in excellent agreement with experiment.

Poly(vinyl chloride)

For PVC, we predict (Table 7.1) that M_c = 10,700, which agrees well with experiment, $M_c \approx$ 11,000, based on $M_c \approx 2M_e$. An orthorhombic unit cell was used, with a planar zigzag conformation [33]. The short formula gives M_c = 12,600 for PVC.

Poly(methyl methacrylate)

For PMMA, we predict (from Table 7.1) that $M_c \approx$ 18,000, in good agreement with $M_c \approx$ 18,400 (using $M_c = 2M_e$). The c-axis length was determined from the isotactic conformation of the chain in an orthorhombic crystal. The short formula gives $M_c \approx$ 25,300 for PMMA, which is consistent with a 3/1 helix. The properties of PMMA are sensitive to the tacticity and conformation. For example, the characteristic ratio has values ranging from 7.4 for syndiotactic PMMA to 11.5 for some conventional PMMA. In the 3/1 helix approximation for the average cross section in the amorphous state, this variation in C_∞ gives M_c values ranging from 22,000 to 34,500. Graessley reports a value of M_c = 31,000 for PMMA with $C_\infty = 8.7$. Wu cites a value of M_e = 9,200 ($M_e \approx \frac{1}{2} M_c$) with $C_\infty = 8.2$, and Aharoni cites M_c = 31,500 with $C_\infty = 7.0$. These variations may be real, since PMMA has many structural forms. The agreement between our model and experiment is not considered to be in jeopardy, since we can account for this range of M_c values once the conformational details are known for a specific type of PMMA. Most commercially available PMMA is not pure and typically consists of a copolymer of methyl and ethyl acrylates (the latter monomer is used to provide thermal stability).

Poly(ethylene oxide)

For PEO, we obtain M_c = 5,000 (Table 7.1), in fair agreement with experiment, M_c = 4,400. In Table 7.1, the bond length $b = 1.51$ was determined from an average of one C–O and two C–C bonds. A 7/2 helix conformation in a monoclinic crystal was used to obtain the c-axis length. The short formula gives poor agreement, with $M_c \approx 8{,}800$ for PEO, since $\alpha^2 j = 1.79$. If we use $\alpha \approx 0.64$ for a 7/2 helix (instead of $\alpha = 0.614$) we obtain $M_c = 4{,}500$, which is closer to the experimental value. An analysis of poly(propylene oxide) gave similar agreement (Table 7.2).

Polycarbonate

For PC (of bisphenol A), we predict that $M_c = 4{,}300$, which is in good agreement with experiment, $M_c = 4{,}800$. The agreement between theory and experiment in this case is gratifying considering the complexity of the PC molecule's structure. In Table 7.1, the bond length $b = 1.43$ was obtained from the average of 4 C–O, 6 C=C, and 2 C–C bonds. The value of $j = 12$ was determined from the sum of (nonparallel) backbone bonds. The short formula should not be used for PC because $\alpha^2 j = 4.4$, and PC is not a simple vinyl polymer with $j = 2$, for which the formula was designed.

The predicted and experimental M_c values are summarized in Table 7.2. We conclude that the entanglement model has ample ability to predict M_c values for random-coil polymers.

Table 7.2 Theoretical and Experimental Critical Entanglement Molecular Weights, M_c

Polymer	M_c from experiment	M_c calculated from theory
polyethylene	4,000	4,200
polystyrene	31,200	32,000
polypropylene	7,000	7,600
poly(vinyl alcohol)	7,500	7,000
poly(vinyl acetate)	24,500	25,200
poly(vinyl chloride)	11,000	10,700
poly(methyl methacrylate)	18,400	18,000
poly(ethylene oxide)	4,400	5,000
poly(propylene oxide)	5,800	5,000
polycarbonate	4,800	4,300

7.4.3 Concentration Dependence of M_c

The concentration dependence of M_c is very important for solvent bonding and fracture of interfaces. A polymer with concentration c in solution can exist in dilute, semidilute, or concentrated states. In dilute solution, the polymer chains in the form of solvated random coils with radius of gyration $R_g \sim M^\nu$ do not overlap. The exponent ν depends on the quality of the solvent, being ⅓ for poor, ½ for theta, and up to 3/5 for good solvents, respectively. With increasing concentration in the dilute region, the random coils overlap eventually at the critical concentration, $c^* \approx M/(N_a V)$, where $V = 4/3\ \pi S^3$ is the volume occupied by the random-coil sphere. Thus we have the molecular weight dependence of c^* as

$$c^* \sim M^{(1-3\nu)} \tag{7.4.18}$$

where the exponent $(1 - 3\nu)$ has a value of 0, $-\frac{1}{2}$, and $-\frac{4}{5}$, for poor, theta, and good solvents, respectively.

When the concentration is in the range $c^* < c < 1$, the chains are in the semidilute region. The static and dynamic properties of polymers in this region have been examined by de Gennes [35, 36] and Daoud *et al.* [37]. Their analysis of polymers in good solvents is represented by the *blob model*, shown in Figure 7.9. Here, a chain is divided into N blobs of average size ξ, known as the *screening length*. Each blob is non-interpenetrating with the other blobs and behaves as a self-avoiding walk (SAW) solvent-swollen chain with exponent $\nu = 3/5$.

The concentration dependence of the screening length may be deduced as follows [35]. When c is equal to c^*, the screening length has the dimensions of the coil radius and therefore $\xi = R_g c^0$; when c is greater than c^*, the screening length becomes independent of the chain radius (and molecular weight) and depends only on concentration, $\xi \sim c^{\beta} R_g{}^0$, where β is an unknown exponent. This permits us to write the scaling law [35]

$$\xi(c) = R_g (c/c^*)^{\beta} \tag{7.4.19}$$

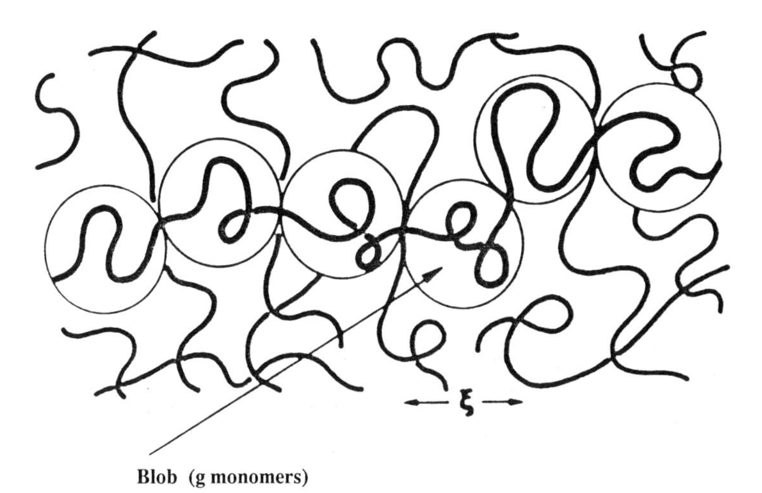

Figure 7.9 Blob model for a chain in the semidilute region (courtesy of P.-G. de Gennes).

where β is the exponent to be determined. Eliminating the molecular weight dependence of $\xi(c)$ at $c > c^*$ in Eq 7.4.19 requires that

$$\beta = \nu/(1-3\nu) \tag{7.4.20}$$

With $\nu = 3/5$, we obtain by substitution $c^* \sim M^{-4/5}$ (from Eq 7.4.18), $R_g \sim M^{3/5}$, and $\beta = -3/4$, and we have the important result

$$\xi(c) \sim c^{-3/4} \tag{7.4.21}$$

Since each blob has the static properties of an SAW, the number of monomers in a blob, g, is related to ξ via

$$\xi \sim g^{3/5} \tag{7.4.22}$$

From the last two equations, it follows that

$$g \sim c^{-5/4} \tag{7.4.23}$$

Since the number of blobs per chain is $N \sim M/g$, N depends on concentration as

$$N(c) \sim c^{5/4} \tag{7.4.24}$$

This result becomes the controlling factor for the concentration dependence of M_c.

We now enquire as to the dependence of the mean-square radius of gyration, $R_g^2(c)$, on concentration in the semidilute region. The question is whether the chain behaves as a simple random walk or an SAW. A chain with N blobs of step length ξ can be described by

$$R_g^2 \sim N^{2\nu}\xi^2 \tag{7.4.25}$$

If the chain behaves as a simple random walk with $\nu = 1/2$, then we have from the concentration dependence of N and ξ that

$$R_g^2(c)/M \sim c^{-1/4} \tag{7.4.26}$$

On the other hand, if the entire swollen chain adopts SAW characteristics with $\nu = 3/5$, then

$$R_g^2(c) \sim c^0 \tag{7.4.27}$$

To address this question, a critical experiment was conducted by Daoud *et al.* [37] using small-angle neutron scattering on deuterated polystyrene in a good solvent (carbon disulfide). They found that in the semidilute region, R_g^2 decreased with increasing concentration of polymer with an exponent of -0.25 ± 0.02, in agreement with Eq 7.4.26. The screening length behaved as $\xi \sim c^{-0.72 \pm 0.06}$, which is in excellent agreement with theory. This confirms their hypothesis that an entangled chain in the semidilute region obeys simple random-walk statistics, while the step length, ξ, has SAW characteristics. Or, when we add solvent to a polymer, the polymer swells, and the radius of gyration increases in accord with $R_g^2(c) \sim c^{-1/4}$. These results are now applied to determine the concentration dependence of M_c.

When the number of bridges, $p(c)/3$, equals the number of chains, $n(c)$, intersecting any plane in the semidilute region, we have from Eq 7.4.1 that a critically connected state exists at

$$3\, n(c)\, a(c) = 1 \tag{7.4.28}$$

where $a(c)$ is the concentration-dependent cross section of a chain. From the blob model, we see that $a(c)$ is controlled by $\xi(c)$ with SAW properties, while $n(c)$ is controlled by $R_g(c)$ with simple random-walk properties. Since $n(c) \sim N_v S(c)$ and $Nv \sim c/M$, it follows that $n(c) \sim M^{-1/2} c^{7/8}$. In terms of the dimensionless concentration $\phi = c/\rho$, the latter relation has the scaling law

$$n(\phi) = n_\infty\, \phi^{7/8} \tag{7.4.29}$$

where $n_\infty \sim M^{-1/2}$ is determined at $\phi = 1$ using Eq 7.4.2.

The chain segment cross section, $a(c)$, is determined from the volume of the chain, $V = N\xi^3$, divided by its length, $L = N\xi$, so that $a(c) \sim \xi^2$, which is the cross-sectional area of a single blob. Therefore, $a(c) \sim c^{-3/2}$, or

$$a(\phi) = a(1)\, \phi^{-3/2} \tag{7.4.30}$$

where $a(1)$ is given by Eq 7.4.4. The assumption in Equation 7.4.30 is that the blobs are close-packed and non-interpenetrating, which are basic assumptions of the blob model.

Substituting for $a(\phi)$ and $n(\phi) \sim M_c^{-1/2}$ in Eq 7.4.28, we obtain the concentration dependence of M_c as

$$M_c(\phi) = 30.89\, (zb/C)^2\, jM_0\, C_\infty\, \phi^{-5/4} \tag{7.4.31}$$

where $\phi^* < \phi < 1$, and all other parameters are determined at $\phi = 1$. This means that when solvent is added to a polymer, its entanglement molecular weight increases (consistent with the blob model) and its entanglement density decreases as the chains swell. Eq 7.4.31 is in excellent agreement with experiments where it is found that $M_c(\phi) \sim \phi^{-x}$, in which the exponent x is approximately equal to or slightly greater than unity.

7.4.4 Entanglement Density and Plateau Modulus

The entanglement density, v_e, is determined by

$$v_e \sim \phi / M_c(\phi) \tag{7.4.32}$$

From Eq 7.4.31, the entanglement density has a concentration dependence in the semidilute and concentrated regions, as

$$v_e(\phi) = v_e(1)\,\phi^{9/4} \tag{7.4.33}$$

where $v_e(1) = \rho N_a / M_e(1)$. This relation is important in solvent bonding and is related to the elastic properties of entanglement networks through the plateau modulus, G_N^0, via

$$G_N^0 \sim v_e R T \tag{7.4.34}$$

where R and T are the gas constant and temperature, respectively. From Eq 7.4.33, the concentration dependence of the plateau modulus is obtained as

$$G_N^0(\phi) = G_N^0(1)\,\phi^{9/4} \tag{7.4.35}$$

When solvent is added to an entangled polymer, ϕ decreases and the stiffness decreases as $\phi^{2.25}$. The reader may have experienced a familiar example of the result of this equation, the softening of chewing gum as the uptake of moisture causes ϕ to decrease, with a resultant decrease in stiffness.

Graessley [26] examined the concentration dependence of G_N^0 for PMMA and *cis*-polyisoprene, and found that G_N^0 was closely proportional to ϕ^2. Ferry [34] reviewed solvent effects on viscosity and plateau modulus. Polystyrenes with a range of molecular weights in solutions of benzyl *n*-butyl phthalate at 100 °C gave exponents that were between 2.2 and 2.3. He notes that G_N^0 solution data for polybutadiene, which do not require temperature reduction for comparison, also give an exponent between 2.2 and 2.3, in agreement with the above model.

7.4.5 Viscosity and Relaxation Times

The zero-shear viscosity, $\eta_0(c)$, is determined as a function of concentration c by the Doi–Edwards theory [38] as

$$\eta_0(c) = G_N^0(c)\, T_r(c) \tag{7.4.36}$$

where $T_r(c)$ is the concentration-dependent reptation time. $T_r(c)$ is given by

$$T_r(c) = \tau_e(c)[M/M_e(c)]^3 \tag{7.4.37}$$

in which $\tau_e(c)$ is the entanglement relaxation time and $M_e(c)$ is the entanglement molecular weight corresponding to the onset of the plateau modulus. We assume that M_e and M_c have the same scaling properties and are related by $M_c \approx 9/4M_e$. A single blob contains g monomers with a concentration dependence $g \sim c^{-5/4}$. The blob relaxation time, $\tau_0(c)$, is determined by Rouse dynamics, so that $\tau_0(c) \sim g^2$ and hence $\tau_0(c) \sim \tau_0(1)c^{-2.5}$. The entanglement relaxation time is similarly controlled by Rouse dynamics, so that $\tau_e(c) = \tau_0(c)N_e^2(c)$, where the number of blobs per entanglement is $N_e(c) \sim M_e(c)/g(c)$. Since $M_c \sim c^{-5/4}$ and $g \sim c^{-5/4}$, then $N_e(c) \sim c^0$, which is correct since $N_e \approx 15$ steps and is independent of concentration: only the blob step length contains the concentration dependence. Thus, τ_0 and τ_e have the same concentration dependence, $c^{-5/2}$. The concentration dependence of the reptation time is obtained from Eq 7.4.37 as

$$T_r(c) \sim T_r(1)\, c^{5/4} \tag{7.4.38}$$

Substituting Eq 7.4.38 and Eq 7.4.37 into Eq 7.4.36 gives the concentration and molecular weight dependence of viscosity as

$$\eta_0(c) = G_N^0\, \tau_e(1)\, [M/M_e(1)]^3\, c^{3.5} \tag{7.4.39}$$

The prediction that $\eta_0(c) \sim c^{3.5}$ is in excellent agreement with experimental data reviewed by Ferry [34]. However, the prediction that $\eta_0 \sim M^3$ does not agree with the generally accepted experimental law, $\eta_0 \sim M^{3.4}$, and has been an unsolved problem for decades. Despite the closeness of the exponents (3.4 and 3.0), the exponent of 3.4 is unacceptable within the reptation framework for the following reason. The reptation time is related to the self-diffusion coefficient, D, and end-to-end vector, R, by [35]

$$T_r = R^2/(3\pi^2 D) \tag{7.4.40}$$

so that $D(c) = D(1)\ c^{-1.5}$. It is firmly established that $R^2 \sim M$, and this requires that $D \sim M^{-2.4}$ in order to have $T_r \sim M^{3.4}$. However, the exponent of -2.4 has not been realized in the majority of diffusion studies, where the data strongly support $D \sim M^{-2}$, which is consistent with $T_r \sim M^3$ in Eq 7.4.40.

7.4.6 The Percolation Correction to Viscosity

Our entanglement model suggests that the relaxation of the network occurs in less than the reptation time. Each stressed chain relaxes by reptation until a critically connected state is reached corresponding to $p_c = 1$. This is physically analogous to the vector percolation relaxation mechanism for networks discussed in Chapter 4. In order for a step-strained network to relax to zero stress, it is not necessary for every bond to relax, but only a certain fraction corresponding to the percolation threshold. The viscosity is determined by the area under the modulus relaxation function. With random bond removal from the network, it is easy to show that the relaxation time at the percolation threshold is shorter than T_r, but the viscosity is still proportional to T_r and M^3. However, with the percolating entanglement model, disentanglement by reptation has an additional molecular weight dependence at the critically connected state via $p_c = ½(p/n - 1)$ (Eq 7.4.1).

The parts of the chain that have reptated from their tubes are nominally in a stress-free state, but exist in an environment of unrelaxed entanglements. The stressed entanglements at early relaxation times are interconnected in a percolating network with some holes due to the relaxed fraction. As relaxation continues, the network develops more holes and finally becomes disconnected at the percolation threshold, $p_c = 1$. The number of stressed bridges per unit area, $p(t)$, is given by

$$p(t) = p_\infty [1 - (t/T_r)^{1/2}] \tag{7.4.41}$$

where the term in square brackets has the same time dependence as the stress relaxation modulus $G(t)$ in the Doi–Edwards function at t less than T_r. The terminal relaxation time τ_c occurs at $p_c = 1$, corresponding to $p(t)/n_\infty = 1$, where the number of chains, n_∞, remains constant. Since the entanglement model gives $p_\infty/n_\infty = (M/M_c)^{1/2}$, the terminal relaxation time is related to the reptation time $T_r \approx \tau_e(M/M_e)^3$ by

$$\tau_c = \tau_e(M/M_e)^3 [1 - (M_c/M)^{1/2}]^2 \tag{7.4.42}$$

When M becomes very large, the terminal relaxation time approaches T_r.

The fraction of the network that relaxes by reptation is $\Phi_r = [1 - (M_c/M)^{1/2}]$, and the fraction relaxing by Rouse–like processes is $\Phi_{RO} = (M_c/M)^{1/2}$. Letting $\eta_r = \Phi r G_N{}^0 \tau_c$ and $\eta_{RO} = \Phi_{RO} G_N{}^0 \tau_{RO}$, where the Rouse time is related to T_r by $\tau_{RO} = 4/27(M_c/M)T_r$, we obtain the zero-shear viscosity as

$$\eta_0 \approx G_N^0 \tau_e (M/M_c)^3 \{[1 - (M_c/M)^{1/2}]^3 + 4/27(M_c/M)^{3/2}\} \tag{7.4.43}$$

Eq 7.4.43 behaves empirically as $\eta_0 \sim M^{3.4}$ for M greater than M_c and converges very slowly ($M/M_c > 100$) to an exponent of 3 with increasing molecular weight, as shown in Figure 7.10. Typically, plastics are tested in the range $M_c < M < 100M_c$, and the exponent of 3.4 appears to be universal.

Doi [39] derived a similar relation using a different physical concept, namely, chain-end fluctuation (CEF). In the CEF model, Doi proposes that Rouse–like fluctuations of the chain ends effectively shorten the tube length and result in a terminal relaxation time similar to that given in Eq 7.4.42. Both models decouple the time dependence of the center-of-mass motion (diffusion) from stress relaxation (viscosity) and produce a pseudo-3.4 exponent. However, the major difference between

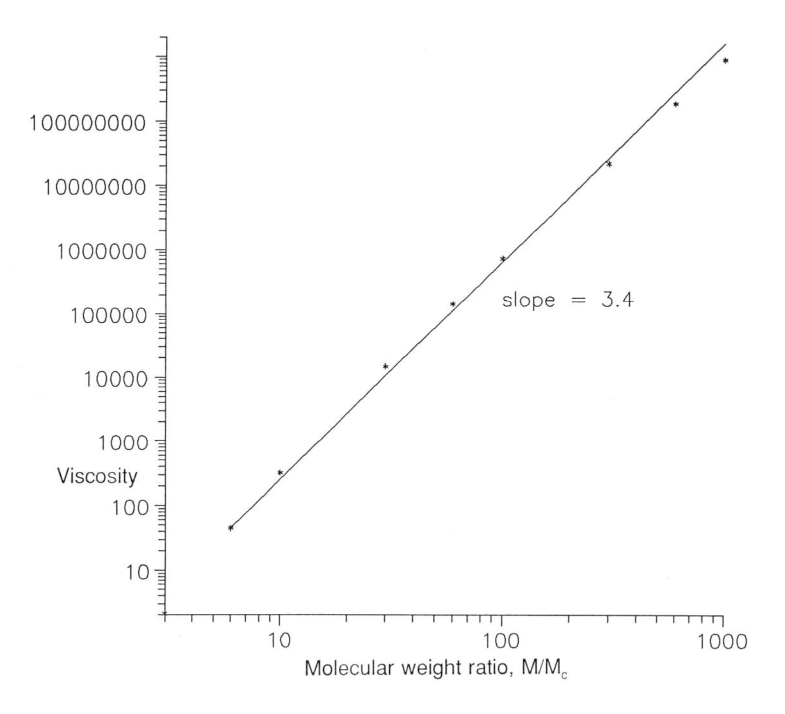

Figure 7.10 Log viscosity versus log M/M_c. A slope of 3.4 is shown through the theoretical data points derived for the percolation correction to viscosity.

our approach and Doi's is that we have a cooperative multichain model of relaxation, while his approach and many others retain the modified single-chain model. The single-chain models require all entanglements to relax with a characteristic time T_r', but networks do not require this and relax earlier at τ_c less than T_r'.

The above analysis on viscosity and diffusion brings up an interesting point: while the dynamics of the chains are the same for diffusion and viscosity, the manifestation of the dynamics leads to slightly different exponents, 3.4 versus 3.0 for relaxation times. This occurs because we can treat diffusion of polymer melts by a single-chain analysis (such as the reptation theory), while relaxation of chains in a network is a cooperative multichain process. Thus, for linear chains, the relaxation time should be less than the diffusion time for a molecule to move a distance of order R_g. This effect becomes more pronounced for multi-arm star polymers, which can stress-relax rapidly by retraction and Rouse–like fluctuations of the arms while the center of mass hardly moves.

The fractal structure of the connected regions in the relaxing network near the percolation threshold zone should have some interesting self-similar effects on the relaxation spectrum.

7.4.7 Analysis of the Diameter of the Reptation Tube

In the reptation model, the tube diameter d_t is controlled by the entanglement spacing, which is proportional to the end-to-end vector, $R_e \sim M_c^{1/2}$, of the entanglement molecular weight, M_e. Since $d_t \sim M_e(\phi)^{1/2}$, we expect the concentration dependence to be

$$d_t \sim \phi^{-5/8} \tag{7.4.45}$$

This result has been confirmed by Richter *et al.* [40] using neutron spin-echo (NSE) analysis of model amorphous polymers. Their NSE studies on polyisoprenes, poly(ethylene–propylene) copolymers, and polybutadienes show that, beyond a characteristic length scale (d_t) and relaxation time (τ_e), the relaxational density fluctuations within a given chain are strongly impeded. The microscopically determined length scales are highly consistent with the existence of an entanglement network, which is the framework for the reptation model.

Our entanglement model predicts that the tube diameter is given in terms of molecular parameters as

$$d_t = 3.93\, C_\infty\, b/\alpha \tag{7.4.46}$$

where $\alpha = C/zjb$. For example, with polyethylene, $C_\infty = 6.7$, $b = 1.54$ Å, and $\alpha = 0.827$, so that $d_t = 49$ Å. For polystyrene, using data from Table 7.1, we obtain $d_t = 86$ Å.

The tube diameter is related to the chain cross-sectional area a by $d_t \sim a$. With increasing temperature, the molecular thermal expansion is anisotropic, and the cross-sectional area increases while the c-axis length typically decreases, as the higher energy torsional angles become more populated. Consequently, C_∞ decreases slightly but the factor $1/\alpha \sim z/C$ increases in proportion to the molecular cross-sectional area thermal expansion. Since the thermal expansion effect is greater than the change in characteristic ratio, the tube diameter should increase with temperature, as noted by Richter *et al.* [40].

7.5 Deformation and Disentanglement

In this section, we explore how entanglement networks can strain harden and disentangle to produce fracture.

7.5.1 Strain Hardening and Fiber Drawing

When a polymer chain in a concentrated melt is subjected to a uniaxial extension, the maximum draw ratio λ_m is determined by

$$\lambda_m = L/R = N^{1/2} \tag{7.5.1}$$

where $L = Nb$ is the extended chain contour length and $R = N^{1/2}b$ is the end-to-end vector. At λ_m, the chain is fully extended and can extend further only by small amounts involving much higher stresses along the backbone bonds of the chain. This is called *strain hardening*, and the resulting melt is in an optimally oriented state. This process is important in fiber or film drawing and plays a very important role in determining the structure of craze fibrils and the deformation zone at crack tips.

When a step strain is rapidly applied to an entangled melt, strain hardening occurs at the level of entanglements. As described by Donald and Kramer [41], who used the rubber elasticity analogy presented by Flory, the strain is taken up by the slack between entanglements, and $\lambda_{sh} \sim M_e^{1/2}$. The contour length between entanglements is determined by $L_e \sim N_e \xi \lambda(\xi)$, where $N_e \approx 16$, $\xi \sim c^{-3/4}$ and $\lambda(\xi)$ is the draw ratio of the blob. When the applied forces are sufficient to straighten out the slack in each blob (containing $g \sim c^{-5/4}$ monomers), it follows that $\lambda(\xi) = gb(1)/\xi$, and $\lambda(\xi) \sim c^{-1/2}$, where $b(1)$ is the bond length of the concentrated state ($c = 1$). This simple assumption may suggest a stress-induced phase transition, since it requires the resulting strain-hardened fibers to have the same local structure in the presence of solvent as in the fully dense state. Thus, substituting for the contour length $L_e \sim N_e \xi \lambda(\xi)$ and $R_e \sim N_e^{1/2} \xi$, we obtain the universal result for the strain-hardening draw ratio, $\lambda_{sh} = L_e/R_e$, as

$$\lambda_{sh}(c) \approx 4\, c^{-1/2} \tag{7.5.2}$$

It is a striking prediction of the entanglement model that all polymer melts strain harden at about the same draw ratio of $\lambda_{sh} \approx 4$ at $c = 1$. Donald and Kramer examined the draw ratio of craze fibrils, λ_f, in several glassy polymers using a densitometry analysis of TEM images. They found fibrillar draw ratios in the range of 2–4. Recently, Kuntz *et al.* [74], using atomic force microscopy (AFM) to study crazing in polystyrene and other glassy polymers, found a maximum draw ratio of about 4, consistent with this model. When solvent is added, the draw ratio increases as $c^{-1/2}$, which has application to fiber manufacturing.

When fibrils are drawn, for example, during solvent spinning of textile fibers in high-elongation flow gradients, entanglements limit the drawing and alignment at the molecular level where it is desired to have fully extended chains. If a good solvent is added to a polymer, the optimal draw ratio is obtained when the concentration-dependent entanglement molecular weight approaches the chain molecular weight, M. Thus, when $M = M_e(1)c_{\mathrm{opt}}^{-5/4}$, the optimal polymer concentration for fiber drawing in terms of molecular weight is

$$c_{\mathrm{opt}} = (M_e/M)^{4/5} \tag{7.5.3}$$

Eq 7.5.3 can also be obtained by letting $\lambda_{sh} = \lambda_m$ and solving for c_{opt}. Note that $\lambda_{sh}(c_{\mathrm{opt}}) = 4(M/M_e)^{2/5}$. When $M \approx 16M_e$, then $c_{\mathrm{opt}} \approx 11\%$ polymer in solution and $\lambda_m \approx 12$, which is less than λ_m determined for the dry state ($c = 1$) using Eq 7.5.1.

7.5.2 Disentanglement and Fracture

How do highly interpenetrated random coil chains disentangle to cause fracture? We now apply the entanglement model to examine disentanglement as a function of a uniaxial draw ratio, λ, in amorphous polymers. The number of chains n crossing a plane normal to the applied strain increases linearly with draw ratio as

$$n(\lambda) = \lambda\, n_\infty \tag{7.5.4}$$

where n_∞ is the number of chains per unit area at $\lambda = 1$. This is the dominant effect in the disentanglement process. The cross-sectional area decreases in an affine manner with increasing draw ratio according to

$$a(\lambda) = a_\infty\, (1 + 1/\lambda^3)^{1/2}/\sqrt{2} \tag{7.5.5}$$

where a_∞ is determined at $\lambda = 1$. At the critical point $3a(\lambda)n(\lambda) = 1$, and from this relation and Eq 7.5.5, the critical draw ratio is determined by

$$\lambda_c^2 + 1/\lambda_c = 2\, M/M_c \tag{7.5.6}$$

Solving this cubic equation, we obtain λ_c in terms of M/M_c as

$$\lambda_c = (8/3\, M/M_c)^{1/2} \cos\{1/3 \cos^{-1} [-1/2\, (3\, M_c/2M)^{3/2}]\} \tag{7.5.7}$$

or

$$\lambda_c = F(M/M_c) \tag{7.5.8}$$

where the function $F(M/M_c)$ depends only on M/M_c. Eq 7.5.8 predicts that $\lambda_c = 1$ when $M/M_c = 1$. Also, $\lambda_c \approx 4$ when $M/M_c \approx 8$, which corresponds to the maximum draw ratio, λ_m, to cause strain hardening between entanglements (Eq 7.5.3). Figure 7.11 shows the dependence of λ_c on M/M_c. The significance of λ_c is that it is the minimum strain to cause fracture by disentanglement, and we thus expect $G_{1c} \sim (\lambda_c - 1)^2$.

When M is much greater than M_c, a useful approximate solution to Eq 7.5.7 gives the critical draw ratio for disentanglement as

$$\lambda_c \approx (2M/M_c)^{1/2} \tag{7.5.9}$$

In terms of the connectivity criterion $3a(\lambda)n(\lambda) = 1$, the approximate solution implies that the cross-sectional area of a chain segment, $a(\lambda)$, does not change with deformation. This is reasonable over most of the data range in Figure 7.11, where the slope is ½, but fails near $M/M_c \approx 1$ where Eq 7.5.9 incorrectly predicts $\lambda_c \approx \sqrt{2}$. This approximation should be remembered later when we make extensive use of Eq 7.5.9.

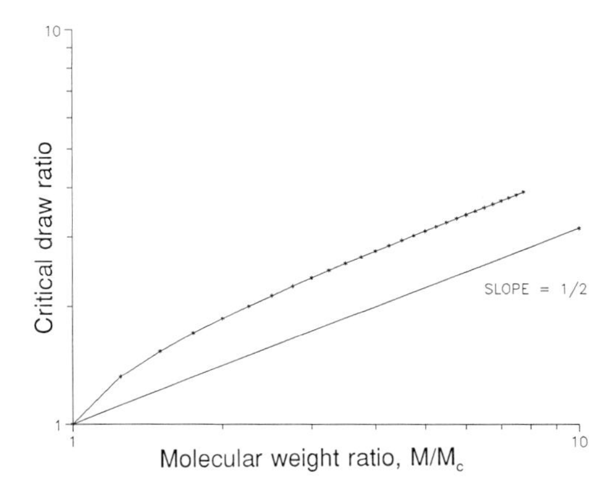

Figure 7.11 Log–log plot of critical draw ratio versus molecular weight ratio, M/M_c. A slope of ½ is indicated.

We believe disentanglement occurs as shown in Figure 7.12, where we depict the response of an entangled chain to a constant (step function) draw ratio, λ, as follows:

A. The average entangled chain with M/M_c greater than 1 is uniaxially deformed to a constant draw ratio, λ. The extension is accommodated by elongation of the random walk (slack) between entanglements so that the end-to-end vector between entanglement points behaves as $R_e(\lambda) = \lambda R_e$. The end-to-end vector R of the whole chain behaves similarly, $R(\lambda) = \lambda R$. The entanglement points deform affinely and the chain stores elastic strain energy. The increase in the primitive path length L between entanglements with stretching behaves as $L(\lambda) = \lambda L_0$, where $L_0 = MR_e/M_e$ is the unperturbed primitive path at $\lambda = 1$. When the stretched path relaxes to the critical length $L_c \approx 2\lambda R_e$, the chains become critically connected and disentangle.

B. Rouse dynamics causes a retraction of the extended chain primitive path length, $L(\lambda)$, and the stored strain energy begins to be released. For an excellent visualization of chain retraction dynamics, see Perkins, Smith, and Chu, "Direct Observation of Tube-Like Motion of a Single Polymer Chain" [76]. As the chain shortens towards its equilibrium path length, it begins to lose entanglements and becomes critically connected at L_c. The time dependence of the retraction process can be approximated as a simple exponential; the stressed fraction of the primitive path $L(t)$ as a function of time is $L(t)/L(\lambda) \approx \exp -t/\tau_{\text{RO}}$, where τ_{RO} is the Rouse relaxation time of the chain.

C. When the chain retracts to a critical length $L_c = 2R_e\lambda$, each chain possesses one bridge and the network becomes critically connected. This state corresponds to the failure time of the entanglement network, τ_f, and is determined by

$$\tau_f \approx \tau_{\text{RO}} \ln (M/M_c) \tag{7.5.10}$$

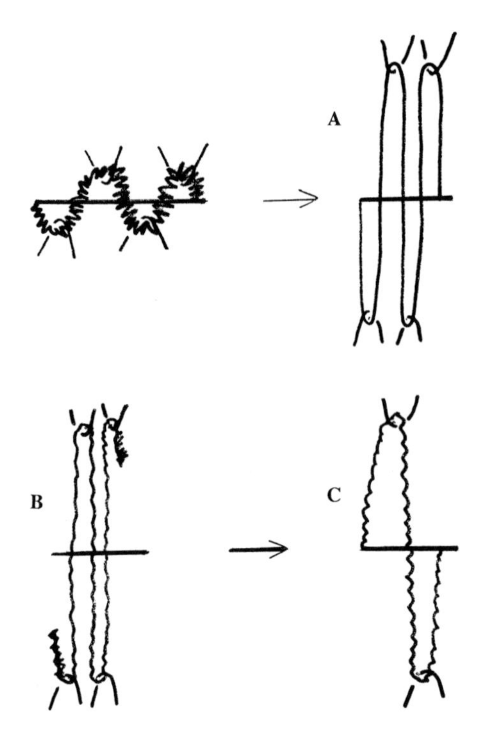

Figure 7.12 Disentanglement mechanism: (A) tightened slack between entanglements; (B) retraction and disentanglement of chain by Rouse relaxation; (C) critically connected state.

If M is approximately equal to M_c, disentanglement is nearly instantaneous but approaches τ_{RO} when $M \approx 8M_c$, the upper bound for chain pullout without bond rupture. Note that the Rouse time, and not the reptation time, plays the dominant role in the disentanglement process. If M is greater than $8M_c$, the chains cannot disentangle completely at the Rouse time and begin to reentangle due to reptation; bond rupture would be necessary to complete the fracture process, and the value of $M^* \approx 8M_c$ sets an upper limit for fracture by disentanglement or chain pullout.

Thus, fracture occurs as the chains are first strained to a critical draw ratio λ_c and mechanical energy $G \sim (\lambda_c - 1)^2$ is stored. The chains relax by Rouse retraction and disentangle if the energy released is sufficient to relax them to the critically connected state. Since $\lambda_c \sim \sqrt{M/M_c}$, we expect the molecular weight dependence of fracture to behave approximately as $G \sim (\sqrt{M/M_c} - 1)^2$ when M is in the range $M_c \leq M \leq 8M_c$.

If the stored strain energy cannot be released by viscous relaxation processes involving disentanglement of the whole chain, bond rupture becomes necessary. This is very important in glassy polymers, where the craze fibrils at the crack tip undergo rupture (discussed in Chapter 8, Section 8.12). The strained entanglement network becomes disconnected by bond rupture when the critical connectivity relation is obeyed, that is, $n = \frac{1}{3} a$. When M is much greater than M_c, n is much smaller than $\frac{1}{3} a$, but n can be increased by breaking N_f bonds, so that the critical condition is met when

$(N_f + n) = \frac{1}{3} a$. Solving for N_f and substituting Eq 7.4.4 for a, we determine the minimum number of broken bonds to cause fracture by

$$N_f = [1 - (M_c / M)^{1/2}]\, C\rho\, N_a / (3\sqrt{2}\, zM_0) \tag{7.5.11}$$

For example, using polystyrene values for C, z, and M_0 (Table 7.1) and a density $\rho \approx 1.0$ g/cm^3, we obtain $N_f \approx 3.0 \times 10^{13}$ $[1 - (M_c/M)^{1/2}]$ /cm^2, in which the numerical coefficient represents the asymptotic number of bridges per unit area (3 crossings per bridge) when M goes to infinity. We do not expect any bond rupture, even for glassy polymers, when $M \approx M_c$, as demonstrated in Chapter 8.

A most interesting point is that the energy to fracture bonds is very small (about 1 J/m^2), and the stored energy released to cause fracture of an entanglement network is about the same whether the chains disentangle or break. This is why the above approach utilizing disentanglement had been very effective in the elucidation of glassy fracture processes, which are dominated by bond rupture rather than chain pullout. In either case, the breaking of the entanglement network is dominated by a length scale that controls both the stored strain energy and the molecular weight dependence of fracture. This point becomes more apparent from welding studies, where experiments support contour length arguments of strength development.

7.5.3 Disentanglement and Fracture of Interfaces

For the welding of amorphous polymer interfaces, the average contour length $<l>$ controls the disentanglement process. Consider entangled amorphous chains of length $L \sim M$ and degree of polymerization $N = M/M_c$. The contour length has contributions from reptation, $L(t/T_r)^{1/2}$, and Rouse dynamics, L_c, which combine approximately via the weighted segmental dynamic contribution (see Chapter 3) as

$$<l> = \tfrac{1}{2} L(t / T_r)^{1/2} + \tfrac{1}{2} L_c[1 - (t / T_r)^{1/2}] \tag{7.5.12}$$

Thus, when $t/T_r \approx 1/N \approx 0$, we find $<l> \approx \frac{1}{2}L_c$, and when $t/T_r = 1$, we have $<l> = \frac{1}{2}L$. Rapid segmental diffusion results in an initial contour length corresponding to $L_c \sim M_c$ at times on the order of the Rouse relaxation time, τ_e, of an entanglement segment.

The critical draw ratio λ_c is approximately equal to $[2<l>/L_c]^{1/2}$, so that we obtain the healing time dependence as

$$\lambda_c(t) \approx [(t/ T_r)^{1/2} (M / M_c - 1) + 1]^{1/2} \tag{7.5.13}$$

where $<l> \leq 8L_c$ and $\lambda_c(t) \leq 4$. When $t = T_r$, we obtain $\lambda_c \approx (M/M_c)^{1/2}$, and when $M = M_c$, we find $\lambda_c(t) \approx 1$.

At fracture, the strain energy density, U_0, is proportional to G_{1c} (Eq 7.2.12). Using $U_0 = E(\lambda_c - 1)^2/2$ (other expressions for U_0 can be used if multiaxial stresses are involved, as discussed later), we have $G_{1c}(t) \sim E(\lambda_c - 1)^2$, and in the approximate form

$$G_{1c}(t) \sim E\{[(t/T_r)^{1/2}(M/M_c - 1) + 1]^{1/2} - 1\}^2 \tag{7.5.14}$$

Figure 7.13 shows a log–log plot of the exact calculation (using Eq 7.5.7) for G_{1c} versus t/T_r. The data are not exactly linear but behave approximately as $G_{1c} \sim t^{0.6}M^{-0.4}$. With increasing molecular weight and with plane-strain conditions at the crack tip (see Chapter 8), the data converge on the pure reptation prediction, $G_{1c} \sim t^{1/2}M^{-1/2}$, or

$$G_{1c}(t) = G_{1c\infty}(t/T_r)^{1/2} \tag{7.5.15}$$

where $G_{1c\infty}$ is the virgin strength when $M \leq 8M_c$.

The welding time τ to achieve complete strength behaves as $\tau \sim M^3$ when $M \leq M^*$, where $M^* \approx 8M_c$. When M is greater than M^*, the welding time is determined from Eq 7.5.12 by $\tau^* \sim M^{*2}M$, and the welding relation is determined by

$$G_{1c}(t) = G_{1c}^*(t/\tau^*)^{1/2} \tag{7.5.16}$$

where $G_{1c}{}^*$ is the maximum strength obtained at M^* and is independent of M. Even though the welding time ($\tau^* \sim M^{*2}M$) is much shorter than $T_r \sim M^3$, the dependence of the welding rate on M remains unaffected since we still have $G_{1c}(t) \sim t^{1/2}M^{-1/2}$. The latter condition is necessary since the diffusing molecules are unaware that an upper limit will be reached before they diffuse a distance equal to their radius of gyration. Thus, chains need not interpenetrate completely to achieve complete strength when M is much greater than M^* and τ^* is much less than T_r. This is a unique and necessary feature of this (contour length) model, and makes it different from all bridge models of welding. The critical stress intensity factor is related to G_{1c} via $K_{1c} \sim \sqrt{G_{1c}}$, and hence we expect the scaling law for welding to be $K_{1c}(t) = K_{1c}{}^*(t/\tau^*)^{1/4}$ when $M \geq M^*$.

7.5.4 Molecular Weight Dependence of Fracture

As t/T_r goes to unity in welding experiments, we recover the molecular weight dependence of the fracture energy in the virgin state. From Eq 7.5.14 we obtain the useful relation

$$G_{1c} \sim M/M_c[1 - (M_c/M)^{1/2}]^2 \tag{7.5.17}$$

in the range $M_c < M < 8M_c$. When M is greater than $8M_c$, we expect $G_{1c} \sim M^0$. It is interesting to note that a plot of log G_{1c} versus log M has an average slope between 2 and 3 in this range of molecular weights, as shown in Figure 7.14. (This accidental slope has led some to postulate that $G_{1c} \sim M^2$, which unfortunately finds misconstrued support from the fact that the energy to pull a chain out behaves as $U_p \sim M^2$). When the fracture energy reaches the plateau, complete disentanglement can be accommodated by bond rupture. We will see in Chapter 8 that bond rupture essentially adds nothing to the total fracture energy.

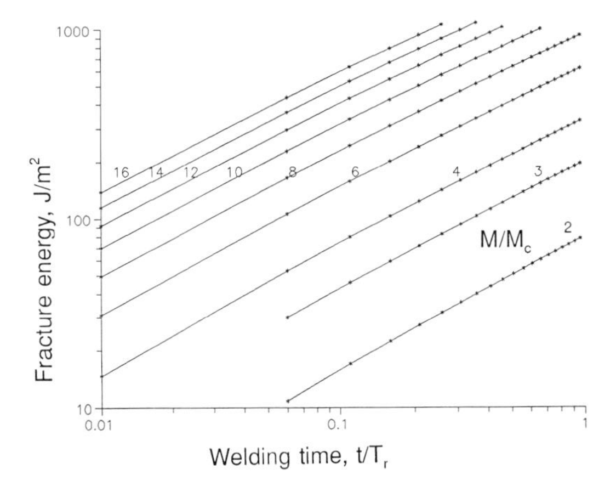

Figure 7.13 Fracture energy versus welding time and molecular weight, M/M_c, for a polymer-polymer interface.

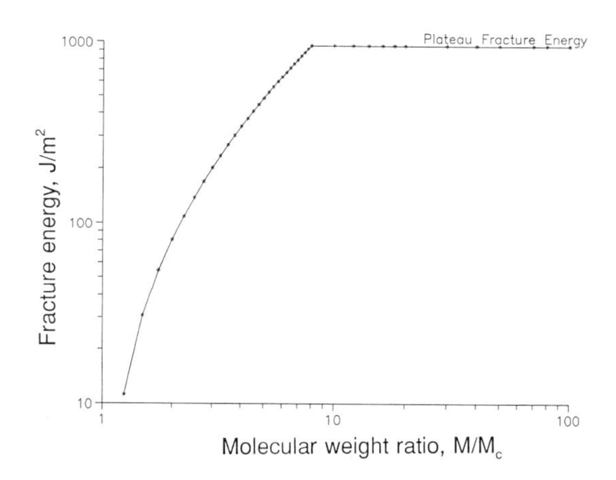

Figure 7.14 Fracture energy versus molecular weight of virgin glassy polymers. The data points are calculated using Eq. 7.5.17.

Finally, the critical stress intensity factor, K_{1c}, fracture stress, σ, and fracture strain, ϵ, are all proportional to $G_{1c}{}^{1/2}$ and hence we have from Eq 7.5.17

$$K_{1c}, \sigma, \epsilon, \sim (M^{1/2} - M_c^{1/2}) \tag{7.5.18}$$

which is in a convenient form for experimental investigation.

7.5.5 Pre-Draw Ratio Dependence of Fracture

We have considered until now the fracture of amorphous polymers that initially were in the unoriented state, that is, they had an initial draw ratio $\lambda_p = 1$. We consider here the interesting case of fracture in a sample that is pre-drawn with $1 \leq \lambda_p \leq \lambda_c$, and the crack propagation direction is normal to the uniaxial draw direction. From the above entanglement analysis, we have for the normal case, $K_{1c} \sim \sqrt{E}(\lambda_c - 1)$, where $\lambda_c \sim \sqrt{(M/M_c)}$. The effect of a pre-draw is to reduce λ_c so that

$$K_{1c} \sim (\lambda_c - \lambda_p) \tag{7.5.19}$$

Thus, when M is greater than $8M_c$, $\lambda_c \approx 4$ and we expect K_{1c} to decrease with increasing λ_p as

$$K_{1c} \sim 1 - \lambda_p/4 \tag{7.5.20}$$

so that K_{1c} goes to zero as λ_p approaches 4. This is perhaps a surprising prediction of the fracture theory and is not intuitively obvious.

Eq 7.5.20 did not take into consideration the effect of draw ratio on the tensile modulus, E, or on the material anisotropy near the crack tip. In the simplest case, if we let $E \sim \lambda_p$, we expect corrections of the type

$$K_{1c} \sim \lambda_p^{1/2}(1 - \lambda_p/4) \tag{7.5.21}$$

This correction predicts that K_{1c} first increases to a maximum at $\lambda_p \approx 1.3$ and then decreases to zero at $\lambda_p = 4$.

Experiments by Lin *et al.* [42] investigated K_{1c} of PMMA as a function of pre-draw ratio. The pre-drawing was done above T_g and the material was then rapidly cooled to trap the draw ratio. They measured K_{1c} using a single-edge notched tensile bar in which the pre-draw direction was parallel to the applied tensile stress. Figure 7.15 shows their K_{1c} versus λ data, which are compared with the above predictions. We see that the experimental data lie between the predictions of Eq 7.5.20 and Eq 7.5.21. The modulus data were not available for exact comparison. The most important agreement with the above theory is that the sample became extremely weak as λ_p approached 4. The experiments of Lin *et al.* could not be readily conducted above $\lambda_p \approx 3.25$.

7.6 Fracture by Chain Pullout

In the last section, we considered the microscopic details of an entanglement model and were able to deduce fracture relations from a disentanglement criterion. Here, we examine the more primitive aspects of chain pullout at an interface and deduce fracture

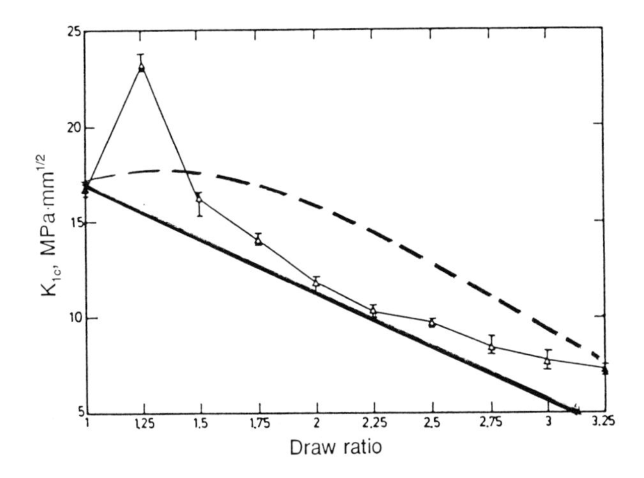

Figure 7.15 K_{1c} versus pre-draw ratio for PMMA (courtesy of Lin *et al.* [42]). The solid line represents Eq 7.5.20 and the dashed line Eq 7.5.21.

criteria that not only have microscopic relevance but also form the basis for a fracture mechanics model of welding and fracture. We begin with the "nail solution".

7.6.1 The Nail Solution

The following practical example of fracture should provide the reader with intuitive insight on the strength of interfaces. Consider two beams of wood (width w, thickness h) in a cantilever beam configuration (Figure 7.16) nailed together by n nails per unit area of penetration length L. The beams have a debonded initial crack length a. What are the fracture energy, G_{1c}, and the critical force, P_{cr}, to pull the beams apart as shown?

The pullout force f for a single nail is proportional to the pullout velocity, $V = dL/dt$, by the dynamic friction relation

$$f = \mu V \tag{7.6.1}$$

where μ is the friction coefficient for the nail segment of length L, and is related to the unit length friction coefficient, μ_0, via $\mu = \mu_0 L$. The pullout force may therefore be written as

$$f = \mu_0 L V \tag{7.6.2}$$

The energy required to pull one nail out at constant velocity is determined from

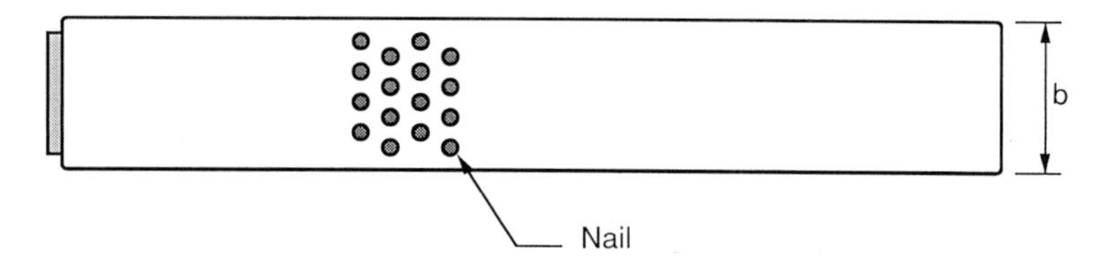

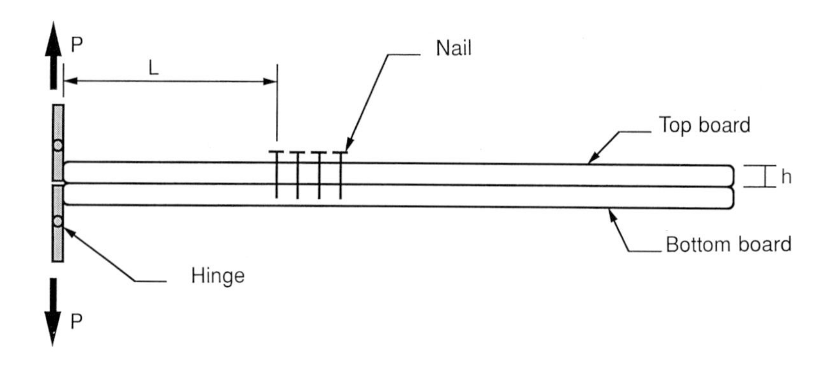

Figure 7.16 Cantilever beams used to measure the fracture energy of nail pullout from wood. Top: plan of beam showing nail heads. Bottom: method of loading beams with a load P (Wool, Bailey, and Friend).

$$U_p = \mu_0 \int_{L=0}^{L} L\, V\, \mathrm{d}L \tag{7.6.3}$$

so that

$$U_p = \tfrac{1}{2} \mu_0\, V L^2 \tag{7.6.4}$$

Thus, $G_{1c} = nU_p$ and we obtain the "nail solution",

$$G_{1c} = \tfrac{1}{2} \mu_0\, V n L^2 \tag{7.6.5}$$

from which we obtain the simple predictions $G_{1c} \sim n$ and $G_{1c} \sim L^2$. Eq 7.6.5 has also been derived by Prentice [43], Evans [44], McLeish *et al.* [45], and Creton *et al.* [46].

To evaluate the nail solution experimentally, we apply Griffith's theory (see Eqs 7.2.1–7.2.5) with the surface energy term $S = awG_{1c}$, so that $\mathrm{d}S/\mathrm{d}a = wG_{1c}$. The strain energy in the beams, U, is determined from simple beam theory by

$$U = P^2 a^3 / (3\, EI) \tag{7.6.6}$$

in which E is the modulus and $I = wh^3/12$ is the moment of inertia. Differentiating U with respect to the crack length a, and equating with dS/da, we obtain

$$G_{1c} = 12\, P_{cr}^2\, a^2/(E w^2 h^3) \tag{7.6.7}$$

Figure 7.17 shows the result of a nail pullout experiment by Wool, Bailey, and Friend [47], which provides support for the linear dependence of G_{1c} on the number of nails per unit area, n. This result is consistent with $P_{cr} \sim n^{1/2}$, which was also noted by Gent *et al.* [48] in fiber pullout experiments. In their work, the force to pull out n fibers from a block of rubber increased as $\sqrt{n}$, and not as n, as might have been intuitively expected. For materials with preexisting cracks, the Griffith expression for the critical stress $\sigma_c \sim n^{1/2}$ (Eq 7.2.5) is

$$\sigma_c = [\mu_0\, E\, V n\, L^2/(2\pi\, a)]^{1/2} \tag{7.6.8}$$

To demonstrate that $G_{1c} \sim L^2$ or $P_{cr} \sim L$, we pulled out nails of different lengths (same diameter) at a constant velocity V of 3 cm/s, and obtained the result shown in Figure 7.18. In this experiment, two nails were pulled out in uniaxial tension; the force

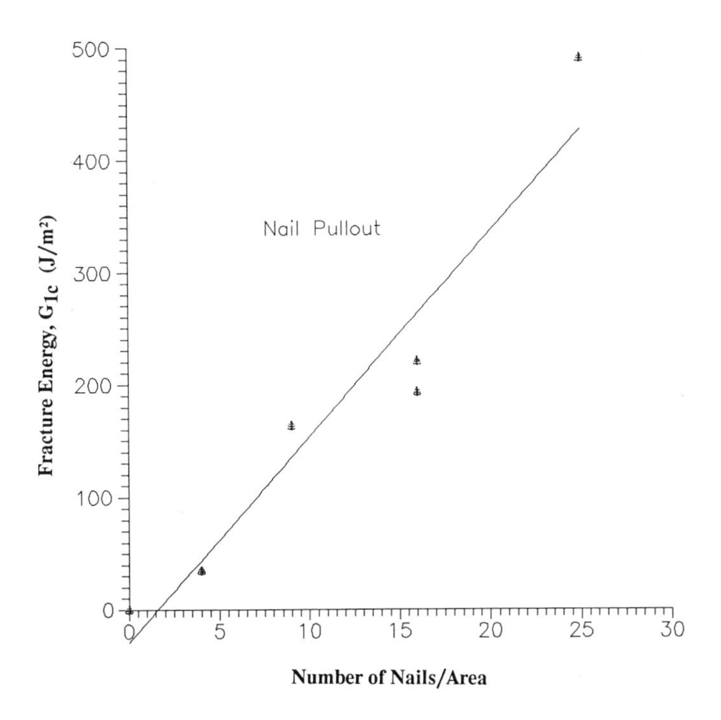

Figure 7.17 Fracture energy versus number of nails per unit area.

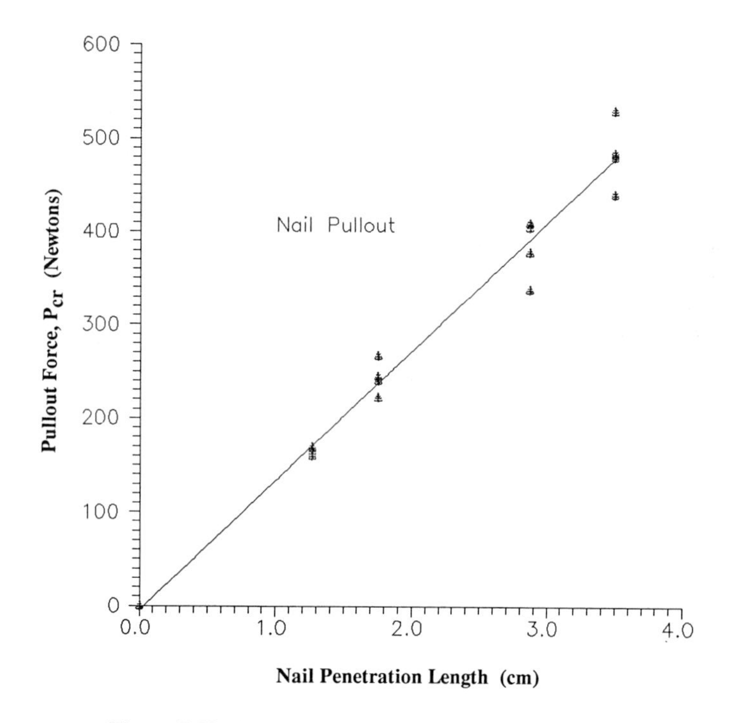

Figure 7.18 Critical fracture load versus length of nails.

could be measured directly and compared to Eq 7.6.2. Clearly, we see the linear dependence of P_{cr} on nail length, L. (One could also deduce that the friction coefficient for a 0.4-cm diameter nail in pine wood was $\mu_0 \approx 9{,}000$ N·s/m^2.)

To improve the analogy between the nail solution and molecular systems, we must add a few additional terms to the fracture energy:

$$G_{1c} = G_{1c}(\text{friction}) + G_{1c}(\text{surface}) + G_{1c}(\text{debond}) \tag{7.6.9}$$

in which $G_{1c}(\text{friction}) \sim nL^2$, as described by Eq 7.6.5; $G_{1c}(\text{surface})$ is the traditional Griffith surface energy term

$$G_{1c}(\text{surface}) \approx 2\,S\,\Gamma \tag{7.6.10}$$

where Γ is the surface energy of the debonded beams and S is the Sapoval number describing surface roughness ($S = 1$ for a perfectly flat fracture surface; see Chapter 4); and

$$G_{1c}(\text{debond}) = \pi\, dnLG_a \tag{7.6.11}$$

where G_a is the energy to debond unit area of the nail surface of length L and diameter d from its matrix. The debonding term is important in fiber–matrix debonding. For example, Gent's group found in fiber pullout experiments [48] that $G \sim L$ for short lengths and then G became independent of L at longer lengths. Thus, the fiber–matrix bonding energy term dominated the friction pullout energy.

We can apply the nail solution to develop our understanding of the behavior of real molecules at interfaces, provided they behave like nails in simple friction pullout. A useful example is the strength of incompatible A/B interfaces reinforced with A–B diblocks in which one end (A) of the diblock is below its entanglement molecular weight and the other end (B) is highly entangled. In that case, Creton *et al.* [46] have shown that the short A end pulls out cleanly from the A side while the long B end remains anchored on the B side. Their data (discussed in Chapter 9) provide strong support for $G_{1c} \sim nL^2$ up to a critical length L_c. In this case, we expect contributions from both friction and the surface interfacial energy term Γ_{AB}, so that

$$G_{1c} = 2S\Gamma_{AB} + 1/2\mu_0 V\Sigma L_A^2 \tag{7.6.12}$$

in which Σ is the areal density of diblock chains.

For rigid rodlike chains, the critical length for chain pullout is determined by the fracture force ($f_b \approx D_0 a_m/2 = \mu_0 V L_c$) required to break the chain via the Morse anharmonic bond approximation (discussed in Chapter 8)

$$L_c = D_0\, a_m/(2\,\mu_0\, V) \tag{7.6.13}$$

where D_0 is the bond dissociation energy and a_m is the Morse anharmonicity constant. Thus, the length dependence of the fracture energy at constant n and V is determined by the scaling law

$$G_{1c} = G_c (L/L_c)^2 \qquad \text{for } L \le L_c \tag{7.6.14}$$

where G_c is the fracture energy at L_c. When L is greater than L_c, if bond rupture occurs, then G_{1c} reaches a plateau value, G_c. When the chains become highly entangled, crazes form and the nail solution no longer applies.

For weak interfaces formed with low molecular weight chains (less than the critical entanglement molecular weight M_c), chain pullout dominates. However, bridge segments of length $l_p \sim \sqrt{L}$ are pulled out, rather than the whole chain. Each chain contains p_c bridges of length l_p, with $p_c l_p = L$ and hence $p_c \sim \sqrt{L}$. The friction contribution to G_{1c} is $\frac{1}{2}\mu_0 V\, np_c\, l_p^2$. In the virgin state, the number of bridges per unit area, P_∞, is constant; we also have the term $np_c = P_\infty$, which is independent of molecular weight, and $l_p^2 \sim M$. Thus, we expect the fracture energy to be determined by an expression of the form

$$G_{1c} = 2S\Gamma + kM \tag{7.6.15}$$

where the constant k is proportional to the pullout velocity and monomer friction coefficient. Fracture by the chain pullout mechanism is favored at low molecular weight [small $l(t)$], low deformation rates (small V), high temperature (low μ_0), short healing times [small $l(t)$], and in the presence of plasticizing agents (low μ_0).

7.7 Dugdale Model of Fracture Mechanics

The Dugdale model [13] was used as a first approximation to determine the molecular aspects of the fracture mechanics. In this model, shown in Figure 7.19, a crack of initial length a_0 propagates through a plastic line zone or craze of length r_p, ahead of the crack tip. The plane-stress field near the crack for a Cartesian volume element, xyz, and angle θ at distance r is given by Rice [14] as

$$\begin{aligned}\sigma_{xx} &= K_1/\sqrt{2\pi r}\cos\theta/2\,(1-\sin\theta/2\sin 3\theta/2)\ +\ldots\\ \sigma_{yy} &= K_1/\sqrt{2\pi r}\cos\theta/2\,(1+\sin\theta/2\sin\theta/2)\ +\ldots\\ \sigma_{xy} &= K_1/\sqrt{2\pi r}\cos\theta/2\ +\ldots\end{aligned} \quad (7.7.1)$$

where K_1 is the stress intensity factor. The length of the plastic zone ahead of the crack tip is determined by

$$r_r = \pi K_{1c}^2/8\sigma_c^2 \quad (7.7.2)$$

where σ_c is the craze or yield stress.

The critical crack opening displacement is given by

$$\delta = K_{1c}^2/\sigma_c E \quad (7.7.3)$$

where E is the tensile modulus corrected for plane stress or plane strain.

Using the approximation that σ_c remains constant within the plastic zone, the critical strain energy release rate, G_{1c}, is determined by

$$G_{1c} = \sigma_c\delta \quad (7.7.4)$$

so that the fracture energy is found as a critical force acting over a critical distance.

7.7.1 Microscopic Aspects of the Dugdale Model at Welding Interfaces

Figure 7.20 shows the essential features of the Dugdale model for a crack propagating through a line deformation zone in styrene–isoprene–styrene (SIS) block copolymer

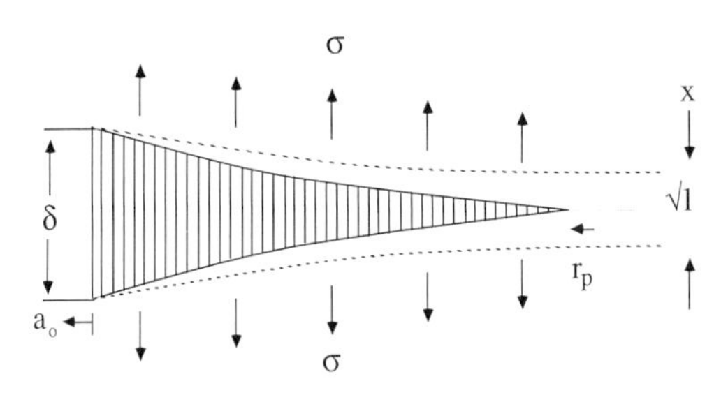

Figure 7.19 Dugdale model for a crack propagating through a partially welded interface, showing the critical crack-opening displacement δ, the critical traction stresses σ, the plastic zone length r_p, and the interface thickness $X \sim \sqrt{l}$.

[49]. The microstructure elucidated by transmission electron microscopy (TEM) consists of styrene spheres (white) with a diameter of about 250 Å in an isoprene matrix (stained black with osmium tetroxide). For the craze zone, $r_p \approx 3$ μm and $\delta \approx 0.2$ μm. The craze propagates in the stress field applied normal to the craze by

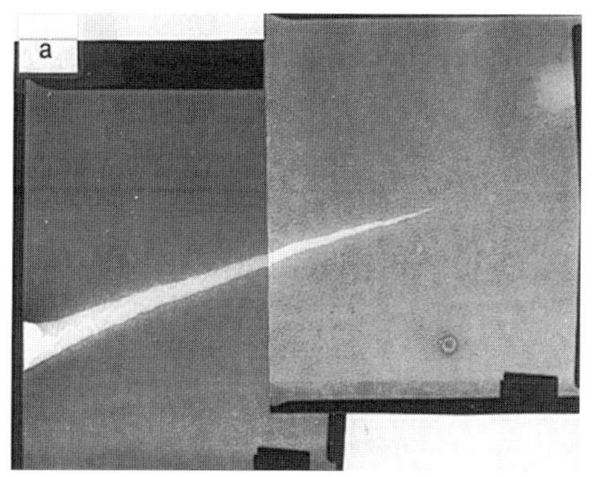

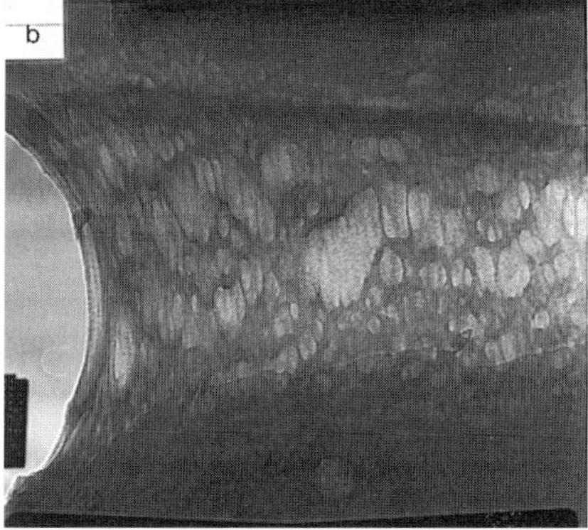

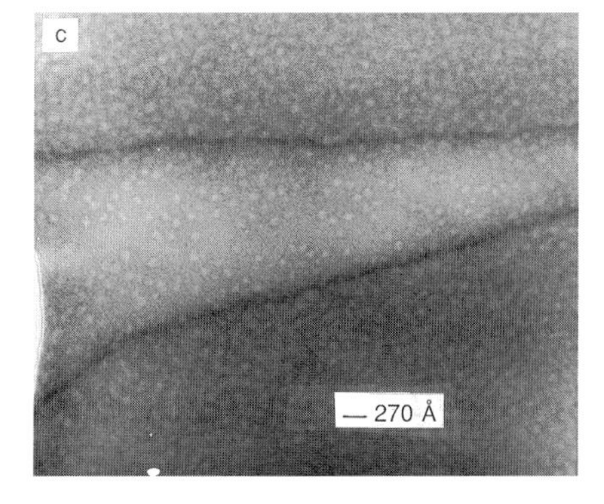

Figure 7.20(a) TEM analysis of the Dugdale-like deformation zone at a crack tip in styrene–isoprene–styrene copolymer. (b) Enlarged detail of the breakdown zone at the crack tip. (c) Detail of the interior of the deformation zone (the white particles represent polystyrene spheres about 250 Å in diameter) (Dolmon, Robertson, and Wool).

incorporating a single styrene sphere into the tip of the zone. As the zone propagates it thickens from 250 Å to 2,000 Å by drawing more material into the zone, and elongates by extending the material already in the zone. At the crack tip, many styrene spheres are drawn into the zone and deformed nonuniformly so that the crack tends to propagate through the center of the zone. Similar mechanisms have been identified for crazes propagating through pure amorphous polystyrene [50, 51]. Thus, a critical stress nucleates the line zone at distance r_p from the crack tip, and the crack-opening displacement is determined by the breakdown of the material drawn into the zone.

For glassy polymers, σ_c can be related to the craze stress, which differs from the shear yielding stress, σ_Y. The molecular weight dependence of σ_c (and δ) in the virgin state is determined by $\sigma_c \sim (M^{1/2} - M_c^{1/2})$ for $M \leq M^*$, above which it becomes independent of molecular weight. The critical crack-opening displacement, δ, is determined by the breakdown of the entanglement network in the deformation zone and the extent to which the zone can thicken, as depicted in Figure 7.20. During welding, the zone consists of interdiffused short chain segments mixed with longer chains with nonequilibrium configurations, and the entangled structure is complex. The crack-opening displacement consists of two components: a component from the virgin material, δ_v, and a component from the diffuse interface, δ_i, with $\delta = \delta_v + \delta_i$. At low welding strengths the contribution from the weaker interface dominates. The crack-opening displacement is then determined by the breakdown of the entanglement network in the interface, by the following approximation

$$\delta \approx \lambda_f\, l(t)^{1/2} \tag{7.7.5}$$

where $\lambda_f \approx 4$ is the constant uniaxial draw ratio to tighten the slack between entanglements.

Substituting for σ and δ, we determine the critical strain energy release rate as

$$G_{1c} \sim t^{1/2} M^{-1/2} \tag{7.7.6}$$

Using the entanglement model from Section 7.5, we determine the molecular weight dependence of the virgin state fracture energy in the range $M_c < M < 8M_c$ as

$$G_{1c}(M) = 0.3\, G_{1c}^* M / M_c\, [1 - (M_c / M)^{1/2}]^2 \tag{7.7.7}$$

where $G_{1c}{}^*$ is the plateau fracture energy occurring at $M \geq 8M_c$.

For welding of glassy polymers, fracture involves both disentanglement and bond rupture mechanisms. The suggestion by de Gennes [52, 53] that molecular bridges could control the strength is similar to the crossing density suggested by Prager and Tirrell [54–56] and Peppas [57], and is also similar to the total number of entanglements per chain per unit area suggested by Kausch [58–62]. In that case, G_{1c} is

$$G_{1c} \sim t^{1/2} M^{-3/2} \tag{7.7.8}$$

$$G_{1c\infty} \sim M^0 \tag{7.7.9}$$

These models predict the correct time dependence of welding, but the molecular weight dependence of the both the welding rate and virgin state is not consistent with experiment.

7.7.2 Craze Model of Fracture

Brown [63] analyzed the fracture of glassy polymers in terms of the craze fibril breakdown at a crack tip. The craze fibrils formed a network with interconnecting members between fibrils. By considering the craze to be an orthotropic elastic material, he derived an expression for the fracture energy in terms of several microscopic parameters, as follows.

$$G_{1c} = (\sigma_f^2 2\pi D/S\lambda^2)(E_2/E_1)^{1/2}(1 - 1/\lambda) \tag{7.7.10}$$

in which σ_f, D, S, λ, E_2, and E_1 are the stress to break the chains in a craze fibril, the fibril diameter, the stress to propagate the craze, the draw ratio of the fibrils, the modulus of the craze in the normal direction, and the modulus of the craze parallel to its propagation direction. To apply this approach to welding, one needs to know the time dependence of each of these variables, which is a formidable requirement. The approach has been used by Brown to examine aspects of the bulk strength and infer relations for diblocks at incompatible interfaces. This relation was applied by de Gennes to welding, so that with chain-end segregation, the number of bridges crossing behaves as $p(t) \sim t^{1/4}M^{-3/4}$, and with $\sigma_f \sim p(t)$, Eq 7.7.10 predicts that $G_{1c} \sim t^{1/2}M^{-3/2}$, which is similar to the previous bridge models. Unfortunately, neutron reflection and SIMS experiments do not support the concept of chain-end segregation.

The fibril fracture stress is determined from the force f to break a single chain segment, multiplied by the areal density of the chain segments, Σ. The force to break a chain can be determined from the anharmonicity of the bonds: for a Morse potential with dissociation energy D_0 and anharmonicity factor a, we obtain [64]

$$f \approx D_0 a/2 \tag{7.7.11}$$

For typical C–C bonds in polymers, we have shown from stress–FTIR and stress–Raman experiments that $D_0 = 5.55 \times 10^{-9}$ N·Å and $a = 1.99$/Å, so that $f \approx 5.52 \times 10^{-9}$ N, which is similar to values obtained by Kausch [58]. From Eq 7.4.4, the cross-sectional area of a chain segment like PS or PMMA is typically about 100 Å^2 and $\sigma_f \approx 5$ GPa.

Using values of $S = 70$ MPa, $\lambda = 3$, $\Sigma = 2.8 \times 10^{17}$/m^2, and $f = 1.4 \times 10^{-9}$ N, Brown determines $G_{1c} \approx 600$ J/m^2 for PMMA, which is reasonably close to experimental values. Using the relation $\sigma_f = f\Sigma$, he argues from Eq 7.7.13 that

$G_{1c} \sim \Sigma^2$, which finds support from his studies of the interfacial strength dependence on the number of A–B diblocks at an A/B incompatible interface, using A = PS and B = PMMA, which we discuss further in Chapter 9.

Raphael and de Gennes [65] examined the joining of two rubber slabs by a single connector molecule of length L (the "one-stitch" problem). The molecule threads only once across the interface. They find in the quasi-static limit of slow fracture that the fracture energy is determined by a Dugdale model expression, $G_{1c} \sim \sigma_0 L$, where σ_0 is the constant stress required to pull out the chain of length L at velocities V less than a characteristic velocity V^*. Above this velocity, the fracture energy is dominated by viscous loss and $G \sim V/V^*$.

Ji and de Gennes [66] extended the one-stitch case to the "many-stitch" problem, where a chain at the interface threads back and forth several times, resembling the minor chains at a symmetric interface. At low velocities, they find results similar to those of the one-stitch case. However, at higher velocities, the situation is more complex, chains cannot pull out and bond scission occurs, giving solutions somewhat similar to that obtained by Brown.

7.8 Saffman–Taylor Fracture

7.8.1 Introduction to Meniscus Instability

Saffman and Taylor examined the penetration of a fluid into a Hele-Shaw cell containing a more viscous liquid [67]. The cell consisted of two parallel glass plates held 0.9 mm apart by strips of rubber to create a channel (12 × 38 cm) for the fluid flow. A pressure gradient was established along the channel by the application of pressure or suction as needed. Air and glycerine were used as the immiscible fluids. The interface between the fluids eventually became unstable and broke up, resulting in the less dense fluid (air) penetrating the more viscous fluid in the form of "fingers". The instability wavelength, Ω, that controls the finger width is described by [67]

$$\Omega = b\pi\,(\Gamma/\eta\, v)^{1/2} \tag{7.8.1}$$

where b is the channel thickness, Γ is the surface tension of the viscous fluid, η is the viscosity of the more viscous fluid, and v is the finger velocity. The quantity $(\eta v/\Gamma)$ is called the *capillary number*, C. High viscosity and velocities result in low Ω values and more profuse fingers.

The Saffman–Taylor meniscus instability model has been applied to craze nucleation and growth phenomena by Kramer [68] and Argon and co-workers [69]. Studies with polystyrene show that the craze propagates by forming fingers at the growth interface. As the air fingers propagate into the stressed glass, they leave behind a wake of fibrillar material that connects the craze surfaces. This mechanism is discussed in more detail with respect to crazing in Chapter 11.

We examined the role of the Saffman–Taylor instability mechanism in a model fracture experiment shown in Figure 7.21 [70–72]. A layer of viscous polymer (polydimethylsiloxane) of thickness b was placed between two glass plates and fractured using the fracture mechanics method in ASTM D 3433 [75]. The fracture energy is determined by

$$G_{1c} = (4P_{max}^2)(3a^2 + h^2)/(Ew^2h^3) \tag{7.8.2}$$

where P is the maximum load before the crack becomes critical and E is the Young's modulus of the glass. The dimensions a, h, w, and b are shown in Figure 7.21. The viscosity η was determined as a function of shear rate R, using

$$\eta = \tau/R \tag{7.8.3}$$

where τ is the shear stress. The room temperature results for the viscosity as a function of shear rate are shown in Figure 7.22. At low shear rates (R less than 0.25/s), the fluid behaves as a classical Newtonian liquid whose viscosity is independent of shear rate, but at higher shear rates, R greater than R^*, we obtain

$$\eta = \eta_0 (R/R^*)^{-\beta} \tag{7.8.4}$$

where the exponent β = 0.74, the zero-shear viscosity η_0 = 10,000 N·s/m^2, and R^* = 0.25/s is the crossover shear rate from Newtonian to non–Newtonian behavior. The magnitude of β is typical for shear-thinning flow of high-molecular-weight polymers. Non–Newtonian behavior sets in at a rate $R^* \approx 1/T_r$. We will see that the non–Newtonian behavior at R greater than R^* plays an important role in determining the fracture energy and microscopic aspects of the deformation zone.

The load–displacement curve and the resulting characteristic Saffman–Taylor (S–T) fracture pattern are shown in Figure 7.23 (obtained from a photocopy of the glass

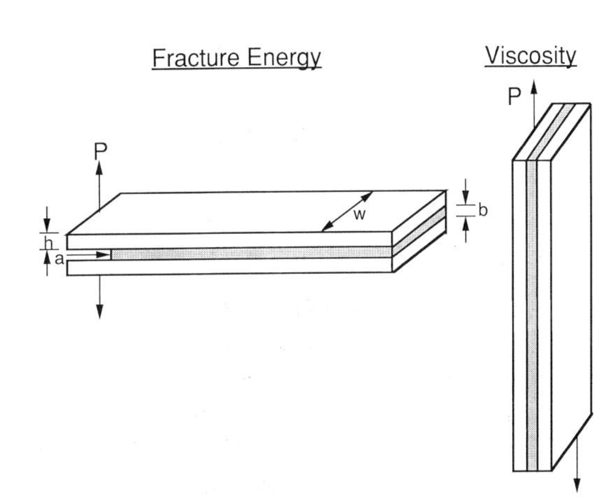

Figure 7.21 Experimental geometry used to investigate fracture energy and shear viscosity of a polymer fluid between two rigid plates (VanLandingham and Wool).

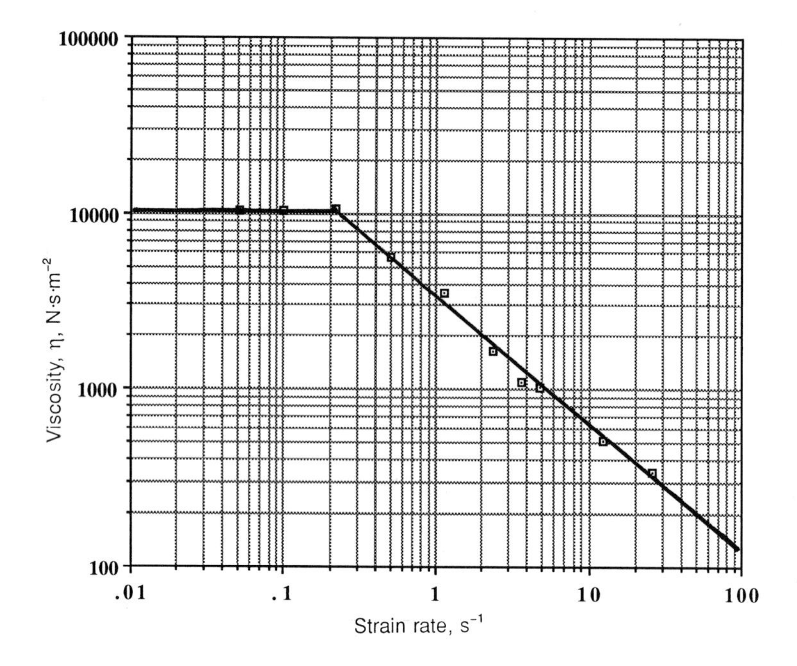

Figure 7.22 Log viscosity versus strain rate (VanLandingham and Wool).

plates after fracture [70]). The pattern has fractal characteristics with multiple branches producing more branches. Initially, instabilities at the air/polymer interface produce many small fingers. These coalesce into larger and fewer branches as the crack propagates. The plates were deformed at a uniaxial deformation rate determined by

$$R = \mathrm{d}(\delta / b)/\mathrm{d}t \tag{7.8.5}$$

where δ is the crack-opening displacement.

7.8.2 Fractal Fracture Pattern

The evolution of the surface fracture pattern in the deformation zone with respect to the load–displacement curve (P-δ) in Figure 7.23 is explained at corresponding numbered positions as follows:

1. Formation of instability and fingers at crack front;
2. Slow stable finger growth to lengths of about 2 mm;
3. Instabilities at finger tips, resulting in termination of most fingers and some branching as crack advances by about 4 mm;
4. Large finger branches forming 5–10 mm ahead of crack tip;
5. At the critical point, fingers propagating 12–15 mm ahead of crack tip.

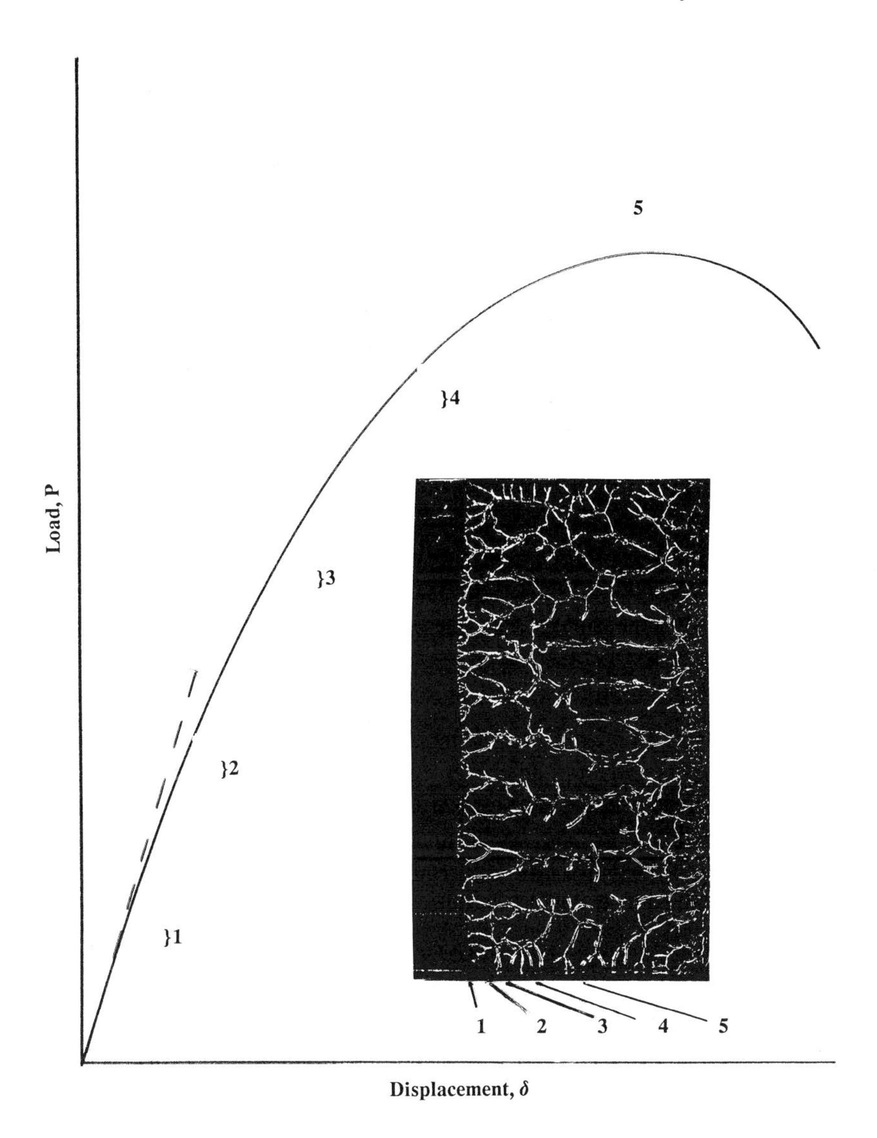

Figure 7.23 Relation between Saffman-Taylor meniscus instability pattern and the load–displacement curve for the fracture experiment. See text for description of numbers.

As the crack is initiated and propagates, the number of fingers and branches decreases in accordance with a "3–1 rule" [70]. The formation and propagation of the fingers represents the evolution of a deformation zone at a crack tip. Nucleation begins in the zone (1) at an applied load that is considerably lower than the fracture load due to the stress concentration at the crack tip. The zone then grows (2–4) with interconnecting fibrils holding the crack surfaces together until the fibrils break and the crack propagates (5) through the deformation zone. The important events in terms of the pattern shown in Figure 7.23 occurred in the first 10 mm of the zone formation, or just the first 20% of the surface features.

The development of the fractal pattern in the deformation zone occurs in three stages, related by a 3–1 rule, where one of three fingers inhibits the growth of the two neighboring fingers and grows to three times the size and spacing of the original set of fingers. This process is repeated and forms a 3–1 deformation zone through which a critical crack propagates. The three stages of development of the deformation zone that control the fracture process occur over the first 10 mm of finger growth, representing a deformation zone length, r_p, of about 18 Ω. This is a very small fraction of the total fracture surface area. The fractal dimension of the pattern in the 3–1 deformation zone before the crack becomes critical is $D_f \approx 1.7$. When the crack becomes critical, the fractal dimension increases to $D_f \approx 2$ as the pattern loses self-similarity.

The critical wavelength, Ω, of the fingers at position (1) can be determined from the S–T relation in Eq 7.8.1 above using $\Gamma = 0.02$ J/m^2, $b = 0.2$ mm, $v = 0.1$ mm/s (by optical observation), and $\eta = 10{,}000$ N·s/m^2, so that $\Omega \approx 0.3$ mm, in good agreement with experiment, where we have 10–20 fingers/cm at the initial unstable interface.

The number of fibrils, N_f, formed per unit surface area bridging the initial crack surfaces before rupture is derived from $N_f = 4/(\pi\Omega^2)$ as

$$N_f = 4\eta\, v/(\pi^3 b^2 \Gamma) \tag{7.8.6}$$

Note that $N_f \sim \eta$. If we model a single deformation zone fibril as a cylindrical rod of initial diameter Ω and fibril draw ratio λ_f (≈ 4), the total fibril surface area, A_f, produced per unit crack surface area is $A_f = 4b\lambda_f^{1/2}/\Omega$, or

$$A_f = 4/\pi\,(\lambda_f\, \eta\, v/\Gamma)^{1/2} \tag{7.8.7}$$

which is independent of the initial polymer layer thickness, b. The fracture energy contribution, G_f, from the surface area only of the fibers is

$$G_f = A_f \Gamma \tag{7.8.8}$$

where Γ is the surface energy of the fibrils in the deformation zone. The surface area increase due to fibrils is dominated by the ratio b/Ω. For crazes, $b \approx 10{,}000$ Å and $\Omega \approx 50$–100 Å, which gives $G_f \approx 30$–60 J/m^2.

7.8.3 Rate and Molecular Weight Effects

The measured fracture energy G_{1c} for the filled polydimethylsiloxane (PDMS, vacuum grease) in this experiment is shown in Figure 7.24 as a function of deformation rate $R = \mathrm{d}(\delta/b)/\mathrm{d}t$. VanLandingham and Wool obtained [70]

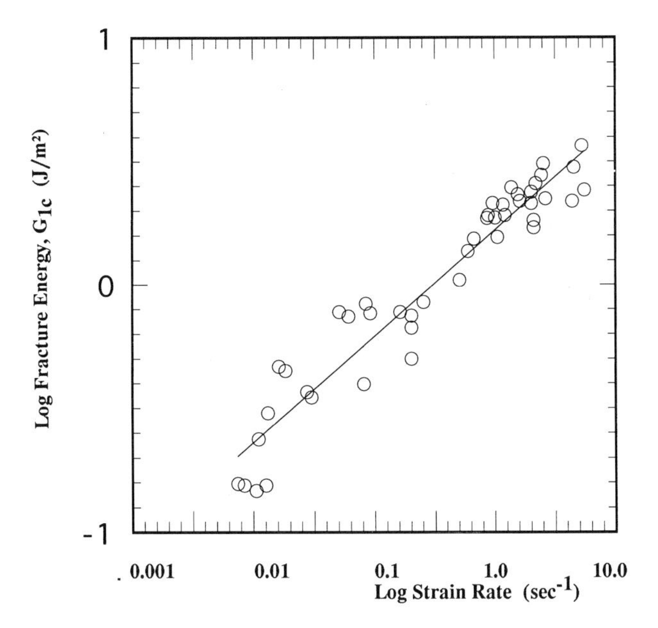

Figure 7.24 Fracture energy versus strain rate for filled PDMS. The solid line has a slope of 0.43 (VanLandingham and Wool).

$$G_{1c} = 1.67 R^{0.43} \qquad (\mathrm{J{\cdot}s/m^2}) \tag{7.8.9}$$

for which the critical load behaved as $P_{cr} \sim R^{0.22}$ [72]. Most of these data were obtained at $R \geq R^*$ in the non–Newtonian region, where we might expect $P_{cr} \sim R^{1-\beta}$. With $\beta = 0.74$ (from Figure 7.22), we predict that $G_{1c} \sim R^{2(1-\beta)} \sim R^{0.52}$.

The effect of the transition from Newtonian to non–Newtonian viscosity (at R^*) on the fracture energy is apparent in Figure 7.25 [71]. Here we examined different molecular weight fractions of pure PDMS using the same Saffman–Taylor fracture configuration. In the Newtonian region at $R < R^*$, the data behaved approximately as $G_{1c} \sim RM^2$, and at $R > R^*$, we obtained $G_{1c} \sim R^x M^y$, where $x \approx 1/7$ and $y \approx 1/2$. In the Newtonian region, one might expect $G_{1c} \sim \eta R$ with a molecular weight dependence of M^3. The transition from Newtonian to non–Newtonian behavior with increasing rate may also involve a change in the Saffman–Taylor fibril breakdown mechanism, possibly from fibril instability ($R < R^*$) to disentanglement ($R > R^*$). The scaling law in the Newtonian region, $G_{1c} \sim RM^2$, is reminiscent of the nail solution (Section 7.7), but this could be accidental since a monolayer of molecules is not controlling the fracture energy.

In addition to the Saffman–Taylor meniscus instability mechanism that causes the characteristic fibrillation, we also observed cavitation due to the negative pressures in the fluid near the crack tip. Cavitation, the consequence of which is formation of

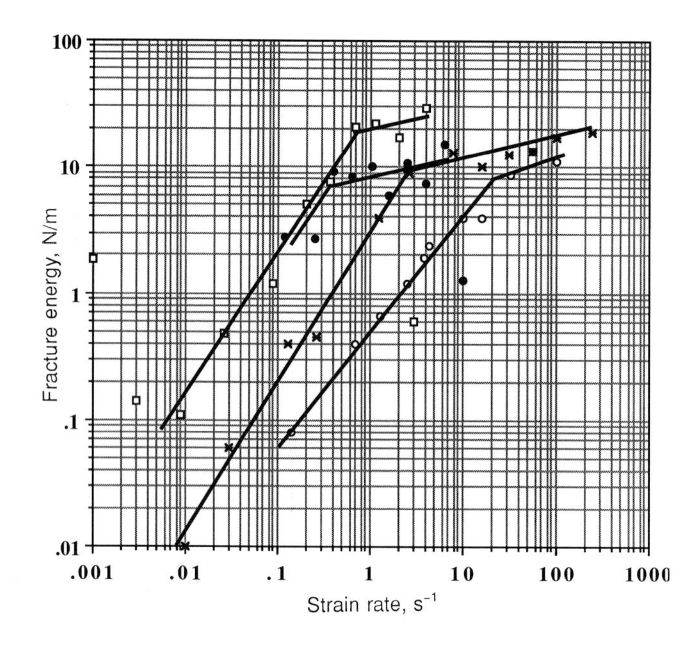

Figure 7.25 Fracture energy versus strain rate for polydimethylsiloxanes of different molecular weights. Key to symbols: ○, *M* of 38,600; ×, *M* of 94,300; •, *M* of 273,500; □, *M* of 369,200 (Wool, McGonigle, and Jackson).

bubbles that expand with increasing load, affects the Saffman–Taylor fibrils and further complicates the structure and the energy dissipation in the deformation zone. We are continuing to study these effects and the transition reported in Figure 7.25.

7.9 References

1. S. S. Schwartz and S. H. Goodman, Eds., *Plastics Materials and Processes*; Van Nostrand Reinhold Company, New York; 1982; Chapter 18, "Fastening and Joining Techniques", p 757.
2. R. P. Wool and K. M. O'Connor, *J. Appl. Phys.* ***52***, 5953 (1981).
3. Y.-H. Kim and R. P. Wool, *Macromolecules* ***16***, 1115 (1983).
4. R. P. Wool, *Rubber Chem. Technol.* ***57*** (2), 307 (1984).
5. R. P. Wool, *J. Elastomers Plast.* ***17*** (Apr), 107 (1985).
6. J. G. Williams, *Fracture Mechanics of Polymers*; Halsted Press, New York; 1984.
7. ASTM D 2979 – 88, Standard Test Method for Pressure-Sensitive Tack of Adhesives Using an Inverted Probe Machine. In *Vol. 15.06, Adhesives, 1993 Annual Book of ASTM Standards*; American Society for Testing and Materials, Philadelphia, PA, 1993.

8. ASTM D 1062 – 92, Standard Test Method for Cleavage Strength of Metal-to-Metal Adhesive Bonds. In *Vol. 15.06, Adhesives, 1993 Annual Book of ASTM Standards*; American Society for Testing and Materials, Philadelphia, PA, 1993.
9. ASTM D 950 – 82 (Reapproved 1987), Standard Test Method for Impact Strength of Adhesive Bonds. In *Vol. 15.06, Adhesives, 1993 Annual Book of ASTM Standards*; American Society for Testing and Materials, Philadelphia, PA, 1993.
10. ASTM D 3807 – 92, Standard Test Method for Strength Properties of Adhesives in Cleavage Peel by Tension Loading (Engineering Plastics-to-Engineering Plastics). In *Vol. 15.06, Adhesives, 1993 Annual Book of ASTM Standards*; American Society for Testing and Materials, Philadelphia, PA, 1993.
11. ASTM D 5041 – 90, Standard Test Method for Fracture Strength in Cleavage of Adhesives in Bonded Joints. In *Vol. 15.06, Adhesives, 1993 Annual Book of ASTM Standards*; American Society for Testing and Materials, Philadelphia, PA, 1993.
12. A. A. Griffith, *Philos. Trans. R. Soc. London, A*, ***221***, 163 (1920).
13. D. S. Dugdale, *J. Mech. Phys. Solids* ***8***, 100 (1960).
14. J. R. Rice, "Mathematical Analysis in the Mechanics of Fracture"; Chapter 3 in *Mathematical Fundamentals*; Vol. II of *Fracture, An Advanced Treatise*, H. Liebowitz, Ed.; Academic Press, New York; 1968; p 192.
15. J. L. Gardon, *J. Appl. Polym. Sci.* ***7***, 643 (1963).
16. A. N. Gent and G. R. Hamed, *J. Adhes.* ***7***, 91 (1975).
17. A. N. Gent and A. G. Thomas, *J. Polymer Sci. Part A-2* ***10***, 571 (1972).
18. R.-J. Chang and A. N. Gent, *J. Polym. Sci., Polym. Phys. Ed.* ***19***, 1619 (1981); *ibid.*, ***19***, 1635 (1981).
19. A. N. Gent, *Int. J. Adhes. Adhes.* ***1*** (April), 175 (1981).
20. A. N. Gent, The 1981 Foundation Lecture, "Strength of Adhesive Bonds"; *Plast. Rubber Int.* ***6*** (4), 151 (1981).
21. G. R. Hamed and C.-H. Shieh, *J. Polym. Sci., Polym. Phys. Ed.* ***21***, 1415 (1983).
22. G. R. Hamed, *Rubber Chem. Technol.* ***54***, 576 (1981).
23. M. L. Williams, *J. Appl. Polym. Sci.* ***13***, 29 (1969).
24. R. P. Wool, *Int. J. Fract.* ***14***, 597 (1978).
25. R. P. Wool, "Crack Healing in Semicrystalline Polymers, Block Copolymers and Filled Elastomers"; chapter in *Adhesion and Adsorption of Polymers*, L.-H. Lee, Ed.; Vol. 12A in series *Polymer Science and Technology*; Plenum Press, New York; 1980; p 341.
26. W. W. Graessley, *The Entanglement Concept in Polymer Rheology*; Vol. 16 in series *Advances in Polymer Science*, H. J. Cantow *et al.*, Eds.; 1982, Springer-Verlag, Berlin 1974; and W. W. Graessley, "Entangled Linear, Branched and Network Polymer Systems—Molecular Theories"; chapter in *Synthesis and Degradation; Rheology and Extrusion*; Vol. 47 in series *Advances in Polymer Science*, H.-J. Cantow *et al.*, Eds.; Springer-Verlag, Berlin; 1982; p 67.
27. S. M. Aharoni, *Macromolecules* ***16***, 1722 (1983).
28. T. A. Kavassalis, and J. Noolandi, *Phys. Rev. Lett.* ***59***, 2674 (1987); T. A. Kavassalis and J. Noolandi, *Macromolecules* ***22***, 2709 (1989).
29. R. P. Wool, B.-L. Yuan, and O. J. McGarel, *Polym. Eng. Sci.* ***29***, 1340 (1989).
30. R. P. Wool, *Macromolecules* ***26***, 1564 (1993).
31. R. P Wool, "Strength and Entanglement Development at Amorphous Polymer Interfaces"; chapter in *Amorphous Polymers and Non-Newtonian Fluids*, C. Dafermos, J. L. Ericksen, and D. Kinderlehrer, Eds.; No. 6 in series *IMA Volumes in Mathematics and Its Applications*, G. R. Sell and H. Weinberger, Eds.; Springer-Verlag, Berlin; 1987; p 169.

32. S. Wu, *Polym. Eng. Sci.* ***30***, 753 (1990).
33. H. Tadokoro, *Structure of Crystalline Polymers*; John Wiley & Sons; New York; 1979.
34. J. D. Ferry, *Viscoelastic Properties of Polymers*; John Wiley & Sons, New York; 3rd ed., 1980.
35. P.-G. de Gennes, *Scaling Concepts in Polymer Physics*; Cornell University Press, Ithaca, NY; 1979.
36. P.-G. de Gennes, *Macromolecules* ***9***, 587 (1976); *ibid.* ***9***, 594 (1976).
37. M. Daoud, J. P. Cotton, B. Farnoux, G. Jannink, G. Sarma, H. Benoit, R. Duplessix, C. Picot, and P.-G. de Gennes, *Macromolecules* ***8***, 804 (1975).
38. M. Doi and S. F. Edwards, *J. Chem. Soc., Faraday Trans. II* ***74***, 1789 (1978).
39. M. Doi, *J. Polym. Sci., Polym. Lett. Ed.* ***19***, 265 (1981).
40. D. Richter, B. Farago, R. Buters, L. Fetters, J. S. Huang, and B. Ewen, *Macromolecules* ***26***, 795 (1993); *ibid.* ***25***, 6156 (1992).
41. A. M. Donald and E. J. Kramer, *J. Mater. Sci.* ***16***, 2977 (1981); A. M. Donald and E. J. Kramer, *J. Mater. Sci.* ***17***, 1871 (1982).
42. C. B. Lin, C. T. Hu, and S. Lee, *Polym. Eng. Sci.* ***33***, 430 (1993).
43. P. Prentice, *Polymer* ***24***, 344 (1983).
44. K. E. Evans, *J. Polym. Sci., Polym. Phys. Ed.* ***25***, 353 (1987).
45. M. McLeish, M. Plummer, and A. Donald, *Polymer* ***30***, 1651 (1989).
46. C. Creton, E. J. Kramer, C.-Y. Hui, and H. R. Brown, *Macromolecules* ***25***, 3075 (1992).
47. R. P. Wool, D. Bailey, and A. Friend, paper submitted for publication.
48. A. N. Gent, G. S. Fielding-Russell, D. I. Livingston, and D. W. Nicholson, *J. Mater. Sci.* ***16***, 949 (1981).
49. L. C. Dolmon, I. M. Robertson, and R. P. Wool, "Deformation of SIS Block Copolymer—A TEM Study", *Bull. Am. Phys. Soc.* ***30*** (3), 488 (1985); L. C. Dolmon, *Deformation of SIS Block Copolymers: A Transmission Electron Microscopy Study*; M.S. Thesis, University of Illinois, Urbana, IL; 1985.
50. B. D. Lauterwasser and E. J. Kramer, *Philos. Mag. A* ***39*** (4), 469 (1979).
51. E. Paredes and E. W. Fischer, *J. Polym. Sci., Polym. Phys. Ed.* ***20***, 929 (1982).
52. P.-G. de Gennes, *C. R. Seances Acad. Sci., Ser. B* ***291***, 219 (1980); *ibid.* ***292***, 1505 (1981).
53. P.-G. de Gennes and L. Leger, *Ann. Rev. Phys. Chem.* ***33***, 49 (1982).
54. D. Adolf, M. Tirrell, and S. Prager, *J. Polym. Sci., Polym. Phys. Ed.* ***23***, 413 (1985).
55. M. Tirrell, D. Adolf, and S. Prager, "Orientation and Motion at a Polymer–Polymer Interface: Interdiffusion of Fluorescent-Labelled Macromolecules"; chapter in *Orienting Polymers: Proceedings of a Workshop Held at the IMA, University of Minnesota, Minneapolis, March 21–26, 1983*, J. L. Ericksen, Ed.; No. 1063 in series *Lecture Notes in Mathematics*, A. Dold and B. Eckmann, Eds.; Springer-Verlag, Berlin; 1984; p 37.
56. S. Prager and M. Tirrell, *J. Chem. Phys.* ***75***, 5194 (1981).
57. A. G. Mikos and N. A. Peppas, *J. Chem. Phys.* ***88***, 1337 (1988).
58. "Fracture Mechanics Studies of Crack Healing"; Chapter 10 in H. H. Kausch, *Polymer Fracture*; Springer-Verlag, Heidelberg; 2nd ed., 1987.
59. K. Jud and H. H. Kausch, *Polym. Bull.* ***1***, 697 (1979); H. H. Kausch and K. Jud, *Plast. Rubber Process. Appl.* ***2***, 265 (1982).
60. D. Petrovska-Delacrétaz and H. H. Kausch, "Interdiffusion of Macromolecules from Surfaces of Different States", *Proceedings of IBM Polymer Symposium, Florence, Italy*, May 10–13 (1989); paper based on the Ph.D. Thesis of D. Petrovska-Delacrétaz, *Effets Mécaniques de la Cicatrisation des Polymères de Structures Différentes*, Thèse No. 866, École Polytechnique Fédérale de Lausanne (1990).

61. K. Jud, H. H. Kausch, and J. G. Williams, *J. Mater. Sci.* ***16***, 204 (1981).
62. H. H. Kausch, *Pure Appl. Chem.* ***55***, 833 (1983).
63. H. Brown, *Macromolecules* ***24***, 2753 (1991).
64. R. P. Wool, R. S. Bretzlaff, B. Y. Li, C. H. Wang, and R. H. Boyd, *J. Polym. Sci., Polym. Phys. Ed.* ***24***, 1039 (1986).
65. E. Raphael and P.-G. de Gennes, *J. Chem. Phys.* ***96***, 4002, (1992).
66. H. Ji and P.-G. de Gennes, *Macromolecules* ***26***, 520 (1993).
67. P. G. Saffman, and G. Taylor, *Proc. R. Soc., Ser. A, Math. Phys. Sci.* ***245***, 312 (1958).
68. E. J. Kramer, "Microscopic and Molecular Fundamentals of Crazing"; chapter in Vol. 52/53 in series *Advances in Polymer Science: Crazing*, H. H. Kausch, Ed., Springer-Verlag, Berlin; 1983; p 1.
69. A. S. Argon and M. M. Salama, *Philos. Mag.* ***36***, 1217 (1977).
70. M. R. VanLandingham and R. P. Wool, "Saffman–Taylor Fracture at a Polymer Liquid/Solid Interface", *Bull. Amer. Phys. Soc.* ***37***, 622 (1992).
71. R. P. Wool, J. McGonigle, and S. Jackson, paper in preparation.
72. M. R. VanLandingham, *Saffman–Taylor Fracture of a Polymer Fluid Between Glass Plates*; M.S. Thesis, University of Illinois, Urbana, IL; 1993.
73. ASTM D 5045 – 91a, Standard Test Methods for Plane–Strain Fracture Toughness and Strain Energy Release Rate of Plastic Materials. In *Vol. 08.03, Plastics (III), 1993 Annual Book of ASTM Standards*; American Society for Testing and Materials, Philadelphia, PA, 1993.
74. A. C.-M. Yang and M. S. Kuntz, *Bull. Am. Phys. Soc.* ***37*** (1), 424 (1992).
75. ASTM D 3433 – 75 (Reapproved 1985), Standard Practice for Fracture Strength in Cleavage of Adhesives in Bonded Joints. In *Vol. 15.06, Adhesives, 1993 Annual Book of ASTM Standards*; American Society for Testing and Materials, Philadelphia, PA, 1993.
76. T. T. Perkins, D. E. Smith, and S. Chu, *Science* ***264***, 819 (6 May 1994).

8 Strength of Symmetric Amorphous Interfaces*

8.1 Introduction

In this chapter we examine a body of experimental data on the strength of symmetric amorphous interfaces, commonly encountered in many practical applications. We consider them in the following order: tack of uncured linear elastomers, welding of polymers, lamination of composites, adhesion, polymer melt processing using compression and injection molding, sintering of powders and pellet resin, and coalescence of latex particles. Our primary interest here is to examine strength between two surfaces that had not previously come into contact. Such is not the case in crack healing, where two fractured surfaces are brought together and which will be considered in detail in Chapter 11. The terms *healing* and *crack healing* were initially applied to the rejoining of fractured surfaces and microvoids [1–3], but we also use it here in conjunction with welding of normal polymer interfaces.

8.2 Stages of Healing at Interfaces

Healing of polymer interfaces was described by Wool and O'Connor [2] in terms of several stages, involving surface rearrangement, surface approach, wetting, diffusion, and randomization. Most of the discussion in the previous chapters focused on the influence of diffusion on mechanical properties. However, several other factors can affect the welding process and are discussed below at the appropriate stages of healing.

8.2.1 Surface Rearrangement

Before bringing the surfaces together, one should consider the roughness or topography of the surface and how it changes with time, temperature, and pressure following contact. In fractured polymers, rearrangement of fibrillar morphology and other such factors can affect the rate of crack healing. Chain-end distributions near the surface

* Dedicated to W. O. Statton

can change as molecules diffuse back into the bulk polymer. Spatial changes of the molecular weight distribution may also occur, for example, where the low-molecular-weight species preferentially migrate to the surface. Chemical reactions, such as oxidation and cross-linking, can occur on the surface and complicate the molecular dynamics of diffusion. Many other processes, for example, molecular orientation changes, can contribute to the stage of surface rearrangement. Each material and each experimental technique usually possesses unique surface rearrangement processes that may need to be quantified.

O'Connor [3] found significant effects of surface rearrangement on the healing rate of lightly cross-linked polybutadiene, so that the fracture energy at constant healing time decreased with increasing surface rearrangement time (prior to contact). This effect can be explained by the chain ends diffusing back into the bulk so that they were not as readily available for diffusion during healing.

8.2.2 Surface Approach

Surface approach takes into consideration the time-dependent contact of the different parts of the surfaces to create the interface. For example, in the healing of cracks, crazes, and voids, contact may be achieved at different locations at different times in the interface depending on the closure mode [1]: slow closure of a double cantilever beam crack would result in different extents of healing along the closed crack. This stage typically contributes as a boundary value problem to the other stages of wetting and diffusion.

8.2.3 Wetting

Wetting at an interface can be time-dependent. For our purposes we provide a brief phenomenological description of wetting to illustrate potential problems in evaluating the time dependence of welding. Figure 8.1 shows a schematic region of the plane of contact of a polymer interface [2]. Due to surface roughness and other factors, good contact and wetting are not achieved instantaneously at all locations. Typically, wetted "pools" are nucleated at random locations at the interface and propagate radially until coalescence and complete wetting are obtained. This problem has been treated as a two-dimensional nucleation and growth process, with the fractional wetted area, $\Phi(t)$, given as [2]

$$\Phi(t) = 1 - \exp(-kt^m) \tag{8.2.1}$$

where k and m are constants that depend on the nucleation function and radial spreading rates.

Contact theories proposed by Anand [4] and others suggest that complete strength may be obtained when the interface has been wetted (at $\Phi = 1$). Others argue that

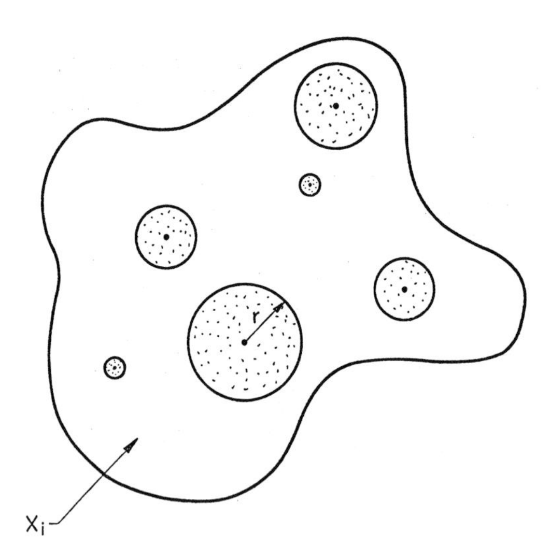

Figure 8.1 Partially wetted domain in a crack interface (Wool and O'Connor).

interdiffusion is necessary for strength development. The time dependence of viscous flow to promote contact and that of interdiffusion may be comparable, since they are subject to the same molecular dynamic processes. We demonstrate in later sections, using both tack experiments and incompatible interface weld strength data (Chapter 9), that diffusion plays a very important role.

8.2.4 Diffusion

The diffusion stage has been discussed in detail (Chapters 2 and 3) with respect to the instantaneous wetting condition. However, in the presence of a time-dependent wetting function, Eq 8.2.1, we see from Figure 8.1 that diffusion has progressed to different extents in different areas of the interface. If the intrinsic diffusion function $H(t)$ (see Chapter 2) as given by

$$H(t) = H_\infty (t/T_r)^{r/4} \tag{8.2.2}$$

does not change its nature with time due to the other stages, then the net diffusion, $H'(t)$, can be expressed as the convolution product

$$H'(t) = \int_{-\infty}^{\tau} H(t-\tau)\, \mathrm{d}\Phi(\tau)/\mathrm{d}\tau \; \mathrm{d}\tau \tag{8.2.3}$$

where τ is a dummy variable on the time axis. The convolution process for typical $\Phi(t)$ functions may mask the time dependence of the intrinsic diffusion function and related mechanical properties. We can solve this part of the problem mathematically by letting the wetting function be a Dirac delta function

$$d\Phi(t)/dt = \delta(\tau) \tag{8.2.4}$$

so that Eq 8.2.3 reduces to $H'(t) = H(t)$. One can attempt this experimentally by obtaining instantaneous wetting of atomically smooth surfaces under moderate contact pressure at temperatures above T_g, the glass transition temperature. However, we have found that the wetting problem can result in difficulty with the analysis of most short-time welding data.

8.2.5 Randomization

The *randomization* stage refers to the equilibration of the nonequilibrium conformations of the chains near the surfaces and, in the case of crack healing and processing, the restoration of the molecular weight distribution and random orientation of chain segments near the interface. The conformational relaxation is of particular importance in the strength development at incompatible interfaces.

The stages of crack healing can have interactive time-dependent functions, so that the welding problem can consist of processes involving five-way convoluted functions, resulting in mechanical properties whose time dependence may not be readily interpretable. Thus, we must take great care in conducting welding experiments designed to critically explore molecular theories of strength development.

8.3 Tack and Green Strength

8.3.1 Introduction

Tack is a term first used in the rubber tire industry that refers to the strength developed between linear amorphous polymers when they are brought into contact for a given time and fractured above the glass transition temperature, T_g. Typically, we are concerned with rubbery materials such as polyisoprene, polybutadiene, and styrene–butadiene rubber, prior to cross-linking via vulcanization or oxidation. Polyethylene and polystyrene melts also fall into this category. *Green strength* is the strength required to obtain cohesive fracture in a single piece of virgin material.

8.3.2 Experiment and Theory: Comparisons of Tack and Green Strength

The disentanglement theory presented in Chapter 7 predicts that the tack or stress at fracture of uncured linear elastomers behaves as

$$\sigma(t) = \sigma_\infty (t/T_r)^{1/4} \tag{8.3.1}$$

so that $\sigma \sim t^{1/4}M^{-1/4}$; and that the green or virgin strength behaves approximately as

$$\sigma_\infty \sim (M^{1/2} - M_c^{1/2}) \tag{8.3.2}$$

In addition, the self-diffusion coefficient should be measurable as

$$D = A/M^2 \tag{8.3.3}$$

where A is a constant derivable from mechanical data. These predictions apply to all amorphous linear polymers whose chain configurations can be described by Gaussian statistics and whose fracture behavior is dominated by chain disentanglement mechanisms.

Figure 8.2 shows tack data obtained by Skewis [5] for several polymer–polymer pairs; we have plotted the tack (units of force applied to a cross-sectional area of 0.4 cm^2) versus $t^{1/4}$. In these experiments, pieces of rubber were pressed together at a constant contact force for a given period of time and then separated with a tensile force. The original data were nonlinear when Skewis plotted them on a $t^{1/2}$ scale (to investigate diffusion mechanisms). The fracture energy behaves as $G_{1c} \sim t^{1/2}$, while the tack behaves as $\sigma \sim t^{1/4}$, as evidenced by the linear plot in Figure 8.2. The SBR/butyl pair is incompatible and its long-time equilibrium strength is considerably weaker than that of either of the symmetric pairs. Similar $t^{1/4}$ results for tack have been reported by Wool and O'Connor [2, 6] for polybutadiene, and by Voyutskii [7] for polyisobutylene.

The molecular weight dependence of tack and green strength was investigated by Forbes and McLeod [8] for fractionated samples of hevea rubber. Their results are shown in Figure 8.3, in which the upper line represents the green strength and the lower line represents the tack stress evaluated at a constant contact time, t_c, of 30 seconds for each molecular weight. The predicted green strength curve was computed using Eq 7.5.18

$$\sigma_\infty = k_1 (M^{1/2} - M_i^{1/2}) \tag{8.3.4}$$

where k_1 = 993 Pa/$M^{1/2}$ and $M_i^{1/2}$ = 284 ($M_i \approx$ 80,000) is the intercept on the $\sigma_\infty = 0$ axis of a plot of σ_∞ versus $M^{1/2}$.

The curve (dashed line) through the tack data in Figure 8.3 was determined from a plot of σ versus $M^{-1/4}$ using

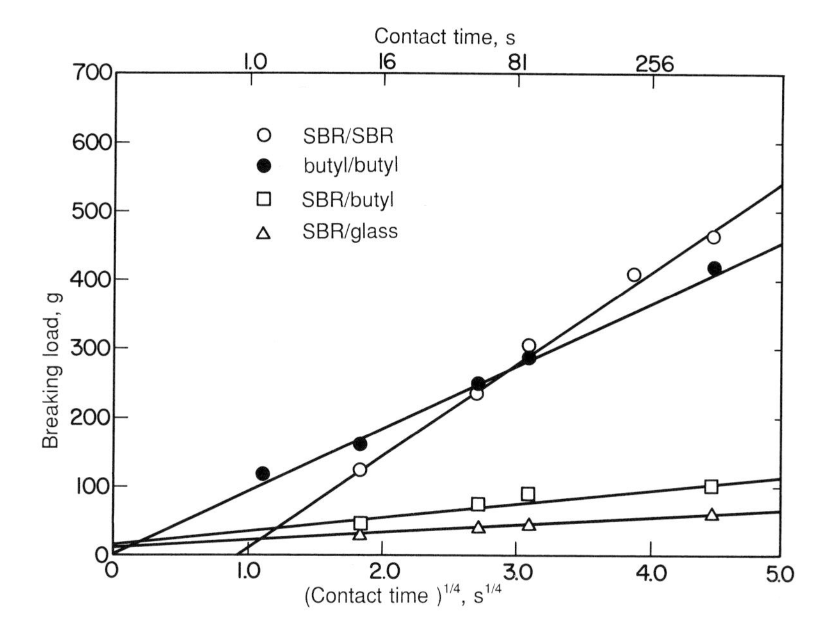

Figure 8.2 The tack or breaking load versus $t^{1/4}$ for polymer–polymer uncured pairs (data of Skewis). Tack measurements were made at a contact load of 1000 g with a contact area of 0.40 cm². Viscosity average molecular weight, M_v, was 260K for SBR and 225K for butyl rubber.

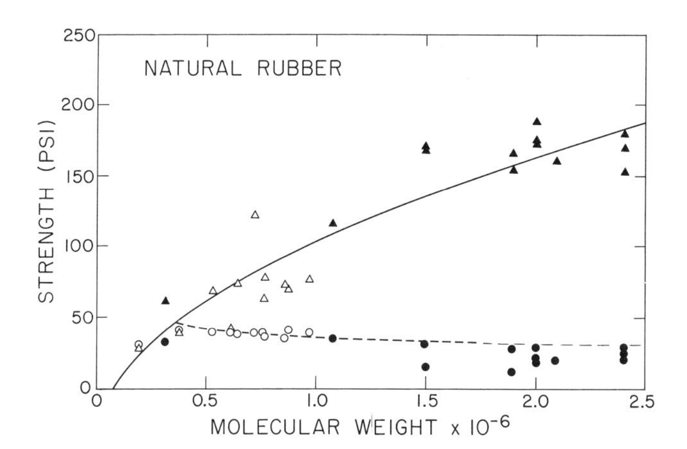

Figure 8.3 Tack (circles) and green strength (triangles) as a function of viscosity average molecular weight for fractionated samples of natural rubber (Forbes and McLeod). The solid line for green strength was obtained using Eq 8.3.4 and the dashed line for tack using Eq 8.3.5. The green strength was evaluated at a uniaxial test speed of 26.67 cm/min at 25 °C, and the tack was evaluated at a constant contact time of 30 seconds for each sample. Filled and unfilled data points refer to unmilled and milled natural rubber, respectively.

$$\sigma = \sigma_0 + \beta M^{-1/4} \tag{8.3.5}$$

where the constant β is 5.29 MPa$\cdot M^{1/4}$ and the intercept σ_0 is 80.5×10^3 Pa. When the tack stress is linear with $t^{1/4}$, as shown for the data in Figure 8.2, then

$$\sigma = k_2 t^{1/4} \tag{8.3.6}$$

Since $\sigma = \sigma_\infty (t/T_r)^{1/4}$, then $k_2 = \sigma_\infty / T_r^{1/4}$. Thus, k_2 is related to the constant β in Eq 8.3.5 by

$$k_2 = \beta / (M^{1/4}\, t_c^{1/4}) \tag{8.3.7}$$

where in this case $t_c = 30$ s.

Considering the difficulties in performing such experiments, the agreement between theory and experiment in Figure 8.3 is satisfactory. If the strength were controlled by bridges, we would expect $\sigma \sim M^{-3/4}$ at constant t_c and $\sigma_\infty \sim M^0$. These predictions are not supported by the data in Figure 8.3. It has been suggested that the tack stress is independent of molecular weight, but in fact it behaves approximately as $M^{-1/4}$, in close agreement with disentanglement theories. This is a weak molecular-weight dependence, but related mechanical properties such as fracture work, G_{1c}, should behave with a stronger molecular weight dependence, as $G_{1c} \sim t^{1/2} M^{-1/2}$ [10].

The relative tack, $R_t = \sigma / \sigma_\infty$, is often measured and compared at constant contact time, t_c, with rheological properties such as the melt viscosity, η. Since $R_t = (t_c/T_r)^{1/4}$, $T_r \sim M^3$, and $\eta \sim M^{3.4}$, we expect that

$$R_t \sim \eta^{-0.22} \tag{8.3.8}$$

I encourage the reader to check this prediction when data become available.

8.3.3 Diffusion Coefficient and Welding Time

In Figure 8.3, the green strength of natural rubber determined by Forbes and McLeod is seen to rise steadily up to a molecular weight of about 1.5×10^6; above that value, the green strength remains fairly constant, and bond rupture is expected to dominate the fracture mechanism by regulating the extent of disentanglement. Given sufficient time, the tack stress increases to the green strength value by continued diffusion. Using tracer diffusion methods, Skewis [9] measured the self-diffusion coefficient of natural rubber at 25 °C as $D = 3 \times 10^{-14}$ cm^2/s with samples with a viscosity average molecular weight, M_v, of 234,000 (Figure 8.4). The polydispersity of his samples was not known.

The self-diffusion coefficient is related to the reptation time, T_r, and end-to-end vector, R, by [11]

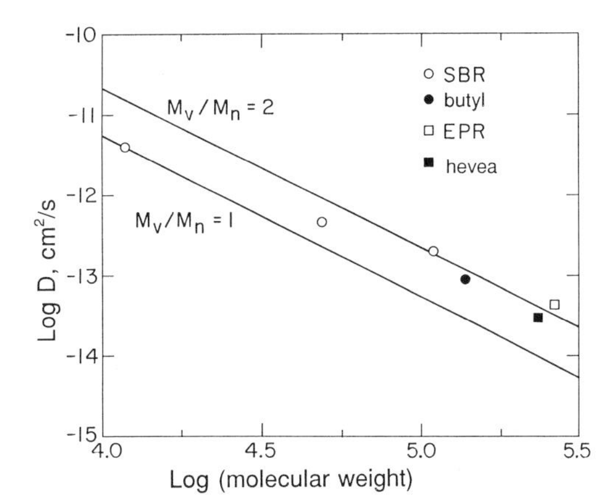

Figure 8.4 Self-diffusion coefficient, D, versus molecular weight for linear uncured rubber materials (data of Skewis).

$$D = R^2/(3\pi^2 T_r) \tag{8.3.9}$$

where $R^2 = C_\infty Mjb^2/M_0$ (Chapter 1). The self-diffusion coefficient can be obtained from welding data as follows. For fracture by disentanglement, the time to achieve complete strength, t_∞, is equal to the reptation time, T_r. Since $T_r = (\sigma_\infty/k_2)^4$, the self-diffusion coefficient is readily obtained as

$$D = (k_2/\sigma_\infty)^4 Mjb^2 C_\infty/(3\pi^2 M_0) \tag{8.3.10}$$

where C_∞ is the characteristic ratio. Using *cis*-polyisoprene ($-CH_2-C(CH_3)=CH-CH_2-$) as a model for the molecular properties of natural rubber, we use Eq 8.3.10 to calculate a value to compare with Skewis's experimental value of D of 3×10^{-14} cm²/s. Letting $M = 234{,}000$, $M_0 = 68$ for polyisoprene (PIP), $b = 1.49$ Å [average of 3 C–C (1.54 Å) and 1 C=C (1.34 Å) bond lengths], $k_2 = 0.103$ MPa/s$^{1/4}$ from Eq 8.3.7, $C_\infty = 4.7$ for *cis*-PIP ($C_\infty = 7.4$ for *trans*-PIP), $j = 4$ backbone bonds per monomer, and $\sigma_\infty = 0.2$ MPa = 29 psi (from Eq 8.3.4; also see Figure 8.4) we obtain $D = 3.4 \times 10^{-14}$ cm²/s. The agreement with experiment is considered to be fortuitous in view of the potential magnification of errors in the term $(k_2/\sigma_\infty)^4$ in Eq 8.3.10. However, we calculated k_2 by averaging over much data, and the green strength can usually be determined fairly accurately.

Thus, for natural rubber we can write Eq 8.3.3 for the molecular weight dependence of the diffusion coefficient at room temperature as

$$D = 1.87 \times 10^{-3}/M^2 \qquad (\text{cm}^2/\text{s}) \tag{8.3.11}$$

Figure 8.4 shows diffusion data for several linear uncured rubber materials as a function of molecular weight. Theoretical lines with slope of −2 are shown, for monodisperse and polydispersity 2 cases. It is interesting to note that the data all fall in the vicinity of the PIP prediction, despite the diversity of polymers examined, namely SBR, butyl rubber, EPR, and hevea rubber.

The healing time ($t_\infty = T_r$) is obtained from Eqs 8.33, 8.3.9, and 8.3.10 as

$$t_\infty = C_\infty j b^2 M^3 / (3\pi^2 M_0 A) \qquad (8.3.12)$$

For natural rubber (NR), using $A = 1.87 \times 10^{-3}$ as in Eq 8.3.11, we obtain

$$t_\infty(\text{NR}) = 1.11 \times 10^{-15} M^3 \quad (\text{s}) \qquad (8.3.13)$$

When M is 300,000, t_∞ is 30 seconds, and in Figure 8.4, this material should be in the green state at $t_c = 30$ s, which is consistent with the data. The healing time required for the samples with M of 10^6 is about 18 minutes.

These results differ from those of Roland and Bohm [12], who found that the time required to achieve maximum autoadhesive strength of polybutadiene was orders of magnitude longer than the time required for the chains to diffuse a distance approximately equal to the end-to-end vector, R. Chain branching and nonuniform wetting at the interface may have contributed to this difference.

Stacer and Schreuder-Stacer [13] investigated the time dependence of autoadhesion in polyisobutylene, $-CH_2-C(CH_3)_2-$. Molecular weights of their samples were in the range 1.5–21 × 10^5, with polydispersities of 1.1 to 3.6. T-peel test configurations were used to evaluate the peel energy, G, as a function of time, molecular weight, temperature, and pressure. Figure 8.5 shows a plot of log G versus log t/a_T for PIB with $M_w = 1.5 \times 10^5$. The contact time was reduced with respect to the time–temperature shift factor, a_T, to create a master plot from data obtained at different temperatures. Several theories were examined. The scaling law for peel energy, $G \sim (t/M)^{1/2}$, predicts a slope of ½, as shown in Figure 8.5. The monodisperse and polydisperse chain interdiffusion theories of Prager and Tirrell [14–16] are compared with experiment in the figure. Their model determines the number of effective crossings per unit area, where an effective crossing involves a minimum interpenetration distance, related to the entanglement molecular weight spacing, which allows a chain segment to participate in energy dissipation processes. The agreement is good for the monodisperse case but less good for the polydisperse model. Both models predict a slower rate of healing (about ½ time decade) than is found experimentally.

The time to achieve complete strength (Eq 8.3.12) for PIB was examined by Stacer for peel adhesion experiments, and reported to be in excellent quantitative agreement with their data. For the PIB sample used in Figure 8.5, $C_\infty = 6.6$, $D = 9.1 \times 10^{-17}$ cm²/s at 25 °C, with $M_v = 150{,}000$, $A = 2.05 \times 10^{-6}$ cm²/s, $b = 1.54$ Å, $M_0 = 56$, and $j = 2$, so that $t_\infty = 3100$ s (log $t_\infty = 3.49$). This value agrees well with the time at which the onset of the plateau strength occurs in Figure 8.5. The welding time equation for PIB at 25 °C is therefore

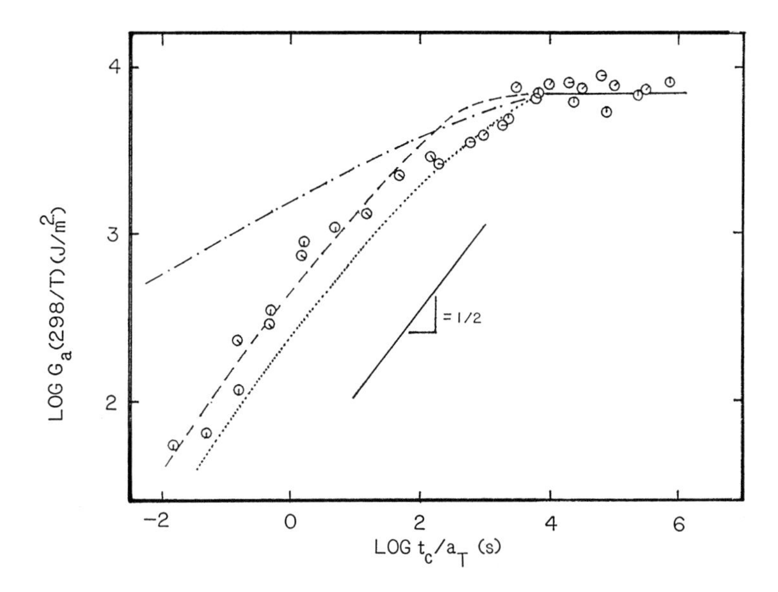

Figure 8.5 Comparison of the predictions of several theoretical models with autoadhesion data for PIB (Stacer and Schreuder-Stacer). Key to symbols: —— (solid line), Wool; ······ (dotted line), Prager & Tirrell, monodisperse; – · – · – · – (dots and dashes), Prager & Tirrell, polydisperse; – – – – (dashed line), first-order kinetics.

$$t_\infty\,(\text{PIB}) = 9.22 \times 10^{-13}\, M^3 \qquad (\text{s}) \tag{8.3.14}$$

where $M \leq M^*$. Thus, with $M = 2.1 \times 10^6$ (the highest molecular weight they tested), the welding time to obtain complete interpenetration is about 100 days. However, the strength is saturated at some value G^* of about 10 kJ/m^2 with molecular weight in the vicinity of about 200,000.

When M is greater than M^*, the welding time τ^* can be estimated from equivalent interdiffused contour lengths $\langle l \rangle$ by

$$\tau^* = (M^*/M)^2\, t_\infty \tag{8.3.15}$$

so that

$$G(t) = G^*\,(t/\tau^*)^{1/2} \tag{8.3.16}$$

For PIB, we obtain $\tau^* = 9.22 \times 10^{-13}\, M^{*2} M$. In this example, with $M = 2.1 \times 10^6$, we obtain $\tau^* = 21.5$ h, rather than 100 days to reach the maximum strength.

8.3.4 Wetting Versus Diffusion

Korenevskaya *et al.* [17] proposed a wetting theory that was expressed as a first-order kinetic process in terms of the fractional peel energy, $G = G(t)/G_\infty$, as

$$G + \ln(1 - G) = -Pt/\eta \tag{8.3.17}$$

where P is the applied pressure and η is the zero-shear viscosity. Figure 8.5 shows that such first-order kinetics provides excellent agreement with the data. Deviations to significantly lower G_{1c} values were observed at short times with the high-molecular-weight PIB samples.

Since the diffusion-controlled healing times are comparable to characteristic relaxation times of viscosity, this work cannot readily distinguish between wetting and diffusion theories. Studies by Wool and O'Connor [2, 6] (Chapter 11) of tack and healing in hydroxy-terminated polybutadiene (HTPB) indicated that the mechanical property recovery due to wetting occurred in a few minutes and was a small fraction of G_∞, while interdiffusion of the partially cross-linked chains took several weeks, during which most of the fracture energy was recovered. The separation of the relaxation times for wetting and diffusion in that case supports the diffusion model.

A more convincing case for the diffusion model is obtained from studies of incompatible amorphous polymer interfaces such as PS/PMMA (Chapter 9). At long contact times, complete wetting or contact has been achieved but the incompatibility allows only limited interdiffusion, with correspondingly low fracture energy. The observed fracture energies of incompatible interfaces [18–20] are consistent with the diffusion model. In these studies, the long-time equilibrium fracture energy, G_∞, increases with the compatibility of the polymer pairs (decreasing Flory–Huggins χ parameter) as $G_\infty \sim 1/\chi$, while the equilibrium thickness behaves as $d_\infty \sim 1/\chi^{1/2}$. Thus, interdiffusion and fracture energy increase together at constant contact area. A possible flaw in this argument is that the incompatibility influences chemical interactions, such as acid–base, at the contact plane, which would affect the wetting theory. However, the contact plane no longer exists as such due to the limited diffusion that has occurred, and the interface becomes fractal with a fractal dimension D_f of about 2.5.

Pressure affects diffusion by decreasing the free volume, which slows the hopping process necessary to achieve Brownian motion and subsequently decreases the self-diffusion coefficient. This effect is more pronounced near the glass transition temperature and is important in polymer melt processing, where large hydrostatic pressures can be encountered. It is not very important in normal tack experiments where the contact pressures are usually much less than a kilobar. However, pressure affects the wetting stage by promoting better contact. Thus, tack measurements should be dependent on pressure to some degree while wetting is occurring, but should be largely independent of pressure when interdiffusion is occurring. Expressions for the pressure dependence of T_r and D have been suggested [2, 6].

8.4 Peel Adhesion

8.4.1 Introduction

Peel adhesion tests are often conducted in the 180° T-peel test configuration, as shown in Figure 8.6 (the material may have an inextensible backing). Following the approach used by Hamed [21] and Gent [22–25] we show a unit volume element (*xyz*) during the course of a peel experiment: before fracture, at fracture, and after fracture. From an energy balance of peel (Chapter 7), the peel force, P, is related to the strain energy per unit volume, U, the sample thickness, h, and width, b, by

$$P = Uhb/2 \tag{8.4.1}$$

where U is the total energy expended in the volume element. U can be estimated by integration of the uniaxial stress–strain curve of the virgin material from zero to the fracture strain. Since $P = \frac{1}{2}bG$ (Eq 7.3.9), for the 180° peel test, where G is the fracture energy per unit area and b is again the sample width, it follows that

$$G = Uh \tag{8.4.2}$$

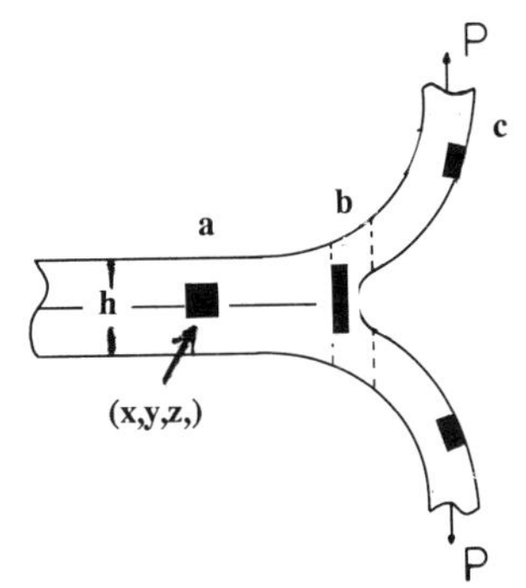

Figure 8.6 Peel adhesion test, showing the fate of an *xyz* element in the interface (a) before, (b) during, and (c) after fracture.

The strain energy density U of a unit volume (*xyz*) of rubber material (Figure 8.6) is given in terms of the three draw ratios λ_x, λ_y, and λ_z, by the Mooney equation

$$U = C_1(\lambda_x^2 + \lambda_y^2 + \lambda_z^2 - 3) + C_2(1/\lambda_x^2 + 1/\lambda_y^2 + 1/\lambda_z^2 - 3) \tag{8.4.3}$$

where C_1 and C_2 are constants. C_1 corresponds to the elastic modulus E from Flory's statistical theory via

$$C_1 = v_e R T/2 = E/2 \tag{8.4.4}$$

in which v_e is the number of entanglement strands per unit volume.

Using the ideal rubber theory in which $C_2 = 0$, we evaluate Eq 8.4.2 in the plane–strain approximation, so that $\lambda_y = \lambda$, $\lambda_x = 1/\lambda^{1/2}$, and $\lambda_z = 1$ (no contraction in the width direction). The dilational strain energy at the crack tip in Figure 8.6 is

$$U = C_1(\lambda^2 + 1/\lambda - 2) \tag{8.4.5}$$

The disentanglement theory (Chapter 7) predicts that fracture occurs at a critical draw ratio λ_c given by Eq 7.5.6, exactly as $\lambda_c^2 + 1/\lambda_c = 2\, M/M_c$, and Eq 8.4.5 then becomes

$$U = 2\, C_1(M/M_c - 1) \tag{8.4.6}$$

The fracture energy for peel adhesion is obtained from the above relations as

$$G = Eh(M/M_c - 1) \tag{8.4.7}$$

which is the green strength.

To obtain the time dependence $G(t) = G_\infty(t/T_r)^{1/2}$, we substitute the contour length for M (Chapter 7) so that

$$G \approx Eh(M/M_c)\,(t/T_r)^{1/2} \tag{8.4.8}$$

Eq 8.4.8 gives characteristic healing slopes of ½ for peel energy, as observed by Tsuji *et al.* [26] and Bhowmick *et al.* [27], and ¼ for tack stress at low peel rates.

8.4.2 Rate Effects on Tack and Green Strength

The effects of strain rate and peeling velocity on tack and green strength have been examined by many investigators [21–27]. For example, Figure 8.7 shows the cohesive fracture energy G obtained by Hamed and Shieh [21] for an SBR elastomer [FRS-146] as a function of T-peel rate R_p and temperature. We see that G increases with both increasing rate and decreasing temperature. The slope of G versus R_p is low at both high and low rates, and at intermediate rates has a slope near ½, but it is constantly changing. Temperature plays a dominant role, and as the temperature approaches T_g (–50 °C), the maximum strength is obtained.

The effects of temperature and rate on G underscore the contribution of viscoelastic processes. To make this point, Hamed and Shieh were able to superimpose the data in Figure 8.7 by horizontal shifts using the William–Landel–Ferry (WLF) equation for the shift factor, a_T. Figure 8.8 shows the master curve obtained by this process, in which log G is plotted versus log $R_p a_T$ at a reference temperature, T_0 of 23 °C. The

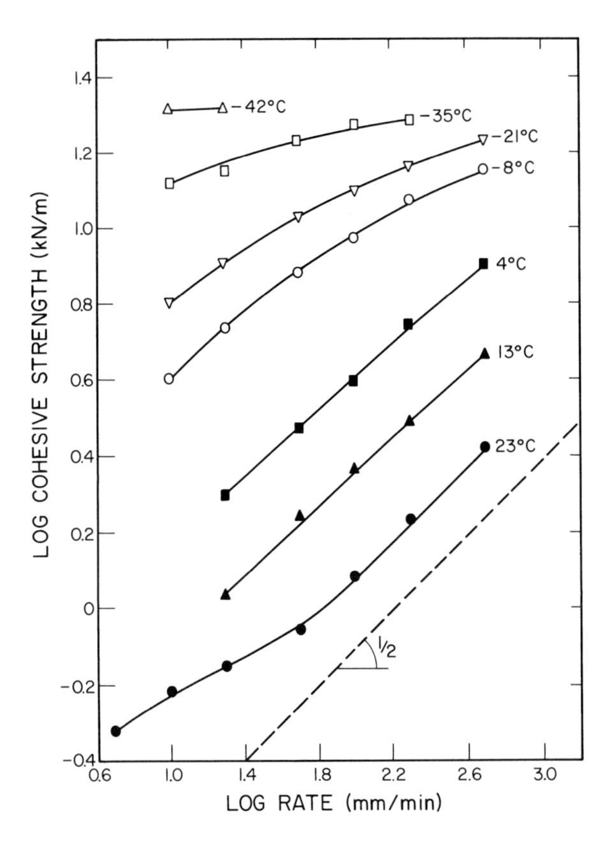

Figure 8.7 Cohesive tear strength of SBR as a function of peeling rate at several constant temperatures (Hamed and Shieh). The dashed line represents the theoretical slope of ½ and was added for comparison purposes. The SBR was a cold emulsion styrene–butadiene random copolymer containing 23% bound styrene and had a Mooney viscosity (ML-4, 100 °C) of 40.

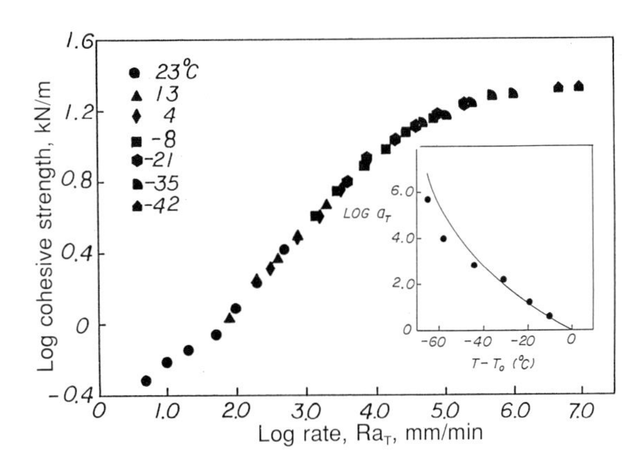

Figure 8.8 Master curve for the data shown in Figure 8.7. The inset gives shift factors versus temperature (Hamed and Shieh).

inset in Figure 8.8 shows the experimental shift factor (points) compared with the WLF prediction (line). Similar results were obtained by Tsuji *et al.* [26] for T-peel tests of polyisobutylene and by Bhowmick *et al.* [27] for adhesive tack and green strength of EPDM rubber.

Chang, Gent, and Lai [22, 23] investigated the effect of peel rate on the fracture energy of lightly cross-linked polybutadiene at temperatures in the range −40 °C to 130 °C. They used both tear (trouser configuration) and T-peel tests. Again, using WLF shift methods, they were able to superimpose data in a master curve of log G versus log $R_p a_T$, as shown in Figure 8.9. At very low rates, they obtained a constant minimum fracture energy G_0, which was found to be proportional to the cross-link density, ν_x, via

$$G_0 = 60\ \nu_x \qquad \mathrm{J/m^2\ (m^3/10^{26})} \tag{8.4.9}$$

The cross-link densities used gave G_0 values in the range of 1–60 J/m². Since the strain energy density is proportional to C_1 in Eq 8.4.3 and $C_1 \sim \nu_x$, this behavior is reasonable. However, at high peel rates, maximum G values were about 3000 J/m², indicating very large viscoelastic effects on fracture. Gent has found that the fracture energy is approximately the product of G_0 and a viscoelastic loss function [23].

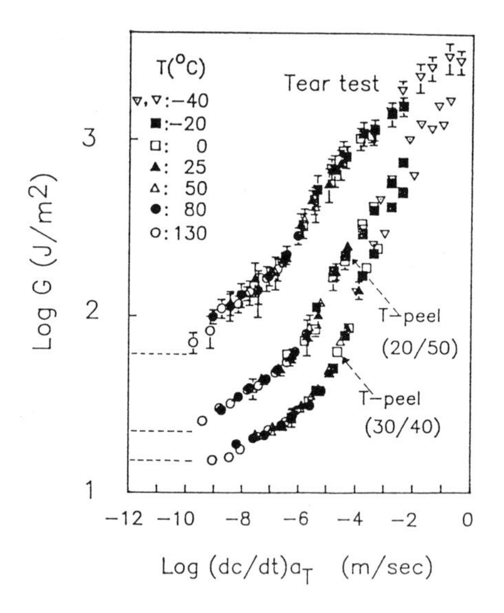

Figure 8.9 Fracture energy versus reduced peel rate for different interlinking densities at −20 °C (peroxide cure) (courtesy of Gent).

We can attempt to understand rate effects on fracture energy by examining viscoelastic contributions to the strain energy function, U, for the volume element xyz in Figure 8.6. The integral of the stress–strain curve for the volume element requires that we have a constitutive relation. For example, let us treat the rubber as a two-parameter Kelvin solid (elastic and viscous elements in parallel) with elastic modulus E and viscosity η, so that the constitutive law for stress σ and strain ϵ is

$$\sigma = \epsilon E + \eta \, \mathrm{d}\epsilon/\mathrm{d}t \tag{8.4.10}$$

In the peel experiment, the rate of peel R_p is related to the strain rate $\mathrm{d}\epsilon/\mathrm{d}t$ by

$$R_p/h = \mathrm{d}\epsilon/\mathrm{d}t \tag{8.4.11}$$

where h is the thickness.

Letting the strain input be described by $\epsilon(t) = Rt$, where $R = \mathrm{d}\epsilon/\mathrm{d}t$ is the constant strain rate, and integrating the stress response from $\epsilon = 0$ to $\epsilon = \epsilon_c$, we obtain

$$U(R) = \tfrac{1}{2} E\epsilon_c^2 + \eta R\epsilon_c \tag{8.4.12}$$

where $\epsilon_c = \lambda_c - 1$ is the critical strain at fracture. This gives the contribution to the fracture work in terms of the stored elastic energy, $\tfrac{1}{2}E\epsilon^2$, and the viscous dissipation term, $\eta R\epsilon_c$. For polymer melts, the viscosity is related to the relaxation time T_r via

$$\eta = E T_r \tag{8.4.13}$$

and Eq 8.4.12 can be rearranged to

$$U(R) = U_0 (1 + 2 T_r R/\epsilon) \tag{8.4.14}$$

where $U_0 = \tfrac{1}{2}E\epsilon_c^2$.

When $2T_rR/\epsilon$ is much greater than 1, the fracture energy becomes the product of the U_0 term and the viscous loss process, as noted by Gent. However, Eq 8.4.14, while predicting the minimum energy at $R = 0$, increases without limit with R. Using the entanglement model, we predict that a maximum energy is obtained at a rate R_∞, related to the Rouse time τ_{RO}^* at $M^* \approx 8M_c$, as

$$R_\infty = 1/\tau_{RO}^* \tag{8.4.15}$$

The Rouse time $\tau_{RO} \sim M^2$ and is related to the reptation time, T_r, by

$$\tau_{RO} = T_r M_e/(3M) \tag{8.4.16}$$

where $M_e = 4/9\ M_c$. Making these substitutions and letting $\lambda_c = 4$, we derive the result for $G(R) = hU(R)$

$$G(R) = G_0(1 + BR/R_\infty) \tag{8.4.17}$$

in which $B \approx 36$ and $G_0 = hU_0$. The factor B is related to the ratio of the maximum energy, G_∞, and the minimum energy, G_0, by

$$B = (G_\infty / G_0) - 1 \tag{8.4.18}$$

We note that in Figure 8.7, G_∞ is approximately 20 kN/m and the lowest value is G_0 of about 0.5 kN/m, giving $B \approx 39$ for Hamed's linear elastomer data. For Gent's lightly cross-linked elastomers in Figure 8.9, B appears to be between 50 and 100.

The time and peel rate dependence of the fracture energy $G(t,R)$ for the Kelvin model during welding has a $t^{1/2}$ dependence for the elastic contribution (via ϵ_c^2) and a $t^{1/4}$ dependence (via ϵ_c). From the above relations we have

$$G(t, R_p) = Eh(M/M_c)(t/T_r)^{1/2} + \eta R_p (M/M_c)^{1/2} (t/T_r)^{1/4} \tag{8.4.19}$$

in which the T-peel rate $R_p = hR$, and E is the plateau modulus of the melt. At low rates (R much less than $1/\tau_{RO}$), $G \sim t^{1/2}$, and at higher rates when the contribution from ηR_p becomes important, we have a mixed time dependence. Thus, the slope of log G versus log t may exhibit a rate dependence, depending on the range. With very high rates (R greater than $1/\tau_{RO}$) at times less than T_r, the ηR_p term dominates and the apparent virgin strength is obtained, even though the interface is only partially welded in a diffusion sense. In Eq 8.4.19, the thickness h is expected to be less than the actual thickness of the peel sample.

Although the above rate model was developed using the constitutive equation for the Kelvin solid, the energy at fracture can generally be partitioned into additive contributions consisting of an elastic term and a viscous dissipation term and gives results for $G(R)$ similar to Eq 8.4.17. This approach gives a sharp cutoff in G versus R as G_∞ is approached, rather than the smooth transition to the G_∞ plateau that would be expected and that is observed experimentally (Figure 8.8). This transition could be incorporated into a more detailed analysis.

The tack problem for lightly cross-linked elastomers has been analyzed by de Gennes [28]. He argues that the fracture energy behaves as $G_{1c} = 2\Gamma E_\infty/E_0$, where 2Γ is the thermodynamic work of adhesion, E_∞ is the high-frequency modulus due to cross-links plus entanglements, and E_o is the zero-frequency modulus, which is small if the cross-link density is low. For uncross-linked linear elastomers, he uses a *viscoelastic trumpet* analysis, where at deformation rates R greater than $1/T_r$, the modulus is dominated by the viscosity, so that $E_\infty \approx \eta R$ and $E_0 = G_N^0$ (plateau entanglement modulus), giving $G_{1c} \approx 2\Gamma\eta R/G_N^0$. This relation is similar in many respects to Eq 8.4.17, in which $\Gamma \sim G_0$. Since $\eta = G_N^0 T_r$, de Gennes obtains the result

$$G_{1c} = 2\Gamma R T_r \tag{8.4.20}$$

8.4.3 Chain Pullout Model

We briefly compare the above viscoelastic model for peel adhesion with the simple model for the energy to pull a chain out from the interface, U_p, given by (Chapter 7)

$$U_p = \tfrac{1}{2}\mu_0\, l(t)^2\, V/b \tag{8.4.21}$$

where μ_0, $l(t)$, V, and b are the monomer friction coefficient, the interpenetration contour length, the velocity, and the statistical segment length, respectively. The stored energy in a chain, U_c, is

$$U_c = \sigma^2\, l(t)\, a_0/2E \tag{8.4.22}$$

in which a_0 is the cross-sectional area of a chain that is being pulled out and E is the elastic modulus. When $U_c = U_p$, the critical stress is obtained as

$$\sigma = [\mu_0 E V\, l(t)/(a_0 b)]^{1/2} \tag{8.4.23}$$

To evaluate this equation at the maximum pullout velocity V^*, we substitute:

$$\mu_0 = kTT_r b/L^3 \tag{8.4.24}$$

$$a_0 L = M/(\rho N_a) \tag{8.4.25}$$

$$V^* = L/\tau_{RO} \tag{8.4.26}$$

$$T_r/\tau_{RO} = 3M/M_e \tag{8.4.27}$$

$$E = G_N^{\circ} = \rho kT/M_e \tag{8.4.28}$$

$$l(t)/L = (t/T_r)^{1/2} \tag{8.4.29}$$

where L, k, T, ρ, and τ_{RO} are the contour length of the whole chain, Boltzmann's constant, temperature, density, and Rouse relaxation time, respectively. This gives the relatively simple result

$$\sigma^* = \sqrt{3}\, E(t/T_r)^{1/4} \tag{8.4.30}$$

where σ^* is the stress at the maximum pullout rate discussed in Section 8.4.2. The factor $\sqrt{3}E$ corresponds to the green state σ_∞ at $t = T_r$. Since E is about 100 psi (0.7 MPa) for many rubbery materials, we predict that σ_∞ is about 170 psi, which is of comparable magnitude to strength values reported in Figure 8.4. The quantitative agreement with experiment is interesting considering the simplicity of the model.

8.4.4 Summary Comments on Tack and Green Strength

Strength development at a polymer–polymer interface was analyzed in terms of the dynamic and static properties of random-coil chains. Interdiffusion of chain segments across the interface was considered to be the controlling factor for tack and green strength of uncured linear elastomers. This concept is similar to that proposed by Voyutskii, but differs markedly from contact theories proposed by Anand and others. In our approach, time-dependent wetting occurs first, followed by interdiffusion. Increasing contact pressure and temperature should promote the establishment of molecular contact (wetting) at the interface up to a saturation point of complete wetting. However, interdiffusion is retarded by increased hydrostatic pressure and enhanced by temperature in the usual thermally activated manner. The effect of pressure on diffusion is to reduce the volume available (depending on the compressibility) for segmental motion and subsequently decrease the diffusion coefficient.

If the tack test is performed at higher temperatures, the average interdiffusion chain segment length increases as $l(T) \sim \exp -Q/2kT$, but the stress required to pull the segment out decreases as $\sigma \sim \exp Q/4kT$, where Q is the activation energy. Therefore, the tack evaluated at constant contact time decreases with increasing temperature for the interdiffusion-controlled process.

The effect of molecular weight on tack at constant contact time, t_c, is to increase the tack according to $\sigma \sim M^{1/2}$ for those samples whose relaxation time T_r is less than t_c, and decrease the tack as $\sigma \sim M^{-1/4}$ for those molecular weights for which T_r is greater than t_c. The tack should reach a maximum at a molecular weight corresponding to $T_r = t_c$. This can be used to determine the self-diffusion coefficient of the chains. At small contact times the results might be complicated by wetting processes. The position of the maximum in a plot of σ versus M is relatively insensitive to the contact time, since $T_r \sim M^3$, and M at the maximum consequently increases as $t_c^{1/3}$. These predictions appear to be in agreement with much experimental data reviewed by Hamed [29] and Rhee [30] but differ in some respects from their own interpretations of the same data.

The above results are not unique to elastomers and many of the predictions will be seen to describe welding and fracture of glassy polymers in Sections 8.5 and 8.6.

8.5 Strength of Glass Interfaces

8.5.1 Introduction

In this section, we examine strength development at symmetric amorphous interfaces where the polymers are welded for a time t, at a temperature above the glass transition temperature T_g, and are then quenched to the glassy state. The glass samples are then fractured by standard fracture mechanics methods where the initial crack is directed along the glass interface. The crack propagates through a craze zone that forms from the interface material. Ideally, the fracture energy is confined to the interfacial region and the rest of the sample deforms elastically.

Figure 8.10 shows the welding method used by Wool and O'Connor to evaluate healing relations for polymer interfaces. Monodisperse-molecular-weight films (or other materials) are bonded to substrates, usually composed of the same polydisperse molecular weight material, and the film surfaces are then welded together. Compact tension and double cantilever beam methods have also been used. The film surfaces to be contacted are first molded against highly polished metal plates and annealed *in vacuo* to assure maximum smoothness. To approximate the instantaneous wetting condition necessary to evaluate the effect of diffusion on welding, the surfaces are wetted at a higher temperature for a short time, t, that is much less than T_r, and are subsequently healed for long times at a constant lower temperature. Under these conditions, the long-time healing data can be extrapolated to zero time.

The stress distribution and the fracture mechanics of the wedge cleavage method have been examined [31, 32]. The wedge is driven into the interface, spreading the

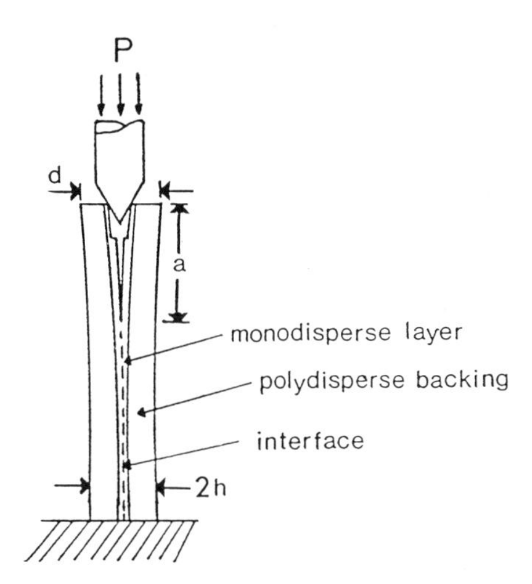

Figure 8.10 Wedge cleavage method of testing strength versus welding time in polymers. Monodisperse-molecular-weight films are bonded to polymer substrates prior to the thermal welding (Wool and O'Connor).

arms apart until the crack of length a begins to grow at a critical value of d, as shown in Figure 8.10. If we model the specimen as an elastic double cantilever beam with each beam partly supported by an elastic foundation, we obtain the critical stress intensity factor, K_{1c}, at constant displacement as [31, 32]

$$K_{1c} = \tfrac{3}{4} E h^{3/2} (d - 2h)\, f(h/a) / a^2 Q \tag{8.5.1}$$

where h is the half-width of the specimen (see Figure 8.10) and d is the separation of the specimen at the upper end. The function $f(h/a)$ is given by

$$f(h/a) = \frac{(1 + 0.64\, h/a)}{1 + 1.92\, h/a + 1.22 (h/a)^2 + 0.39 (h/a)^3} \tag{8.5.2}$$

and the factor Q is a correction for the extra compliance due to the presence of the end slot [32], $Q = 1 + [(h/h')^3 - 1]\,(l_s/a)^3$, where l_s and h' are the length and half-width of the slot, respectively. The fracture energy is obtained from Eq 8.5.1 via the relation $G_{1c} = K_{1c}^2/E$.

The advantage of the wedge cleavage technique is that the crack propagates in a stable manner, so a single sample can yield multiple data points and the significance of the statistics is improved. We can obtain the weld strength as a function of welding time by alternating crack propagation with healing treatments so that each measurement corresponds to crack advance into a previously nonfractured part of the interface.

8.5.2 Time and Molecular Weight Dependence of Welding

We used wedge cleavage methods to examine the time and molecular weight dependence of PS/PS interfaces. To overcome the time-dependent wetting stage, we designed a two-stage experiment where we welded at high temperature (138 °C) for short times (30 seconds) and then at the normal welding temperature (120 °C) for long times. In this case, the stress intensity factor K_{1c} is expressed as [3, 33]

$$K_{1c} = (K_W^4 + \alpha t)^{1/4} \tag{8.5.3}$$

where K_W is the initial wetted strength after 30 seconds and α is a constant. Figure 8.11 shows the behavior of this function for two-stage welding on a plot of K_{1c} versus $t^{1/4}$. The data taken at long times, when αt is much greater than K_W, can be extrapolated to $t = 0$ as $K_{1c} \sim t^{1/4}$, and the slope gives the constant α. Alternatively, Eq 8.5.3 can be linearized by rearranging to

$$K_{1c}' = (K_{1c}^4 - K_W^4)^{1/4} = (\alpha t)^{1/4} \tag{8.5.4}$$

where K_{1c}' is the effective strength corrected for wetting.

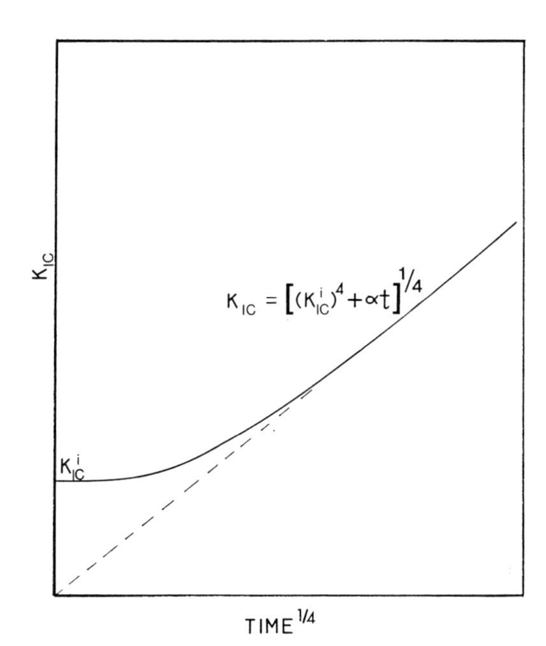

Figure 8.11 Schematic behavior of K_{1c} versus $t^{1/4}$ for two-stage welding (Wool and O'Connor).

Figure 8.12 shows results obtained by O'Connor [3, 33] for welding monodisperse-molecular-weight polystyrene interfaces at 120 °C with an initial 30-second weld at 138 °C. The data trend is similar to that shown schematically in Figure 8.11, where the long-time data tend to be linear and can be extrapolated to zero time. Figure 8.13 shows the same data plotted according to Eq 8.5.4. The K_{1c} values have a linear dependence on $t^{1/4}$, and we see a clear effect of molecular weight on the healing rate: the higher molecular weight gives the slower healing rate. The best fit to the data for four molecular weights, of 142K, 217K, 330K, and 524K gives

$$K_{1c} \sim t^{1/4} M^{-x} \tag{8.5.5}$$

where x is in the range 0.31 to 0.37 with an optimal fit at $x = 0.35$.

The scaling law for welding is $K_{1c} = K_{1c\infty}(t/T_r)^{1/4}$, where $K_{1c\infty} \sim (M^{1/2} - M_c^{1/2})$, and it follows that the molecular weight exponent, x, is determined by

$$x = ¼ - \log[1 - (M_c/M)^{1/2}] \tag{8.5.6}$$

In the asymptotic limit, $x = ¼$, but for molecular weights with $M/M_c \approx 10$, then $x \approx 0.4$, which is in very good agreement with experimental values. This result is highly supportive of the relation between contour length and strength. Since $G_{1c} \sim K_{1c}^2$, these data give $G_{1c} \sim t^{1/2}M^{-0.7}$, where the molecular weight exponent (0.7) is closer to that for the chain disentanglement model (0.5) than to that for the bridge fracture

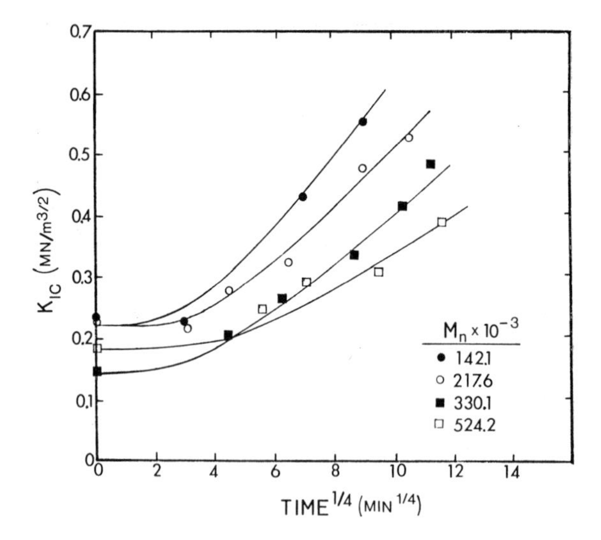

Figure 8.12 K_{1c} versus $t^{1/4}$ for two-stage welding of polystyrene of different narrow-fraction molecular weights, taken with the wedge cleavage technique (Wool and O'Connor).

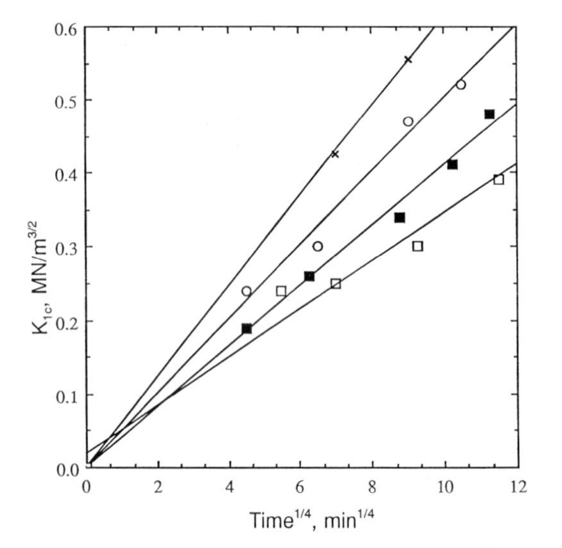

Figure 8.13 K_{1c} versus $t^{1/4}$ wedge cleavage data corrected for single-stage welding (Eq 8.5.4) of polystyrene interfaces (data from Figure 8.12).

model (1.5). McGarel repeated some of these wedge cleavage experiments with monodisperse polystyrenes and found similar results [45].

Using improved welding techniques to eliminate wetting problems, Wool and McGarel tested monodisperse PS/PS interfaces as compact tension specimens. The vacuum welding chamber shown in Figure 8.14 had a piston that applied normal pressure to the polymer slabs being welded. This ensured the removal of trapped air

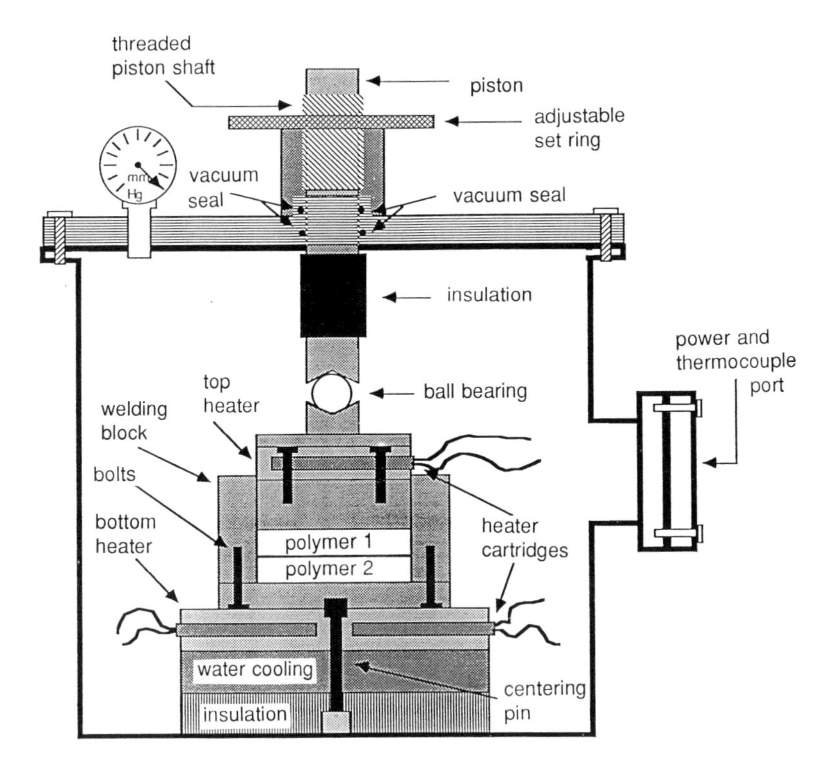

Figure 8.14 Polymer vacuum welding chamber (Foster and Wool).

and promoted excellent contact at the interface. The sample was also contained in a sealed welding block to prevent melt leakage or sample distortion during welding under pressure. Once the slabs were welded for a time t, they were cut into compact tension specimens, with the crack on the interface plane.

Figure 8.15 shows G_{1c} versus $t^{1/2}$ for welding PS/PS interfaces with monodisperse molecular weights, M, of 207K, 600K, and 900K, at 120 °C. The data are reasonably linear on a $t^{1/2}$ scale (log–log plots also give a slope of ½) with the greatest scatter seen for the 900K sample. Again we note that the sample with the lowest molecular weight has the fastest healing rate. To compare contour length with bridge welding models, we plotted the data as G_{1c} versus $t^{1/2}M^{-1/2}$ in Figure 8.16(a) and G_{1c} versus $t^{1/2}M^{-3/2}$ in Figure 8.16(b), and enquired about superposition of data. It is clear that the contour length model gives the better superposition in Figure 8.16(a), similar to the wedge cleavage data. These experiments give support to the relation that $G_{1c}(t) \sim l(t)$, where $l(t)$ is the average contour length, and to the relation that $K_{1c} \sim X(t)$, where $X(t)$ is the average monomer diffusion depth corresponding to the characteristic width of the interface.

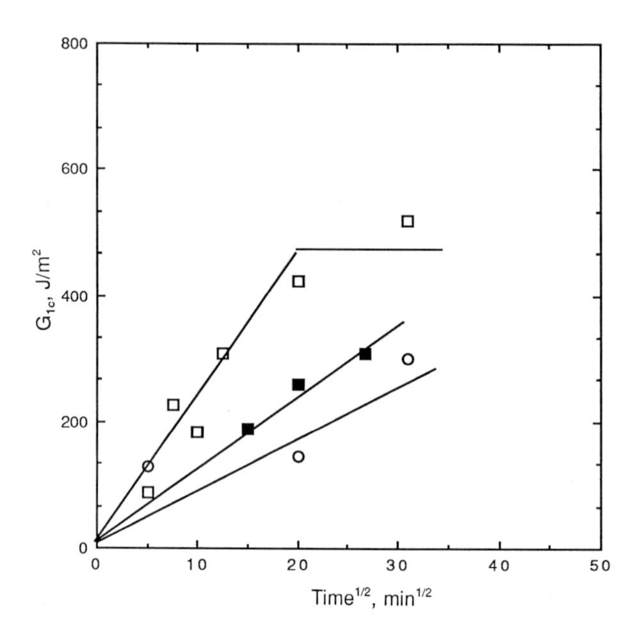

Figure 8.15 G_{1c} versus $t^{1/2}$ welding data for PS interfaces (monodisperse molecular weight) obtained with compact tension methods. Key to symbols: □ M = 207,000; ■ M = 600,000; ○ M = 900,000 (McGarel and Wool).

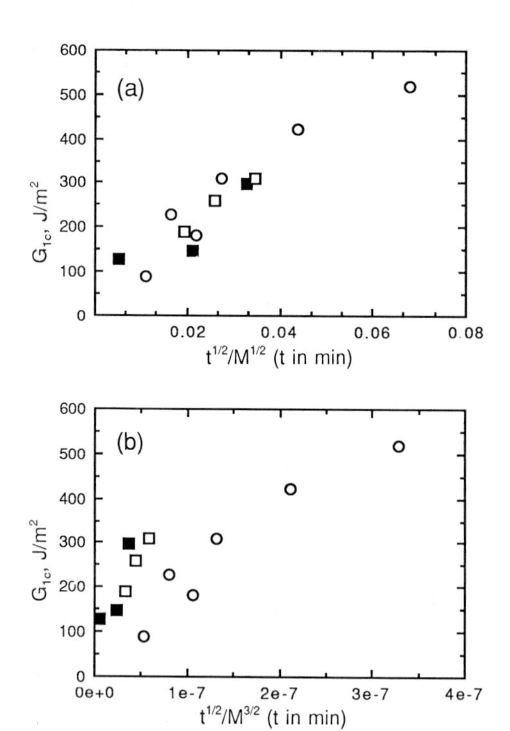

Figure 8.16 Test of scaling laws for welding of PS interfaces. (a) Contour length model, G_{1c} versus $t^{1/2}M^{-1/2}$; (b) Bridges model, G_{1c} versus $t^{1/2}M^{-3/2}$. Key to symbols: ○ M = 207,000; □ M = 600,000; ■ M = 900,000. Data from Figure 8.15 (McGarel and Wool).

8.6 Molecular Weight Dependence of Fracture of Glassy Polymers

8.6.1 Fracture Energy Versus Molecular Weight

Figure 8.17 shows the virgin fracture energy, G_{1c}, of polystyrene as a function of molecular weight M (monodisperse fractions) [34, 35]. Three regions are quite distinct, the first when M is less than M_c, the second when M is between M_c and M^*, and the third when M is greater than M^*, where M^* is approximately $8M_c$. An abrupt change in fracture behavior is observed in the vicinity of M_c of about 32,000 that parallels changes in the zero-shear viscosity behavior. The G_{1c} data versus molecular weight are shown in Table 8.1.

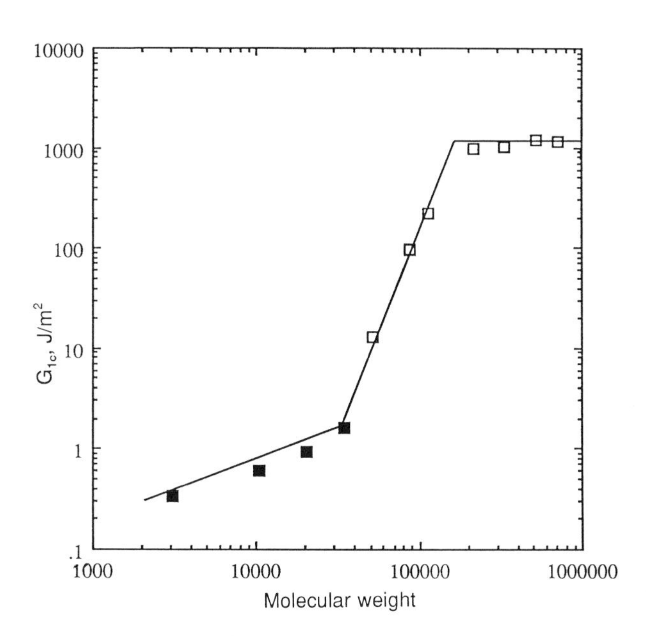

Figure 8.17 G_{1c} versus molecular weight for polystyrene in the virgin state. Data of Robertson (M less than 40,000, ▪ filled squares) and O'Connor and Wool (▫ open squares).

8.6.2 Fracture at M Less Than M_c

For M less than M_c, the data obtained by Robertson [36] using wedge cleavage methods (Table 8.1) indicate that the material is very fragile, with G_{1c} values of less than 1 J/m^2. From Chapter 7, we expect the fracture energy to be governed by the nail

Table 8.1 G_{1c} Versus M_n for Polystyrene

M_n	G_{1c}, J/m^2	M_n	G_{1c}, J/m^2
3,100	0.34	114,921	220.1
10,300	0.60	217,600	980.79
20,500	0.94	330,070	1040.52
34,500	1.64	524,220	1220.66
52,050	12.9	713,950	1160.45
88,340	97.6		

solution, so that $G_{1c} \sim pl_p^2$, where p is the number of nails of length $l_p \sim M^{1/2}$. Since p is independent of M, we expect the linear relation $G_{1c} \sim M$. The fracture energy at M less than M_c is well described by the linear equation (Eq 7.6.15)

$$G_{1c} = 0.24 + 3.44 \cdot 10^{-5} M \qquad (\text{J/m}^2) \tag{8.6.1}$$

As M decreases, the fracture energy approaches the Griffith limit, $G_{1c0} = 2S\Gamma$, where Γ is the surface energy and S is the Sapoval number describing the surface roughness ($S = 1$ is a flat surface without distortion). For PS at room temperature, $2\Gamma = 0.08$ J/m^2, which yields a surface roughness, S, of 3.

The same fracture data at M less than M_c can also be described by a best-fit power law

$$G_{1c} = 4.67 \cdot 10^{-3} M^{0.53} \qquad (\text{J/m}^2) \tag{8.6.2}$$

Kramer has suggested that G_{1c} depends on $M^{1/2}$ [37], based on the random-coil radius $R_g \sim M^{1/2}$ controlling the crack-opening displacement at constant craze stress, σ_c, giving

$$G_{1c} = \sigma_c(\lambda_f - 1) R_g \tag{8.6.3}$$

where $\lambda_f \approx 4$ is the craze fibril extension ratio at the crack tip. Experiments by Lauterwasser and Kramer [38] on PS crazes indicate that the craze stress has a value of about 25 MPa at the base, rising to about 35 MPa at the craze tip, resulting in an average craze stress, σ_c, of about 30 MPa. Using densitometry measurements on TEM images of craze fibrils at M greater than M_c, they measure the extension ratio as $\lambda_f \approx 4$. This gives the critical crack-opening displacement $\delta = 3R_g$, and substituting for the molecular weight dependence of R_g, they obtain

$$\delta = 2.1\, M^{1/2} \qquad (\text{Å}) \tag{8.6.4}$$

Using the Dugdale model with $G_{1c} = \sigma\delta$, Kramer predicts

$$G_{1c} = 0.0063\, M^{1/2} \qquad (\text{J/m}^2) \tag{8.6.5}$$

which is in close agreement with the experimentally derived power law, $G_{1c} = 0.0047M^{0.53}$.

The result, $G_{1c} \sim M$, which is consistent with chain pullout arguments, provides a slightly better fit of the data in the molecular weight range M less than M_c. However, Kramer's approach gives a better extrapolation to the ideal Griffith limit, $G_{1c} = 2\Gamma = 0.08$ J/m^2 (S = 1), near the monomer molecular weight M_0, and is in excellent quantitative agreement with the experimental data. We can attempt to reconcile the two models by applying Kramer's analysis to the chain pullout model as follows. For chain pullout, the critical crack-opening displacement in the absence of an entanglement network (M less than M_c) can be given as the average chain contour length L, as

$$\delta = L/4 \tag{8.6.6}$$

The factor of 4 comes about from the fact that at the crack plane, the maximum a chain needs to pull out is $L/2$ and the minimum is zero, so that we have a distribution of chain pullout lengths with an average of about $L/4$.

The contour length is determined by

$$L = R^2/b \tag{8.6.7}$$

where R is the end-to-end vector and b is the statistical bond length. For PS, $R = 0.456\,M$ Å and b = 6.5 Å, which gives

$$\delta = 0.0175\,M \qquad (\text{Å}) \tag{8.6.8}$$

With Kramer's value of σ_c = 30 MPa, the fracture energy for chain pullout is determined from $G_{1c} = \sigma_c\delta$ as

$$G_{1c} = 0.24 + 5.26 \cdot 10^{-5}\,M \qquad (\text{J/m}^2) \tag{8.6.9}$$

in which the extrapolated Griffith limit (0.24) has been added. The factor of 5.26×10^{-5} is in pretty good agreement with the experimental factor of 3.44×10^{-5}. A potential problem with this analysis is that we require σ_c to remain constant in this molecular weight range.

8.6.3 Fracture at M Greater Than M_c

When M is in the molecular weight range between M_c and M^*, which in this case means $M^* \approx 8M_c \approx 250{,}000$, G_{1c} increases from 1 to 1000 J/m^2, as shown in Figure 8.17. The welding theory extrapolated to the virgin state predicts that the fracture energy can be described approximately (Chapter 7) by

$$G_{1c} \approx (0.3\, G_{1c}^{*}\, M / M_c)\, [1 - (M_c / M)^{1/2}]^2 \tag{8.6.10}$$

where $G_{1c}{}^*$, approximately 1000 J/m^2, is the upper plateau in Figure 8.17. Values calculated with this equation are compared with experiment in Table 8.2, using M_c = 30,000. A slightly better fit is obtained if M_c = 33,000.

Table 8.2 G_{1c} Versus Molecular Weight for Polystyrene

M_n	G_{1c} (Expt), J/m^2	G_{1c} (Theory), J/m^2
34,500	1.64	2
52,100	12.9	30
88,300	97.6	133
115,000	220	274
218,000	980	896
330,000	1040	1000
524,000	1220	1000

The critical stress intensity factor, $K_{1c} = (G_{1c}E)^{1/2}$, is derived from Eq 8.6.10 as $K_{1c} \sim (M^{1/2} - M_c{}^{1/2})$. Figure 8.18 shows K_{1c} versus $M_n{}^{1/2}$ data obtained by Wool and O'Connor for both polydisperse (M_w/M_n = 1.85) and near-monodisperse PS samples fractured by compact tension [2, 3]. The monodisperse molecular weight data can be described at M less than $8M_c$ by

$$K_{1c} = 0.006\, (M^{1/2} - M_c^{1/2}) \qquad (\mathrm{MPa\,m}^{-1/2}) \tag{8.6.11}$$

where the intercept at $K_{1c} = 0$ gives $M_c{}^{1/2}$ = 188.45, or M_c = 35,500. The polydisperse samples have the same slope (0.006) but show a higher strength, presumably due to the higher molecular weight tail in the distribution and the plasticizing action of the low molecular weights. The separation between the two sets of data is about $\sqrt{(M_w/M_n)} \approx \sqrt{2}$, which suggests that the weight average M_w, rather than the number average M_n, would better describe most fracture data.

Prentice [39], Evans [40], and McLeish [41] have also argued that the fracture energy should depend on molecular weight via the chain pullout mechanism, so that $G_{1c} \sim n_\infty L^2$, where $L \sim M$ is the contour length of the chain. They assume that n_∞ is independent of molecular weight (actually, $n_\infty \sim M^{-1/2}$) and suggest that $G_{1c} \sim M^2$. Figure 8.19 shows G_{1c} versus M_v for PMMA on a log–log plot with a slope of 2.5. Similar plots for PC also have a slope of 2.5. We suspect that this slope is due to the inhomogeneous nature of G_{1c} versus M, and the data are more justifiably represented by relations similar to Eq 8.6.10 above. If the relation $G_{1c} \sim L^2$ were correct, one would expect the time dependence of welding to be $G_{1c}(t) \sim t/M$, but this is not consistent with the body of welding data.

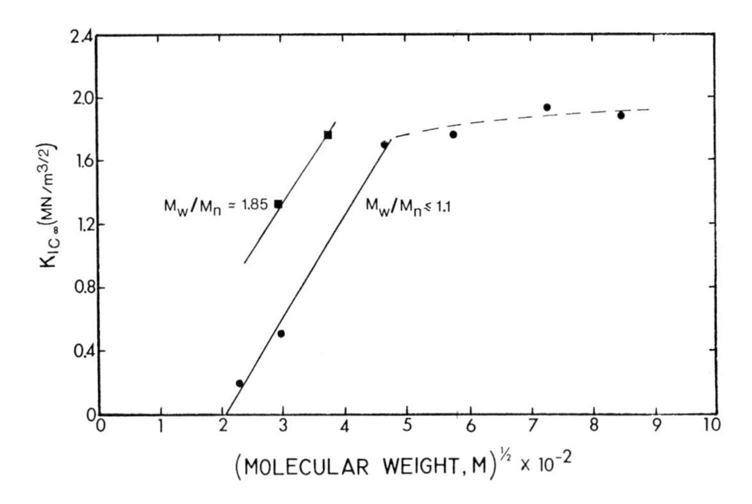

Figure 8.18 K_{1c} versus $M^{1/2}$ for PS (Wool and O'Connor).

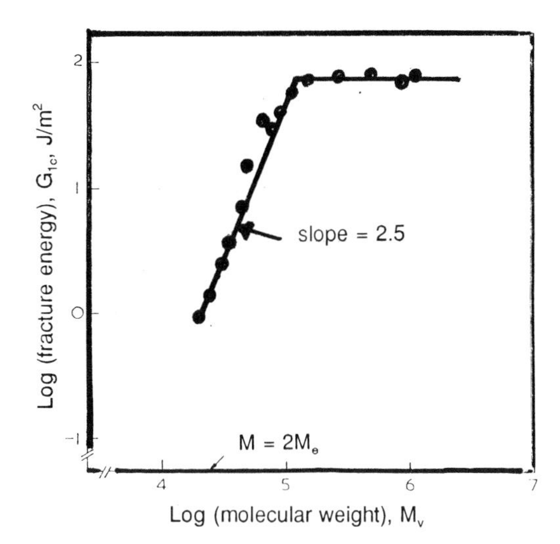

Figure 8.19 Log fracture surface energy versus log molecular weight for PMMA (courtesy of P. Prentice).

8.6.4 Tensile Fracture Properties Versus Molecular Weight

The tensile properties of monodisperse and polydisperse PS samples were investigated by McCormick, Brower, and Kin [42]. The PS tensile samples ("dog bones") were melt-processed by injection molding; their fracture properties are shown in Table 8.3.

Table 8.3 Tensile Test Data for Injection-Molded Polystyrene

M_n	$M_n^{1/2}$	Tensile Strength, psi	Modulus, psi × 10^{-5}	Elongation, %
		Anionic Polystyrene		
40,000	200	a	a	a
66,300	257	1,430	4.6	–
95,700	309	3,710	4.52	0.90
116,000	341	5,570	4.04	1.30
131,000	362	6,540	4.49	1.60
159,000	399	7,020	4.60	2.20
189,000	435	7,100	4.48	1.90
240,000	490	7,920	4.62	3.30
		Isothermal Polystyrene		
49,000	221	1,580	5.14	0.3
99,000	315	6,250	4.50	1.5
110,000	332	6,980	4.51	1.9
130,000	361	7,760	4.52	2.4
161,000	401	8,070	4.75	2.8
157,000	396	8,170	4.64	2.4

a - Too brittle for testing

Figure 8.20 shows our analysis of the uniaxial tensile fracture stress, σ_∞, plotted as a function of $M_n^{1/2}$ for both mono- and polydisperse samples [43]. The data are reasonably linear up to M_n of about 131,000 and reach a plateau at higher molecular weight. The monodisperse PS strength in the linear region has a least squares correlation coefficient $r^2 = 0.996$ (1 = perfect linear fit) and is described by

$$\sigma_\infty = 49.19\,(M^{1/2} - M_c^{1/2}) \qquad \text{(psi)} \tag{8.6.12}$$

where $M_c = 52{,}700$ is the intercept at $\sigma_\infty = 0$.

The polydisperse data are described similarly in the linear region ($r^2 = 1.00$) by

$$\sigma_\infty = 49.00\,(M^{1/2} - M_c^{1/2}) \qquad \text{(psi)} \tag{8.6.13}$$

where $M_c = 35{,}000$. Again the mono- and polydisperse data sets are offset by a factor of about $\sqrt{2}$ on the $M^{1/2}$ axis. However, the front factors (about 49) are the same.

The fracture strain, ϵ_∞, of the virgin injection-molded monodisperse-molecular-weight samples behaves in the linear region (Figure 8.21) as

$$\epsilon_\infty = 0.013\,(M^{1/2} - M_c^{1/2}) \qquad (\%) \tag{8.6.14}$$

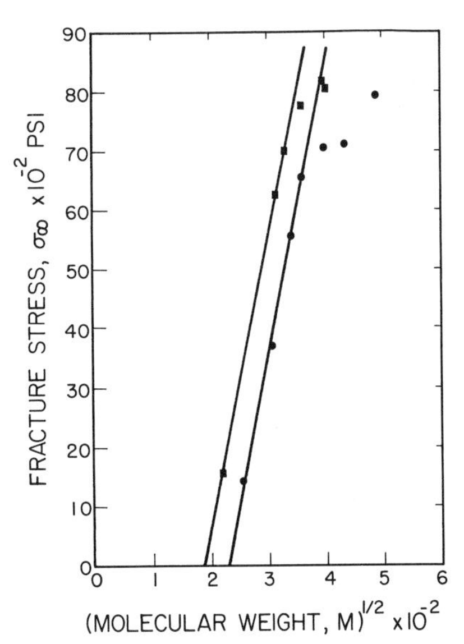

Figure 8.20 Fracture stress versus $M^{1/2}$ for injection-molded PS.
Key to symbols:
• (circles), monodisperse molecular weight data;
▪ (squares), polydisperse data.
(Data of McCormick *et al.*).

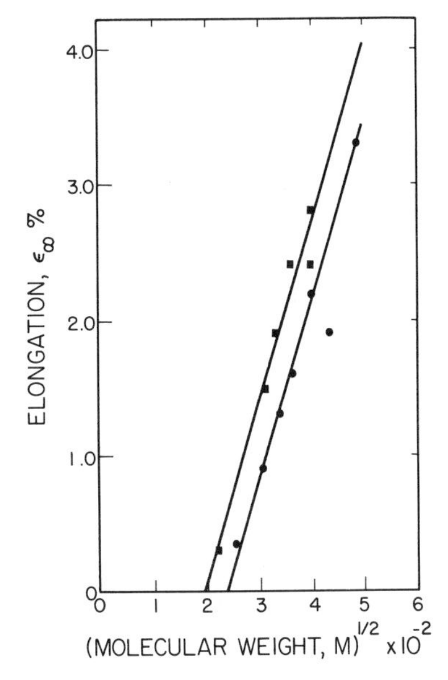

Figure 8.21 Fracture strain versus $M^{1/2}$ for injection-molded PS.
Key to symbols:
• (circles), monodisperse molecular weight data;
▪ (squares), polydisperse data.
(Data of McCormick *et al.* [42]).

where M_c = 55,500 and r^2 = 0.996. The polydisperse data have the same slope of 0.013 and M_c = 38,300, with r^2 = 0.967. The ratio of the slopes for fracture stress and strain, 49/0.00013 ≈ 3.8 × 10^5 psi, should be approximately equal to the Young's modulus E, as given in Table 8.3.

For elastic brittle fracture, the total fracture work, W (area under the stress–strain curve), is well approximated by $W = \sigma\epsilon/2$ as

$$W = 3.2 \cdot 10^{-3} (M^{1/2} - M_c^{1/2})^2 \qquad \text{(psi)} \tag{8.6.15}$$

where M is less than M^*. The units of stress (psi) in this case correspond to energy per unit volume, similar to strain energy density.

8.6.5 Dugdale Model Parameters Versus Molecular Weight

The critical energy to propagate a crack through a craze zone is given by $G_{1c} = \sigma_c\delta$, where σ_c is the stress necessary to form the craze and δ is the critical crack-opening displacement. Both σ_c and δ should show a dependence on M and M_c that behaves approximately as $\sigma_c \sim \delta \sim (M^{1/2} - M_c^{1/2})$. Also, the critical deformation zone length r_c in the Dugdale model should exhibit a similar molecular weight dependence. Some support for these predictions is provided by measurements by Pitman and Ward of σ_c, δ, and r_c as functions of M_n and M_w during fracture of polycarbonate [44]. In Figure 8.22, the craze length, crack-opening displacement, and craze stress all show similar behavior with increasing molecular weight. Each Dugdale parameter upon extrapolation reaches zero at some finite M_c value (about 4,000 for M_n and 10,000 for M_w plots), similar to our earlier analyses for G_{1c}, K_{1c}, σ_∞, ϵ_∞, and W. The tests were conducted on different PC samples at −30 °C, and the Dugdale parameters were determined by optical interference fringe analysis. Knowing σ_c and δ by independent measurements, Pitman and Ward were able to test the Dugdale theory by calculating $G_{1c} = \sigma_c\delta$ and comparing the result with the measured values shown in Figure 8.23. Excellent agreement was obtained. The data are consistent with $G_{1c} \sim M$, as in Eq 8.6.10, and a plot of log G_{1c} versus log M gives a slope of 2.5.

8.6.6 Effects of Pressure on Welding

The effect of the applied pressure (0–2 MPa) on welding of monodisperse samples of polystyrene was studied by McGarel and Wool [45]. PS blocks 52 × 52 × 3.5 mm were welded at different hydrostatic pressures for 30 minutes at 115 °C using the vacuum welding chamber shown in Figure 8.14. The blocks were cut into compact tension fracture mechanics specimens, the notch was fatigue sharpened at 10 Hz to produce a starter crack, and the specimens were fractured at a rate of 0.25 mm/min at room temperature. When no pressure was applied during welding, the blocks pulled apart due to thermal contraction upon cooling. The effect of pressure on G_{1c} at constant welding time is shown in Figure 8.24. The fracture energy of the interface is seen to decrease from 385 J/m^2 to 327 J/m^2 with increasing pressure up to 2 MPa.

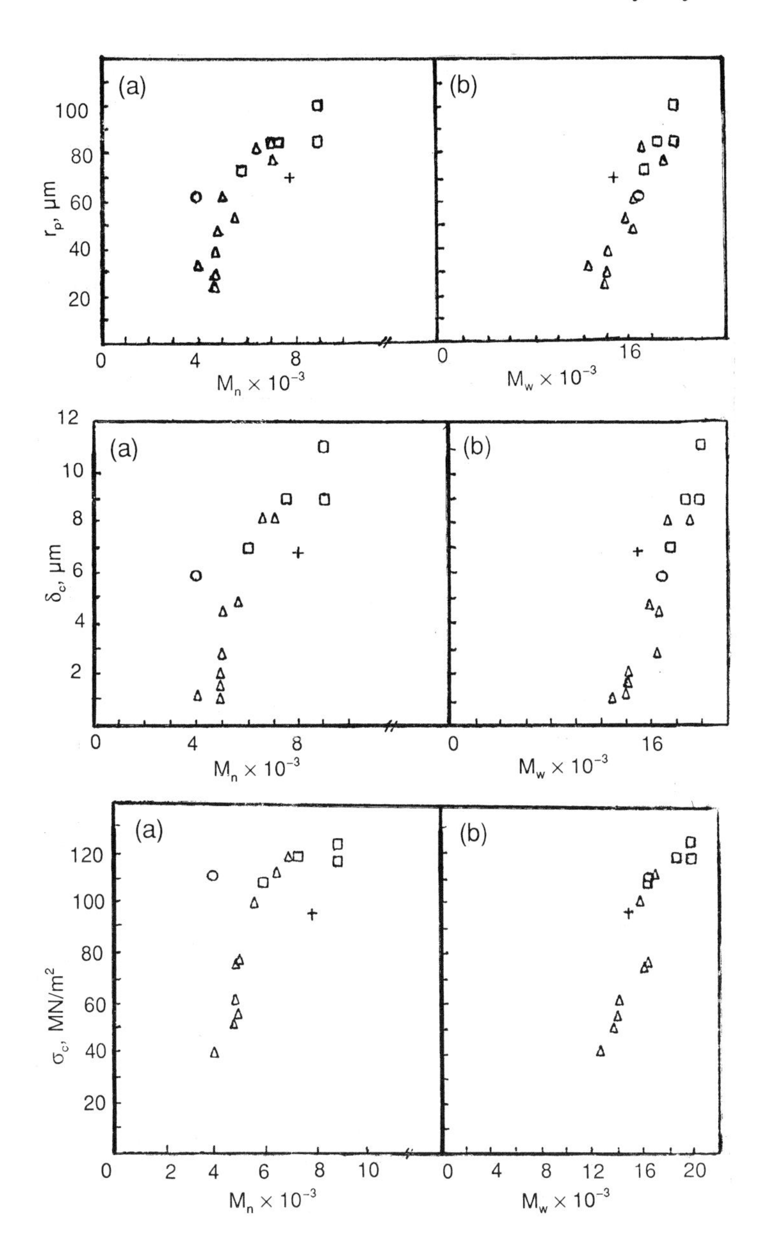

Figure 8.22 Dugdale fracture parameters versus molecular weight for polycarbonate.
Top: plastic zone length r_p versus M.
Middle: critical crack-opening displacement δ versus M.
Bottom: critical stress σ_c versus M.
(a) Dugdale parameters versus number average molecular weight;
(b) the same parameters versus weight average molecular weight.
(Courtesy of Pitman and Ward).

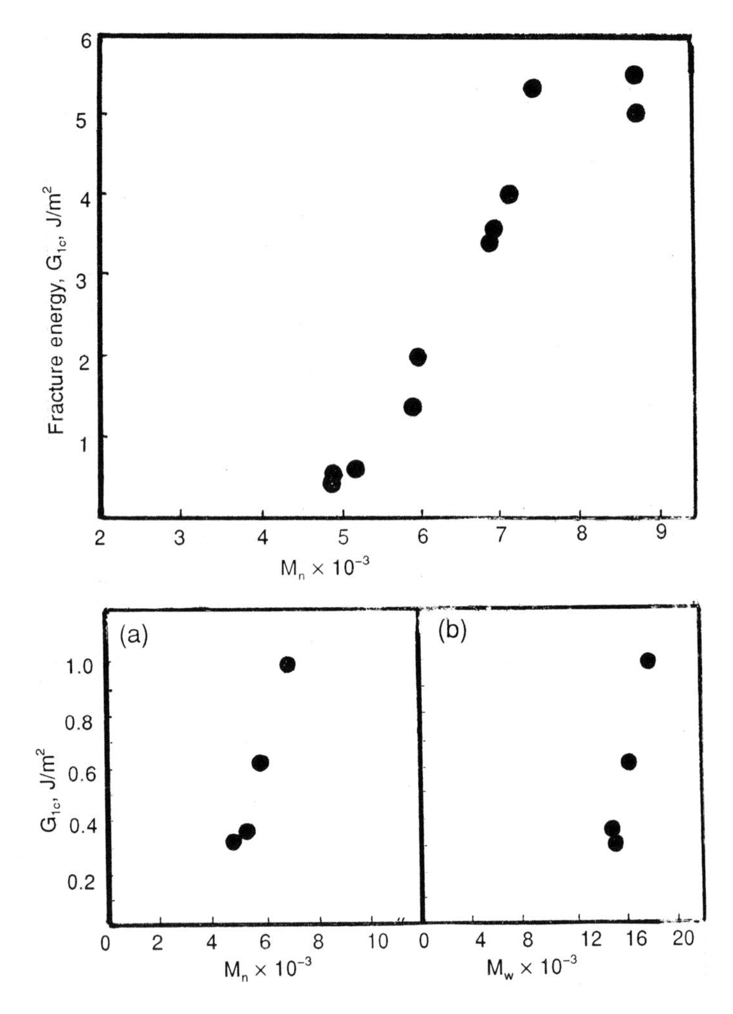

Figure 8.23 Top: measured strain energy release rate, G_{cm}, versus number average molecular weight, Lexan sheet; Makrolon 2803. Bottom: measured strain energy release rate in plane strain, G_{1c}, versus (a) number average molecular weight, (b) weight average molecular weight for Lexan sheet (courtesy of Pitman and Ward).

The weld pressure influences the strength development in two ways. First, a certain amount of pressure is necessary to promote intimate contact and wetting of the polymer surfaces. Once good contact has been achieved, the effect of pressures in the range 0.3–2.8 MPa (50–400 psi) appears negligible. At very high pressures, the glass transition temperature of PS increases by about 20 °C per kilobar and this could have a substantial effect on welding rates at low weld temperatures.

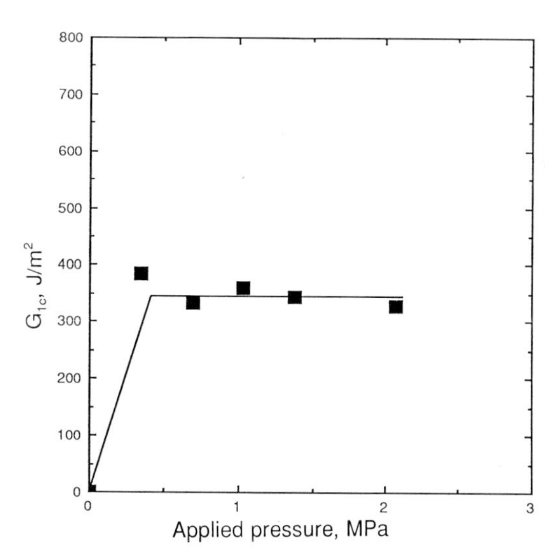

Figure 8.24 Effect of pressure on fracture energy for partially welded PS at constant time (McGarel and Wool).

The pressure dependence of the diffusion coefficient, D, is related to pressure, P, by

$$\partial \ln D / \partial P)_T = -V_a / kT \tag{8.6.16}$$

where V_a is the activation volume. Since the welding time, t_w, is related to the diffusion relaxation time via $t_w = R^2/(3\pi^2 D)$, the dependence of the welding time on P, T, and M can be described in the Arrhenius approximation by

$$t_w \sim M^3 \exp[(Q + PV_a)/kT] \tag{8.6.17}$$

where Q is the activation energy. Thus, as P increases, the ratio of $t_w(P)$ to $t_w(0)$ increases exponentially as

$$t_w(P)/t_w(0) \approx \exp PV_a/kT \tag{8.6.18}$$

If T_g increases by about 20 °C per 100 MPa applied pressure, we infer that $V_a \approx 0.05Q/100$ MPa and the relative change in G_{1c} is determined by

$$G_{1c}(P)/G_{1c}(0) \approx \exp - PV_a/2kT \tag{8.6.19}$$

For example, if Q = 100 kcal/mol, P = 2 MPa, and T = 115 °C = 388 K, then $PV_a = 0.1$ and we obtain $G_{1c}(2)/G_{1c}(0) \approx 0.9$ at constant welding time. This result is consistent with the data shown in Figure 8.24, where an initial value of $G_{1c} \approx 380$ J/m^2 is calculated to decrease to $G_{1c} \approx 342$ J/m^2 with application of 2 MPa.

The exact dependence of pressure on welding time can be determined from the equation of state for a specific melt. When T is much greater than T_g, the melt becomes incompressible and V_a is effectively zero. The approach given above is considered to be empirical, but it serves to make the point that the application of pressure at polymer interfaces may result in counterintuitive negative effects on the weld strength.

8.7 Lap Shear Welding

8.7.1 Introduction to Lap Shear

Previous studies of strength development at interfaces most often used Mode I tensile fracture mechanics techniques, such as wedge cleavage, compact tension, and double cantilever beam methods. Another technique based on ASTM D 3163 [74] utilizes a lap shear joint (shown in Figure 8.25) to measure the development of strength across an interface. The method is a standard test of adhesive strength and contains a large contribution of Mode II (shear) deformation. However, it is not well characterized theoretically and yields only one data point per sample, like the compact tension fracture method. The purpose of this section is to describe the lap joint procedure and present data to demonstrate that this technique is useful for evaluating the time dependence of strength development at a polymer–polymer interface.

8.7.2 Experimental Methods for Lap Shear

Kline and Wool used a lap shear test piece (shown in Figure 8.25) to study strength development during welding of PS interfaces [46]. The surfaces had previously not been in contact and care was taken to ensure rapid wetting of the interface. Lap shear samples were welded for a given time at constant temperatures above T_g and fractured at room temperature.

A set of six stresses, the normal stresses σ_{xx}, σ_{yy}, and σ_{zz}, and the shear stresses τ_{xy}, τ_{yz}, and τ_{xz}, describes the stress state in the sample (Figure 8.26). However, the load cell recorded only those forces transmitted through the y direction. Therefore the normal stresses, σ_{xx} and σ_{zz}, have no effect upon the failure load except as experienced through their respective shear stresses. If one assumes that there is no relative displacement in either the x or z direction, one concludes that only shear stresses in the yz plane are included in the recorded failure load, P_f. Hence, one arrives at the expression for the failure load

$$P_f = A_{xz}\sigma_{yy} + A_{yz}\tau_{yz} \tag{8.7.1}$$

An order of magnitude analysis yields the shear failure stress

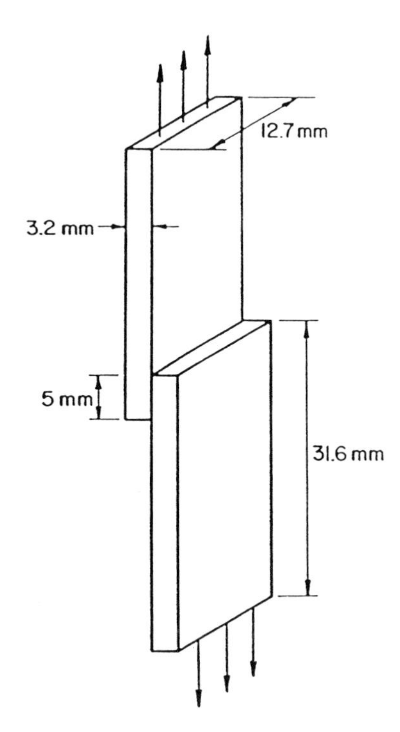

Figure 8.25 Lap shear joint test piece used to test the weld strength of polymers (Kline and Wool).

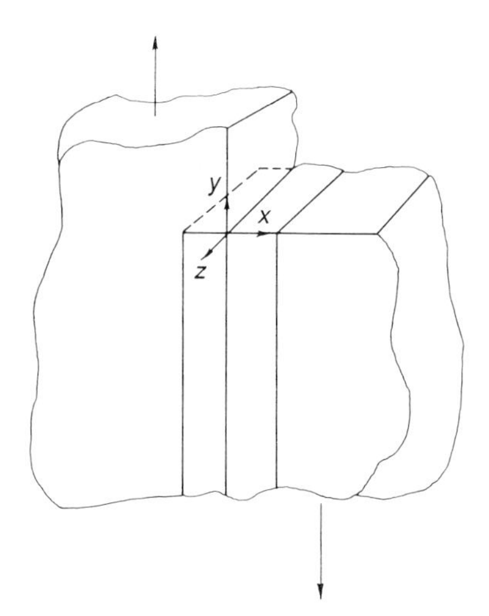

Figure 8.26 The coordinate system for the lap shear joint.

$$\tau_f = P_f / A_{yz} \tag{8.7.2}$$

where A_{yz} is an experimental parameter (welded area) that normalizes the failure load.

8.7.3 Lap Shear of Symmetric Interfaces

Lap shear experiments were done using a commercial grade of polystyrene (PS) from Dow Chemical with M_n = 142,000 and a polydispersity of 1.85 [46]. Figure 8.27 shows a plot of the (average of 15 samples) shear failure stress, τ_f, versus time to the fourth power at each constant temperature. The linear response of τ_f with $t^{1/4}$ is seen at each temperature and provides further support for diffusion-controlled strength at polymer–polymer interfaces.

The time to achieve complete healing, T_h, can be estimated from the upper plateau in Figure 8.27. At 391.3 K (118.1 °C), $T_h \sim 256$ min for the polydisperse PS samples with M_n = 142,000 and M_w = 262,000. This time can be compared with related times measured by other techniques. Lee and Wool [47], using FTIR dichroism techniques, measured the relaxation time, T_r, of uniaxially oriented monodisperse PS with M_n = 233,000 as T_r = 215 min. Green and Kramer [48] report a self-diffusion coefficient for monodisperse PS with M_n = 255,000 as $D = 5.8 \times 10^{-17}$ cm²/s at 125 °C. Whitlow and Wool [49] observed similar diffusion coefficients using SIMS (Chapter 5). Shifting to 118 °C using the Vogel–Fulcher relation [48]

$$\log D/T = A - B/(T - T_\infty) \tag{8.7.3}$$

where $B = 710$, $T_\infty = 49$ °C and $A = -9.49$, then $D(118\ °\text{C}) = 6.42 \times 10^{-18}$ cm²/s. With $T_r = R^2/(3\pi^2 D)$, the time required for the center of mass to diffuse a distance equal to

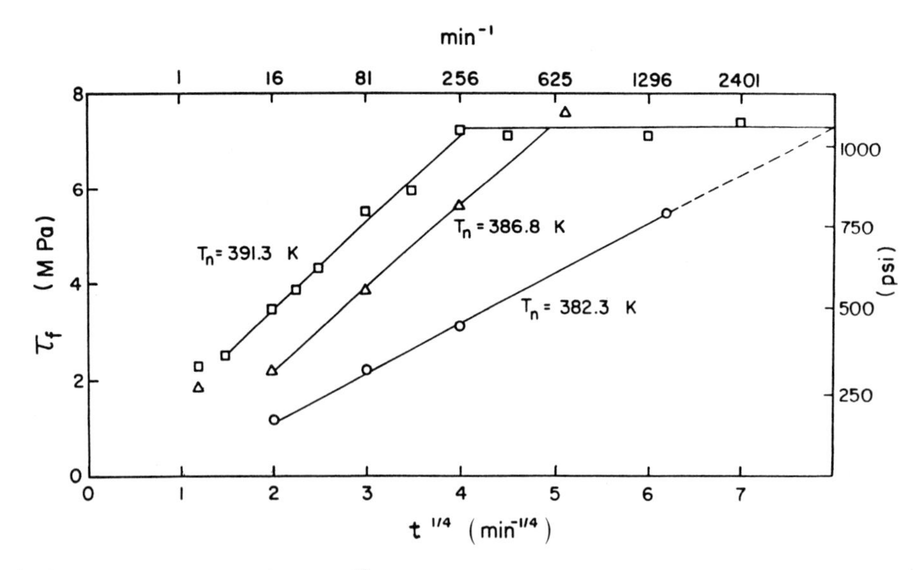

Figure 8.27 Shear stress versus $t^{1/4}$ for lap welding of polystyrene at several temperatures (Kline and Wool).

the end-to-end vector is obtained as $T_r = 1000$ min, which is about four times longer than the time required to achieve complete strength of the lap shear joint with $M_w = 262{,}000$. On the other hand, with $T_r \sim M^3$ and $D \sim M^{-2}$, the relaxation time for the chains with $M_n = 143{,}000$ at 118 °C is about 180 minutes, which is in close agreement with the lap shear result. One should keep in mind when evaluating healing times from the reptation theory that at $t = T_r$, the center of mass of chains in the bulk diffuses a distance related to the end-to-end vector, while the average monomer interpenetration distance at the polymer interface, X, is roughly equivalent to the radius of gyration ($X = 0.81R_g$).

The temperature dependence of welding is obtained from the temperature dependence of T_r via Eq 8.7.3. The shear stresses at temperatures T_1 and T_2 and the same welding time are related by

$$\tau(T_1)/\tau(T_2) = (T_1/T_2)^{1/4}\, 10^{1/4\,B[1/(T_2-T_\infty)-1/(T_1-T_\infty)]} \tag{8.7.4}$$

Since the term $(T_1/T_2)^{1/4} \approx 1$, Eq 8.7.4 can be simplified to

$$\tau(T_1)/\tau(T_2) \approx 10^{1/4\,B[1/(T_2-T_\infty)-1/(T_1-T_\infty)]} \tag{8.7.5}$$

The factor of ¼ in the exponent is derived from the same exponent in the welding relation for shear stress, $\tau = \tau_\infty(t/T_r)^{1/4}$.

The change in slope with temperature can also be described by the William–Landel–Ferry (WLF) theory of thermal activation near T_g. At $T = 113.5$ °C, the activation energy, E_a, of 93.2 kcal/mol, is predicted from a WLF analysis using the parameters for PS, $c_1 = 13.7$, $c_2 = 50$, and $T_g = 100$ °C. This agrees well with $E_a = 96.1$ kcal/mol obtained from a plot of the superposition shift factor versus $1/T$ in Figure 8.27.

Exercise

Compare Equation 8.7.5 with the actual data shown for the temperature dependence of lap shear in Figure 8.27. Does the equation provide an adequate description of the temperature dependence of welding?

8.8 Melt Processing of Internal Weld Lines

8.8.1 Introduction to Internal Weld Lines

Internal weld lines are encountered in melt processing of pellet and powder resin. For example, a double-gated plate mold has a weld line in the center, assuming equal flow through the gates. Other cases involve sintering of powder, compression molding and injection molding of pellets, and coalescence of latex particles.

8.8.2 Compression Molding of Pellets

Consider the compression molding of resin pellets, where the interfaces and weld lines form when the pellet surfaces make contact in the mold (Figure 8.28). With increasing time and pressure, the interfaces become wet, diffusion occurs, and the strength of the sample increases to its virgin strength if sufficient time for diffusion is allowed.

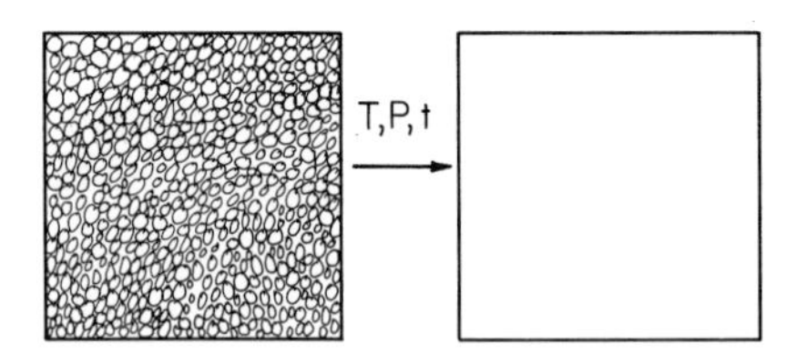

Figure 8.28 Schematic of powder/pellet interface welding during polymer melt processing. With temperature, pressure, and time, the pellet interfaces heal to form the plastic product.

Wool and O'Connor [2] studied the time dependence of compression molding using PS pellets of the same molecular weight as described above in the lap shear experiments. PS plates were compression-molded in a Carver press at 127 °C ± 6 °C, with pressures of 5.5 MPa (800 psi) for varying processing times, t_p. The plates were withdrawn from the mold, quenched, cut into tensile "dog bone" samples, and fractured at room temperature in uniaxial tension. Weak interfaces within a sample provided a source of crack initiation.

Figure 8.29 shows the fracture stress (average of 6 samples) plotted against the one-fourth power of the processing time. The prediction of $\sigma(t) \sim t^{1/4}$ is again noted to be satisfactory. The processing time to reach the virgin state near $\sigma_\infty \sim$ 6,000 psi (41.4 MPa) was between 10 and 20 minutes, estimated from the saturation of the strength at longer times. Using Whitlow and Wool's diffusion data at 125 °C (Chapter 5), the reptation time for pellets with M_n = 143,000 is about 20 minutes, which is in reasonable agreement with the observed processing time.

Tests that used pre-notched specimens of molded pellet samples to study fracture mechanics as a function of welding time could be sensitive to the position of the notch and the length of the pre-crack compared to the pellet diameter. Tiegi and co-workers [50–52] examined the fracture toughness, K_{1c}, of PMMA and SAN pellets as a function of compression molding temperature at constant molding time (30 minutes) and pressure (8 MPa). They found that when the molding was done at lower temperatures, the fracture morphology was substantially intergranular (between the pellets), and at higher temperatures, the fracture morphology was transgranular (through the pellets). The K_{1c} values for the high-temperature transgranular morphologies were consistently higher than those for the intergranular low-temperature case. However, as healing neared completion, they noted a small drop in K_{1c} that was due to the damage zone becoming more focused on the crack plane. With partially healed pellets, considerable collateral damage can occur in the vicinity of the crack plane and raise the K_{1c} values.

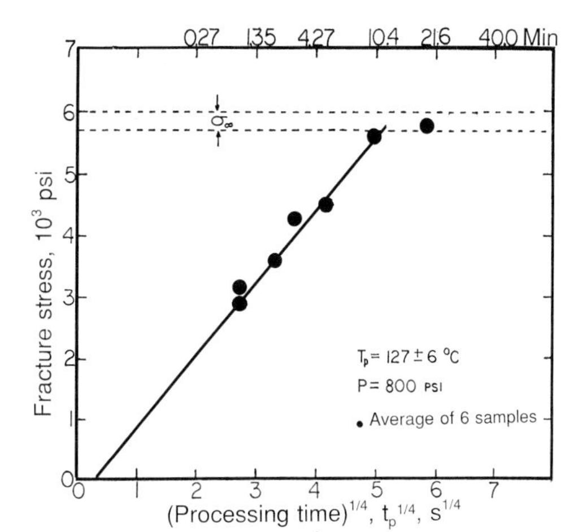

Figure 8.29 Fracture stress versus processing time ($t^{1/4}$) for compression molding of polystyrene pellets. "Dog bone" samples were prepared from compression-molded pellets (Wool and O'Connor).

McCormick *et al.* have determined the molecular weight dependence of the virgin strength, σ_∞, of compression-molded PS pellets using similar sample preparation and fracture conditions [42]. Their data are shown in Table 8.4. Figure 8.30 [2] shows σ_∞ versus $M^{1/2}$ for monodisperse- and polydisperse-molecular-weight PS. The monodisperse data are described by the linear equation ($r^2 = 0.974$)

$$\sigma_\infty = 33.40\,(M^{1/2} - M_c^{1/2}) \tag{8.8.1}$$

where the intercept at $\sigma = 0$ is $M_c^{1/2} = 245$, so that $M_c = 60{,}000$. For the polydisperse fracture data, the front factor is about the same (33.02 psi, or 0.23 MPa) and $M_c = 34{,}000$, with a correlation coefficient $r^2 = 0.936$. We note that the slope for injection molding (49 psi; see Section 8.6.4) is greater than that for compression molding (33 psi). This may be due to orientation effects for injection-molded samples, or to the existence of more severe flaws in compression-molded pellets.

The strain at fracture, ϵ_∞, was described in a similar manner for the monodisperse data by (Figure 8.31)

$$\epsilon_\infty = 0.009\,(M^{1/2} - M_c^{1/2}) \qquad (\%) \tag{8.8.2}$$

with $M_c = 65{,}000$ and correlation coefficient $r^2 = 0.978$. The polydisperse data had a front factor of 0.008% and $M_c = 34{,}200$ with $r^2 = 0.938$. The front factor for the monodisperse data (0.009 in Eq 8.8.2) is smaller than that for the injection-molded samples (0.013 in Eq 8.6.14).

The mechanical properties for the compression-molded pellets (Figures 8.30 and 8.31) are seen to increase linearly in accordance with virgin state predictions up to $M \sim 160{,}000$, above which they decrease slightly with increasing M. The data were obtained from samples compression-molded at a constant time of 1 minute at 170 °C

Table 8.4 Tensile Test Data for Compression-Molded Polystyrene

M_n	$M_n^{1/2}$	Tensile Strength, psi	Modulus, psi × 10^{-5}	Elongation, %
		Anionic Polystyrene		
40,000	200	a	a	a
66,300	257	440	–	0.06
95,700	309	1,980	4.38	0.43
116,000	341	3,000	4.45	0.73
131,000	362	4,380	4.48	1.03
159,000	399	4,910	4.72	1.20
189,000	435	5,150	4.47	1.17
240,000	490	4,660	4.50	1.06
		Isothermal Polystyrene		
49,000	221	970	5.11	0.23
99,000	315	5,000	4.48	1.18
110,000	332	4,880	4.38	1.18
130,000	361	5,310	4.38	1.26
161,000	401	5,280	4.30	1.34
157,000	396	5,310	4.38	1.33

a – Too brittle for testing

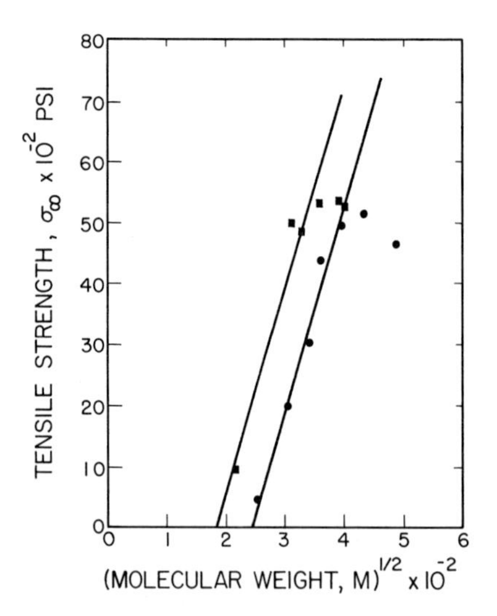

Figure 8.30 Tensile strength versus the square root of the molecular weight of compression-molded polystyrene (pellets). Molding occurred at 170 °C for 1 minute. Monodisperse- (circles) and polydisperse-molecular-weight (squares) pellets were used. (Data of McCormick *et al.*)

for all materials. We have suggested that, under the processing conditions used by McCormick *et al.*, samples with M less than 160,000 had reached their virgin state at t of 1 minute, but samples with M greater than 160,000 were not completely healed at 1 minute, and the fracture stress should decrease as observed according to $\sigma \sim M^{-1/4}$

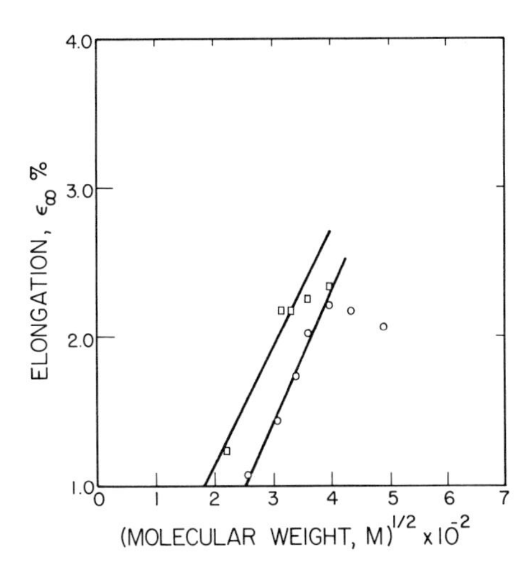

Figure 8.31 (Data from Figure 8.30) Strain versus $M^{1/2}$ for compression-molded pellets.

[2]. The reptation time for M = 160,000 is about 1 second at 170 °C, and the time of 1 minute used in this experiment presumably includes warm-up time in the mold in addition to diffusion times. However, the maximum in strength with increasing M at constant contact time is expected, and is similar to the tack and green strength data presented in Figure 8.3.

Strength development at internal weld lines in injection-molded specimens is expected to behave similarly, although some complications are expected with interfaces involving oriented molecules at the weld line. However, it is interesting to note that orientation relaxation occurs at about the same rate as normal diffusion. Orientation relaxation may contribute to interdiffusion at the weld line in a similar manner as for unoriented surfaces, but problems are anticipated with barrier effects produced by oriented layers that could result in anisotropic diffusion rates.

Comment on De-Welding

De-welding occurs when an interface is heated and, instead of gaining strength, it weakens. This interesting phenomenon may occur, for example, after compression molding of pellets in a Carver press to form a transparent glass plate. Considerable flow and elastic deformation of individual pellets are required when the pressure is applied at T greater than T_g. When the healing time, t_h, is less than the relaxation time of the deformed pellets, T_r, the pellets when quenched into the glassy state still retain considerable elastic energy [proportional to $1 - (t_h/T_r)^{1/2}$]. Optically, the pellet interfaces have disappeared and the glass appears clear as the virgin glass. (The residual strain orientation in the pellets would be visible in crossed polars because of orientation.) However, since t_h is less than T_r, the interfaces are weak and, if the plate is heated in the vicinity of T_g, the stored elastic energy is released, the interfaces are de-welded, and the pellets reappear. In this case, the glass loses its transparency (because of light scattering from the interfaces) and becomes more fragile. This situation can occur at other weld interfaces, depending on the processing conditions.

8.9 Healing of Latex Particles

8.9.1 Latex Paints

Latexes are used as water-soluble paints and coatings, and comprise about 20% of all synthetic polymers. A latex consists of spherical particles, 50–500 nm in diameter, suspended in an aqueous solution. The particles are made by emulsion polymerization and typically are uniform in size but have a broad range of molecular weights within each particle. Coalescence of these particles has recently been treated as a welding problem by Winnick and co-workers [53] and Sperling and co-workers [54–56].

The formation of latex films is shown schematically in Figure 8.32. First, the water evaporates from the aqueous latex, leaving behind the close-packed particles. The surface of the particles (chain ends, surfactants) becomes important as the particles come into contact, pack into face-centered cubic structures, deform, and then diffuse together to produce a mechanically rigid film. There are several important differences between latex film formation and welding in the usual sense. The surface of latex particles contains charged chain-end groups and could impede diffusion [54] if it repels the other surfaces. However, the presence of the chain ends on the surface could enhance the diffusion rate, as discussed in Chapter 2. Some particle diameters are smaller than the end-to-end vector R of the chains, and this could provide an additional entropic driving force for a chain to escape from its confined surroundings. Also, the effect of the surface could be to promote molecular weight segregation within the particle.

8.9.2 Fluorescence Studies of Latex Particle Interdiffusion

The interdiffusion of particles as an important stage in the formation of a latex film has recently been verified by direct energy transfer (DET) with fluorescent active labeled chromophores [53], and by SANS studies of deuterated PS latex particles in a normal PS latex [54]. To perform the DET experiment, Winnick and co-workers placed donor molecules in one poly(butyl methyl acrylate) (PBMA) particle and acceptor molecules in the others [53]. The particle diameter was 3370 Å and M_w was 7.6×10^4. As interdiffusion proceeded at 70 °C (T_g + 40 °C), the volume fraction of mixing, f_m, increased. They found that the number of spectroscopically monitored donor–acceptor complexes increased initially as $f_m \sim t^{1/2}$, up to about 60% mixing in 100 minutes, and then slowed down, as shown in Figure 8.33. The diffusion coefficient was determined as $D = 1.3 \times 10^{-15}$ cm^2/s, 6.6×10^{-15} cm^2/s, and 3.05×10^{-14} cm^2/s, at 70 °C, 80 °C, and 90 °C, respectively. The apparent activation energy for diffusion is about 39 kcal/mol.

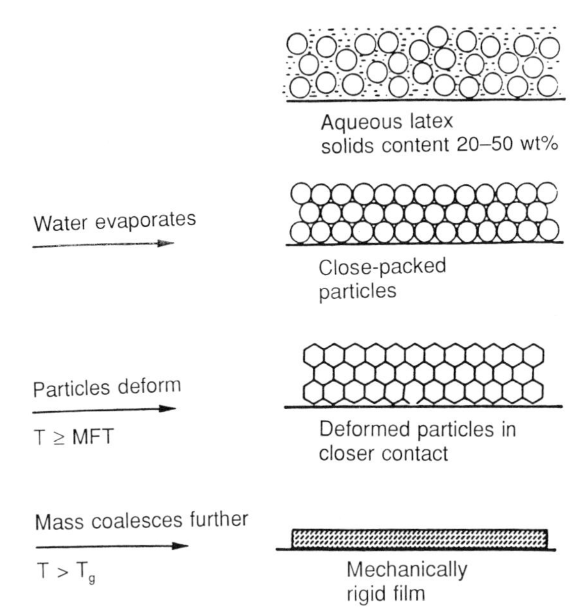

Figure 8.32 Schematic depiction of latex film formation. (Note: MFT is the mechanically rigid film temperature.)

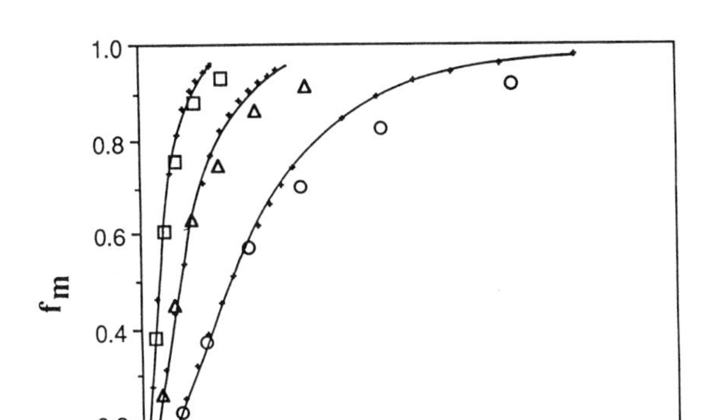

Figure 8.33 Volume fraction of mixing, f_m versus $t^{1/2}$ (courtesy of Wang and Winnick).

We compute the relaxation time for the Winnick experiment using $T_r = R^2/(3\pi^2 D)$. With $R \approx 160$ Å, and $D = 1.3 \times 10^{-15}$ cm^2/s at 70 °C, we obtain $T_r \approx 1$ min, which is typical for these molecular weights at (T_g + 40 °C), and we conclude that most of the data were obtained at t much greater than T_r. In that case, the observation that $f_m \sim t^{1/2}$ is to be expected for normal Fickian diffusion at t greater than T_r. One also expects

$f_m \sim t^{1/2}$ at t less than T_r when the chain ends are initially on the surfaces of the latex particles. The data in Figure 8.33 were found to behave as $f_m \sim t^{1/4}$ over the entire time range. This is more likely due to polydispersity effects, where the small molecules diffuse first at a faster rate than the longer molecules.

8.9.3 SANS Studies of Latex Interdiffusion

In an elegant series of experiments, Sperling and co-workers [54–56] investigated small-angle neutron scattering from deuterated latex particles as they became welded with protonated particles. They examined monodisperse-molecular-weight particles with M_n of 150,000 in uniformly sized spherical particles, 600–700 Å. The radius of gyration of the particles was determined by Guinier plots of SANS data at various annealing times, and the interpenetration depth, d, was determined by subtraction of the initial particle diameter.

Figure 8.34 shows d versus annealing time on a log–log plot at temperatures of 145 °C, 135 °C, and 125 °C. The data fall on lines of different slopes above and below T_r: at t less than T_r, they find good agreement with $d \sim t^{1/4}$, and at t greater than T_r, they have $d \sim t^{1/2}$. The transition in slope is similar to that observed by SIMS, SNR, and NRA analysis in Chapters 5 and 6. Knowing T_r from the diffusion analysis, they find that the slope transition occurs near T_r in each case. For their samples, the crossover depth, $d_\infty \approx 0.81R_g$, is predicted to occur at about 86 Å. In Figure 8.34, we note that $d_\infty \approx 50$ Å, which is smaller by about 30–40 Å than predicted. We suspect that this discrepancy may be due to the SANS measurement and subtraction technique.

The tensile properties of films made from annealed latex particles were determined as a function of interdiffusion depth. Figure 8.35 shows the tensile strength (TS) versus d. As expected, the increase in TS with d is reasonably linear before saturation occurs near $d \approx 100$ Å, which is consistent with TS $\sim t^{1/4}$. Sperling and co-workers argue that the tensile strength first goes through a maximum before finally reaching a plateau value. This is most likely due to the extent of transgranular versus intergranular fracture modes with increasing annealing time, similar to the Tiegi experiments with pellets discussed in Section 8.8.2.

8.10 Fatigue of Welded Interfaces

8.10.1 Introduction

When a symmetric interface is partially welded, the critical fracture energy behaves as $G_{1c} \sim t^{1/2}$. However, many plastic materials are subjected to loads less than the critical load. In such cases the crack of length a can advance in a stable manner by a small amount da, until a critical-sized crack forms and the interface fractures. We are

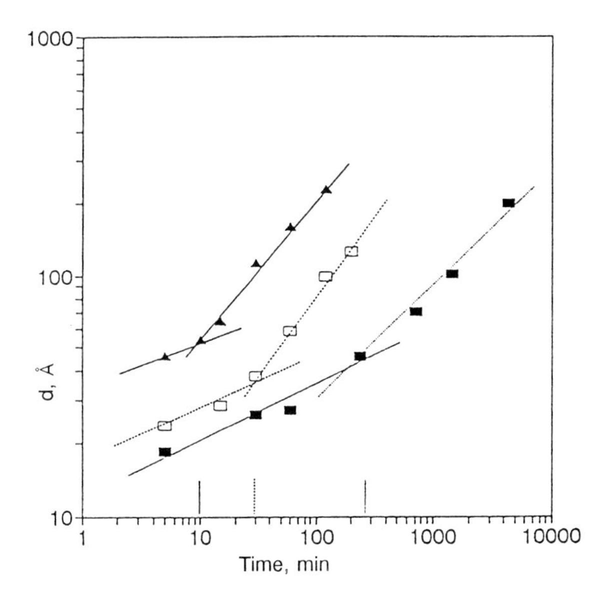

Figure 8.34 Log–log plot of interpenetration depth, *d*, versus annealing time for latex film formation at temperatures of 145 °C ▲ (triangles), 135 °C □ (open squares), and 125 °C ■ (closed squares) (courtesy of Sperling *et al.*).

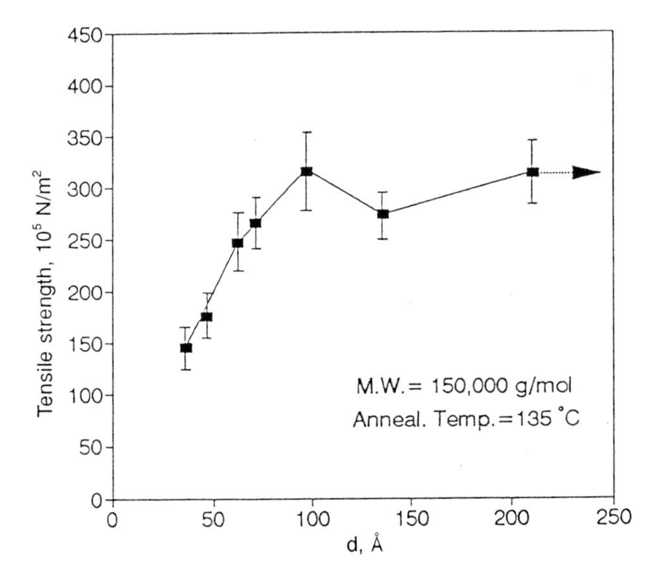

Figure 8.35 Tensile strength versus interpenetration depth (Sperling *et al.*).

interested in the case where a partially healed interface is subjected to cyclic fatigue loads of N cycles. The measured variable in this study is the incremental increase in crack length per cycle, da/dN, known as the *fatigue crack propagation rate, FCP*. When the interface is at full strength, da/dN should be at a minimum, since the maximum resistance is being experienced. However, if the interface is only partially healed, da/dN may be large and the lifetime of the interface correspondingly short. What penalties are to be paid for a partially healed interface? This question is very important for laminated composites and internal weld lines in melt-processed polymers. The answer is surprising.

8.10.2 Experimental Methods for Fatigue Welding

The effect of welding conditions on the fatigue crack propagation (FCP) rate, da/dN, along a welded PS/PS interface was investigated by Yuan and Wool [34, 57] using the DCB (double cantilever beams) test configuration shown in Figure 8.36. These experiments provide information on the expected lifetime of welded or laminated parts as a function of the time of welding. The goal is to determine how the FCP rate of a partially healed interface compares with that of the fully healed state. The fatigue tests were conducted in tension–tension displacement, δ, control mode with a haversine wave form at frequency 1 Hz and ratio $R = \delta_{min}/\delta_{max} = 0.33$. The fatigue crack length, a, was measured and the applied strain energy release value, G_1, was calculated using the following expression by Wang [58]

$$G_1 = 9\,(EI/a^4 b)\,\delta^2[1 + 15/4\,\alpha(\delta/a) + (189/64\,\alpha^2 + 6/7)(\delta/a)^2] \tag{8.10.1}$$

where E is the modulus, I is the moment of inertia, and $\alpha = (d + h/2)/a$. The FCP rate is usually a power law in G and can be described by the Paris law

$$da/dN = A\,G^m \tag{8.10.2}$$

where A and m are constants.

8.10.3 Fatigue Healing of Symmetric Interfaces

Figure 8.37 shows the FCP rate (da/dN) of PS as a function of ΔG at several constant healing times from 10 to 165 minutes. The welding was done at 112 °C on PS samples with $M_n = 133{,}000$ and $M_w = 303{,}000$. Two sets of data are shown for each healing time, and the reproducibility is seen to be good for any data set. At low ΔG values (about 150 J/m^2), the FCP rate is not strongly dependent on either ΔG or healing time. With increasing ΔG (above 200 J/m^2), the FCP rate develops a strong dependence on ΔG and t with Paris law exponents $m \approx 2$ (Eq 8.10.2) for each welding time.

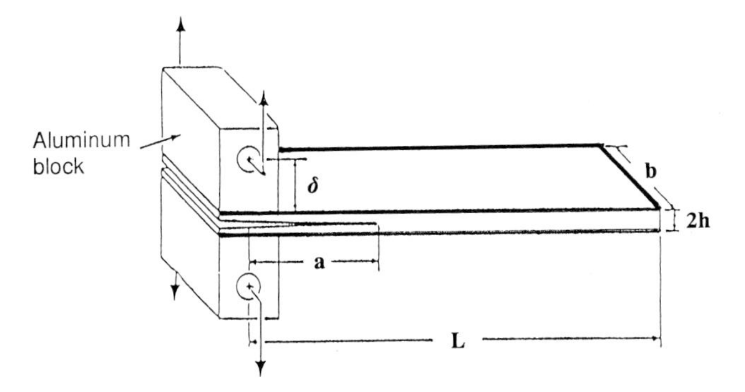

Figure 8.36 DCB specimen for the fatigue testing of welded polystyrene plates with b = 12.7 mm, d = 12.7 mm and $2h$ = 6.6 mm (Yuan and Wool).

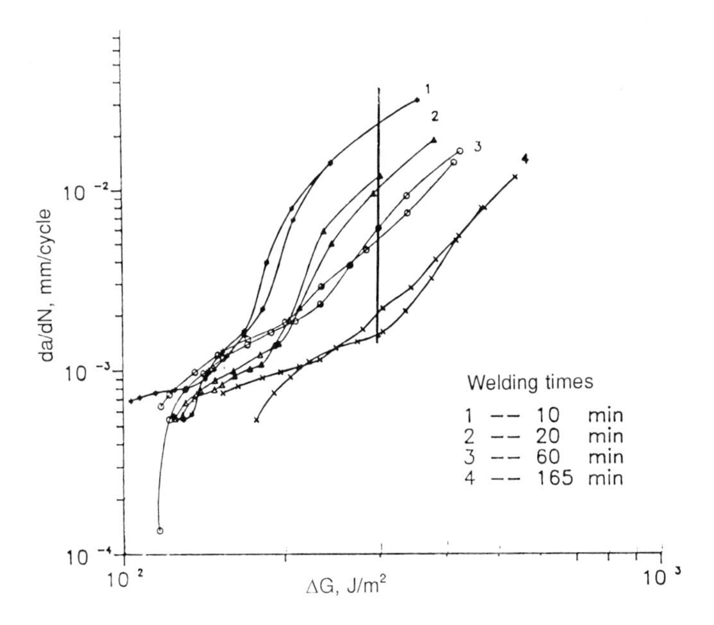

Figure 8.37 Fatigue crack propagation rate, da/dN, versus crack propagation energy ΔG, for welding of PS interfaces at 10–165 min (Yuan and Wool).

A cross-plot of the fatigue crack propagation data, da/dN, at constant crack propagation energy, ΔG, (vertical line in Figure 8.37) reveals the effect of welding time on da/dN. The results are shown in Figure 8.38 for welding temperatures of 108 °C and 117 °C and ΔG of 240 J/m^2. A strong effect of welding time was noted and the results can be expressed as a power law

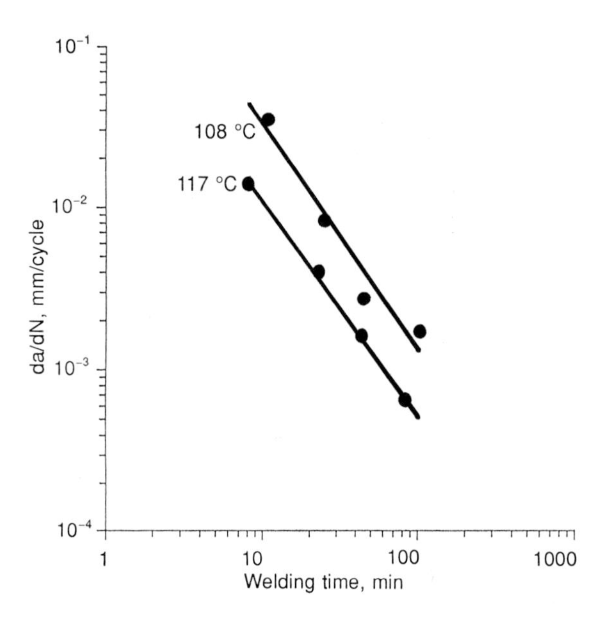

Figure 8.38 Fatigue crack propagation FCP rate, da/dN, versus welding time of PS interfaces at 108 °C and 117 °C (Yuan and Wool).

$$(\mathrm{d}a/\mathrm{d}N)_{\Delta G} = A(T)\, t^{-1.3} \tag{8.10.3}$$

where $A(T)$ is a temperature-dependent prefactor. Thus, as the strength of the interface increases with welding time, the FCP rate decreases with a strong power-law dependence, and the failure time, t_f, for a crack to reach a critical length increases with welding time as $t_f \sim t^{1.3}$.

The exponent of −1.3 can be compared with the chain disentanglement model discussed in Chapter 7. With the Dugdale model, the incremental increase in the crack length, da, is proportional to the increase in the plastic zone, r_p, so that

$$\mathrm{d}a \sim r_p \sim K^2/\sigma_c^2 \tag{8.10.4}$$

Thus, when the stress intensity factor K is applied at the crack tip, the plastic zone advances by $r_p \sim 1/\sigma_c^2$, where $\sigma_c \sim l(t)^{1/2}$ is the stress necessary to propagate the deformation zone. The time required to disentangle the chains, t_p, is proportional to dN via

$$\mathrm{d}N \sim t_p \sim \mu_0\, l(t)^2/\sigma_c \tag{8.10.5}$$

which gives $\mathrm{d}a/\mathrm{d}N \sim X^{-5}$, where X is the average interface thickness. Substituting for the welding time via $X = X_\infty(t/T_r)^{1/4}$, and $X_\infty \sim M^{1/2}$, we have

$$da(t)/dN \sim K^2 M^{-5/2} (t/T_r)^{-5/4} \qquad (8.10.6)$$

The exponent of −1.25 agrees well with the experimental value of −1.3 from Figure 8.38. Compared to other healing exponents of ¼ and ½, this larger negative exponent bears the message that there is a high penalty to pay in fatigue for partially welded interfaces.

8.10.4 Molecular Weight Dependence of Fatigue

The molecular weight dependence of da/dN (at constant welding time) is $da/dN \sim M^{5/4}$, where the exponent is positive (which is correct since the higher molecular weight materials heal less rapidly and therefore have higher da/dN values). High-molecular-weight polymers are often used for their good fatigue resistance, for example, UHMWPE in artificial hip joints. However, this real advantage may turn to a disadvantage if the weld is not properly formed.

The time dependence of da/dN during welding suggests that at $t = T_r$, or when the virgin state is obtained, the molecular weight dependence is determined by

$$(da/dN)_G = A\,(M^{-5/2} - M_c'^{-5/2}) \qquad (8.10.7)$$

where A is a constant. In a manner analogous to that of the fracture energy approaching zero near M_c, the FCP should increase to infinity at an M_c value corresponding to critical crack propagation at the applied K_1 value. Hence M_c' in Eq 8.10.7 depends on K_1 by a relation similar to Eq 8.6.11, so that M_c' is greater than M_c. Figure 8.39 shows da/dN versus molecular weight data of Rimnac *et al.* [59] for PVC. The solid line corresponds to Eq 8.10.7 in which A is 4.5×10^8 mm/cycle and M_c' is 60,000. The strong molecular weight dependence is similar to that predicted by Eq 8.10.7. We also deduce that the critical stress intensity factor for PVC is given by $K_{1c} \approx 0.005\,(M^{1/2} - M_c^{1/2})$ MPa·m$^{1/2}$, where M_c is 11,000.

The effects of a partially welded interface on the FCP rate are considerable, as shown in Figure 8.38. Many plastic parts are designed with regard to M^*, so that G_{1c} (see Figure 8.17) is a maximum at minimum melt viscosity. However, it is clear that fatigue lifetimes continue to increase at M much greater than M^*, as can be deduced from Figure 8.39. These results are of importance for composite lamination of thermoplastic matrices.

We can compare the molecular weight dependence of da/dN with craze growth behavior. It is interesting to note that when Kramer and Berger [60] examined the craze thickening velocity v_c as a function of molecular weight, they found that $v_c \sim M^{-5/2}$. The correspondence between these results may be reasonable under stable crack–craze growth conditions. Similarly, a large dependence of craze (length) growth on molecular weight was observed by McGarel and Wool (discussed in Chapter 11). All of these phenomena suggest a strong viscous flow contribution in craze growth

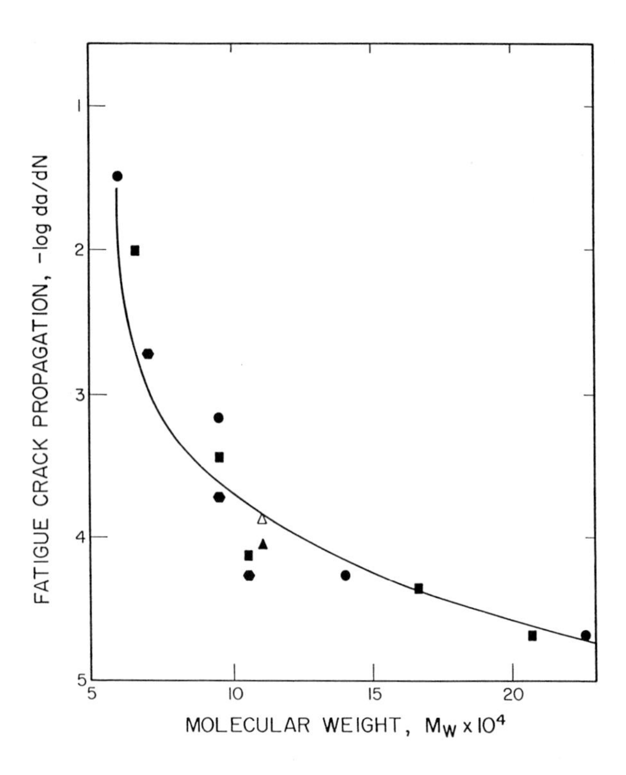

Figure 8.39 Fatigue crack propagation rates versus molecular weight of PVC (Rimnac *et al.*). The line is calculated from disentanglement theory by $da/dN = A/(M^{5/2} - M_c^{5/2})$ (Eq 8.10.7), where $A = 4.5 \times 108$ mm $M^{5/2}$/cycle and $M_c = 60{,}000$ is the molecular weight at which critical crack propagation occurs at $K = 0.7$ MPa·m$^{1/2}$.

behavior where the time to flow or draw craze fibrils is related to characteristic times of order M^3.

Fatigue crack healing is also an important concept for fiber-filled composites. Klosterman and Wool [61] found that fatigue damage involving fiber–matrix failure and matrix cracking in glass-fiber-filled thermoplastics could be healed at temperatures above T_g. As a result, the fatigue crack propagation rate decreased with increasing healing time in the damage zone until the virgin-state FCP rate was obtained. Furthermore, we found that we could obtain complete crack healing with highly cross-linked epoxies (Chapter 12). These studies suggest that the lifetime of complex composites in practice could be considerably increased by periodic healing treatments. The possibility also exists that selected radiation, such as microwaves, and other treatments, such as solvents, could be used to excite healing processes in specific regions of the composite.

8.11 Welding Applications

We now consider the symmetric interface theories in the context of real-life applications, involving vibrational welding, non-isothermal welding and composite interfaces.

8.11.1 Relation of Thermal to Vibrational Welding

Vibrational welds can be effected in thermoplastics over a frequency range 25–400 Hz, (Stokes [62–65]). Typical welding pressures are 130–2000 psi (1–14 MPa), with 0.3–1.6 mm weld amplitudes. The process can be divided into four phases, which are independent of the thermoplastic. In Phase 1, the weld interface is heated by Coulombic friction, which ends when the interface begins to melt; the penetration, which is the moving together of the separate pieces due to lateral flow out of the interface, is essentially zero during this phase. With continued energy input, the interface begins to melt and flow, and penetration begins in Phase 2 up to a steady-state lateral flow. Phase 3 is the steady-state flow condition during which the penetration increases linearly with time so that the material is melting at the same rate as it is flowing out of the lateral surfaces. The molten film is about 0.1–0.24 mm, and the temperature is very close to ambient within about 1.0 mm of the solid–liquid interface [62]. Phase 4 occurs when the vibratory motion is stopped, and the interface solidifies.

The role of diffusion in vibrational welding is apparent in Phases 2–4. However, since the diffusion necessary to promote maximum strength involves distances on the order of 100 Å, the large amount of local melting (0.1 mm) and flow may be unnecessary. Obviously, the extent to which Phase 3 steady-state flow is sustained determines the extent of lateral flow and local deformation of the weld line. For each thermoplastic, the temperature rise at the end of Phase 2 could be used to determine the healing time from the diffusion analysis, and this could be used to minimize Phase 3 flow processes. For optimal component integrity, we would like to minimize contributions from Phases 2 and 3 by heating the interface in Phase 1 to cause melting, and then solidifying in Phase 4. The pressure history could also be tailored to promote melting in Phase 1, and reduced in Phases 2–4 to minimize flow. If reduced pressures were used in vibrational welding the interdiffusion process would not be affected if good contact had been achieved. The fact that flow occurs during Phase 3 at the interface presents new opportunities for mechanical interlocking, and this could be important for incompatible interfaces.

8.11.2 Non-Isothermal Welding of Composites

Bastien and Gillespie [66] considered the welding of polyether–ether–ketone (PEEK) thermoplastic composite laminates with an amorphous polyetherimide (PEI) interlayer between the laminate plies. The PEEK had commingled graphite fibers and woven fabric in the laminates. The PEI layer was first bonded to the PEEK laminate surfaces during the initial consolidation. During welding in a high-temperature autoclave, the PEI surfaces made contact to form a symmetric PEI/PEI interface. Welding was done at temperatures in the range 250–280 °C, which are above the T_g (210 °C) of PEI but below the melting point (344 °C) of the semicrystalline PEEK.

To fabricate the large composite panels, 16 quasi-isotropic laminates were molded [±45/0/90] at a pressure of 1.38 MPa using a non-isothermal thermal history (shown in Figure 8.40). Here, the temperature, T, was raised from room temperature, and healing commenced when T was above T_g and continued at a faster rate until the upper limit used in the isothermal process (250 °C) was reached. They analyzed the non-isothermal part by using an incremented version of the minor chain healing model (Chapter 2). In the isothermal case, the lap shear stress, σ, of the laminates is given by

$$\sigma = \sigma_\infty (t/T_r)^{1/4} \tag{8.11.1}$$

where σ_∞ is the virgin strength, and the stress ratio is related to the average contour length $l(t)$ via

$$\sigma/\sigma_\infty = [l(t)/L]^{1/2} \tag{8.11.2}$$

in which L is the tube length.

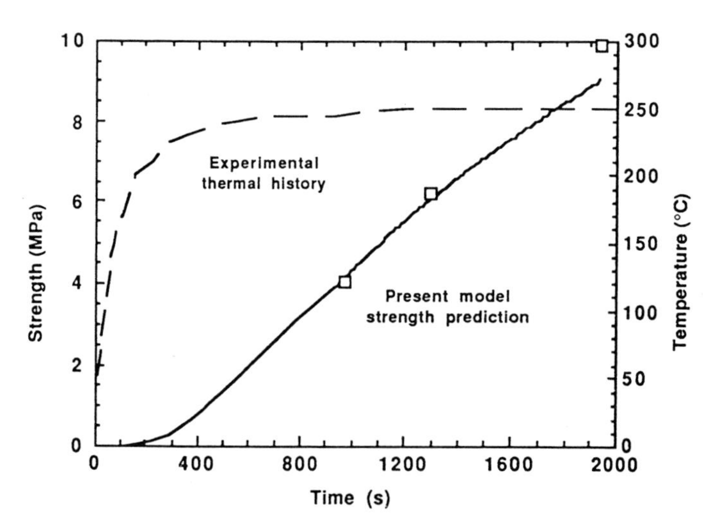

Figure 8.40 Thermal history and corresponding strength predictions of lap shear compression molded PEI interfaces at 250 °C (courtesy of L. Bastien and J. Gillespie).

For non-isothermal histories, they divided the temperature profiles into constant temperature steps of duration $\delta t_i = (t_{i+1} - t_i)$, in which the reptation time, T_r^*, is temperature dependent, and then summed the increments in the following manner

$$\sigma / \sigma_\infty = [\Sigma (t_{i+1}^{1/2} - t_i^{1/2}) / T_r^{*1/2}]^{1/2} \tag{8.11.3}$$

The temperature dependence of the reptation time was determined using an Arrhenius relation

$$T_r^* = B \exp(A/T) \tag{8.11.4}$$

where ln $B = -105.6$ s$^{1/4}$ and $A = 59728$ K^{-1}.

Figure 8.40 compares the observed strength ratio with the model predictions for the experimental thermal history (up to 250 °C). Similar agreement with theory and experiment was obtained for temperatures up to 270 °C and 280 °C. Bastien and Gillespie found that their model was in very good agreement with the composite fabrication process and allowed them to optimize the toughness and processing time by choosing the correct thermal history. In principle, their incremental solution for non-isothermal histories can be applied to any symmetric interface once the temperature dependence of T_r is known.

8.11.3 Autoadhesion at Composite Interfaces

The autoadhesion of polysulfone (PSU) thermoplastic composite resin (Udel P1700) was examined by Loos *et al.* [67]. This resin is impregnated with unidirectional fibers and is used as a prepreg material for composite fabrication. They were interested in optimizing the processing parameters and used our tack theory (Chapter 7) as a basis for evaluating their results. The tensile test assembly used to evaluate the strength of the PSU/PSU interface as a function of contact time and temperature is shown in Figure 8.41. This method provided a constant contact area surrounded by a radial pre-crack generated by the nonbonding kapton spacers. The samples were tested at constant strain rates and temperatures **above** T_g (194 °C) of the resin.

The autoadhesive load, σ, of the PSU/PSU interface test is shown as a function of $t^{1/4}$ in Figure 8.42. Loos *et al.* found that the welding theory describes their results in several important respects: (a) good agreement was obtained with $\sigma \sim t^{1/4}$; (b) the welding time, T_r, decreased in a predictable manner with increasing temperature; (c) the virgin strength, $\sigma_\infty(T)$, decreased with increasing temperature.

They defined the degree of autoadhesion as

$$f_{AU} = \sigma(T,t) / \sigma_\infty(T) \tag{8.11.5}$$

with $0 \leq f_{AU} \leq 1$. Eq 8.11.5 was used by Loos *et al.* to form a master curve of the autoadhesion data as a function of T and t using

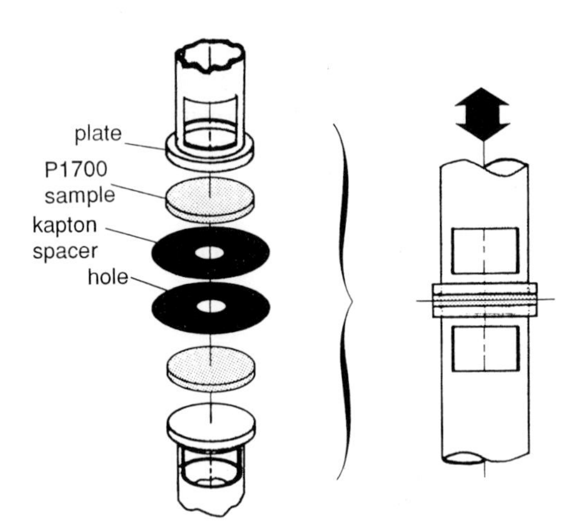

Figure 8.41 Schematic diagram of the interfacial tensile test assembly for polysulfone interfaces (courtesy of A. Loos *et al.*).

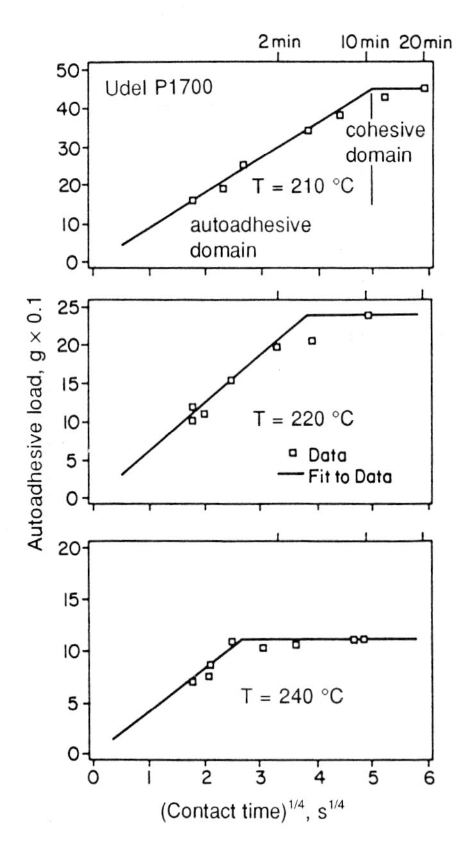

Figure 8.42 Autoadhesive load as a function of the fourth root of contact time for polysulfone resins at different temperatures (courtesy of A. Loos *et al.*).

$$f_{AU} = K(T)\, t^{1/4} \tag{8.11.6}$$

where $K = K_0 \exp(-E_a/kT)$, in which k is Boltzmann's constant. The Arrhenius constants K_0 and E_a were determined from a plot of ln $K(T)$ versus $1/T$ that gave $E_a = 6.0902 \times 10^{-20}$ J and $K_0 = 1922$. These relations provided excellent guides for optimizing the composite fabrication history.

8.12 Broken Bonds During Fracture

8.12.1 Introduction

We now explore the issue of broken bonds during fracture. When a crack propagates through a glassy polymer, the chains may pull out or they may fracture. We wish to know how many bonds are broken per unit fracture surface area, N_f, and how N_f is affected by molecular weight, molecular weight distribution, and temperature. Knowing N_f for a fracture event, we can evaluate the relevant contribution of bond rupture and chain disentanglement to the fracture energy G_{1c}. This problem has been examined at Urbana by Rockhill [68], Willett and O'Connor [69], and Paulson [70, 71]. The method used is a slice–fracture experiment in which a microtome blade slices through a polymer of known molecular weight, as shown in Figure 8.43. The change in the molecular weight of the slice gives the exact number of broken bonds due to slicing. The slicing is done in a manner that simulates brittle fracture, where the blade is preceded by a craze at its tip and fracture occurs by the breakdown of the craze. This point can be verified by examination of TEM images of fracture surfaces obtained from compact tension specimens that prove indistinguishable from sliced surfaces.

In a slice–fracture experiment, the number of bonds broken per unit area, N_f, can be found exactly as follows [68]

$$N_f = \frac{\rho N_a h}{2}\left(\frac{1}{M} - \frac{1}{M_i}\right) \tag{8.12.1}$$

where M_i, M, ρ, N_a, and h are the virgin monodisperse number average molecular weight, number average molecular weight after slicing, density of the sample, Avogadro's number, and thickness of the sample slice, respectively. The molecular weights and molecular weight distributions were obtained through size exclusion chromatography (SEC), also known as gel permeation chromatography (GPC).

Scanning electron micrographs provide evidence that the microtome technique was controlled to simulate a crack propagating through a craze. Figure 8.44 compares the surface of a microtomed slice with that of a compact tension sample, at the same magnification. We see that they are fairly indistinguishable.

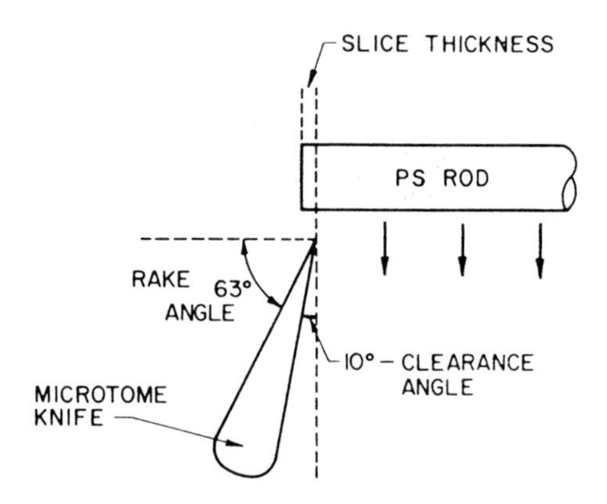

Figure 8.43 The slicing method used by Wool and Rockhill to determine the number of broken bonds per unit area. The microtome blade is stationary and the polystyrene sample is sliced as it moves up and down in a temperature-controlled environment.

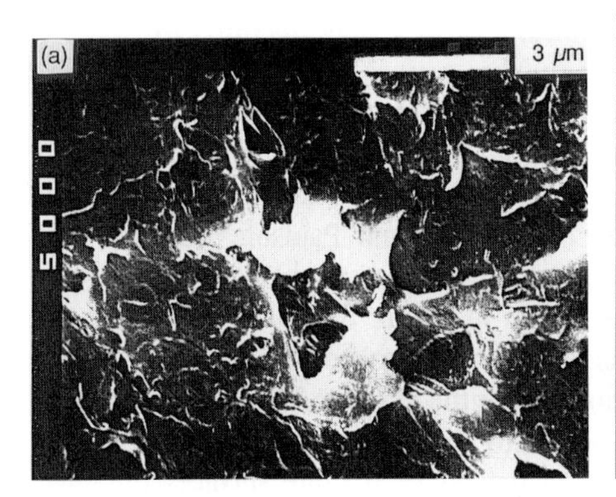

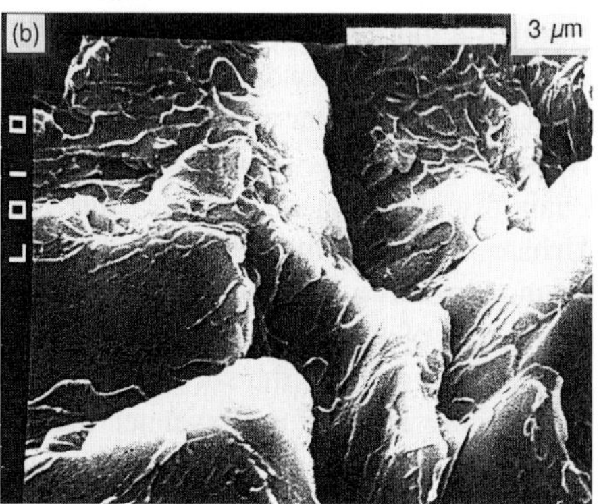

Figure 8.44 Scanning electron micrographs of (a) a compact tension fracture surface and (b) a microtomed slice; the bars represent 3 μm (Paulson and Wool).

8.12.2 Number of Broken Bonds, N_f

Representative SEC spectra of virgin and sliced material are shown in Figure 8.45. Monodisperse polymers have a narrow spectrum, excellent for showing the low-molecular-weight tail created as chains are broken. The spectrum shown is for the 172,000 number average molecular weight sample. The peak height is lower but the peak has not moved along the axis due to the abundance of 172,000-length chains still present in the slice. The number average molecular weight of the degraded sample has shifted to 154,000 due to the increase in short chains (seen in the low-molecular-weight tail). In all cases, the low-molecular-weight tail did not extend below M_c, indicating that fracture by chain pullout occurs near M_c. Results for polyethylene were similar. The low-molecular-weight region extends smoothly from the peak molecular

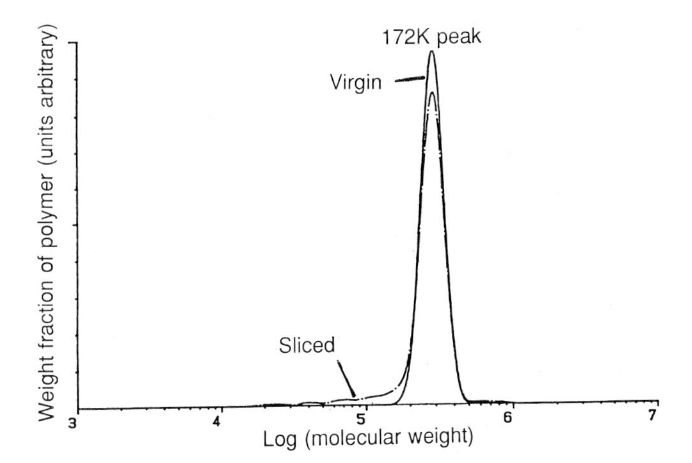

Figure 8.45 Size exclusion chromatography spectra for virgin monodisperse polymer (M = 172,000), and sliced polymer (M = 154,000). The slice thickness of 5 μm was obtained at room temperature (Paulson and Wool).

weight, M_n, to M_c, indicating the formation of all length species, which is to be expected if the chain breaks at random locations. The absence of a high-molecular-weight tail is also indicative of little or no postfracture free-radical interactions.

The decrease in the number average molecular weight for all molecular weights tested at room temperature is shown in Table 8.5. Percent decreases show that longer chains undergo more relative degradation, although the number of broken bonds as a function of area, N_f, shows little change with molecular weight. The $M^{-1/2}$ dependence of the number of chains, n, at an interface may explain this observation. As the molecular weight increases, the number of chains in a given area decreases. Therefore, if the number of breaks per unit area is a constant, independent of molecular weight, each longer chain must undergo more breaks per chain, as seen in Table 8.5.

The room-temperature samples had an average of 6.8×10^{13} breaks per cm^2. Sperling and co-workers [72, 73] recently obtained similar results using a dentist's burr to drill polystyrene surfaces [56]. This result may be compared with the number of bridges intersecting a unit area, $p = 1/a$, and the number of bonds required to break an entanglement network, N_f. From Chapter 7 (Eq 7.5.11) we have the critically connected state existing when $(N_f + n)3a = 1$, so that

$$N_f = (C\rho N_a)/(\sqrt{2}\,zM_0)\,[1 - (M_c/M)^{1/2}] \tag{8.12.2a}$$

or

$$N_f = N_{f\infty}\,[1 - (M_c/M)^{1/2}] \tag{8.12.2b}$$

Table 8.5 Room-Temperature Molecular Weight Changes and N_f

Number average virgin molecular weight	Room-temperature degraded molecular weight	Percent change	N_f, 10^{-13} cm^{-2}
91,500	84,400	7.76	13.84
93,600	90,200	3.63	4.76
167,000	143,000	14.37	15.22
172,000	154,000	10.47	7.21
274,000	210,000	23.36	17.04
282,000	230,000	18.44	11.56
395,000	279,000	29.37	16.09
432,000	361,000	16.44	6.34
765,000	615,000	19.61	4.27
877,000	618,000	29.53	6.93
929,000	626,000	32.62	6.69

where z, M_0, C, ρ, and N_a are the number of monomers per chain per c-axis length, the monomer molecular weight, the c-axis length of a unit cell, the density, and Avogadro's number, respectively. This equation gives $N_{f\infty} = 3 \times 10^{13}/\text{cm}^2$, in which the front factor $N_{f\infty} = \frac{1}{3}\,a$. When $M = M_c$, $N_f = 0$, which is consistent with all SEC and dilute solution viscosity data. When M becomes very large, N_f approaches $3 \times 10^{13}/\text{cm}^2$, which is in close agreement with the experimental results in Table 8.5. The number of crossings $p = 9 \times 10^{13}/\text{cm}^2$ also agrees well with experimental data. The chain fracture results strongly suggest that the entanglement network in the craze zone in glassy polymers fractures primarily by a combination of bond rupture and chain pullout.

If we assign an energy of $D_0 = 80$ kcal/mol to each bond rupture, the contribution of bond fracture, G_f, to the total fracture energy is obtained as

$$G_f = N_f D_0 J / N_a \approx 0.4 \text{ J/m}^2 \tag{8.12.3}$$

where $J = 4.18$ J/cal. Since G_{1c} for virgin PS samples is about 1000 J/m^2 (see Figure 8.17), bond rupture contributes a fraction 0.0004 to the fracture energy, or 0.04%. Thus, to an excellent approximation, bond rupture contributes essentially nothing to the magnitude of the fracture energy. However, it plays a major role in determining the extent of plastic deformation in the craze zone at the crack tip. In the deformation zone, the traction stress, σ_c, draws fresh material into crazes, and the crack-opening displacement increases until the competing processes of bond rupture and disentanglement stop the craze thickening at the crack tip.

8.12.3 Temperature Dependence of Bond Fracture

The virgin and degraded number average molecular weights for the thermal testing are shown in Table 8.6. N_f values calculated for each of the four molecular weights at the four temperatures are summarized. Temperature effects on N_f are free of molecular weight dependence, as the values illustrate; this molecular weight independence makes it possible to average the values of N_f at each temperature.

Table 8.6 Molecular Weight Changes and N_f Versus Temperature

Virgin number average molecular weight	Temperature °C	Degraded number average molecular weight	M_n change	N_f, 10^{-13} cm^{-2}
91,500	25	84,400	7,100	13.84
	45	83,300	8,200	16.19
	65	85,600	5,900	11.34
	85	85,900	5,600	10.72
167,000	25	143,000	24,000	15.22
	45	144,000	23,000	14.48
	65	151,000	16,000	9.61
	85	149,000	18,000	10.95
274,000	25	210,000	64,000	17.04
	45	219,000	55,000	14.04
	65	222,000	52,000	13.10
	85	239,000	35,000	8.19
395,000	25	279,000	116,000	16.19
	45	301,000	94,000	12.09
	65	310,000	85,000	10.61
	85	304,000	91,000	11.59

Figure 8.46 shows the effect of temperature on the average N_f values. As the temperature increases, N_f decreases for all molecular weights. The graph can be extrapolated to zero fracture events to find the temperature where no scission takes place, as well as to determine N_f at 0 K. We obtain $N_f = 0$ at 192.8 °C, which is approximately equal to $T_g + 90$ °C. At this temperature we expect the chains with M less than $8M_c$ to fully disentangle and pull out without bond rupture. This should be relevant to tack and green strength analysis. Extrapolating to the other extreme, at 0 K

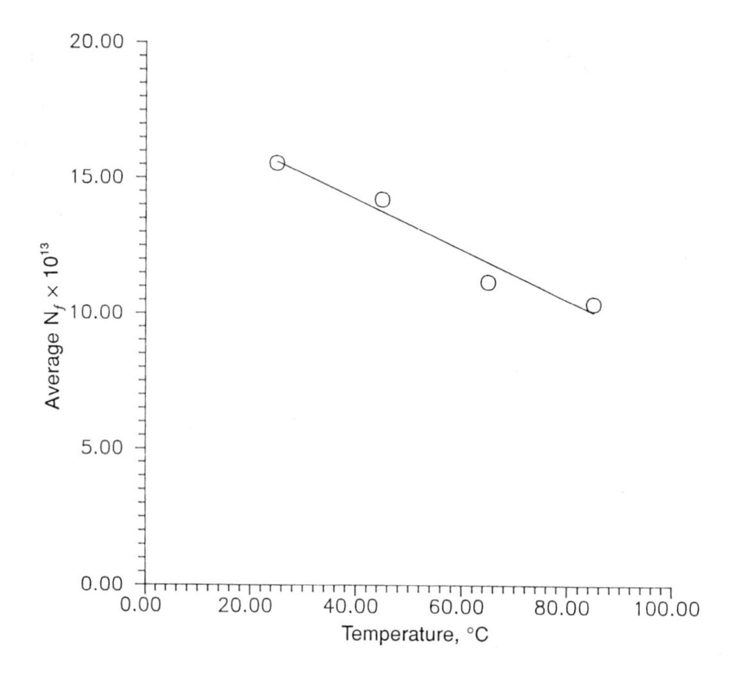

Figure 8.46 Number of broken bonds, N_f, versus temperature. The N_f values were averaged for each molecular weight.

(−273 °C), we obtain $N_f = 43.3 \times 10^{13}$ scissions/cm^2. These extrapolations must be viewed with some scepticism, given the relatively small change in N_f for a 60 °C change in temperature, and it may be more appropriate to say that the data show little change in N_f over this temperature range. When M is much greater than M^*, the entanglement network has little option but to fracture, even though chain pullout is favored at higher temperatures.

8.12.4 Relation Between Fracture Stress and Broken Bonds

In the high-molecular-weight limit (M much larger than $8M_c$), the fracture stress in the craze at the crack tip, σ_c, can be deduced from the critical energy relation

$$U_c \geq U_f \tag{8.12.4}$$

where U_c is the strain energy density stored in the craze and U_f is the energy to fracture the bonds. The stored strain energy in the craze is determined by

$$U_c = \sigma_c^2 / 2\, E_c \tag{8.12.5}$$

where E_c is the modulus of the craze in the drawing direction. The energy due to bond rupture of an entanglement network consisting of N_{ev} entanglements per unit volume is

$$U_f = D_0\, f\, N_{ev} \tag{8.12.6}$$

where D_0 is the bond fracture energy, $f \approx 1/2$ is the percolation fraction of bonds that must be broken to cause fracture in the network (discussed in Chapter 4), and $N_{ev} = \rho/M_c$ is the entanglement density. In this approach, the strain energy is first stored in the craze fibril entanglement network, and we enquire if this energy is sufficient to break $f\rho/M_c$ bonds when it is released at a critical strain energy density. Substituting Eqs 8.12.5 and 8.12.6 in Eq 8.12.4 and solving for σ_c, we obtain

$$\sigma_c = (2\, f\, E_c D_0\, \rho / M_c)^{1/2} \tag{8.12.7}$$

This equation predicts that the fracture stress increases with the square root of the number of bonds to be broken, and is inversely proportional to $M_c^{1/2}$, as expected from the entanglement model and noted by Wu (discussed in Chapter 7). The behavior of the fracture stress with modulus and bond energy is analogous to the Griffith equation (7.2.5), via $\sigma \sim \sqrt{EG}$.

Equation 8.12.7 for fracture stress can be further simplified if we assume that $2f \approx 1$, and $\sigma^* = \sigma_c/\lambda$, where σ^* is the applied stress at the craze/glass fibril interface and λ is the fibril draw ratio. The resulting solution for the fracture stress of polymers with M much greater than M^* is

$$\sigma^* \approx (E D_0\, \rho / M_c)^{1/2} / \lambda \tag{8.12.8}$$

The bond dissociation energy D_0 is about 80 kcal/mol for a C–C bond. This value varies within a range of ± 10 kcal/mol for individual polymers, depending on the inductive and polar nature of the substituent groups; electron donors increase D_0 while electron attractors decrease D_0. For polystyrene, $E = 2$ GPa, $D_0 \approx 80$ kcal/mol, $\lambda = 4$, $\rho = 1.2$ g/cm^3, $M_c = 30{,}000$, and we obtain $\sigma^* \approx 37$ MPa (5,400 psi), which agrees well with reported values (Table 8.4). Polycarbonate, with a much lower M_c of 4,800 and E of 3 GPa is expected to have a fracture stress of about 110 MPa (see Figure 8.23).

The bond fracture stress, σ_b, is determined by

$$\sigma_b = D_0\, a_m / (2\, a \cos\Theta) \tag{8.12.9}$$

where $a_m = 1.99/\text{Å}$ is the Morse parameter and cos Θ is the angle between the bond and the applied stress (typically, cos $\Theta = 1$). Using values for polystyrene with

a = 111 Å^2 and D_0 = 80 kcal/mol, we obtain $\sigma_b \approx$ 5 GPa. We can relate the macroscopic fracture stress to the chain fracture stress by substituting for D_0 as

$$\sigma^* \approx [2\, aE\sigma_b \rho / (M_c\, a_m)]^{1/2} / \lambda \tag{8.12.10}$$

From the above relations, we deduce the result for the fracture energy $G_{1c} \sim \sigma^{*2}$ as

$$G_{1c} \sim D_0 / M_c \tag{8.12.11}$$

Since D_0 is essentially constant for linear polymers, this relation highlights the role of M_c in fracture processes. Thus, we would expect polycarbonate to have a fracture energy about six times that of polystyrene, or $G_{1c} \approx$ 6 kJ/m^2 (Figure 8.23).

When applied to healing interfaces, the number of bonds to be broken is $n(t)\,l(t)$ in an interface of volume $V \sim X(t)$. In this case, Eq 8.12.4 is used as

$$\sigma^2 X(t) / 2\, E \approx l(t)\, n(t)\, D_0 / M_c \tag{8.12.12}$$

Since $n(t)/X(t) = 1/M$, we obtain $\sigma(t) \sim [l(t)/M]^{1/2}$, or

$$\sigma(t) = \sigma^* (t/\tau)^{1/4} \tag{8.12.13}$$

where the welding time $\tau \sim M^{*2}M$. Hence, $\sigma(t) \sim t^{1/4}M^{-1/4}$, which is identical to the chain pullout case. Thus, the scaling laws for welding are effectively independent of the details of fracture, since both chain pullout and bond rupture mechanisms depend on the same interdiffused contour length. The stored energy in the chain at critical conditions is released either by pullout or bond rupture.

As a final comment in this chapter, we enquire into the dependence of G_{1c} on the number of chains, Σ, at an interface. Since $G_{1c} \sim \langle l \rangle$, and $\langle l \rangle \sim \Sigma^2 M^2$, Eq 8.12.13 yields

$$G_{1c} = G_{1c}^* (\Sigma / \Sigma^*)^2 \tag{8.12.14}$$

where G_{1c}^* is the fracture energy corresponding to a saturation number of chains $\Sigma^* \sim M^{*-1/2}$. This relation has significance for the use of compatibilizers at incompatible interfaces and coupling agents at polymer–metal interfaces. It will be examined more closely using classical experiments of Brown and co-workers in the next chapter.

8.13 References

1. R. P. Wool, "Crack Healing in Semicrystalline Polymers, Block Copolymers and Filled Elastomers"; chapter in *Adhesion and Adsorption of Polymers*, L.-H. Lee, Ed.; Vol. 12A in series *Polymer Science and Technology*; Plenum Press, New York; 1980; p 341.

2. R. P. Wool and K. M. O'Connor, *J. Appl. Phys.* ***52***, 5953 (1981).
3. K. M. O'Connor, *Crack Healing in Polymers (Fracture, Reptation, Tack)*; Ph.D. Thesis, University of Illinois, Urbana, IL; 1984.
4. J. N. Anand, *Adhesion* ***1***, 31 (1969).
5. J. D. Skewis, *Rubber Chem. Technol.* ***39***, 217 (1966).
6. R. P. Wool and K. M. O'Connor, *J. Polym. Sci., Polym. Lett. Ed.* ***20***, 7 (1982).
7. S. S. Voyutskii, *Autohesion and Adhesion of High Polymers*; Vol. 4 in series *Polymer Reviews*, H. F. Mark and E. H. Immergut, Eds.; Wiley–Interscience, New York; 1963.
8. W. G. Forbes and L. A. McLeod, *Trans. Inst. Rubber Ind.* ***30*** (5), 154 (1958).
9. J. D. Skewis, private communication.
10. R. P. Wool, *Rubber Chem. Technol.* ***57*** (2), 307 (1984).
11. P.-G. de Gennes, *J. Chem. Phys.* ***55***, 572 (1971).
12. C. M. Roland and G. G. A. Bohm, *Macromolecules* ***18***, 1310 (1985).
13. R. G. Stacer and H. L. Schreuder-Stacer, *Int. J. Fract.* ***39***, 201 (1989).
14. S. Prager and M. Tirrell, *J. Chem. Phys.* ***75***, 5194 (1981).
15. D. Adolf, M. Tirrell, and S. Prager, *J. Polym. Sci., Polym. Phys. Ed.* ***23***, 413 (1985).
16. M. Tirrell, D. Adolf, and S. Prager, "Orientation and Motion at a Polymer–Polymer Interface: Interdiffusion of Fluorescent-Labelled Macromolecules"; chapter in *Orienting Polymers: Proceedings of a Workshop Held at the IMA, University of Minnesota, Minneapolis, March 21–26, 1983*, J. L. Ericksen, Ed.; No. 1063 in series *Lecture Notes in Mathematics*, A. Dold and B. Eckmann, Eds.; Springer-Verlag, Berlin; 1984; p 37.
17. N. S. Korenevskaya, V. V. Laurent'ev, S. M. Yagnyatinskaya, V. G. Rayevskii, and S. S. Voyutskii, *Polym. Sci. USSR (Engl. Transl.)* ***8***, 1372 (1966).
18. K. L. Foster and R. P. Wool, *Macromolecules* ***24***, 1397 (1991).
19. J. L. Willett, K. M. O'Connor, and R. P. Wool, "Mechanical Properties of Polymer–Polymer Welds: Time and Molecular Weight Dependence"; *Polym. Prepr. (Am. Chem. Soc., Div. Polym. Chem.)* ***26*** (2), 123 (1985).
20. J. L. Willett and R. P. Wool, *Macromolecules* ***26***, 5336 (1993).
21. G. R. Hamed and C.-H. Shieh, *J. Polym. Sci., Polym. Phys. Ed.* ***21***, 1415 (1983).
22. R.-J. Chang, A. N. Gent, and S.-M. Lai, "Effect of Interfacial Bonds on the Strength of Adhesion", *PMSE Preprints; Proceedings of the American Chemical Society Division of Polymeric Materials: Science and Engineering* ***67***, 41 (1992).
23. R. J. Chang and A. N. Gent, *J. Polym. Sci., Polym. Phys Ed.* ***19***, 1619 (1981).
24. A. N. Gent, The 1981 Foundation Lecture, "Strength of Adhesive Bonds"; *Plast. Rubber Int.* ***6*** (4), 151 (1981).
25. A. N. Gent and A. G. Thomas, *J. Polymer Sci. Part A-2* ***10***, 571 (1972).
26. T. Tsuji, M. Masuoko, and K. Nakao, "Superposition of Peel Rate, Temperature and Molecular Weight for T Peel Strength of Polyisobutylene"; chapter in *Adhesion and Adsorption of Polymers*, L.-H. Lee, Ed.; Vol. 12A in series *Polymer Science and Technology*; Plenum Press, New York; 1980; p 439.
27. A. K. Bhowmick, P. P. De, and A. K. Bhattacharyya, *Polym. Eng. Sci.* ***27***, 1195 (1987).
28. P.-G. de Gennes, *C. R. Acad. Sci., Ser. 2,* ***312***, 1415 (1991).
29. G. R. Hamed, *Rubber Chem. Technol.* ***54***, 576 (1981).
30. C. K. Rhee and J. C. Andries, *Rubber Chem. Technol.* ***54***, 101 (1981).
31. R. Guernsey and J. Gilman, *Exp. Mech.* ***1***, 50 (1961).
32. M. F. Kanninen, *Int. J. Fract.* ***9***, 83 (1973).
33. K. M. O'Connor and R. P. Wool, *Bull. Amer. Phys. Soc.* ***30*** (3), 389 (1985).

34. R. P. Wool, B.-L. Yuan, and O. J. McGarel, *Polym. Eng. Sci.* ***29***, 1340 (1989).
35. R. P. Wool, "Welding, Tack and Green Strength of Polymers"; chapter in *Fundamentals of Adhesion*, L.-H. Lee, Ed.; Plenum Press, New York; 1991.
36. Robertson, R. E., "The Fracture Energy of Low Molecular Weight Fractions of Polystyrene"; chapter in *Toughness and Brittleness of Plastics*, R. D. Deanin and A. O. Crugnola, Eds.; Vol. 154 in Advances in Chemistry Series; American Chemical Society, Washington, DC; 1976; p 89.
37. E. J. Kramer, *J. Mater. Sci.* ***14***, 1381 (1978).
38. B. D. Lauterwasser and E. J. Kramer, *Philos. Mag. A* ***39***, 469 (1979).
39. P. Prentice, *Polymer* ***24***, 344 (1983).
40. K. E. Evans, *J. Polym. Sci., Polym. Phys. Ed.* ***25***, 353 (1987).
41. M. McLeish, M. Plummer, and A. Donald, *Polymer* ***30***, 1651 (1989).
42. H. W. McCormick, F. M. Brower, and L. Kin, *J. Polym. Sci.* ***39***, 87 (1959).
43. R. P. Wool, "Strength Development at a Symmetric Polymer–Polymer Interface"; In *Advances in Rheology 3. Polymers*, B. Mena, A. Garcia-Rejon, C. Rangel-Nafaile, Eds.; Proc. IXth Int. Congr. Rheology; Univ. Nacional Autónoma de México, Mexico, 1984; p 573.
44. G. L. Pitman and I. M. Ward, *Polymer* ***20***, 895 (1979).
45. O. J. McGarel and R. P. Wool, "Welding of Symmetric Amorphous Polymer Interfaces", *Bull. Am. Phys. Soc.* ***34*** (3), 939 (1989).
46. D. B. Kline and R. P. Wool, *Polym. Eng. Sci.* ***28***, 52 (1988).
47. A. Lee and R. P. Wool, *Macromolecules* ***19***, 1063 (1986).
48. P. F. Green and E. J. Kramer, *Macromolecules* ***19***, 1108 (1986).
49. S. J. Whitlow and R. P. Wool, *Macromolecules* ***24***, 5926 (1991).
50. F. Danusso, G. Tiegi, and A. Lestingi, *Polym. Commun.* ***27***, 56 (1986).
51. F. Danusso, G. Tiegi, and P. Botto, *Polym. Commun.* ***26***, 221 (1985).
52. G. Tiegi, private communication of unpublished results at the Politechnico di Milano, Italia, 1984.
53. Y. Wang and M. A. Winnick, *J. Phys. Chem.* ***97***, 2507 (1993).
54. J. N. Yoo, L. H. Sperling, C. J. Glinka, and A. Klein, *Macromolecules* ***23***, 3962 (1990).
55. N. Mohammadi, A. Klein, and L. H. Sperling, "Elucidation of the Fracture Micromechanisms of Glassy Polymers via Healing and Fracture of a Polyinterface System", *PMSE Preprints; Proceedings of the American Chemical Society Division of Polymeric Materials: Science and Engineering* ***67***, 268 (1992).
56. L. H. Sperling, Paper presented at the Annual Symposium on Latex Particles, Lehigh University, 1993.
57. B.-L. Yuan and R. P. Wool, "Fatigue Studies of Healing Interfaces", *Bull. Am. Phys. Soc.* ***34*** (3), 939 (1989); R. P. Wool and B.-L. Yuan, paper in press.
58. S. S. Wang and A. Miyase, *J. Compos. Mater.* ***20***, 439 (1986).
59. C. M. Rimnac, J. A. Manson, R. W. Hertzberg, S. M. Webler, and M. D. Skibo, *Macromol. Sci.-Phys., B* ***19***, 351 (1981).
60. E. J. Kramer and L. L. Berger, *Adv. Polym. Sci.* ***91/92***, 1 (1990).
61. D. H. Klosterman and R. P. Wool, *Bull. Am. Phys. Soc.* ***27*** (3) 361 (1982).
62. V. K. Stokes, *Polym. Eng. Sci.* ***28***, 718 (1988).
63. V. K. Stokes, *Polym. Eng. Sci.* ***28***, 728 (1988).
64. V. K. Stokes, *Polym. Eng. Sci.* ***28***, 989 (1988).
65. V. K. Stokes, *Polym. Eng. Sci.* ***28***, 998 (1988).
66. L. J. Bastien and J. W. Gillespie, Jr., *Polym. Eng. Sci.* ***31***, 1720 (1991).

67. A. C. Loos, J. C., Howes, and P. H. Dara, "Thermoplastic Matrix Composite Processing Model"; in *Adhesion Science Review*, H. F. Brinson, J. P. Wightman, and T. C. Ward, Eds.; Commonwealth Press Inc., Radford, VA; 1987; p 263.
68. R. P. Wool and A. T. Rockhill, *J. Macromol. Sci. Phys. B* ***20***, 85 (1981).
69. J. L. Willett, K. M. O'Connor, and R. P. Wool, *J. Polym. Sci., Part B: Polym. Phys.* ***24***, 2583 (1986).
70. K. Paulson, *Molecular Fracture in Glassy Polymers*; M.S. Thesis, University of Illinois, Urbana, IL; 1990.
71. K. Paulson and R. P. Wool, paper in press.
72. N. Mohammadi, A. Klein, and L. H. Sperling, *Macromolecules* ***26***, 1019 (1993).
73. M. Sambasivam, A. Klein, and L. H. Sperling, *Macromolecules*, in press.
74. ASTM D 3163 – 92, Standard Test Method for Determining Strength of Adhesively Bonded Rigid Plastic Lap-Shear Joints in Shear by Tension Loading. In *Vol. 15.06, Adhesives, 1993 Annual Book of ASTM Standards*; American Society for Testing and Materials, Philadelphia, PA, 1993.

9 Strength of Incompatible Amorphous Interfaces*

9.1 Introduction

The structure and strength of symmetric A/A interfaces have received considerable study, both theoretical [1–5] and experimental [6–12] (Chapter 8). Asymmetric compatible interfaces have also been studied theoretically [13–16] and experimentally [6, 7, 17, 18]. Incompatible A/B interfaces have received more attention [19–38], largely because of their natural abundance and numerous practical applications in blends.

In this chapter, we first examine the fracture energy, G_{1c}, of model asymmetric amorphous interfaces as a function of welding time and temperature using a wedge cleavage fracture mechanics technique. Second, we analyze the fracture surfaces by XPS and SEM to determine the locus of failure and attempt to elucidate the mechanism of fracture. Third, we relate the strength of the interface to its structure with the microscopic analysis discussed in Chapters 7 and 8. Several families of A/B interfaces are explored: polystyrene/poly(methyl methacrylate) (PS/PMMA) with varying PS molecular weight; poly(styrene–co-acrylonitrile)/poly(methyl methacrylate) (PSAN/PMMA) with varying acrylonitrile (AN) content; and poly(styrene–co-acrylonitrile)/polycarbonate (PSAN/PC) with varying AN content in the PSAN. Finally, we examine the role of diblock (A–B) compatibilizers on the strength of A/B interfaces.

9.1.1 Theoretical Aspects of Incompatible Interfaces

The concentration profile of an incompatible polymer interface, as determined by Helfand and co-workers [19, 20], is shown in Figure 9.1. Monomer units from one side diffuse to the other side until an equilibrium distribution is reached having a characteristic interpenetration depth d_∞. The greater the amount of mixing or chain interpenetration at the interface, the greater the strength developed.

The interfacial mixing of an incompatible pair of polymers is believed to depend on two opposing forces (see Chapter 1). The first force is repulsion of chain segments, in the presence of incompatible chain segments, due to an interaction energy; the second force is entropic in nature, and allows chains near the interface with reduced

---

* Dedicated to E. Helfand and E. J. Kramer

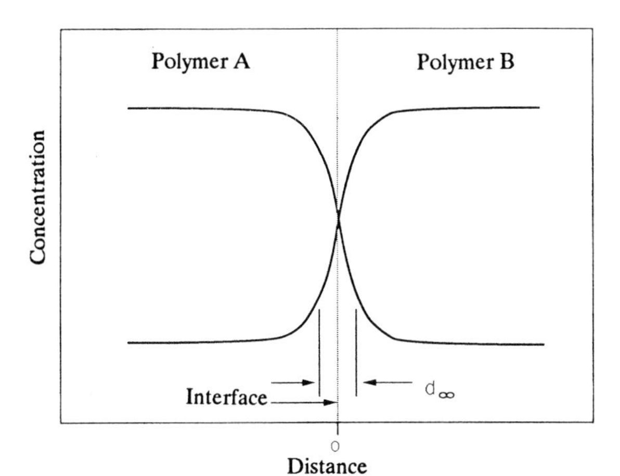

Figure 9.1 Concentration profile of an A/B incompatible polymer interface, as determined by Helfand.

entropy, which may lie in unfavorable conformations, to increase their entropy by crossing the interface. Helfand and co-workers [19, 20] examined the theory of the incompatible interface between two immiscible polymers of infinite molecular weight and described the diffusion of one polymer species into another by solving the diffusion equation of a random walk in a potential field due to the presence of an incompatible monomer. In particular, they determined an expression for the equilibrium interpenetration zone, d_∞, or depth of a polymer chain with respect to the Flory–Huggins interaction parameter, χ

$$d_\infty = 2b/(6\chi)^{1/2} \tag{9.1.1}$$

where b is the statistical segment step length and is taken as 6.5 Å for PS and PMMA.

To relate the interface structure to mechanical strength, we compare the structure and strengths of the incompatible interface with the partially diffused symmetric interface. We assume that the strength that exists at d_∞ can be related to the strength that exists at a corresponding average monomer interpenetration distance $X(t)$ for the partially diffused healing symmetric interface. The fracture energy of the symmetric interface is given by [5, 8, 9]

$$G(t) = G^*_{1c}(t/T_r)^{1/2} \tag{9.1.2}$$

where $G(t)$ is the fracture energy at time t and $G_{1c}{}^*$ is the maximum fracture energy attained at the reptation time, T_r. This relationship assumes that chain disentanglement dominates the fracture mechanism. When chain rupture dominates, a similar equation applies, except that T_r becomes the time at which the maximum strength $G_{1c}{}^*$ is attained.

The average monomer interpenetration distance, X, of the symmetric interface is given by (see Chapter 2)

$$X(t) = X_\infty (t/T_r)^{1/4} \tag{9.1.3}$$

where X_∞ is the interpenetration distance attained at T_r, related to the radius of gyration, R_g, by [44]

$$X_\infty \approx R_g \tag{9.1.4}$$

The strength of the interface is related to the average interpenetration of the minor chain contour length, $l(t)$, via $G(t) \sim l(t)$. Since $l(t) \sim X(t)^2$, then

$$G(t) \sim X(t)^2 \tag{9.1.5}$$

X is related to the equilibrium thickness, d_∞, by

$$d_\infty = 2X \tag{9.1.6}$$

which gives the fracture energy of the A/B interface as

$$G_{1c} \sim d_\infty^2 \tag{9.1.7}$$

From the above relations it follows that

$$d_\infty \approx 2R_g (G_{1c}/G_{1c}^*)^{1/2} \tag{9.1.8}$$

and also

$$G_{1c} = G_{1c}^* (d_\infty/2R_g)^2 \tag{9.1.9}$$

These equations allow one to relate the strength of the partially interpenetrated incompatible interface to the structure and strength of the partially and fully diffused interface. The utility of these equations is discussed later.

The temperature dependence of the interaction parameter χ can be written as [45]

$$\chi(T) = A/T + B \tag{9.1.10}$$

with A and B to be determined experimentally. Substituting in Eq 9.1.1 and 9.1.7,

$$G_{1c}(T) \sim 1/(A/T + B) \tag{9.1.11}$$

For Case I, where A/T is much greater than B, we find G_{1c} varies reversibly with T as

$$G_{1c}(T) \sim T \tag{9.1.12}$$

while for Case II (A/T much smaller than B), $G_{1c}(T)$ is independent of temperature:

$$G_{1c}(T) \sim T^0 \tag{9.1.13}$$

These relations for G_{1c} and d_∞ are examined in later sections.

9.1.2 Fractal Structure of Incompatible Interfaces

In Chapter 4 we discussed, using gradient percolation arguments, how the diffusion field at an interface could be divided into a connected and a nonconnected region. The diffusion front is fractal and represents the frontier between the connected and nonconnected regions of the interface. With incompatible A/B interfaces, the nonconnected region is negligible, and only the connected region contributes to the interface structure. In three dimensions, especially at short diffusion distances, the entire polymer-diffused region contributes to the fractal frontier. Thus, the equilibrium diffusion field of incompatible A molecules into the B side would be viewed as a "dancing" fractal on the B side, with fractal dimension D of about 2.5.

In the absence of the nonconnected region, the fractal structure should be readily observable by scattering methods. Hashimoto *et al.* [40] examined the space–time organization of the late-stage spinodal decomposition of a binary mixture of perdeuterated polybutadiene and polyisoprene near the critical point, using SANS and light-scattering techniques. They observed that the structure factor versus inverse scattering length in the interface region has a characteristic slope of 2.5 independent of time. They believe that this "intrinsic nonuniversality" can be explained in terms of the fractal arguments in Chapter 4 [77].

The number of particles, N_f, on the fractal polymer interface is determined at diffusion distances less than R_g by

$$N_f = nL \tag{9.1.14}$$

where n is the number of chains per unit area of length L. For incompatible interfaces, n is constant, $L \sim d_\infty^2$, and hence $N_f \sim d_\infty^2$. The nail solution (Chapter 7) for weakly entangled polymers gives the fracture energy as $G_{1c} \sim nL$, or $G_{1c} \sim L$ when n is constant. This gives the intuitively appealing solution that the fracture energy is proportional to the number of A–B contacts in the incompatible interface. Thus, with $G_{1c} \sim N_f$, we obtain $G_{1c} \sim d_\infty^2$, as in Eq. 9.1.7.

The Sapoval number, S, which describes the interface roughness, is related to the equilibrium depth d_∞ of incompatible interfaces by

$$S = S_0(d_\infty / d_0)^2 \tag{9.1.15}$$

where S_0 is the interface roughness at diffusion depth d_0. If $S_0 = 1$ at $d_0 \approx b$ (bond length), then $S \approx d_\infty^2/b^2$. Hence, for weakly entangled polymers, we expect the fracture energy to be related to the interface roughness via

$$G_{1c} \approx 2\Gamma (d_{\infty} / b)^2 \tag{9.1.16}$$

where Γ is the surface tension when $S = 1$. For example, if $d_{\infty} \approx 20$ Å, $b = 1.5$ Å, and $\Gamma = 0.04$ J/m^2, one obtains $G_{1c} \approx 14$ J/m^2 with a Sapoval number S of 177.

When the diffusion depth is much greater than the entanglement radius R_{ge}, crazing of entangled chains is the dominant mechanism, the nail solution no longer applies, and the fracture energy is related to the interface roughness via $G_{1c} \sim S/n$, and Eq 9.1.9 is applicable, where again $G_{1c} \sim d_{\infty}^2$.

9.2 Experimental Methods for Incompatible Interfaces

9.2.1 Welding Methods for Incompatible Interfaces

The welding process consists of placing two molded A/B plates together at room temperature and quickly heating them to the desired welding temperature in the welding block shown in Figure 9.2 [32]. This is an aluminum box designed to prevent the 53-mm-square polymer samples from deforming excessively (less than 3% thickness change) at temperatures above T_g; it unbolts to allow removal of samples

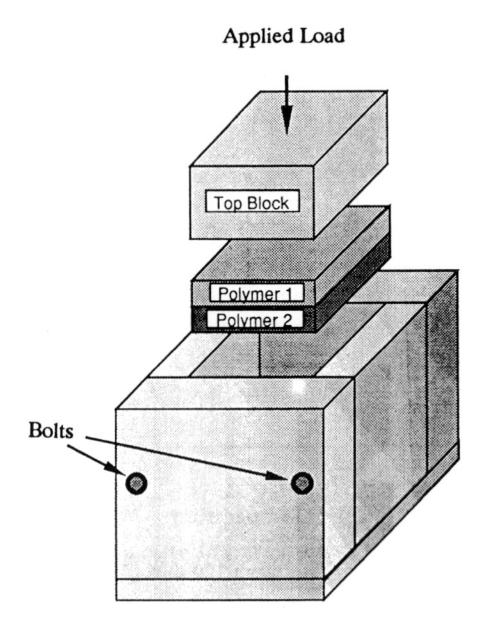

Figure 9.2 The welding block for A/B interfaces (Foster and Wool).

without damage to the weak interface. The welding block is placed either into a preheated Carver press, or into the welding chamber (Figure 8.19 in Chapter 8).

After welding had proceeded for a controlled time, temperature, and pressure, the A/B plates were cut into wedge cleavage fracture mechanics specimens of thickness, 6.55 mm ± 0.14 mm; width, 8.66 mm ± 0.35 mm; and length, 52.8 mm. A 10-mm crack was initiated with a razor blade, and a groove for the wedge, 0.76 mm thick and 2.90 mm deep, was centered and cut on the top interface. These samples were then dried for at least 24 hours in the vacuum oven at 80 °C before being tested.

9.2.2 X-Ray Photoelectron Spectroscopy

Analysis of the fracture surface helps determine whether fracture is *adhesive* or *cohesive*. We posit that a fracture surface that is the result of chain disentanglement and chain pullout has failed by a primarily adhesive process, and we do not expect to find A molecules on the B surface or *vice versa*. In adhesive failure, fracture occurs at the original interface, but, in cohesive failure, the locus of fracture depends on the asymmetry of the deformation zone and typically occurs some distance into the weaker matrix, rather than at the original interface plane. In cohesive failure, therefore, the locus of failure determines the extent to which we observe A molecules on the B side, or B molecules on the A side.

X-ray photoelectron spectroscopy (XPS) makes use of the photoelectric effect. Low-energy X-rays irradiate the surface of a sample, and the kinetic energy of the ejected electrons is analyzed. The kinetic energy is the difference between the X-ray photon energy and the ejected electrons' binding energy from specific atoms. The measured binding energy is determined by the elemental composition of the sample. One then obtains a spectrum of electron intensity versus energy.

For PS/PMMA, we were concerned with the atomic ratios of the O_{1s} (540 eV) to C_{1s} (285 eV) atomic constituents. If PMMA were transferred to the PS side, a strong O_{1s} signal should develop, whereas if PS were transferred to PMMA, a decrease in the O_{1s} signal should occur. We determined atomic ratios of oxygen to carbon (O/C) and nitrogen to carbon (N/C) by measuring the peak areas A_i for each element i = O, C, N, and converting the area ratios to atomic ratios using atomic sensitivity factors S_i. Each S_i is a measure of the cross section of that element in the X-ray electron ejection process, relative to fluorine ($S_F = 1$). We obtained atomic ratios using the relations

$$\frac{\mathrm{N}}{\mathrm{C}} = \frac{(A_N/S_N)}{(A_C/S_C)} \tag{9.2.1}$$

$$\frac{\mathrm{O}}{\mathrm{C}} = \frac{(A_O/S_O)}{(A_C/S_C)} \tag{9.2.2}$$

By comparing atomic ratios of the pure (as-molded) polymer surfaces to those of the fracture surfaces, we can sensitively determine the composition of the surface. Since the XPS electron escape depths are low, the composition analysis is conveniently confined to a depth of a few nanometers, which is a typical equilibrium depth of incompatible interfaces.

9.2.3 Microscopic Analysis

A section of approximately 1 cm^2, on one side of a fractured wedge cleavage sample, was glued to an aluminum base. The sides of the sample were coated with a conductive adhesive, DAG 154, to minimize charging in the SEM. The adhesive was dried for 24 hours, and the surfaces were sputter-coated with gold for 2 minutes, resulting in a gold coating of approximately 150–200 Å thickness.

We used the technique of Cho and Gent [46] to examine the fracture surface of the PMMA by dyeing PS remaining on the surface. Cyclohexane is a good solvent for PS but a poor solvent for PMMA, so a red dye dissolved in cyclohexane penetrates PS fragments and shows up as a red fracture surface on the PMMA side [32]. After a 5-minute wash in isooctane (a poor solvent for PS and PMMA but a good solvent for the dye), the samples were air dried and examined with an optical microscope.

9.2.4 Fracture Methods

We used a wedge cleavage (WC) geometry (Figure 9.3), sometimes called a modified double cantilever beam (DCB), to measure the critical strain energy release rate, G_{1c}. A wedge is driven downward into the interface at 0.254 mm/min until a crack is observed advancing. Once an advancing crack is observed, the motion of the wedge is stopped and the crack is allowed to advance at constant displacement (that is, with no further change in m) until the stress at the crack tip is low enough so that cracking stops. This usually takes about 15 minutes. The displacement, m, is measured to $\pm$ 0.01 mm, the sample is removed, and the crack length, a, is measured to $\pm$ 0.1 mm. This procedure was repeated as many as 15 times per sample, with the crack therefore becoming ever longer, resulting in approximately 40 measurements per welding experiment. In practice, we have chosen to use the G_{1c} arrest rather than the true critical strain initiation value. The length of the sample outside the grips is slightly different for each sample. In order to better compare one sample to another, we defined a relative crack length as the ratio of the arrested crack length to the length of the specimen outside the grips, H, shown in Figure 9.3.

The fracture mechanics of the wedge cleavage method was analyzed by Kanninen [47]. He derived an expression for G_{1c} with a double cantilever beam geometry on an elastic foundation and obtained

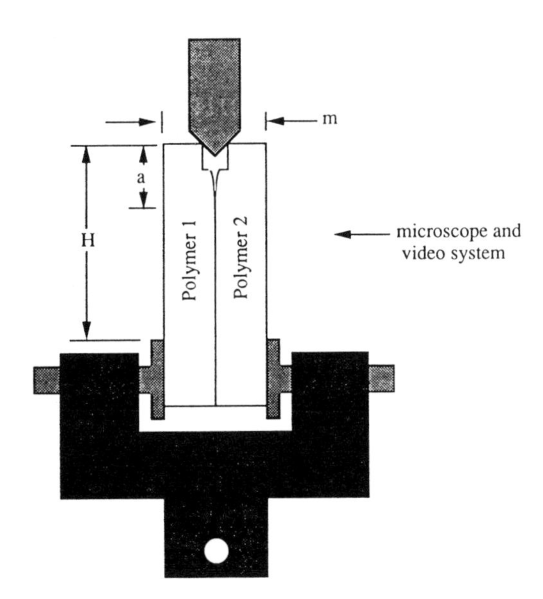

Figure 9.3 Arrangement and geometry of the wedge cleavage experiment.

$$G_{1c} = \frac{3}{16}\frac{(m-2h)^2 Eh^3}{a^4}\left[1+0.64\left(\frac{h}{a}\right)\right]^{-4} \tag{9.2.3}$$

where m is the separation of the outside top surface of the sample (displacement), $2h$ is the total sample thickness, E is Young's elastic modulus, and a is the crack length. Thus, the crack length a and the displacement $(m - 2h)$ due to the wedge are measured and used in Eq 9.2.3 to obtain G_{1c}. This method does not guarantee that the crack propagates along the interface. If one material has a lower craze stress than the other, the deformation zone becomes asymmetric and the crack may run off the interface plane. Asymmetric WC methods that use methods suggested by Brown [48, 49] are discussed in Section 9.7.

9.2.5 Fracture of Bimaterial Interfaces

Williams [50] examined fracture at the interface between two dissimilar materials with moduli E_1 and E_2, and Poisson's ratios v_1 and v_2. In a DCB or WC configuration, he expects the strain energy release rate, G_{1c}, to be determined to a first approximation by addition of the individual crack-opening displacements, so that the modulus E for the homogeneous case (1 = 2) in Eq 9.2.3 is replaced by

$$E = \tfrac{1}{2}(1/E_1 + 1/E_2)^{-1} \tag{9.2.4}$$

With some approximations (see [50]), an examination of the stress fields at the crack tip results in a relation between G_1 and the stress intensity factor, K_1, as

$$G_1 = \tfrac{1}{2} K_1^2 (1/E_1 + 1/E_2)(2/\alpha - 1/\alpha^2) \tag{9.2.5}$$

The α in Eq 9.2.5 is related to differences in the moduli and Poisson's ratios, via

$$\alpha = (E_1/E_2 + 1)/(\phi + 1) \tag{9.2.6}$$

where

$$\phi = 1 + \tfrac{1}{2} E_2 [(1 + v_1)/E_1 - (1 + v_2)/E_2] \tag{9.2.7}$$

With symmetrical loading of dissimilar materials, a shear stress $\tau_{0,r}$ results, in the plane of the interface, which is given in terms of distance r from the crack tip as [50]

$$\tau_{0,r} = -\left(\frac{1-\alpha}{\alpha}\right) \frac{K_1}{(2\pi r)^{1/2}} \tag{9.2.8}$$

In the symmetric case with $E_1 = E_2$ and $v_1 = v_2$, $\phi = 1$ and $\alpha = 1$, the shear stress vanishes, and Eq 9.2.3 is valid. The following two cases give further insight into the dissimilar fracture approximation with symmetrical loading (equal thickness of beams).

Case i: $E_2/E_1 = \beta$; $v_1 = v_2$; $\phi = 1$.

The shear stress at the interface plane $\tau_{0,r}$ (Eq 9.2.8) is related to the modulus ratio β and the maximum stress σ_{yy} for Mode I opening (Eq 7.7.1) by

$$\tau_{0,r} = [(\beta - 1)/(\beta + 1)]\, \sigma_{yy} \tag{9.2.9}$$

For example, a modulus ratio of $\beta = 1.1$ (10% difference) results in an additional shear stress at the interface plane that is about 9% of the opening stress. With $\beta = 2$, $\tau_{0,r} = \sigma_{yy}/3$, which is considerable. The combination of a large shear stress with the tensile σ_{yy} component could have a resultant that would drive the crack off the interface and into the bulk material.

Case ii: $E_2 = E_1$; $\beta = 1$; $v_1 > v_2$.

In this case, the shear stress simplifies to

$$\tau_{0,r} = \tfrac{1}{4} (v_2 - v_1)\, \sigma_{yy} \tag{9.2.10}$$

Since the difference in Poisson's ratios is on the order of 0.1, the shear stress is quite small. We conclude that the effect of a difference in Poisson's ratios is much smaller than that of a difference in moduli.

The complex stress field and mode mixing at a bimaterial interface with different moduli and Poisson's ratios have been the subjects of many studies and have been reviewed by Suo and Hutchinson [54]. There is no solution yet to the problem of coupling the evolution of an asymmetric deformation zone with the mode-mixing stress field due to asymmetry of E and v, so as to predict the fracture energy and locus of failure. The locus of failure (side 1 or 2) and mode of failure (cohesive or adhesive) appear to depend on several factors, such as the differences in craze stress ($\sigma_2 - \sigma_1$), modulus ($E_2 - E_1$), sample thickness ($h_2 - h_1$), and strength of interface [S (strong) or W (weak)]. In Table 9.1, we summarize the possibilities for the failure mode using small differences (positive, if the stress, modulus, or thickness of 2 is greater than that of 1; negative if *vice versa*) in various combinations of these variables.

9.3 PS/PMMA Interfaces

9.3.1 Strength of PS/PMMA Interfaces

The results of wedge cleavage fracture experiments on PS/PMMA interfaces are shown in Figure 9.4 [32]. Here, G_{1c} is plotted as a function of the relative crack length, a/H. These results show the typical experimental reproducibility between samples prepared, welded, and cut under similar conditions, in this case from two welding experiments at 140 °C for 5000 seconds. Four samples were from one welding block and three were from a different block. The data show good reproducibility, with a standard

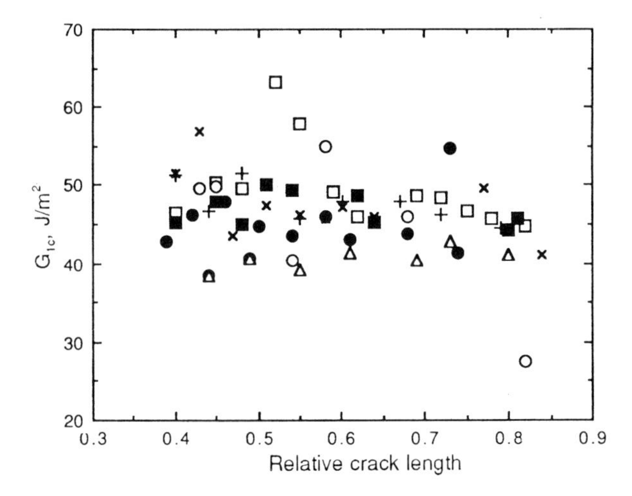

Figure 9.4 Results of wedge cleavage fracture experiments on PS/PMMA interfaces (Foster and Wool).

Table 9.1 Factors Affecting the Mode of Failure and the Locus of Failure in Asymmetric Interfaces

Craze stress, $\sigma_2 - \sigma_1$	Modulus, $E_2 - E_1$	Thickness, $h_2 - h_1$	Interface, strong/weak	Mode, cohesive/ad	Locus, 0,1,2
0	0	0	S	A	0
0	0	0	W	A	0
+	0	0	S	C	1
+	0	0	W	A	0
0	+	0	S	C	1
0	+	0	W	A	0
0	0	+	S	C	1
0	0	+	W	A	0
+	+	0	S	C	1
+	+	0	W	C	1
+	–	0	S	A	0
+	–	0	W	A	0
+	0	+	S	C	1
+	0	+	W	C	1
0	+	+	S	C	1
0	+	+	W	A	0
0	+	–	S	A	0
0	+	–	W	A	0
+	+	+	S	C	1
+	+	+	W	C	1
+	+	–	S	C	1
+	+	–	W	A	0
+	–	+	S	C	1
+	–	+	W	A	0
+	–	–	S	C	2
+	–	–	W	A	0

deviation within 10–15% of the measured value, both along the crack length of the sample and in comparison to neighboring samples from the same welding sheet, as well as from block to block. This indicates uniform if not complete wetting, and shows the expected scatter from a fracture experiment. More important, the G_{1c} values were found to be independent of the relative crack length, which validates the use of Eq 9.2.3 for these experiments. The average G_{1c} value is 45.5 ± 4.1 J/m^2. Samples welded for 600 seconds had an average strength of 42.7 ± 4.9 J/m^2, and the scatter in these data was generally greater.

These values are significantly lower than the virgin values (G^*) for either the PS or PMMA. G^* has been measured with a variety of fracture geometries [31, 50, 51, 52]; its value depends on molecular weight up to some upper limiting molecular weight, M* (Chapter 8). For PS this critical molecular weight is 100,000–250,000, depending on the fracture test [9, 52]. At higher rates generally used with the Charpy impact and compact tension geometries, the strength above the critical molecular weight is approximately 600–1000 J/m^2. Our results are also much higher than predicted by the Griffith theory of fracture, where G_{1c} is twice the surface energy ($\Gamma_{PS} \approx 0.04$ J/m^2).

A comparison of the magnitude of the plateau G_{1c} values of Figure 9.4 with theoretical predictions of the work of adhesion, W_{12}, provides further support for the interpenetration hypothesis. The thermodynamic work of adhesion, W_{12}, is given as

$$W_{12} = \Gamma_1 + \Gamma_2 - \Gamma_{12} \tag{9.3.1}$$

where Γ_1 and Γ_2 are the surface tensions of phase 1 and 2, and Γ_{12} is the interfacial tension. Substitution of appropriate values for PS and PMMA into Eq 9.3.1 gives W_{12} of 0.08 J/m^2 at 25 °C. The thermodynamic work of adhesion is about 500 times smaller than the measured G_{1c} values. Analyses based on van der Waals dispersion forces between two infinite plates [36] or on acid–base interactions [55, 56] gave similar results [36].

Recently, Tirrell and co-workers [53], using the surfaces forces apparatus (SFA) designed by Israelachvili, provided the first direct evidence for the correctness of Eq 9.3.1. The SFA was used to determine the work of adhesion, W_{12}, for a symmetric PET/PET interface and for fluorinated PET/normal PET interfaces, and W_{12} was compared to the surface tension values, Γ.

It is clear that the above classical estimates of the work of adhesion do not adequately describe the plateau strengths shown in Figure 9.4. While they are not conclusive proof of chain interpenetration at the PS/PMMA interface, our results support the interpenetration hypothesis, which predicts that

$$G_{1c} \sim 1/\chi \tag{9.3.2}$$

This relation is further explored in the next section.

9.3.2 Temperature Dependence of PS/PMMA Strength

G_{1c} of a series of PS/PMMA samples (Figure 9.5) welded at four different temperatures between 120 °C and 150 °C was independent of temperature, with an average value of 42.5 ± 4.5 J/m^2 [32]. This accords with Case II, Eq 9.1.13. Furthermore, by comparing G_{1c} at two different temperatures, we can predict the difference in χ over the temperature range (T_1, T_2) from

$$G_{T_1}/G_{T_2} = \chi_{T_2}/\chi_{T_1} \tag{9.3.3}$$

Our experimental results would cause us to predict little or no change in χ. Russell *et al.* measured χ for a PS/PMMA diblock system within this temperature range [57, 58]. They obtained

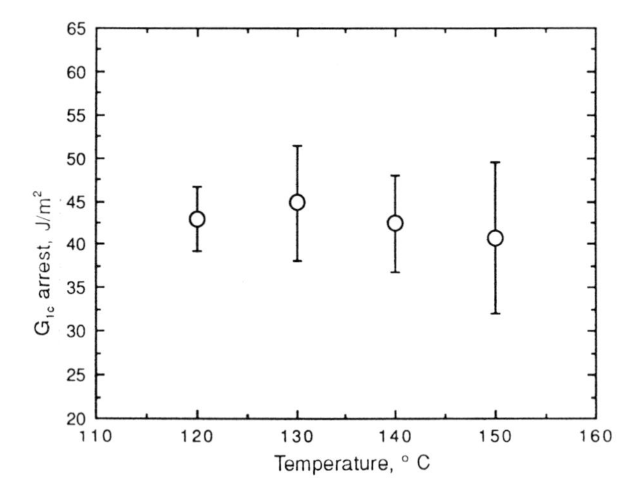

Figure 9.5 G_{1c} of a series of PS/PMMA samples welded at four different temperatures between 120 °C and 150 °C (Foster and Wool).

$$\chi = 0.0284 + 3.902/T \tag{9.3.4}$$

where T is the temperature in degrees Kelvin. The Flory–Huggins interaction parameter, χ, changes by only 2% in the temperature range of 120–150 °C. This change is well within the experimental value we determined for the value of G_{1c} of 42.5 ± 4.5 J/m^2.

9.3.3 Thickness of PS/PMMA Interfaces

We can calculate the interpenetration depth as a function of temperature by substituting Eq 9.3.4 into Eq 9.1.1. We find this depth is between 27.1 and 27.4 Å for the temperature range of 120–150 °C. Fernandez, Higgins, *et al.* [59] report, on the basis of neutron reflection experiments, that the interfacial thickness of a PS and PMMA sample heated for 6 hours at 120 °C was no greater than 20 ± 5 Å. However, Russell *et al.*, also using neutron reflection techniques, obtained $d_\infty \approx 50$ Å for this temperature range [57, 58].

As shown earlier, we can combine the PS/PMMA fracture results with the virgin PS fracture results and calculate d_∞ with Eq 9.1.8. By substituting the fracture strength of virgin PS, measured with compact tension, of $G_{1c}^* = 1000$ J/m^2 at a number average molecular weight, M_n, of 250,000, and radius of gyration R_g = 138 Å, we obtain $d_\infty \approx 58$ Å.

9.3.4 Molecular Weight Dependence of PS/PMMA Strength

The fracture energy of PS/PMMA interfaces welded at 132 °C is shown in Figure 9.6 as a function of welding time for three PS molecular weight distributions, listed in Table 9.2 [33].

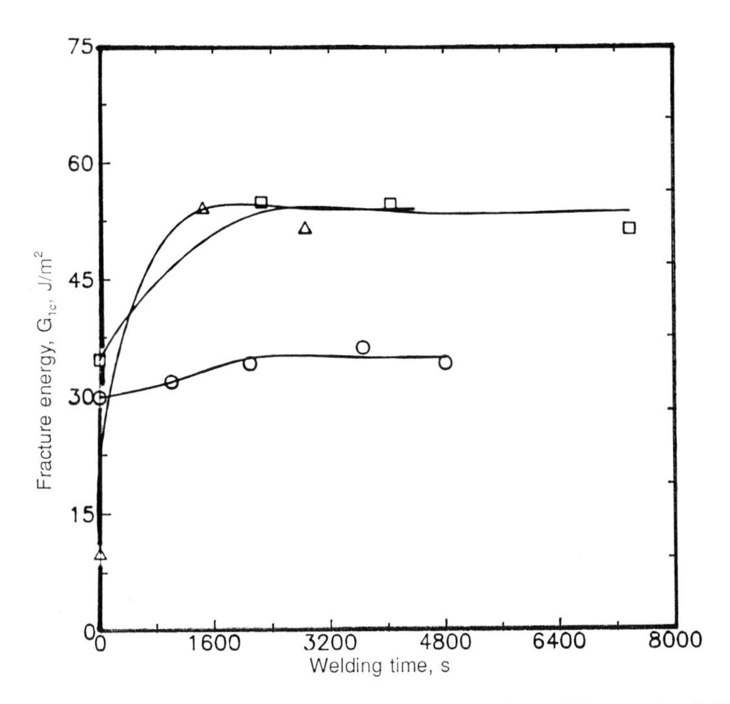

Figure 9.6 The effect of the molecular weight of PS on the weld strength of PS/PMMA interfaces. Key: ○ (circles), M_n = 132,000, P = 2.00; □ (squares), M_n = 128,000, P = 1.05; △ (triangles), M_n = 575,000, P = 1.06. T_w is 132 °C. (Willett and Wool).

Table 9.2 Properties of PS and PMMA

	M_n	M_w	Modulus, psi	T_g
PS	132,000	302,000	390,000	105 °C
PS	128,000	134,000	390,000	106 °C
PS	575,000	609,500	390,000	106 °C
PMMA	62,000	120,000	390,000	108 °C

The results are typical of all the incompatible pairs studied. The strengths at zero welding time are due to the initial wetting treatment. With time, the strength increases to a plateau G_{1c} value (about 50 J/m^2) and remains constant thereafter. The equilibrium strength is about a tenth of the fracture energy (500–1000 J/m^2) of either bulk polymer, but much greater than the work of adhesion, discussed below.

No difference is seen (Figure 9.6) between the plateau strength of the monodisperse molecular weight (128K and 575K) PS samples. However, the polydisperse (132K) PS sample has a slightly lower strength, possibly due to molecular weight segregation effects at the interface. The time to attain the plateau strength is about 1600 seconds, and appears not to depend on molecular weight. These experiments, unlike our symmetric interface studies, were not designed to analyze the dynamics of strength development, but a comment is appropriate. The dynamics of strength development at interfaces is controlled by two convoluted time-dependent functions, wetting and diffusion [8]. Unless we are careful to minimize the wetting function, the contribution of the diffusion function is masked by the convolution process.

For symmetric PS/PS interfaces in SIMS experiments, we have shown in Chapter 5 that the interface thickness, d (note that $d = 2X$), increases with time t as

$$d(t) = 2R_g(t/T_r)^{1/4} \tag{9.3.5}$$

If we assume that the interface dynamics is largely dominated by reptation, then the interface thickness d_∞ develops at a time t_∞ derived from Eq 9.3.5, as

$$t_\infty = (d_\infty/2R_g)^4\, T_r \tag{9.3.6}$$

Since $R_g \sim M^{1/2}$ and $d_\infty \sim M^0/\chi^{1/2}$, the molecular weight dependence of the reptation-controlled equilibration time is $t_\infty \sim M$. If this were indeed the case, we would expect a stronger molecular weight dependence of the time to achieve the plateau strength in Figure 9.6. Also, substituting typical values for $d_\infty \approx 50$ Å, $R_g \approx 100$ Å and $T_r \approx 3000$ s for $M = 132{,}000$ at 125 °C, we obtain $t_\infty \approx 10$ s, which is much shorter than the observed welding time of about 2,000 seconds.

If d_∞ is of the order of the tube diameter d_t, the diffusion is dominated by Rouse dynamics of the entanglement segments of molecular weight M_e. In that case, Eq 9.3.6 is replaced by $t_\infty = (d_\infty/d_t)^4\tau_e$. Here, the tube diameter $d_t \sim M_e^{1/2}$, $\tau_e \sim M_e^2$, and $d_\infty \sim 1/\sqrt{\chi}$, so that $t_\infty \sim 1/\chi^2$, which is independent of molecular weight. This result is more consistent with the Helfand theory and limited experimental observations in which no molecular weight dependence was observed. Thus, interpenetration can be achieved by small fluctuations of entanglement lengths near the interface, as discussed in Chapter 3. The entanglement relaxation time is related to the reptation time by

$$\tau_e = (M_e/M)^3\, T_r/3 \tag{9.3.7}$$

Thus, if $M_e = 18{,}000$, $M = 132{,}000$ and $T_r = 3000$ s, then $\tau_e \approx 2.5$ s at 125 °C. This time typically is less than the experimental wetting times.

The case of shear at the interface between two incompatible polymers A and B was examined by Brochard-Wyart, de Gennes, and Troian [60]. If A and B are not entangled (d_∞ is smaller than $N_e^{1/2}$), they expect a weak Rouse friction with viscosity η_{RO} that is size dependent, as

$$\eta_{RO} = \eta_1 (d_\infty / b)^2 \tag{9.3.8}$$

where b is the length of a statistical segment and η_1 is the viscosity of monomers. For very weak interfaces at $d_\infty \approx b$, $\eta_{RO} \rightarrow \eta_1$, and when $d_\infty \approx N_e^{1/2} b$, $\eta_r \rightarrow \eta_1 N_e$, where N_e is the number of monomers in the entanglement molecular weight. If V is the sliding velocity, the stress can be represented by

$$\sigma = \eta_r V / d_\infty \tag{9.3.9}$$

where V/d_∞ is the shear rate. Combining this with Eq 9.3.8, we predict that $\sigma \sim d_\infty$, and the energy to create unit fracture surface area in shear is $G_{III} \sim \sigma^2$, which gives

$$G_{III} \sim d_\infty^2 \tag{9.3.10}$$

This is similar to the Mode I fracture case for disentanglement discussed above.

The probability for entanglements, f, in this region is given by de Gennes as

$$f = \exp(-N_e \chi) \tag{9.3.11}$$

The entanglements should suppress the slippage whenever $N_e \chi$ is greater than ln $(N^3/N_e^{5/2})$ [60].

9.3.5 XPS Analysis of PS/PMMA Interfaces

While the experimental results of the previous section are in good agreement with Helfand's model of incompatible polymer interfaces, a complete understanding of the mechanical properties of these interfaces requires knowledge of the actual interface breakdown process and subsequent crack growth. If fracture is dominated by chain pullout, one might expect the interpenetrated region to completely disentangle during fracture, resulting in chemically pure surfaces, that is, with no residue of one polymer on the other surface. Analysis of the fracture surface chemistries reveals whether fracture is adhesive (at the interface) or cohesive (in one or the other bulk), or a combination of the two modes.

Spectra from 250 eV to 550 eV for PMMA fracture surfaces welded to PS at 125 °C and 140 °C are shown in Figure 9.7. The C_{1s} peak is near 285 eV and the O_{1s} peak is near 540 eV. The O_{1s} peaks of the fracture surfaces are clearly reduced relative to C_{1s} peaks, indicating an increase in carbon content. O/C ratios for the spectra of the two welded samples are 0.26 and 0.28. In contrast, the O/C ratio of as-molded PMMA,

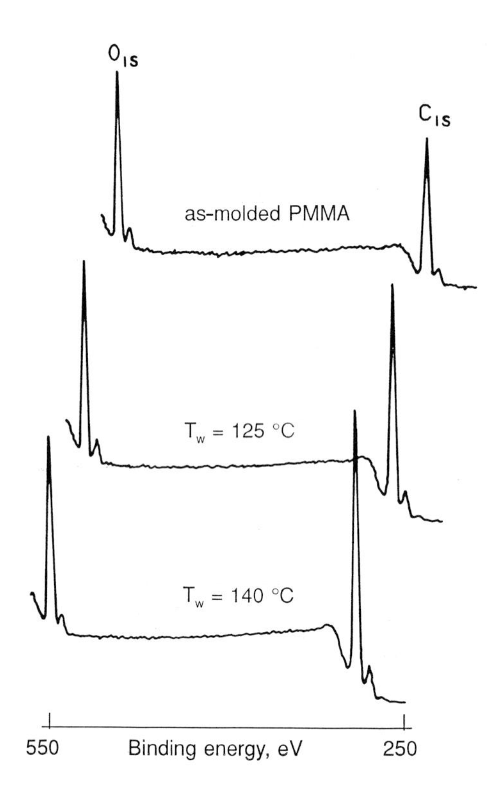

Figure 9.7 XPS spectra of PMMA fracture surfaces welded to PS: (top) as-molded PMMA; (middle) T_w = 125 °C; (bottom) T_w = 140 °C. The binding energy range is 250 to 550 eV (Willett and Wool).

shown as the top curve in Figure 9.7, is 0.38, in good agreement with the theoretical value of 0.40 (2 oxygen atoms and 5 carbon atoms per monomer).

O/C ratios measured on the corresponding PS fracture surfaces are on the order of 0.01 or less [33]. The PS fracture surfaces are indistinguishable from as-molded PS surfaces. Foster and Wool [32] obtained similar results in a related study of PS/PMMA interfaces. These results show that crack growth at PS/PMMA interfaces is both cohesive and adhesive in nature. Cohesive fracture occurs only in the PS side to any extent that XPS analysis can detect. If the fracture were completely adhesive, we would expect an O/C ratio of about 0.38 for the PMMA side; if the fracture were completely cohesive (in PS), we would expect O/C ≈ 0. Thus, the approximate relative contribution of each failure mode, as estimated by the ratio 0.38/0.28 ≈ 0.74, suggests that the fracture process is more adhesive than cohesive (about 3:1) for a PS/PMMA interface.

Fracture energies measured at the PS/PMMA interfaces are in the range of 40 to 50 J/m^2. The work of adhesion at PS/PMMA interfaces is on the order of 0.01 J/m^2 [33]. Bulk G_{1c} values for PS and PMMA are 500 to 1000 J/m^2, depending on the technique used in the measurement. These differences, coupled with the surface structures observed, suggest limited interdiffusion has occurred. It is unclear, however, how much the cohesive fracture of the PS contributes to the measured fracture energies.

9.3.6 SEM Analysis of PS/PMMA Interfaces

Representative SEM images of a PS fracture surface welded to PMMA at 125 °C and 140 °C are shown in Figure 9.8. The most striking feature of these and other PS surfaces is their similarity, despite their different welding conditions and molecular weights. All surfaces shown are characterized by a series of lines, hundreds of micrometers in length with a well-defined spacing of roughly 50 μm, which are perpendicular to the direction of crack growth. There was very little effect of either welding temperature or molecular weight on the surface features of fractured interfaces.

At higher magnification, these lines are seen to be composed of several ridges of highly deformed material, as shown in Figure 9.9. These ridges exhibited birefringence

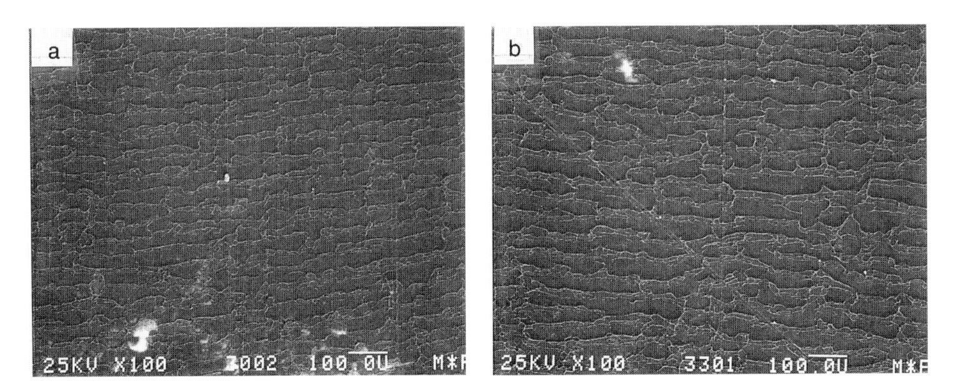

Figure 9.8 SEM micrographs of PS fracture surfaces welded to PMMA: (a) T_w = 125 °C; (b) T_w = 140 °C; bar is 100 μm (Willett and Wool).

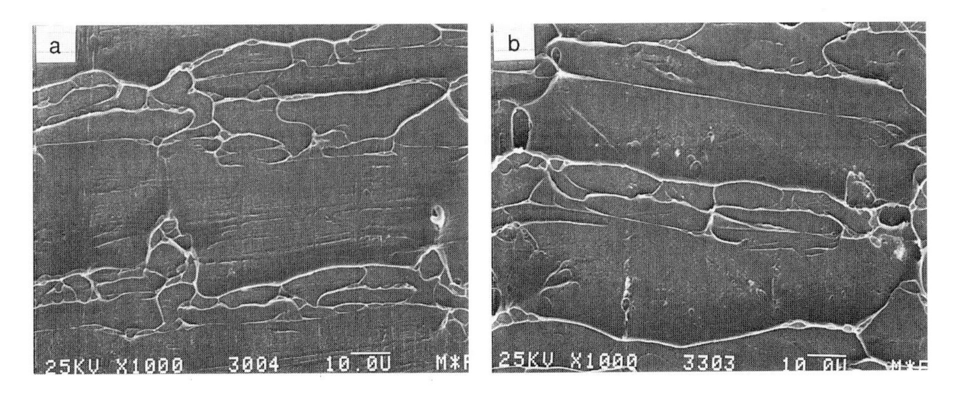

Figure 9.9 Higher magnification SEM micrographs of PS fracture surfaces welded to PMMA: (a) T_w = 125 °C; (b) T_w = 140 °C; bar is 10 μm (Willett and Wool).

when viewed through crossed polars. Since the XPS analysis showed PS fracture surfaces are pure PS, it is clear that these ridges are highly deformed PS and not PMMA residues. These features are indicative of discontinuous "stick–slip" crack growth, a type of unstable crack growth common in polymers loaded at very slow rates [61]. The structures of Figure 9.9 suggest the sticking phase involves the deformation and rupture of the elongated strands, and the slipping phase leads to the relatively featureless regions between the deformed ridges.

On the other hand, the corresponding PMMA fracture surfaces are quite different, as shown in Figure 9.10. In each case, PS residue is evident; the rest of the surface in each photo is rather featureless. The ratio of the featureless to ridge material in Figure 9.10 is about 3:1, which is consistent with the XPS data. However, the resulting fracture energy (about 50 J/m^2), computed in terms of virgin strength/adhesive fracture ratios, does not scale with the ratio of adhesive/cohesive area contributions, probably because the cohesive energy contribution (associated with the ridges) is not the same as that in the virgin material. Comparison of the corresponding PS and PMMA surfaces also indicates that good wetting was achieved.

Examination of the PMMA surfaces at higher magnification suggests the residues of the polydisperse PS occur when the crack leaves the interface, grows in the PS, then returns to the interface. It appears the residues correspond to regions where crack growth was "slip"-like, while the "stick" mode resulted in the featureless regions.

Changing the molecular weight of the PS led to different residue geometry [33]. The features of the 128,000-M_n PS, while showing line-like correlations seen in the polydisperse material, were observed to be more discrete, being in the form of small islands 5 to 10 μm across. When the PS molecular weight is increased to 575,000, the residue on the PMMA is in the form of randomly distributed islands of PS ranging from 1 to more than 10 μm across.

It is clear from these figures that the weld strength of PS/PMMA interfaces is derived from at least two components: the breakdown of the PS/PMMA interface, and

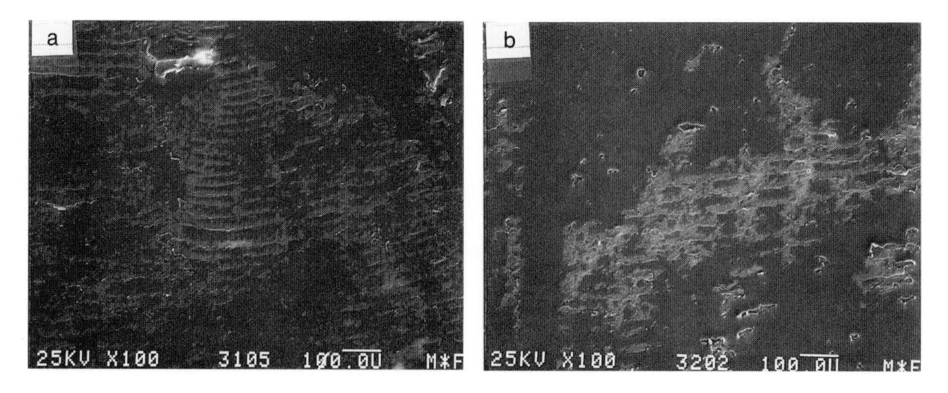

Figure 9.10 SEM micrographs of PMMA fracture surfaces welded to PS: (a) T_w = 125 °C; (b) T_w = 140 °C; bar is 100 μm (Willett and Wool).

the fracture of bulk PS. The first process gives rise to the ridge features seen in Figure 9.9 and the featureless regions of the PMMA surfaces. The second one leaves the PS residues seen on the PMMA fracture surfaces. The similarity between the PS fracture surfaces suggests the same crack growth mechanism is dominant in all cases.

On the basis of the surface analysis, we conclude that there is probably a combination of adhesive and cohesive failure, and that the cohesive failure occurs very close to the interface on the PS side only. This is probably due to the lower craze stress of PS in comparison to PMMA. A surface analysis with XPS determined that some PS was transferred to the PMMA side as a result of the fracture, indicating some cohesive fracture, but no PMMA was transferred to the PS side. An optical dye study found some evidence of polystyrene transferred onto the fracture surface of PMMA, with less than 20% coverage of crazed fracture zones testing positive for PS [32].

9.4 PSAN/PMMA Interfaces

9.4.1 Strength of PSAN/PMMA Interfaces

We have seen that we can vary the interaction parameter, χ, by changing the welding temperature. We now describe experiments in which we change χ chemically, using a copolymer as one of the components of the interface. This approach gives us another way to investigate the role played by χ in the determination of the weld strength of incompatible polymer interfaces [33].

The materials chosen for this study were the same PMMA used in the PS study, and three PSANs of various AN content. Their characteristics are given in Table 9.3.

Table 9.3 Properties of PSAN

% AN	M_n	Modulus, psi	T_g
5.7	136,000	397,000	106
23	75,000	439,000	108
37	68,000	469,000	109

The interaction parameter χ for a copolymer–homopolymer pair can be written [35]

$$\chi = \beta\chi_{13} + (1-\beta)\chi_{23} - \beta(1-\beta)\chi_{12} \tag{9.4.1}$$

where χ_{ij} is the interaction parameter between monomers i and j, 1 and 2 are the comonomers, 3 is the homopolymer, and β is the mole fraction of monomer 1 in the copolymer. Eq 9.4.1 therefore expresses χ in terms of a random mixture of the two polymers.

The quadratic relationship between χ and β has two important consequences. First, χ has a minimum at β^* given by

$$\beta^* = 0.5\,(\chi_{12} + \chi_{23} - \chi_{13})/\chi_{12} \tag{9.4.2}$$

which implies that the composition at β^* has the greatest weld strength. Second, if χ_{12} is greater than $(\chi_{23} + \chi_{13})^2$, the minimum χ is negative, and a range of compositions exists where the two materials are compatible. In this composition range, weld strength is controlled by center-of-mass interdiffusion, which leads to fracture strengths on the order of bulk strengths.

If we substitute the values of χ that Kammer and co-workers found [35] for PMMA, PAN, and PS, we obtain the following relation

$$\chi = (624\beta^2 - 727\beta + 193)/T \tag{9.4.3}$$

where β is the mole fraction of styrene in the copolymer. [The numerical coefficients include all the terms indicated in Eq 9.4.1 for the interaction between each pair of monomers (S/MMA, S/AN, AN/MMA)]. A plot of Eq 9.4.3 is shown in Figure 9.11. Differentiating Eq 9.4.3 with respect to β, one finds the minimum value of χ is at $\beta^* = 0.58$. This value corresponds to a weight percent AN of 27%. The minimum value of χ is –0.05 at 125 °C, which predicts compatibility; in fact, the range of composition that gives compatibility is $0.41 < \beta < 0.75$, which corresponds to AN contents of 15% to 42%. Therefore, the copolymers with 23% AN and 37% AN are predicted to be compatible with PMMA, while the 5.7% AN copolymer should be incompatible, but have a higher weld strength than pure PS with PMMA.

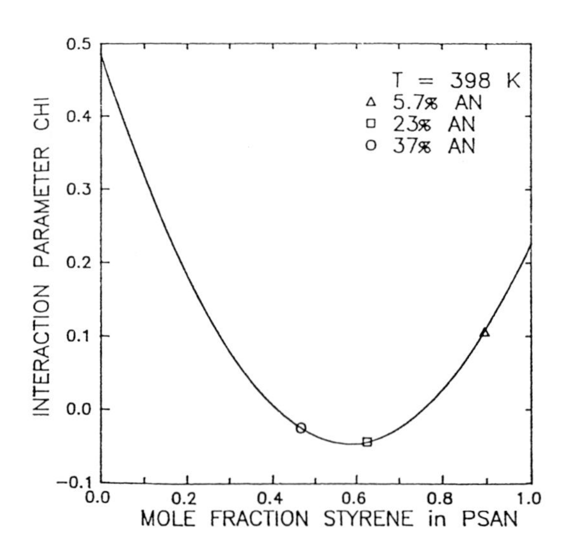

Figure 9.11 The interaction parameter χ for PSAN/PMMA as a function of the mole fraction of styrene in the PSAN. T = 125 °C (Willett and Wool).

The results of welding experiments at 140 °C for PSAN/PMMA interfaces with different AN content are shown in Figure 9.12. The same general features are seen as for PS/PMMA welding. While the 5.7% AN/PMMA pair behaves as expected relative to PS, the 37% AN copolymer's behavior indicates that it too is incompatible with PMMA, in contradiction with Eq 9.4.3. One way to explain this departure from theory is to question the random mixing assumption used in deriving the equation. This assumption neglects any configurational or comonomer sequence effects on χ. Such effects have been studied and shown to affect the validity of Eq 9.4.3 [62, 63]. Since the 37% AN copolymer has a nearly 1:1 molar ratio of styrene to acrylonitrile monomers (see Figure 9.11), these effects are probably important and therefore lead to an incorrect calculation of χ.

The 23% AN copolymer, on the other hand, exhibits completely different welding behavior with PMMA, as shown in Figure 9.13. Not only are the values of G_{1c} much greater at both temperatures, but no plateau seems apparent even after 6000 seconds. These results are consistent with the prediction of compatibility for this pair. If we use the minor chain model for polymer welding and assume a chain-pullout-dominated fracture mechanism, we predict that G_{1c} depends on t as $t^{1/2}$. We see that a plot of G_{1c} against the square root of welding time exhibits reasonable linearity. Similar behavior has been observed by Kausch and co-workers [6, 7] in crack-healing experiments with PMMA and a 25% AN copolymer, which also form a compatible pair (Chapter 12).

The incompatibility–compatibility transition with AN content is confirmed by the optical appearance of cast films of PSAN/PMMA mixtures. We prepared 50/50 (w/w) solutions in methylene chloride at a concentration of 1.0 g/dl. Films were cast after 24 hours of dissolution time. The 23% AN copolymer/PMMA blend formed a clear film with no visible phase separation after being annealed at 125 °C and 140 °C. The other three films (0.0%, 5.7%, and 37% AN) exhibited phase separation in the form of

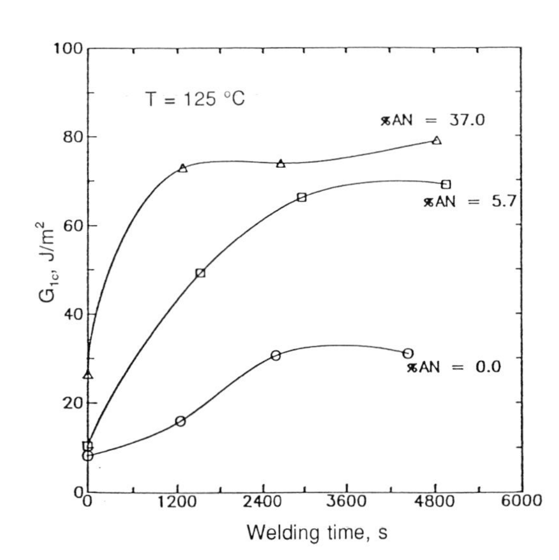

Figure 9.12 Weld strength for three PSAN/PMMA interfaces as a function of time. T_w = 125 °C (Willett and Wool)

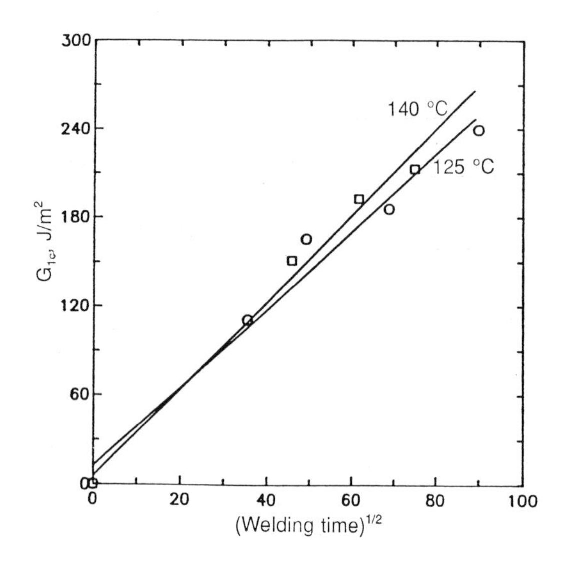

Figure 9.13 Weld strength versus the square root of time for the compatible PSAN/PMMA pair (Willett and Wool).

spherical droplets; annealing did not visibly alter their appearance. These results are in agreement with other studies of the compatibility of PSAN/PMMA blends [38].

9.4.2 XPS Analysis of PSAN/PMMA Interfaces

XPS spectra of the PMMA fracture surfaces welded to PSAN with different AN content at 125 °C are shown in Figure 9.14. N_{1s} peaks near 400 eV are apparent on each PMMA welded fracture surface. The O_{1s} peak on the PMMA welded to PSAN 1 (5.7% AN) is greatly reduced, and virtually no O_{1s} signal is measured on the surface welded to PSAN 2 (23% AN). The N/C values for these spectra are 0.013 and 0.072, respectively. The O/C values are 0.065 and 0.009, substantially less than the as-molded PMMA value of 0.38 (top curve). By comparison, the N/C values for as-molded PSAN 1 and PSAN 2 (not shown in Figure 9.14) are 0.015 and 0.068, and the theoretical values are 0.016 and 0.061, respectively. These results show that fracture of welded PSAN 1/PMMA and PSAN 2/PMMA interfaces is almost purely **cohesive**. With PSAN 3 (37% AN) at 125 °C, nitrogen is detectable (Figure 9.14), but to a much smaller extent than with PSAN 1 and PSAN 2. The O/C value is 0.36, and the N/C value is 0.013. The as-molded N/C value for PSAN 3 (not shown in Figure 9.14) is 0.101; the theoretical value is 0.106.

When the welding temperature is increased to 140 °C, similar results are obtained. N/C values from the PMMA fracture surfaces are 0.012, 0.069, and 0.021 for PSAN 1, PSAN 2, and PSAN 3, respectively. O/C values for these PMMA surfaces are 0.007, 0.010, and 0.333. PSAN 1 and PSAN 2 exhibit purely cohesive failure, while at both temperatures PSAN 3 exhibits a combination of cohesive and adhesive failure.

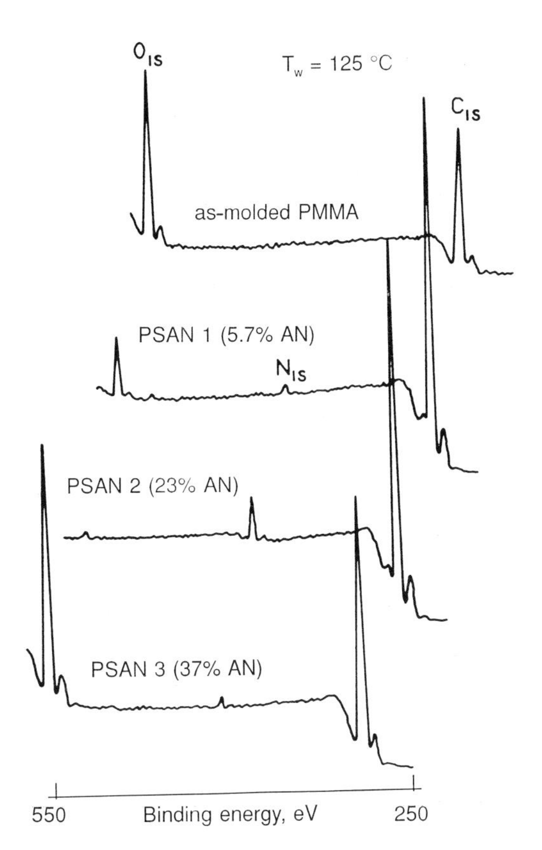

Figure 9.14 XPS spectra of PMMA fracture surfaces molded to PSAN copolymers, for as-molded PMMA; 5.7% AN; 23% AN; 37% AN. T_w = 125 °C (Willett and Wool).

Analysis of the PSAN fracture surfaces indicates that no measurable cohesive fracture occurred in the PMMA. In all three cases, the N/C values measured on PSAN fracture surfaces were equal to the as-molded values, within experimental error. O/C values of 0.01 to 0.03 measured on as-molded PSAN surfaces as well as on PSAN fracture surfaces are due to oxidation of acrylonitrile groups during molding.

The presence of cohesive fracture at all the interfaces studied, particularly with PSAN 1 and PSAN 2, indicates the interface fracture process can be dominated by deformation processes occurring in the bulk phases. Therefore, any comparison between models of interface strength and experimental measurements must account for the role of cohesive fracture. Clearly, interfaces between incompatible polymers can attain sufficient strength to shift the locus of fracture to one of the bulk materials.

9.4.3 SEM Analysis of PSAN/PMMA Interfaces

The corresponding fracture surfaces of the fractured PSAN/PMMA strongly resembled each other, which was not the case for the fractured PS/PMMA discussed in Section

9.3. We saw in Section 9.4.2 that two of the PSAN/PMMA compositions showed cohesive fracture behavior, indicating fracture did not occur at the interface, so we are not surprised to find that the members of each pair of these PSAN/PMMA fracture surfaces closely resemble each other. The fracture surfaces of the 5.7% PSAN/PMMA interfaces showed matching parallel lines 10–20 μm apart and aligned perpendicular to the crack propagation direction (similar to Figure 9.8). This periodicity is indicative of a stick–slip type of crack growth (as discussed in Section 9.3.6). The fracture surfaces of the compatible PSAN/PMMA (23% AN) pair, with a strength of about 200 J/m^2, were similar in appearance, which is not surprising given the completely cohesive fracture that occurred. Again, stick–slip features with a periodicity of roughly 10 μm were observed.

The surfaces of the PSAN/PMMA interfaces with 37% AN are markedly different from the others, and are shown in Figure 9.15. Crack growth in this case was quite discontinuous, snapping rapidly for several hundred micrometers, then growing slowly for a time, then snapping again. This behavior gave rise to the banded structure seen in Figure 9.15. Extensive deformation is seen in the slow growth bands. This deformation becomes much finer in detail as the slow band grows, and finally the crack snaps again, giving a featureless PMMA surface. The fractal-like features seen on the PMMA are PSAN fibrillar edges that were ripped away from the other surface during fracture [33]. In addition, the corresponding bands on the PSAN have the appearance of troughs, indicating that the crack grew into the PSAN.

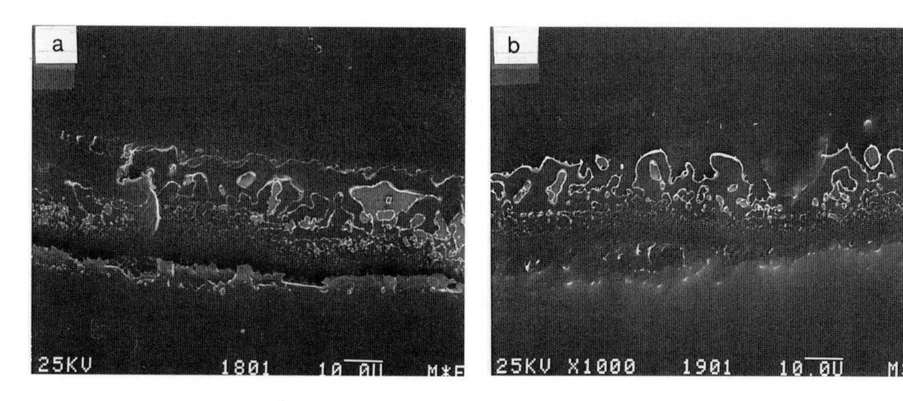

Figure 9.15 SEM micrographs of PSAN/PMMA fracture surfaces: (a) PSAN; (b) PMMA. The AN content is 37%; bar is 10 μm (Willett and Wool).

The influence of both welding temperature and the interaction parameter χ on the strength of PSAN/PMMA interfaces is largely consistent with the interpenetration scheme. However, microscopic analysis of the resulting fracture surfaces indicates fracture is cohesive in two of the three cases studied, which does not accord so well with the interpenetration hypothesis.

9.5 PSAN/PC Interfaces

9.5.1 Strength of PSAN/PC Interfaces

The third polymer pair we considered in these welding experiments was PSAN/PC. This pair is of practical importance as PC/ABS blends have been the subject of some interest in recent years. The same three PSAN copolymers described in Section 9.4 were used in this study; their characteristics are given in Table 9.3. The PC had a molecular weight M_n of about 28,000, a T_g of 150 °C, and a value of Young's modulus of 280,000 psi. The modulus ratio, β, was about 1.5 in this case, and according to Eq 9.2.9, there is an additional shear component in the interface, $\tau_{0,r} \approx 0.17\ \sigma_{yy}$. Of particular interest in this system is the large discrepancy between the values of T_g of the two types of polymers. This difference allows welding to be studied in two regions: between the glass transitions of the two polymers (a 40 °C difference), and above T_g of the PC. Only welding at or below T_g of the PC is discussed here.

As in the PSAN/PMMA case, the AN content allows us to change χ in the PSAN/PC interface without changing the temperature. Therefore, [35, 37]

$$\chi = 556\beta^2 - 697\beta + 314)/T \tag{9.5.1}$$

with β again being the styrene mole fraction in the PSAN. A plot of this function is given in Figure 9.16. In this case, χ is nowhere negative, so we expect all three PSAN copolymers to behave in the incompatible fashion observed in the previous systems. The minimum value of χ occurs at $\beta^* = 0.63$, which corresponds to a copolymer of 23% AN and should give rise to the highest weld strength with PC.

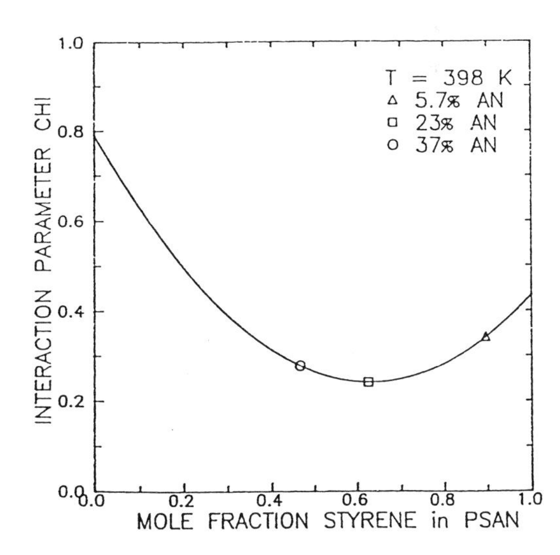

Figure 9.16 The interaction parameter χ for PSAN/PC interfaces versus the mole fraction of styrene in the PSAN. T = 125 °C (Willett and Wool).

Welding curves generated at 125 °C for PSAN/PC interfaces are shown in Figure 9.17. Again we see the familiar shape for incompatible polymer interfaces, with strengths of the same order of magnitude as in the other two systems. These results indicate a certain amount of mixing is taking place, even though the PC is still more than 20 ° below its T_g. This mixing may be a result of a plasticizing action of the more fluid PSAN on the PC during welding.

The composition of the PSAN has a marked effect on the ultimate weld strength, as evidenced by the fact the 23% AN copolymer has a weld strength three times larger than that of the 5.7% AN. This composition dependence clearly indicates the role that the interaction parameter plays in determining the weld strength of polymer interfaces. Similar results were obtained near T_g at 149 °C.

The temperature and composition results are condensed in Figure 9.18. At each temperature, the 23% AN copolymer exhibits the maximum ultimate strength. This composition is also the one which Eq 9.5.1 predicts will have the highest strength. In addition, since from Eq 9.5.1 we calculated that χ(37% AN) is less than χ(5.7% AN), we expect the 37% AN copolymer to exhibit the higher strength of the two; this is indeed observed. A plot of the ultimate weld strength versus $1/\chi$ gives a good linear correlation. Although the calculation of χ involves several assumptions, the correlation between the calculated values and the plateau strength indicates the importance of χ in determining the ultimate strength of incompatible polymer interfaces.

The effect on composition dependence seen for this system is similar to the results of Keitz and co-workers [30], who observed a maximum in fracture strength of PSAN/PC interfaces, measured by lap shear, at an AN composition of about 24%. Their welding conditions were given as 8 minutes at 190 °C, under slight pressure. We note that the optimum PSAN composition applies for welding both above and below T_g of the PC.

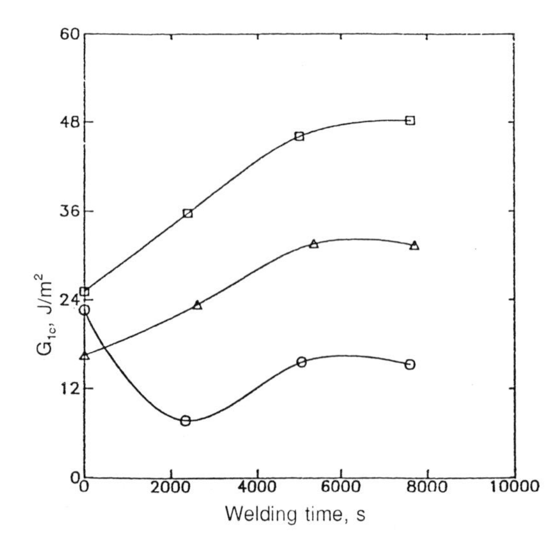

Figure 9.17 Weld strength as a function of time for PSAN/PC interfaces. T_w = 125 °C. Key to symbols: ○, 5.7% AN; □, 23% AN; △ 37% AN (Willett and Wool).

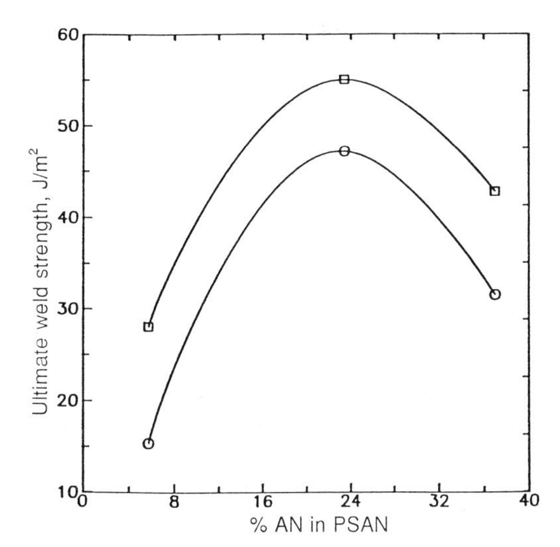

Figure 9.18 The effect of copolymer composition on the plateau strength of PSAN/PC interfaces.
Key to symbols: ○ 125 °C; □ 149 °C (Willett and Wool).

Keitz *et al.* [30] and Mendelson [64] have shown that, when melt-blended at temperatures in the neighborhood of 220 °C, PC and PSANs of various composition are in fact partially miscible. By measuring the change in T_g of each material when blended, relative to the T_g of the pure phase, they were able to calculate the composition of each phase. For several different blend ratios of PC and PSAN, the peak weight fraction of PC in the PSAN phase always occurred at the PSAN composition that gave the highest lap shear strength; the weight fraction of PSAN in PC was somewhat insensitive to the AN content. Typical maximum values for PC in PSAN were 15 to 20%, while only 10% or less by weight PSAN was measured in the PC phase.

9.5.2 XPS Analysis of PSAN/PC Interfaces

Figure 9.19 illustrates the XPS spectra obtained from PC fracture surfaces welded to PSAN at 125 °C, below the T_g of PC (149 °C). As in the PMMA case, an N_{1s} signal is detected for all three PSANs. A decrease in the O/C ratio is also evident relative to the value for as-molded PC. These results indicate cohesive fracture occurred even though the welding temperature was lower than T_g of the PC. Similar results were obtained for PC fracture surfaces welded at T_g (149 °C).

Figure 9.20 shows the N/C ratios plotted against the acrylonitrile content of the PSAN copolymers. At both welding temperatures, the maximum N/C ratio corresponds to PSAN 2 (23% AN), followed by PSAN 3 and PSAN 1. The measured fracture energies follow the same trend, with PSAN 2 having the largest interface strength with PC. In addition, the correlation between the acrylonitrile content and either the N/C ratio or the weld strength accords with the interaction parameter calculations, which

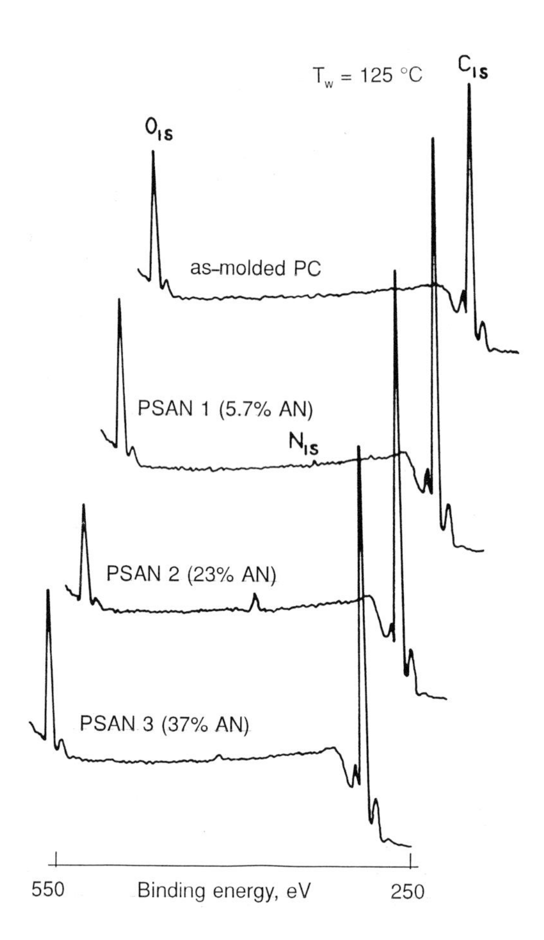

Figure 9.19 XPS spectra of PS fracture surfaces welded to PSAN, for as-molded PC; 5.7% AN; 23% AN; 37% AN. T_w = 125 °C (Willett and Wool).

exhibit a broad minimum near 25% AN. As in the PSAN/PMMA case, cohesive failure in the PSAN/PC system limits the interpretation of the fracture energy to strictly interface interpenetration terms.

9.5.3 SEM Analysis of PSAN/PC Interfaces

Although this pair of polymers is different from the other two systems, since welding took place at or below T_g of one of the components, the PSAN surfaces show characteristics similar to those seen for the PS/PMMA and PSAN/PMMA interfaces. Figure 9.21 illustrates the features seen on the PSAN and PC fracture surfaces. The presence of structures hundreds of micrometers in length with spacings of 2 or 3 μm is indicative of the familiar stick–slip type of crack growth seen in the previous cases. The similarity of the three PSAN fracture surfaces suggests the same underlying mechanism is in operation at all three interfaces during crack growth.

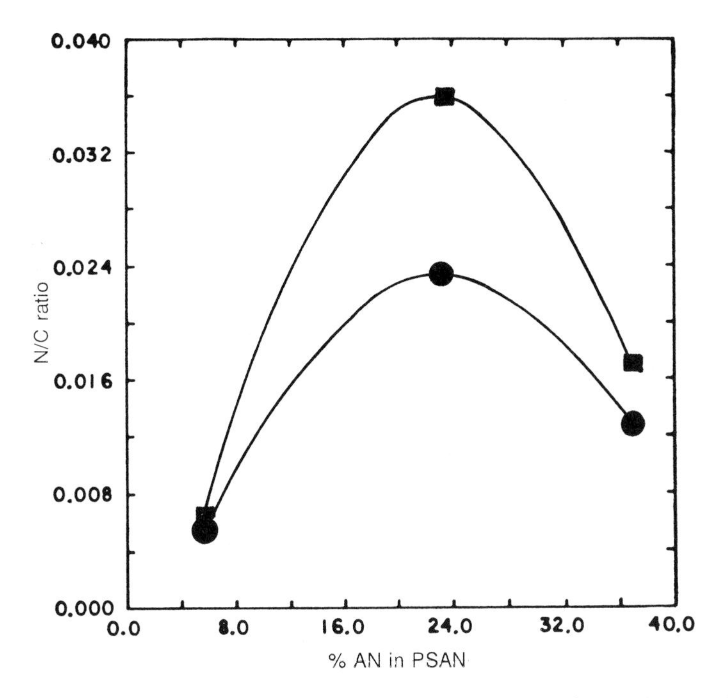

Figure 9.20 Nitrogen/carbon ratios versus %AN measured on PC fracture surfaces of PSAN/PC. Key to symbols: • (circles), 125 °C; ▪ (squares), 149 °C (Willett and Wool).

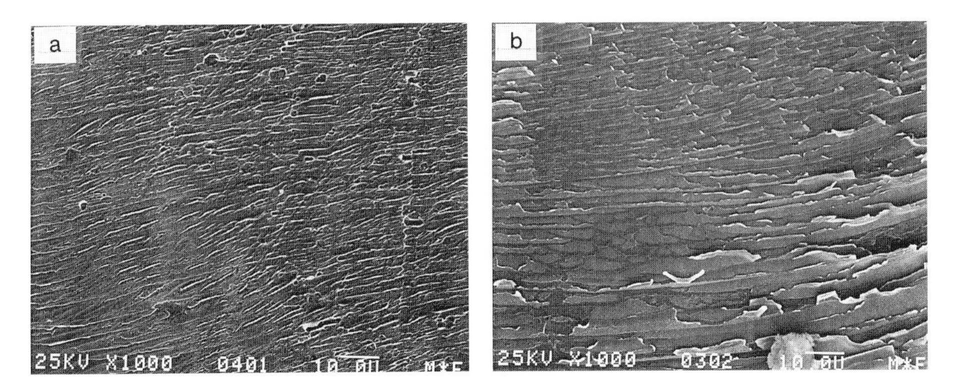

Figure 9.21 SEM micrographs of PSAN/PC fracture surfaces: (a) PSAN; (b) PC. The AN content is 23%; bar is 10 μm (Willett and Wool).

The PC surfaces welded to the 5.7% AN and 37% AN copolymers were relatively featureless, while the PC welded to the 23% AN copolymer has large regions of residual copolymer, as shown in Figure 9.21. In Section 9.5.2 we saw that this pair gave the largest N/C ratio for each welding temperature, in addition to having the highest weld strength. The source of the N_{1s} signal in Figure 9.21 is clearly the residual PSAN.

The orientation of the features seen on the PC surface of Figure 9.21 is quite interesting. The tops of the ridges of PSAN point in the direction from which the crack came, not in the direction of crack growth. Clearly, then, they are not due to the crack simply running off the interface, as was seen in the polydisperse PS/PMMA fracture surfaces. We show in the next section that these features are due to the growth of crazes ahead of the crack before fracture occurs.

The micrographs presented in this section show that the paired interfaces that failed in a largely adhesive manner had very dissimilar fracture surfaces. Only when fracture was cohesive, in two of the PSAN/PMMA pairs, were the fracture surfaces symmetric.

9.6 Role of Crazes in the Fracture of Incompatible Interfaces

9.6.1 Crazes Near Crack Tip

The use of cleavage geometries to characterize isotropic materials is hampered by the fact that cracks run to the specimen surface once they start growing [65–68]. This behavior is due to the interaction between the bending force acting on the crack surfaces and the tensile force normal to the median plane of the sample. The net result is a maximum tensile stress whose normal makes an angle of roughly 80° with the median plane [68]. It is this plane in which fracture occurs, via craze growth and breakdown in glassy polymers like PS and PMMA, and thus the crack once begun immediately runs away from the middle of an isotropic sample. Various techniques have been developed to avoid this problem in cleavage specimens [65–68].

The fracture of the incompatible polymer interfaces in this study is a result of the interplay between three forces, tensile, bending and compressive. When a cleavage specimen is wedge loaded, an additional compressive force acts in the median plane of the sample in the opposite direction to that of the bending force. Therefore, it shifts the direction of the maximum tensile stress, and according to its magnitude, the plane of fracture is also changed.

The importance of the asymmetric properties on the deformation and fracture process is illustrated by the presence of crazes ahead of the crack, which form in the subcritical stage. Figure 9.22 illustrates this phenomenon for two systems, the 23% AN PSAN/PC and the 5.7% AN PSAN/PMMA. In both cases, the crazes begin and grow in the PSAN phase, and attain a length of roughly 20 μm by the time fracture occurs. The crazes do not begin at the interface and grow out into the bulk, but rather begin in a plane roughly 45° to the crack growth direction. Their subsequent curvilinear growth reflects the evolution of the stress field as loading continues.

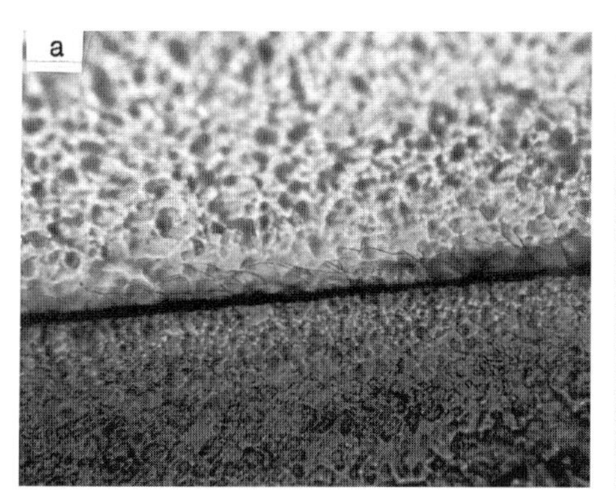

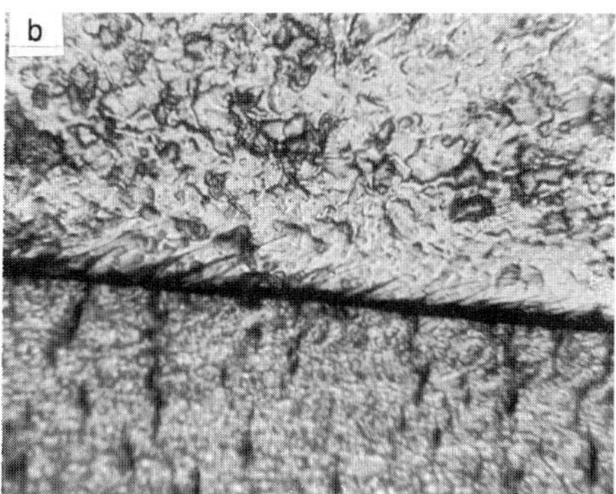

Figure 9.22 Craze formation ahead of crack at incompatible polymer interfaces: (a) PSAN (top)/PMMA (bottom) (5.7% AN); (b) PSAN (top)/PC (bottom) (23% AN). In (a), the crack grows from left to right; in (b), it grows from right to left. Magnification of both photomicrographs is 70× (Willett and Wool).

The tendency of the crazes to become more parallel to the interface is due to the increasing magnitude of the bending force as loading continues and the beams of the sample are spread farther apart. The net effect is that the maximum tensile stress axis more closely approaches the perpendicular to the interface plane. Since the interface is weaker than either bulk, it begins to break down, and crack growth is initiated if the stress is high enough. Stopping the wedge motion at this point allows the system to dissipate energy via crack growth.

If the tensile stress is not great enough to initiate crack growth, further loading increases the bending stress still more. Finally, a stage is reached at which the stress distribution is similar to that of an isotropic sample, and the crack runs away from the interface, once it has begun. This type of failure did indeed occur occasionally with the PSAN/PMMA interfaces that exhibited the highest strengths, with the result that the sample was destroyed.

The presence of crazes in front of the crack tip provides an explanation for the features seen on the PC fracture surfaces welded to the 23% AN PSAN (Figure 9.21). Once crack growth started, the crazes ahead of the crack provided paths for it to leave the interface. As the crack jumped between the crazes, it tore through the uncrazed regions between crazes before returning to the interface, leaving behind the ridges observed.

9.6.2 Role of Entanglements in Crazing

Although we saw no direct evidence of off-axis crazes in the interfaces other than 23% AN PSAN, our understanding of the role of crazes in the fracture process is deepened when we consider the relative tendency of each material to form crazes. Kramer and co-workers have shown that the entanglement density determines whether an amorphous polymer crazes or forms deformation zones [41, 42]. PS, with a relatively low entanglement density, readily crazes, whereas PC, with a much higher density, tends to shear yield. Table 9.4 lists the entanglement densities of each of the materials used in this work (from [39] and [43]). It is significant to note that in all cases, a residue of the polymer with the lower entanglement density is found on the fracture surface of its welding partner, regardless of differences in moduli and glass transition temperatures. The only exception to this statement is the 37% AN PSAN/PMMA pair, but some uncertainty exists in the correct value of M_e for this copolymer. Also of significance is the fact that where crazing was observed, it always happened in the side that had the lower entanglement density (Figure 9.22). This is consistent with the craze stress behaving as $\sigma_c \sim M_c^{-1/2}$, as discussed in Chapter 7.

Table 9.4 Entanglement Densities ν_e (cm^{-3})

Polymer	M_e	ν_e (10^{-19})
PS	19,100	3.3
PSAN (5.7% AN)	17,100	3.7
PSAN (23% AN)	11,000	5.9
PSAN (37% AN)	6,100	10.6
PMMA	8,800	8.2
PC	2,500	29.0

The effect of entanglement density on crazing is one possible explanation for the discrepancy in the normalized N/C ratios of PC fracture surfaces and ultimate weld strength. In Section 9.5, we saw that the PC surface welded to the 5.7% AN PSAN had more PSAN residue than did the PC surface welded to the 37% AN PSAN, even though its interfacial strength was lower. Since the former copolymer crazes more readily, because of a lower entanglement density, craze growth before fracture may produce more pathways for cohesive fracture compared with the latter copolymer. The net result is then a greater amount of the lower AN content copolymer on the PC fracture surface. Three-point bend tests confirmed this relative crazing behavior.

Robertson observed crazes at PS/PMMA interfaces when the two materials were solvent bonded [69]. He concluded that the crazes were primarily in the PMMA, although he could not confirm that. He surmised that "good molecular contact" is all that is needed to achieve an interface whose resistance to crazing is the same as those of the bulk materials. It is clear from the results above, however, that for thermally welded PS/PMMA interfaces, the PS undergoes more deformation during the fracture process. In addition, the combination of low G_{1c} value and residual PS on the PMMA surface indicates that the stress distribution and its evolution also play an important role in what type of bulk deformation occurs. Therefore we can conclude that while "good molecular contact" may be sufficient to induce plastic deformation in the adjacent bulk, it is not sufficient to raise the interface strength to that of the bulk.

9.6.3 Good's Theory of Dissimilar Interfaces

Good [70] examined fracture of dissimilar materials (1 and 2) using the Griffith–Irwin crack theory. He looked at differences in elastic moduli ($\Delta E = E_1 - E_2$) and bulk fracture energy ($\Delta G = G_1 - G_2$), the presence or absence of weak or strong interfacial forces, the presence of flaws, and the thickness (δ_1 or δ_2) of the zone in which the energy dissipation occurred. He categorized the type of fracture (adhesive/cohesive) by the signs of ΔE and ΔG, and the strength of the interface. He predicts:

1. True interfacial fracture (adhesive) occurs only when the interfacial strength is low, and ΔG and ΔE have opposite signs.
2. If ΔG and ΔE are of the same sign, cohesive failure is likely to be the mode of fracture even if the interfacial force is weaker than the bulk cohesive forces.
3. If the interfacial forces are moderately strong, cohesive failure occurs regardless of the signs or relative magnitudes of ΔE and ΔG. The exact locus of failure in this case is determined by ΔE and ΔG.

When combined with the interpenetration picture of incompatible interfaces, these predictions are fairly accurate in describing the behavior observed in all the systems discussed in this work. Weak interfacial forces, corresponding to poor interpenetration, lead to adhesive failure regardless of the signs of ΔE and ΔG; the PS/PMMA interfaces fall into this category. Since fracture is predominantly adhesive at these interfaces, the measured fracture energies directly reflect the role of interpenetration.

Strong interfacial forces lead to cohesive failure for two PSAN/PMMA interfaces. The N/C ratios measured on the PMMA fracture surfaces show that the cohesive failure occurred in the PSAN phase in both cases. Although one would expect, from Good's theory, that failure would occur in the PMMA since it has lower E and G values, the fact that it occurs in the PSAN is most likely because of a larger size of inherent flaws in the PSAN, which leads to preferential crazing in the PSAN. This effect is probably due to the lower entanglement density in the PSANs, as discussed in the next section. The higher strength of these interfaces is due to increased interpenetration, resulting from lower χ values relative to PS. Finally, the PSAN/PC interfaces fall into the same category as PS/PMMA. Here, one would intuitively expect low interfacial forces since welding took place at or below T_g of the PC. In addition, ΔE and ΔG for these materials have opposite signs. This combination gives the conditions that Good predicts are needed for true interfacial failure. For all three of the copolymer combinations, fracture was predominantly adhesive.

9.7 Compatibilizers at Incompatible Interfaces

9.7.1 Diblock Copolymers at Incompatible Interfaces

Creton, Kramer, Hui, Brown, and co-workers examined the role of A–B diblocks on strength development at incompatible A/B interfaces in an elegant study, using the system shown in Figure 9.23 [71–76]. Here, the diblocks are segregated to the interface, the A block on the A side and the B block on the B side. An interfacial region of thickness d_∞ forms. The interfacial thickness for PS/PMMA interfaces with PS–PMMA diblocks was examined by Russell *et al.* [57, 58] using neutron reflection. They found that the equilibrium PS/PMMA interface increased from 50 Å to 75 Å with the addition of diblock. If $G_{1c} \sim d_\infty^2$ holds for the diblock case, an increase in d_∞ by 50% can result in a substantial increase in strength, from 45 J/m^2 to about 100 J/m^2.

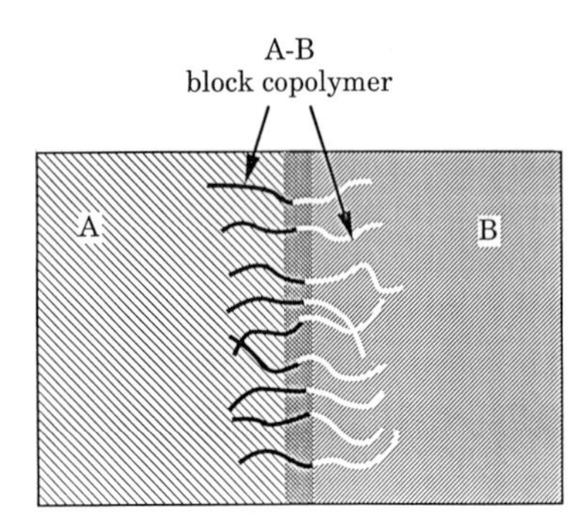

Figure 9.23 Schematic drawing of a layer of A/B block copolymer chains segregated at the interface between A and B homopolymers (courtesy of Creton *et al.* [71]).

In the diblock study by Creton *et al.* [71], several parameters were varied independently: Σ, the number of A–B diblock chains per unit area; N_A, the length of the A block; N_B, the length of the B block; and the χ parameter. In the system they examined A was polystyrene; B was poly(2-vinylpyridine) (PVP); and the diblock was PS–PVP in which the length of the PVP block, N_{PVP}, was varied systematically, and N_{PS} was greater than 280. The main purposes of this study were to explore the effects of entanglements on G_{1c} by varying the PVP block length above and below its M_e value, and to evaluate the relation between G_{1c} and Σ.

9.7.2 Asymmetric Fracture Tests

Figure 9.24 shows the asymmetric DCB cleavage test used by Creton *et al.* [71] to measure fracture toughness. The arrangement is asymmetric, to compensate for the difference in elastic moduli, E_1 and E_2, and differences in craze stress and craze thickening rate, as discussed in the last section. Detailed discussions of this method are in references by Brown [48, 49] and Evans and co-workers [72]. If the arm thickness is controlled and the stored strain energy is balanced, the crack can be directed along the interface. Otherwise, deviation of the crack from the interface and the generation of secondary crazes near the crack tip can give higher apparent strength values.

With a wedge cleavage specimen of different moduli E_1 and E_2, and arm thicknesses h_1 and h_2, as shown in Figure 9.2.4, Kanninen's equation for the equal-arm wedge cleavage (Eq 9.2.3) is modified to the following relation [71],

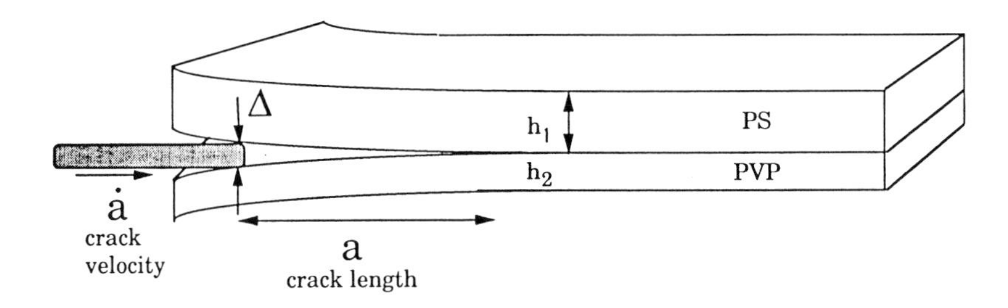

Figure 9.24 Asymmetric wedge cleavage test used by Brown [48, 49].

$$G_{1c} = \frac{3\delta^2 E_1 h_1^3 E_2 h_2^3}{8a^4}\left[\frac{(E_1 h_1^3 C_2^2 + E_2 h_2^3 C_1^2)}{(E_1 h_1^3 C_2^2 + E_2 h_2^3 C_1^2)^2}\right] \tag{9.7.1}$$

where δ is the wedge opening displacement, a is the crack length, and the constants C_1 and C_2 are given by $C_1 = 1 + 0.64\ h_1/a$, and $C_2 = 1 + 0.64\ h_2/a$.

The degree of asymmetry is characterized in terms of the ratio of the crack-opening displacement of the PVP side, Δ_{PVP}, to the total crack-opening displacement Δ, as Δ_{PVP}/Δ. Figure 9.25 shows the effect of asymmetry on the measured fracture energy of PS/PVP interfaces without the diblocks. For the experiments discussed below, the asymmetry value of 0.65 was used (where G_{1c} was at its minimum value). It is interesting to note that the minimum strength is $G_{1c}(0) \approx 1$ J/m^2 for the PS/PVP interface. This interface is very weak and is therefore ideal for evaluating the effects of diblocks on strength.

9.7.3 Effect of Diblock Molecular Weight on G_{1c}

Creton *et al.* measured G_{1c} versus the number of diblock chains per unit area, Σ, for the PS–PVP diblock 1300–173 [71] at a PS/PMMA interface. In this case, the PS block was above its entanglement molecular weight but the PVP was below $N_{e\,PVP}$. The strength was found to increase from $G_{1c}(0) \approx 1.2$ J/m^2 (no diblocks) to a maximum G_{1c} of about 4 J/m^2. The strength remained constant beyond Σ_∞ of about 0.1 chains/nm^2, which corresponds to a saturated brush of diblock across the interface. For all diblock fracture tests below $N_{e\,PVP}$ (about 250), the maximum fracture energy was about 5–10 J/m^2, which is in agreement with data for pure polystyrene in the molecular weight range 14,000–28,000 (Chapter 8), which is also below its M_c (30,000). FRES analysis of the fracture surfaces indicated that the PVP blocks pulled out cleanly from the PVP side while the PS blocks remained anchored. These results are consistent with the nail solution, $G_{1c} = \frac{1}{2}\mu_0 V \Sigma L^2 + G_{Ic}(0)$ (Eq 7.6.12), in which $G_{Ic}(0) = 2S\,\Gamma_{AB}$.

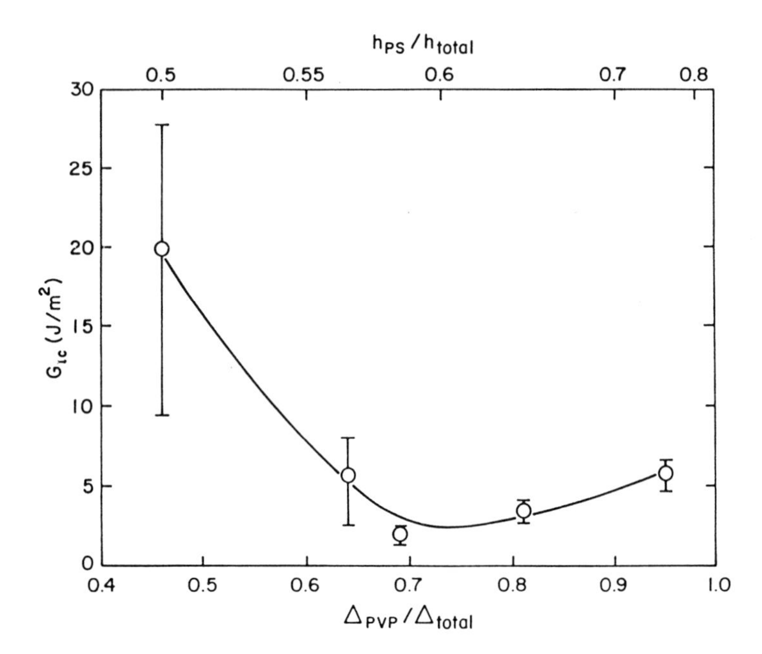

Figure 9.25 Fracture energy as a function of relative thickness (upper abscissa) and of the degree of beam thickness asymmetry, Δ_{PVP}/Δ (lower abscissa; see text) in the wedge cleavage test in Figure 9.24.

When the PVP block was longer than its entanglement length, a large increase in G_{1c} was observed, with values up to 100 J/m^2. Figure 9.26 shows the effect of increasing N_{PVP} on the fracture energy per chain, G_{1c}/Σ. As entanglements are created the fracture energy rises rapidly, analogous to the effect of molecular weight on G_{1c} of symmetric interfaces discussed in Chapter 8. Two fracture mechanisms could be distinguished by Creton *et al.* above N_{ePVP}. At low values of Σ, G_{1c} increased slowly with Σ, as shown in Figure 9.27, and fracture occurred by bond rupture at the PS–PVP junction in the diblock. At larger values of Σ, the interface fracture began with the formation of a stable craze ahead of the crack, followed by rupture of the craze fibrils, resulting in a much higher G_{1c} value, (about 120 J/m^2). At low values of Σ, the fracture energy behaves as

$$G_{1c} \sim \Sigma^2 \qquad \text{for } N_{PVP} >> N_{ePVP} \tag{9.7.2}$$

and at higher values of Σ, the slope approaches 1 as G_{1c} approaches saturation. The change in slope at high Σ values may be due to changes in the packing factor of diblock chains at the interface.

We can make an analogy between the symmetric PS/PS interface with M greater than M_c and this study as follows. Although microscopically the interpenetrated

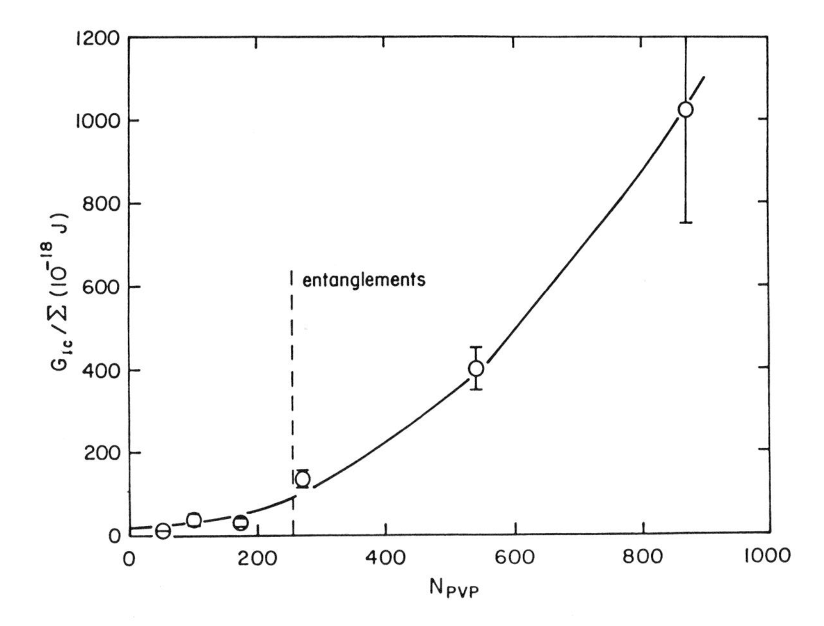

Figure 9.26 Fracture toughness per chain as a function of the degree of polymerization of the PVP block. The full line is drawn to guide the eye (courtesy of Creton *et al.* [71]).

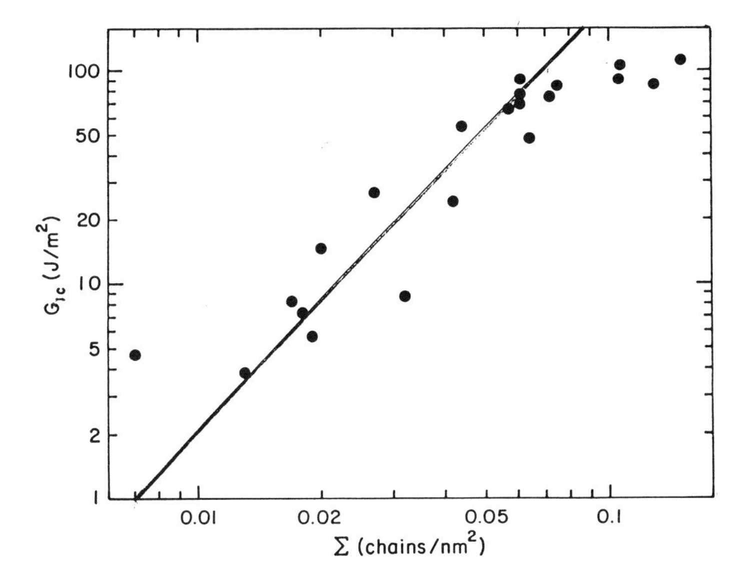

Figure 9.27 Fracture energy, G_{1c} for the 800 PS/870 PVP diblock copolymer as a function of the number of diblocks per unit area of interface. The solid line has a slope of 2 on this log-log plot (courtesy of Creton *et al.* [71]).

random-coil structure of the symmetric interface is very different from the PS/PVP interface with diblocks (Figure 9.23), the crazing, bond rupture, and disentanglement behavior may provide a basis for comparison. At short healing times, the minor chains of interpenetration contour length $\langle l \rangle$ represent the PVP chain segments in the diblock, and the number of minor chains intersecting the interface, n, is equivalent to the areal density of diblock chains, Σ.

From Chapters 2 and 7, we have the dependence of G_{1c} on $\langle l \rangle$, n, and M as

$$G_{1c}(l) = G_\infty \langle l \rangle / L, \quad G_c(n) = G_\infty (n/n_\infty)^2, \text{ and } \quad G_\infty(M) \sim G^*(M/M_c)[1 - (M_c/M)^{1/2}]^2$$

This gives the dependence of G_{1c} on Σ and N_{PVP} as

$$G_{1c}(\Sigma) = G_\infty (\Sigma / \Sigma_\infty)^2 \tag{9.7.3}$$

where G_∞ is the fracture energy at saturation (Σ_∞). G_∞ is determined by

$$G_\infty (N_{PVP}) \sim G^* (N_{PVP} / N_{ePVP}) [1 - (N_{ePVP} / N_{PVP})^{1/2}]^2 \tag{9.7.4}$$

in which G^* is the maximum fracture energy at $\Sigma = \Sigma_\infty$ and N_{PVP} much larger than N_{ePVP}. Eq 9.7.4 well describes the data in Figure 9.26. Since this function is inhomogeneous, it does not behave like a power law of the type $G_\infty \sim N_{PVP}{}^\beta$, where β is some exponent, except at very high N_{PVP} values where β is about 1. At lower β values near N_{ePVP}, the apparent exponent β is about 2–3.

Thus, the welding model predicts that $G_{1c} \sim \Sigma^2$ at constant N_{PVP}, which is in excellent agreement with the data in Figure 9.27, where the solid line has a slope of 2 at low values of Σ (less than 0.06/nm^2). In this range of Σ values, we assume that the PVP segments have random-walk-like conformations, $\langle l \rangle$ and n are related by $\langle l \rangle / L = (n/n_\infty)^2$, and the packing factor of chains is controlled by $n/M^{1/2} = 1$. With increasing Σ, the diblock chains become more crowded and extended in a manner resembling a brush. For extended chains, and also for unentangled chains at N_{PVP} less than N_{ePVP}, the dependence on Σ changes to a linear dependence

$$G_{1c} = G_\infty \Sigma / \Sigma_\infty \tag{9.7.5}$$

which is in good agreement with experiment.

9.7.4 Effect of Diblock Layer Thickness on G_{1c}

When A–B symmetric diblock chains of length N are applied as a layer of thickness h between two slabs A and B of incompatible polymers, we enquire as to the effect of h and N on G_{1c}. Using equation 9.7.3, we have $\Sigma \sim h/N$ and $\Sigma_\infty \sim N^{-1/2}$, so that

$$G_{1c} \sim G_\infty h^2 / N \tag{9.7.6}$$

where G_∞ is the maximum fracture energy attainable. For a given molecular weight, the optimal layer thickness h_∞ is obtained from Eq 9.7.6 at $G_{1c} = G_\infty$ as

$$h_\infty \sim N^{1/2} \tag{9.7.7}$$

Thus, Eq 9.7.6 may be given as the dimensionally correct scaling law

$$G_{1c}(h) = G_\infty (h/h_\infty)^2 \tag{9.7.8}$$

Results of experiments by Brown and co-workers [75, 76] can be compared with these relations. They examined the effect of molecular weight N and thickness h of PS–PMMA diblocks on the fracture energy of PS/PMMA interfaces. Their results are shown in Figure 9.28, where G_{1c} is plotted versus copolymer film thickness for diblocks with molecular weights in the range 84K to 900 K. We observe a significant effect of diblock thickness on G_{1c}. For each molecular weight, the strength increases from an initial value of about 15 J/m^2 to a saturation value of $G_\infty \approx 600$ J/m^2. Although the data have considerable scatter, if we analyze the points in the range $G_{1c} < 500$ J/m^2 and $h < 50$ nm, we obtain the following results.

At $G_{1c} = 300$ J/m^2, a crossplot of h versus N gives

$$h_{300} = 0.043\, N^{0.51} \tag{9.7.9}$$

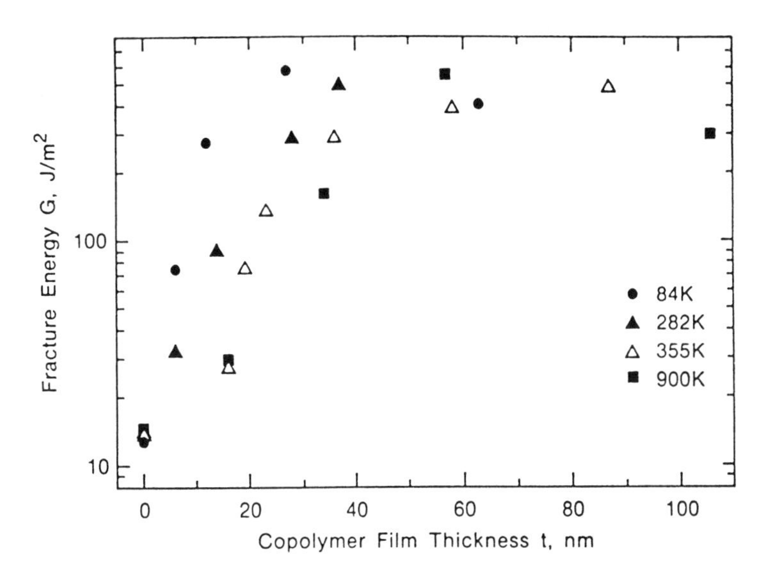

Figure 9.28 Interface fracture energy versus thickness h of the PS/PMMA diblock layer. The symmetric diblock molecular weights are indicated (courtesy of Brown *et al.* [75]).

which is in good agreement with Eq 9.7.7.

We proceed to evaluate the prediction $G_{1c} \sim h^2$. A plot of log G_{1c} versus log h gives approximate slopes of 2.4 (84K), 1.63 (282K), 1.92 (355K), and 2.39 (900K); the average slope is 2.08. The approximation involves our ability to "eyeball" the data from Figure 9.28 (see the Exercise below). Thus, combining Eq 9.7.9 with Eq 9.7.8, we obtain the quantitative solution for G_{1c} as a function of h and N as

$$G_{1c} \approx 162 \times 10^3 h^2/N \qquad (9.7.10)$$

which applies in the range G_{1c} less than 500 J/m^2.

Brown and co-workers note that when the annealing temperature is increased at constant welding time (30 minutes), the fracture energy decreases, which they attribute to a lamellar type of ordering within the diblock phase. Thus, the optimal properties are obtained when the diblock chains remain as random coils and promote crazing through entanglements.

Exercise

Analyze the experimental data of Brown et al. (Figure 9.28) for the effect of PS–PMMA diblock layer thickness h and molecular weight N on G_{1c}, and determine the validity of the scaling law $G_{1c} = G_\infty (h/h_\infty)^2$, where $h_\infty \sim N^{1/2}$.

9.8 References

1. S. Prager and M. Tirrell, *J. Chem. Phys.* ***75***, 5194 (1981).
2. D. Adolf, M. Tirrell, and S. Prager, *J. Polym. Sci., Polym. Phys. Ed.* ***23***, 413 (1985).
3. Y.-H. Kim and R. P. Wool, *Macromolecules* ***16***, 1115 (1983).
4. P.-G. de Gennes, *C. R. Hebd. Seances Acad. Sci., Ser. B,* ***291***, 219 (1980).
5. R. P. Wool, *Rubber Chem. Technol.* ***57*** (2), 307 (1984).
6. K. Jud and H. H. Kausch, *Polym. Bull.* ***1***, 697 (1979).
7. K. Jud, H. H. Kausch, and J. G. Williams, *J. Mater. Sci.* ***16***, 204 (1981).
8. R. P. Wool and K. M. O'Connor, *J. Appl. Phys.* ***52***, 5953 (1981); R. P. Wool and K. M. O'Connor, *J. Polym. Sci., Polym. Lett. Ed.* ***20***, 7 (1982).
9. R. P. Wool, B.-L. Yuan, and O. J. McGarel, *Polym. Eng. Sci.* ***29***, 1340 (1989).
10. C. M. Roland and G. G. Bohm, *J. Appl. Phys.* ***29***, 3803 (1984).
11. J. D. Skewis, *Rubber Chem. Technol.* ***39***, 217 (1966).
12. W. G. Forbes and L. A. McLeod, *Trans. Inst. Rubber Ind.* ***30*** (5), 154 (1958).
13. P.-G. de Gennes, *C. R. Seances Acad. Sci., Ser. 2,* ***292***, 1505 (1981).
14. F. Brochard, J. Louffroy, and P. Levinson, *Macromolecules* ***16***, 1683 (1983).
15. G. Foley and C. Cohen, *J. Polym. Sci., Polym. Phys. Ed.* ***25***, 2027 (1987).
16. E. J. Kramer, P. F. Green, and C. J. Palmstrom, *Polymer* ***25***, 473 (1983).
17. H. H. Kausch, *Polymer Fracture*; Springer-Verlag, Berlin; 2nd ed., 1987.
18. D. R. Paul and S. Newman, Eds., *Polymer Blends Vol. 1*; Academic Press, New York; 1978.

19. E. Helfand and Y. Tagami, *J. Chem. Phys.* ***56***, 3592 (1972); E. Helfand and A. M. Sapse, *J. Chem. Phys.* ***62***, 1327 (1975).
20. E. Helfand, *Macromolecules* ***25***, 1676 (1992).
21. H.-W. Kammer, *Z. Phys. Chem. (Leipzig)* ***258***, 1149 (1977).
22. K. Binder and H. L. Frisch, *Macromolecules* ***17***, 2928 (1984).
23. A. D. McLaren, *J. Polym. Sci.* ***3***, 652 (1948).
24. S. S. Voyutskii, A. I. Shapalova, and A. P. Pisarenko, *Colloid J. USSR (Engl. Transl.)* ***19***, 279 (1957).
25. S. S. Voyutskii, *Rubber Chem. Technol.* ***33***, 748 (1960).
26. S. S. Voyutskii and V. L. Vakula, *J. Appl. Polym. Sci.* ***7***, 475 (1963).
27. Y. Inegar and D. E. Erickson, *J. Appl. Polym. Sci.* ***11***, 2311 (1967).
28. S. S. Voyutskii, *Autohesion and Adhesion of High Polymers*; Vol. 4 in series *Polymer Reviews*, H. F. Mark and E. H. Immergut, Eds.; Wiley–Interscience, New York; 1963.
29. S. S. Voyutskii, S. M. Yagnyatinskaya, L. Ya. Kaplunova and N. L. Garetovskaya, *Rubber Age* ***105*** (2), 37 (1973).
30. J. D. Keitz, J. W. Barlow, and D. R. Paul, *J. Appl. Polym. Sci.* ***29***, 3131 (1984).
31. Robertson, R. E., "The Fracture Energy of Low Molecular Weight Fractions of Polystyrene"; chapter in *Toughness and Brittleness of Plastics*, R. D. Deanin and A. O. Crugnola, Eds.; Vol. 154 in Advances in Chemistry Series; American Chemical Society, Washington, DC; 1976; p 89.
32. K. L. Foster and R. P. Wool, *Macromolecules* ***24***, 1397 (1991).
33. J. L. Willett and R. P. Wool, *Macromolecules* ***26***, 5336 (1993).
34. S. Wu, *Polymer Interface and Adhesion*; Marcel Dekker, New York; 1982.
35. J. Kressler, H. W. Kammer, and K. Klosterman, *Polym. Bull.* ***15***, 113 (1986).
36. J. L. Willett, *Strength Development at Incompatible Polymer Interfaces*; Ph.D. Thesis, University of Illinois, Urbana, IL; 1988.
37. M. Suess, J. Kressler, and H. W. Kammer, *Polymer* ***28***, 957 (1987).
38. M. E. Fowler, J. W. Barlow, and D. R. Paul, *Polymer* ***28***, 1177 (1987).
39. S. Wu, *Polymer* ***28***, 1144 (1987).
40. T. Hashimoto, H. Jinnai, H. Hasegawa and C. C. Han, *Physica A (Amsterdam)* ***204***, 261 (1994); and private communication with T. Hashimoto, Kyoto, Nov. 1993.
41. E. J. Kramer, *Polym. Eng. Sci.* ***24***, 761 (1984).
42. A. M. Donald and E. J. Kramer, *J. Mater. Sci.* ***16***, 1967 (1981); *ibid.* ***16***, 2977 (1981); *ibid.* ***17***, 1871 (1982).
43. J. D. Ferry, *Viscoelastic Properties of Polymers*; John Wiley & Sons, New York; 3rd ed., 1980.
44. H. Zhang and R. P. Wool, *Macromolecules* ***22***, 3018 (1989).
45. P. J. Flory, *Principles of Polymer Chemistry*; Cornell University Press, Ithaca, NY; 1953.
46. K. Cho and A. N. Gent, *J. Adhes.* ***25***, 109 (1988).
47. M. F. Kanninen, *Int. J. Fract.* ***9***, 83 (1973).
48. H. R. Brown, *Macromolecules* ***24***, 2752 (1991).
49. H. R. Brown, *J. Mater. Sci.* ***25***, 2791 (1990).
50. J. G. Williams, *Fracture Mechanics of Polymers*; Halsted Press, New York; 1984.
51. L. J. Broutman and F. J. McGarry, *J. Appl. Polym. Sci.* ***9***, 589 (1965); *ibid.* ***9***, 609 (1965).
52. R. Greco and G. Ragosta, *Plast. Rubber Process. Appl.* ***7***, 163 (1987).
53. W. M. Merrill, A. V. Pocius, B. V. Thakker, and M. Tirrell, *Langmuir* ***7***, 1975 (1991).
54. Z. Suo and J. W. Hutchinson, *Int. J. Fract.* ***43***, 10 (1990).

55. R. S. Drago, G. C. Vogel, and T. E. Needham, *J. Am. Chem. Soc.* ***93***, 6014 (1971).
56. F. M. Fowkes, D. O. Tischler, J. A. Wolfe, L. A. Lannigan, C. M. Ademu-John, and M. J. Halliwell, *J. Polym. Sci., Polym. Chem. Ed.* ***22***, 547 (1984).
57. T. P. Russell, A. Menelle, W. A. Hamilton, G. S. Smith, S. K. Satija, and C. F. Majkrzak, *Macromolecules* ***24***, 5721 (1991).
58. T. P. Russell, *Mater. Sci. Rep.* ***5***, 171 (1990).
59. M. L. Fernandez, J. S. Higgins, J. Penfold, R. C. Ward, C. Shackleton, and D. Walsh, *Polymer* ***29***, 1923 (1988).
60. F. Brochard-Wyart, P.-G. de Gennes, and S. Troian, *C. R. Acad. Sci., Ser. 2,* ***310***, 1169 (1990).
61. A. J. Kinlock, *Met. Sci.* ***14***, 305 (1980).
62. A. C. Balazs, I. C. Sanchez, I. R. Epstein, F. E. Karasz, and W. J. MacKnight, *Macromolecules* ***18***, 2188 (1985).
63. A. C. Balazs, F. E. Karasz, W. J. MacKnight, H. Ueda, and I. C. Sanchez, *Macromolecules* ***18***, 2784 (1985).
64. R. A. Mendelson, *J. Polym. Sci., Polym. Phys. Ed.* ***23***, 1975 (1985).
65. L. J. Broutman and F. J. McGarry, *J. Appl. Polym. Sci.* ***9***, 589 (1965).
66. J. J. Benbow and F. C. Roesler, *Proc. Phys. Soc. (London) Ser. B*, ***70***, 201 (1957).
67. J. P. Berry, *J. Appl. Phys.* ***34***, 62 (1963).
68. P. P. Gillis and J. J. Gilman, *J. Appl. Phys.* ***35***, 647 (1964); R. Guernsey and J. J. Gilman, *Exp. Mech.* ***1***, 50 (1961).
69. R. E. Robertson, *J. Adhes.* ***4***, 1 (1972).
70. R. J. Good, *J. Adhes.* ***4***, 133 (1972).
71. C. Creton, E. J. Kramer, C.-Y. Hui, and H. R. Brown, *Macromolecules* ***25***, 3075 (1992).
72. P. G. Charalambides, H. C. Cao, J. Lund, and A. G. Evans, *Mech. Mater.* ***8***, 269 (1990).
73. C. Creton and E. J. Kramer, *Macromolecules* ***24***, 1846 (1991).
74. K. Cho, H. R. Brown, and D. C. Miller, *J. Polym. Sci., Polym. Phys. Ed.* ***28***, 1699 (1990).
75. H. R. Brown, K. Char, V. R. Deline, and P. F. Green, *Macromolecules* ***26***, 4155 (1993).
76. K. Char, H. R. Brown, and V. R. Deline, *Macromolecules* ***26***, 4164 (1993).
77. T. Hashimoto (Kyoto), private communication.

10 Strength of Incompatible Semicrystalline Interfaces*

10.1 Introduction to Incompatible Semicrystalline Interfaces

In this chapter we examine mechanisms of strength development at incompatible crystallizable interfaces. Polyethylene and polypropylene are used as model materials. Relevant topics from Chapter 8 and Chapter 9 on symmetric and incompatible interfaces are integrated into the discussion [1–13]. New factors involved in the strength development of crystallizable amorphous interfaces are considered, in addition to "local crystallization" effects proposed by Galeski [14–16]. In his scenario, volume contraction due to crystallization of spherulites in the vicinity of the interface is important and can result in a breakup of the original interface plane with a subsequent influx of melt across the interface. Crystallization of the influxes can produce a mechanically interlocked interface with a fracture energy considerably greater than that predicted by our model for incompatible amorphous interfaces without crystallization.

Isotactic polypropylene (PP) and high density polyethylene (HDPE) are thermodynamically incompatible. Both theory and experiment show that local segmental interdiffusion can occur as long as the temperature is high enough to permit a significant degree of mobility. Theoretical predictions of the interface thickness, d_∞, in the range of 50–100 Å, were made by Helfand [8], Kammer [17], and Shilov [18]. My purpose here is to observe how the interface forms and strength develops between the two polymers after contact is made in the melt and crystallization proceeds under isothermal and controlled cooling rate thermal histories. The results have application to recycling of mixed plastics, polymer blends, and composite laminates.

10.2 Experimental Techniques

10.2.1 Interface Preparation and Thermal History

The properties of the polymers studied by Yuan and Wool are given in Table 10.1 [10]. Linear low density polyethylene (LLDPE) and HDPE were used to investigate

---

* Dedicated to J. D. Hoffman

Table 10.1 Test Materials

Material	Code	MI*	T_m, °C	$M_w \times 10^{-5}$	Appearance
PP	Shell 5520	5.0	161	2.0	Pellet
HDPE	Chemplex 5853	1.3	131	1.3	Pellet
LLDPE	Grsin 7144	20.0	122		Powder
LLDPE	Grsin 7147	50.0	121		Powder

* MI = melt index, a measure of flow related inversely to melt viscosity

structure development in the interface, and HDPE was used for the butt welding mechanical studies. The PP/PE interface was prepared by laminating plates of PP and PE together in a Carver hot press with varying contact time and temperature. The details of the sample preparation methods are in the references [10].

The thermal histories are shown in Figure 10.1. PP/PE interface strength was evaluated with a butt joint specimen at room temperature, prepared as shown in Figure 10.2.

10.3 Structure Development at Semicrystalline Interfaces

The studies of amorphous polymer interfaces discussed in Chapters 7 and 8, as well as the crystallization behavior of semicrystalline polymers (Chapter 1), lead us to expect interface formation of PP/PE to occur in five steps, shown in Figure 10.3. The steps—(1) surface rearrangement, (2) wetting, (3) interdiffusion, (4) crystallization, and (5) solidification—are discussed individually in the sections that follow.

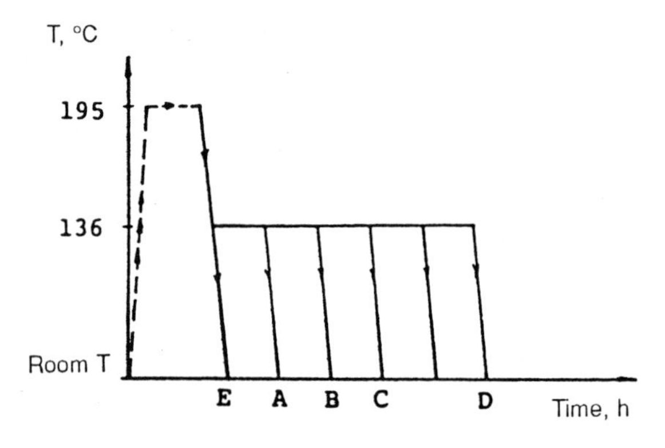

Figure 10.1 Thermal history for the PP/PE laminate isothermally crystallized at 136 °C for times (A) 0.5 h, (B) 1.0 h, (C) 1.5 h, (D) 2.5 h, and (E) rapid cooling at 10 °C/min.

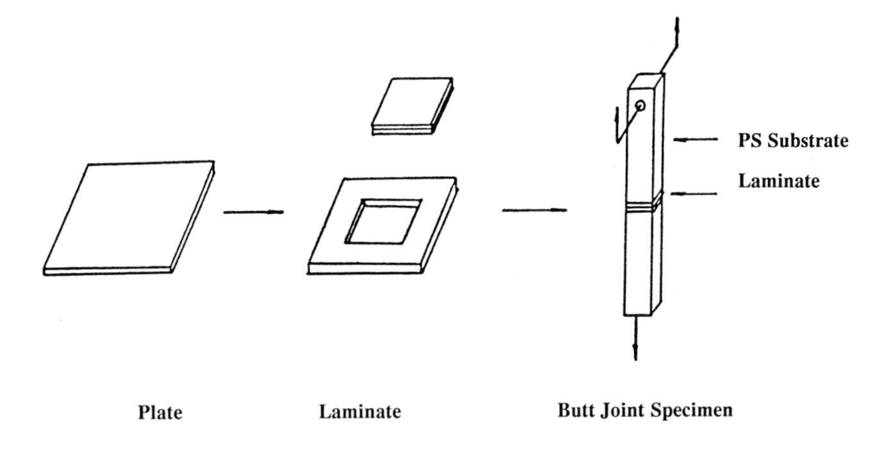

Figure 10.2 Preparation of the PP/PE butt joint specimen.

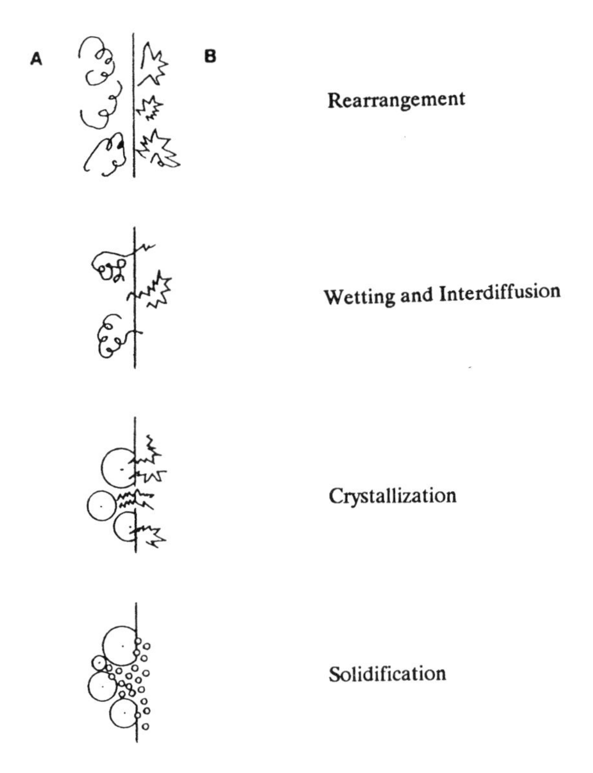

Figure 10.3 Schematic representation of the stages of interface formation of an A/B polymer pair during crystallization.

10.3.1 Surface Rearrangement

Surface rearrangement phenomena occur at the surfaces both before and after contact has been made to form the polymer–polymer interface. We are primarily concerned with structure formation or relaxation near the surface; chain-end distribution functions; surface segregation of impurities, low-molecular-weight species, and nucleating agents; and nonequilibrium chain configurations. With regard to whether memory of crystal nuclei persists in the melt, we know from Rault's data that the molten PP/PE laminate is in an equilibrium state (with loss of previous crystallization history) after it has been held at 195 °C for 30 minutes [19]. The relaxation time for PP with an M_w value of 10^5 is around 30 minutes if the previous crystallization was done at a cooling rate of 20 °C/min. The conditions for making the plates and laminates in our experiments were similar to these conditions.

Surface rearrangement effects also involve the nonequilibrium configurations of the melt chains due to the reflecting boundary condition of the surface "compressing" the normal Gaussian coil shape of the amorphous chains that lie within a distance of the radius of gyration of the surface. The reduced entropy of the surface layer of chains acts as a driving force to permit limited interdiffusion at incompatible interfaces. Rouse–like relaxation of the compressed configurations has been shown (Chapter 3) to cause rapid interdiffusion up to distances of the order of the entanglement spacing in polymer melts.

The spatial distribution function for chain ends is affected by the chemical nature of the chain ends and the molecular weight distribution near the surface. Melt crystallization results in several new phase processes and segregation processes, which lead to surface rearrangement and need to be considered as a function of thermal history.

10.3.2 Wetting

This step involves the establishment of contact of PE with PP at distances of the order of the van der Waals radius. Surface roughness and impurities play an important role in determining the rate of wetting and contact of the surfaces to form the interface. Usually the spreading coefficient, F_{ij}, is used to estimate the wettability of phase i spreading on phase j [20], as

$$F_{ij} = \Gamma_j - \Gamma_i - \Gamma_{ij} \tag{10.3.1}$$

where Γ is the surface tension. The following values of the surface tension at 140 °C [20] were substituted in Eq 10.3.1 to calculate F: Γ_{PE} = 28.8 dyne/cm; Γ_{PP} = 23.1 dyne/cm ; $\Gamma_{PP/PE}$ = 1.1 dyne/cm, and consequently, $F_{PP/PE}$ = + 4.6 dyne/cm.

From a thermodynamic point of view, since F is positive, we infer that PP can spread on the PE melt and that good contact can be achieved at the interface. Therefore, interdiffusion can subsequently occur in the wetted areas. If the time dependence of wetting is comparable to the time dependence of interdiffusion, then the extent of diffusion is determined by the convolution product of wetting and diffusion functions (Chapter 8).

10.3.3 Interdiffusion

If we use the Flory–Huggins solution theory applied to a binary polymer blend system [22], we can find the critical condition for miscibility in such a system in terms of the interaction parameter χ (Chapter 1), as

$$\chi_{cr} = \tfrac{1}{2}(1/\sqrt{N_a} + 1/\sqrt{N_b})^2 \tag{10.3.2}$$

so that

$$\chi_{cr} = 2/N \tag{10.3.3}$$

when $N_a = N_b = N$. The effect of molecular weight on χ_{cr} is based on Eq 10.3.2; the higher the molecular weight, the lower the χ_{cr} value, and the greater the incompatibility. The χ_{cr} for the PP/PE system studied with a value of M_w of 10^5 (for PE) is 0.00056, calculated by Eq 10.3.3. Polymer pairs with χ greater than χ_{cr} should be incompatible.

Helfand and co-workers developed a theory to predict the interfacial properties between two incompatible polymers [8], as discussed in Chapter 9. The equilibrium interdiffused thickness of the interface at long contact times, d_∞, was derived as

$$d_\infty = 2b/(6\chi_{ab})^{1/2} \tag{10.3.4}$$

and

$$\chi_{ab} = V/RT(\delta_a - \delta_b)^2 \tag{10.3.5}$$

where b is the effective bond length of a statistical segment in the melt, with typical values in the range 5–7 Å, δ is the solubility parameter expressed in units of $J^{1/2}/cm^{2/3}$, V is the molar volume (cm^3/mol), R is the gas constant (cal/mol·K) and T is the absolute temperature (K). The data needed for the calculation of χ_{ab} at 140 °C in this study are available from reference [29] and are as follows: $\delta_{PP} = 16$, $\delta_{PE} = 17$, $V = 32.8$, $R = 1.98$ and $T = 413$ K. The calculated values of χ_{ab} and d_∞ are 0.011 and 54 Å, respectively. The calculated values show that χ_{ab} is much larger than the value of χ_{cr} (0.011 >> 0.00056); the PP/PE polymer pair used here should represent an incompatible interface in the melt.

The end-to-end distance, R, and radius of gyration, R_g, of PE chains in the melt are calculated from Eqs 1.1.1 and 1.1.2 using $C_\infty = 6.7$ [23] and $M = 10^5$: the results are $R = 337$ Å and $R_g = 138$ Å. When the PE molecules diffuse across the interface to an equilibrium depth of about 27 Å (corresponding to $d_\infty/2$), this is about one-fourth of the distance necessary to achieve complete strength at a diffusion distance X_∞ of 0.81 R_g. The entanglement radius R_{ge} is about 48 Å for both PP and PE, so a diffusion depth of 27 Å is not expected to produce much entanglement between PP and PE.

Thus, a thin layer (about 30 Å) of mixed PP and PE forms in the melt prior to crystallization. When crystallization occurs, the fate of the layer is sensitive to the crystallization mechanism, with opportunities for either further phase separation induced by crystallization, or entrapment via entanglements in the amorphous component of the semicrystalline structure.

10.3.4 Crystallization

Both PP and PE are semicrystalline polymers with a spherulitic structure. Voids and defects are induced by the density change due to crystallization occurring in the melt; for example, the volume contraction for PP could reach 10%. If there is a difference in melting points and crystallization rates, it is possible for one polymer to crystallize while the other is still in the liquid phase. The volume contractions associated with the random nucleation of spherulites near the interface have the effect of pulling the other polymer in the liquid phase across the interface and creating interspherulitic influxes. This effect, known as *local crystallization* [14–16], often produces a mechanically interlocked interface. Intraspherulitic influxes also develop and contribute to the strength of the interface.

The surface morphology of a semicrystalline polymer has been shown to be dependent on the nature of the mold surface, the cooling rate, and processing conditions [14, 20, 21, 26]. Because of differences in both melting points and crystallization rates, the PP/PE laminate can be formed with either the PP crystallized against the PE melt or the PE crystallized against the PP melt. Consequently, the interface structure and strength change with the thermal history. Also, it is well known that melt-cooled PP has a three-phase system, namely amorphous, smectic, and crystalline phases; these may play a role in determining the macroscopic properties of the interface region [27, 28].

10.3.5 Solidification

Partial solidification of the PP/PE laminate first occurs with crystallization, and the final solid interface is obtained by further cooling of the sample. The interface morphology, and in turn the interface strength, depend on the two steps. If the PE influxes have fully developed during slow cooling, further cooling does not have a significant effect on the interface morphology. In contrast, when cooling from the melt to room temperature is rapid, the interface morphology is determined by the relative crystallization rates on both sides of the laminate. These two cases were studied in our experiments and are discussed in the next section.

10.4 Structure of Polypropylene/Polyethylene Interfaces

10.4.1 Isothermally Crystallized PP/HDPE and PP/LLDPE

The laminates crystallized at 136 °C were subjected to varying crystallization times to produce specimens with different interface structure. In SEM photos of etched sections cut perpendicular to the interface (Figure 10.4), we see that the interface shape, which is flat if crystallization time is short, assumes an irregular wave shape with increasing crystallization time; some pear-like interface shapes even occur. The distortion of the interface plane obviously leads to higher fracture energies, which are expected to be proportional to the surface roughness.

The PP surface of a fractured PP/HDPE interface was examined with SEM, and the results are shown in Figure 10.5. With increasing time, the size of the spherulites at the interface increases and their number decreases appreciably. The interface structure is composed of large spherulites that grew slowly under isothermal conditions, and many small spherulites that are the result of quenching the interface at the end of the crystallization time (see thermal history in Figure 10.1). The single large spherulite shown in Figure 10.5(A) is one of the larger PP spherulites. Nucleating agents are an

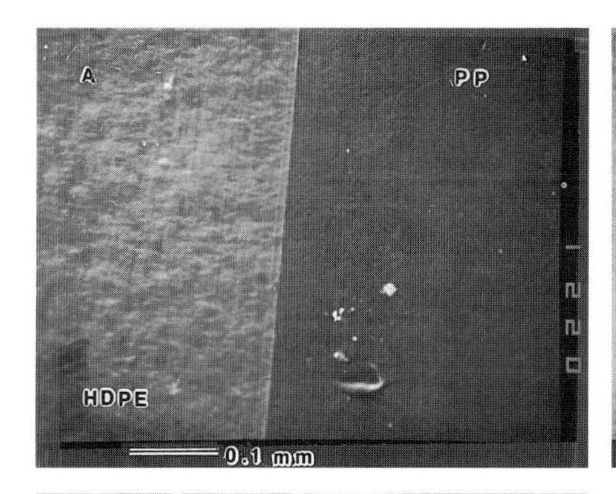

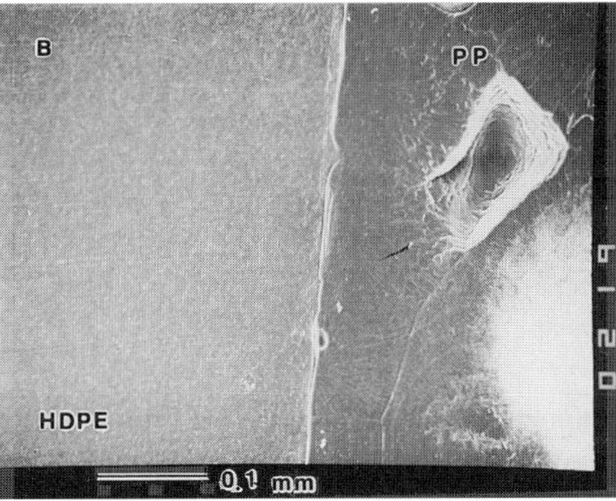

Figure 10.4 SEM photos of a PP/HDPE interface crystallized isothermally at 136 °C for the following times: (A) 0.5 h, (B) 1.0 h, (C) 1.5 h. The interface shape changes from being initially flat (A), to an irregular wave shape (C). The surfaces were etched for 30 min (Yuan and Wool).

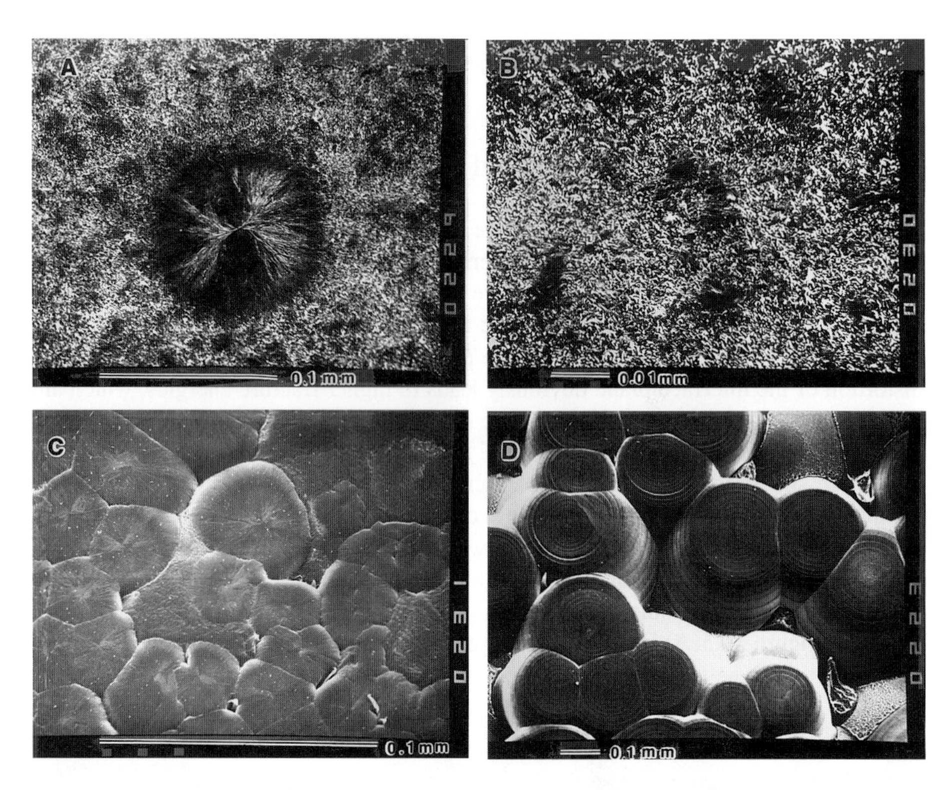

Figure 10.5 SEM micrographs of the PP surface of a fractured PP/HDPE interface with the following crystallization times at 136 °C: (A) 0.5 h, showing one spherulite and fine fibers; (B) high magnification of fibers in (A); (C) 1.0 h, showing many more spherulites growing near the interface; (D) 1.5 h, well-developed spherulites with "clean" shell debonded surfaces and interstitial areas due to polymer segregation and volume contraction (Yuan and Wool).

important variable in the determination of the interface structure. A high nucleation density results in structures similar to that shown in Figure 10.5(B), whereas stronger interfaces form with fewer nuclei and spherulites, Figure 10.5(C,D).

The material remaining in the interstices between spherulites is clearly visible. For example, Figure 10.6 shows that adhesive fracture occurred in the PP/LLDPE interface plane, with extensive pullout and fracture of (PE) fibrils. Here, the LLDPE surface shows both the deformed fibrils and the "imprint" of the PP spherulites. The LLDPE spherulites are much smaller than the imprint size. The fractured surfaces were well matched with each other and became more undulating with increasing crystallization time. Well-defined spheres, holes, and highly drawn (PE) fibers appear on the PP side (Figures 10.5 and 10.6). Differential scanning calorimetry (DSC) measurements [10] support the hypothesis that fracture remnants of the HDPE influxes resided in the PP side. IR spectroscopy could also be used to investigate this aspect of the fracture process.

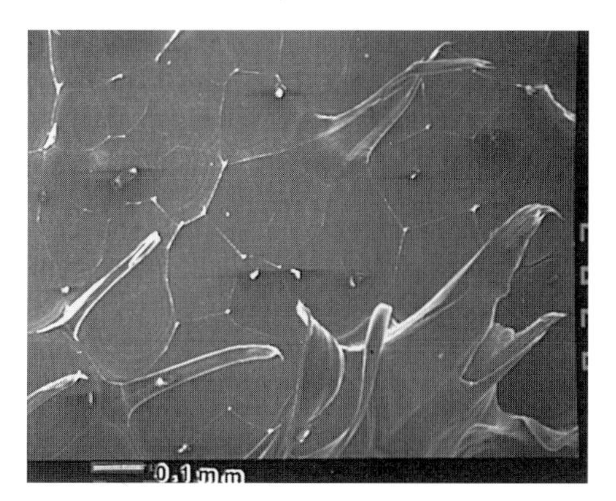

Figure 10.6 LLDPE surface of a fractured PP/LLDPE interface. The fibrils on the surface are due to pullout of the PE influx material that was incorporated into the PP side during isothermal crystallization (Yuan and Wool).

10.4.2 Strength of PP/PE Interfaces

Results of the butt strength tests of the PP/HDPE interface indicate that the strength increases with increasing crystallization time, as shown in Figure 10.7. The yield strength of HDPE is about 20 MPa, which is about twice the strength of the interface formed at 1.5 hours. Although the comparison of fracture stresses is complicated by the fracture mechanism differences, the results are highly supportive of a considerable enhancement of the interface strength due to mechanical interlocking. The ratio of the fracture stresses is much greater than the proportion of influx material in a unit of cross-sectional area that one could deduce from Figure 10.5 or Figure 10.6. Thus, we expect contributions to strength from intraspherulitic entrapment as well as from the

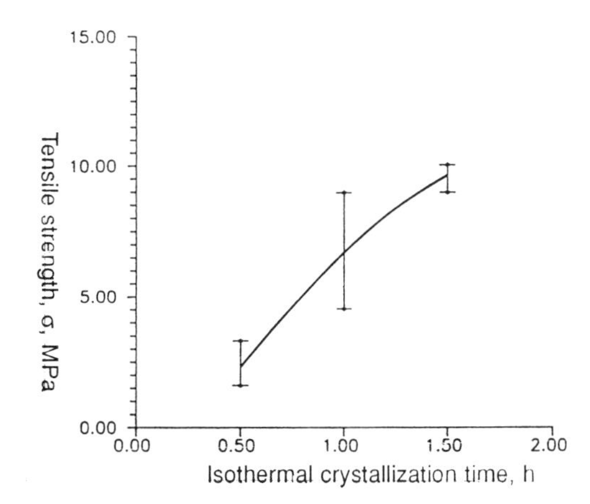

Figure 10.7 Butt tensile strength of the PP/HDPE laminate as a function of isothermal crystallization time at 136 °C (Yuan and Wool).

interspherulitic influxes. Our data are insufficient to determine the relative contribution of each, but the intraspherulitic contribution is not considered to be as significant as the interspherulitic. This judgment is based on the fact that rapidly quenched interfaces, which are mechanically very weak, have ample opportunity for intraspherulitic entrapment but do not develop interspherulitic influxes.

An estimate of the effect of influxes on the interface strength, σ, and fracture energy, G, can be approximated for the fully crystallized interface using simple geometric arguments as follows. We let the interface be composed of PP spherulites of radius r, in which the interstitial area is completely occupied by PE influx fibrils. The centers of the circles representing the contiguous spherulites are separated by a distance $2r$. The fracture energy is given by

$$G \approx N_f G_f \tag{10.4.1}$$

where N_f is the number of influxes per unit area and G_f is the energy to fracture a single influx fibril. We assume that the energy to fracture the average fibril is simply related to its area A_f via

$$G_f = G_0 A_f \tag{10.4.2}$$

where G_0 is the fracture energy of pure PE in the virgin state. We are also assuming here that the influx fibrils are sufficiently interlocked that they fracture rather than pull out. The area of an average fibril is determined from the interstitial area between four contiguous circles as

$$A_f = r^2 (4 - \pi) \tag{10.4.3}$$

The number of fibrils per unit area is

$$N_f = 1/(4 r^2) \tag{10.4.4}$$

Substituting for N_f and A_f in the above relations, we obtain the fracture energy due to influxes as

$$G = 0.22\, G_0 \tag{10.4.5}$$

which is independent of the spherulite radius. To compare with our fracture experiments, we have the strength of the interface σ, which is a function of the square root of G, so that

$$\sigma = 0.46\, \sigma_0 \tag{10.4.6}$$

where σ_0 is the tensile strength of the pure PE material. For HDPE, σ_0 values are in the range 3,200–4,500 psi (22–31 MPa), giving an interface strength, σ, on the order of 10–14 MPa. The maximum values reported in Figure 10.7 for the PP/HDPE interface are consistent with this very simple analysis.

The role of heterogeneous nuclei (impurity particles) at the interface is unexpected. For example, they can migrate across the interface from the PP side to the PE side during the wetting step at 195 °C for 30 minutes [14–16]. It is possible that when the PP crystallizes first, the PP side could be "open" in regions without solidifying for a certain time, due to the depletion of nuclei. This would facilitate the formation of the PE influxes when the laminate crystallizes at 136 °C.

When the crystallization time of the PP/HDPE interface is around half an hour (Figure 10.5), the PP spherulites initiated near the interface are still too small to reach the interface. A planar appearance, with a few sporadic disk-like spherulites, was seen by SEM in the fractured interface after further cooling for solidification. With a higher magnification (1500×), a number of very short and fine fibers (3–5 μm) covering the entire surface were revealed. The limited interdiffusion at the incompatible polymer melt interface, resulting in the entanglements between the two polymers, is thought to be one of the mechanisms in the development of interface strength (Figure 10.5) and is thought to facilitate the influxes of fluid.

In contrast with the profuse fine-scale fibers appearing everywhere on the fracture surface of the short-time–crystallized laminate, longer and thicker PE fibrillar remnants are clearly seen in the interstices of the PP spherulites of the long-time–crystallized laminate. However, only a few fine fibers can be seen on the "shell" of PP spherulites (Figure 10.5). The withdrawal or segregation of the PE chains from the PP side during the slow crystallization process may have contributed to the lack of fine fibers on the surface. This is consistent with rapid crystallization causing entrapment of PE, and slow crystallization resulting in further phase separation from the interdiffused layer in the melt.

The formation of the influxes to depths of several micrometers should be considered when concentration profiles are evaluated in terms of diffusion mechanisms at crystallizing interfaces. Obviously, the contribution of influxes could give misleading results if it is superimposed on diffusion-controlled concentration profiles. Many experiments of this type are reported in the literature without consideration for the effective breakup of the interface.

We can summarize the results on isothermally crystallized PP/PE interfaces by stating that the origin of the interface strength may come from two sources. One is the entanglement of the PP and PE chains in the melt to a depth of about 50 Å, which results in a fine fibrillated fracture morphology (Figure 10.8). The other is the mechanical interlocking caused by the PE "flow", resulting in influxes, which should be more effective the longer the crystallization time (Figure 10.9).

When the PP spherulites have grown to where their outer edges impinge on each other, their size no longer increases with time and the interface strength should level off. In our experiments, it seems that the interface strength tends to level off at around 1.5 h at 136 °C. The maximum strength at this point is about one-third to one-half of the tensile strength of the virgin PE and is considerably stronger than the interface formed without the development of influxes.

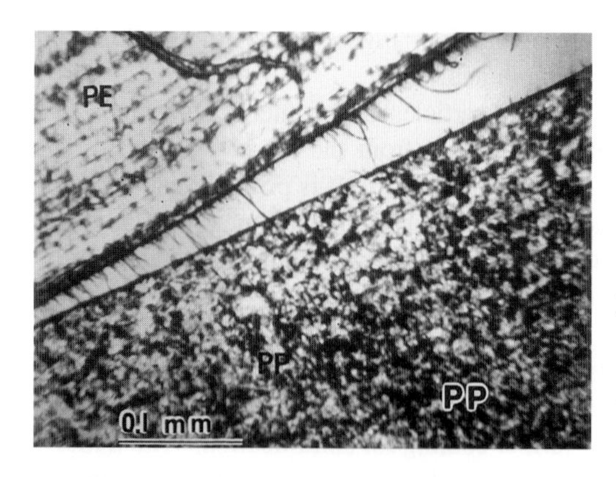

Figure 10.8 Optical micrograph of the PP/LLDPE fractured interface, which was obtained by rapid cooling from 195 °C in the hot press. Note the thin fibrils on the fracture surfaces (Yuan and Wool).

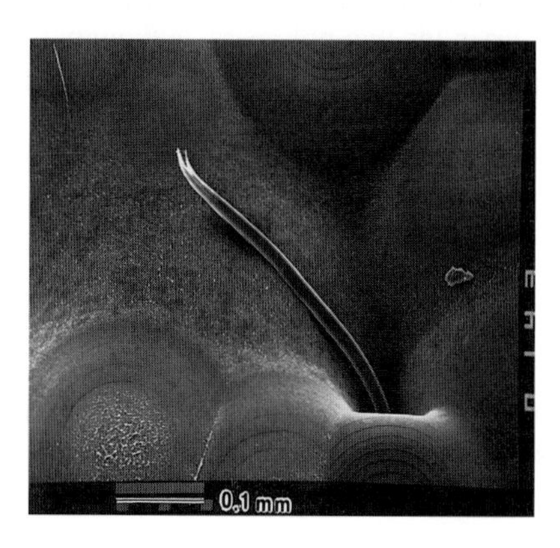

Figure 10.9 SEM photo of the PP surface of a fractured PP/HDPE interface showing a highly drawn PE fiber (Yuan and Wool).

The force transferred to the PE melt during the crystallization of PP is dependent on the location of the PE fluid with respect to the occluded PP melt and the growing PP spherulites, as shown in Figure 10.10. As the PP molecules crystallize, a driving force towards the spherulites is created from the local volume contraction and is transferred to the PE through melt entanglements. If the PP spherulites crystallize on the interface plane, the direction of the resultant force exerted on the PE is largely parallel to the interface (Figure 10.10, Case A). Therefore, the PE molecules have difficulty with "flow" into the PP matrix. The quenched and short-time–crystallized laminate gives many spherulitic nuclei appearing right on the border, which either lets the interface "close" to influxes, or minimizes the pulling force on PE as mentioned above. The other situation (Figure 10.10, Case B) is similar to the slow isothermal crystallization case. The PP melt is surrounded by the PP spherulites nucleated away from the interface, and the PE influxes can be formed.

The structure of the interface was the same regardless of which kind of polyethylene (HDPE or LLDPE) we used in these isothermal crystallization experiments. We can expect that during isothermal crystallization, PEs with different viscosity

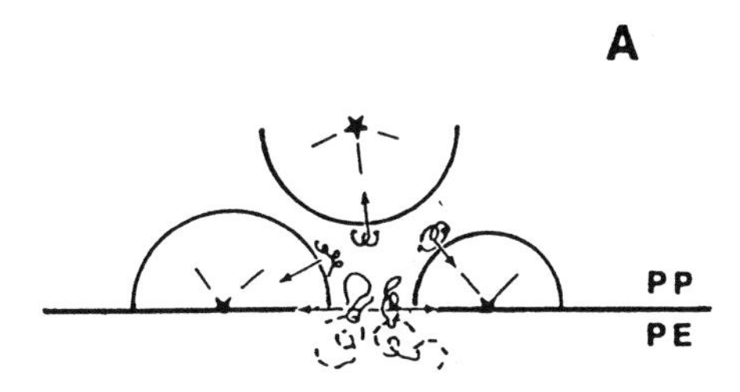

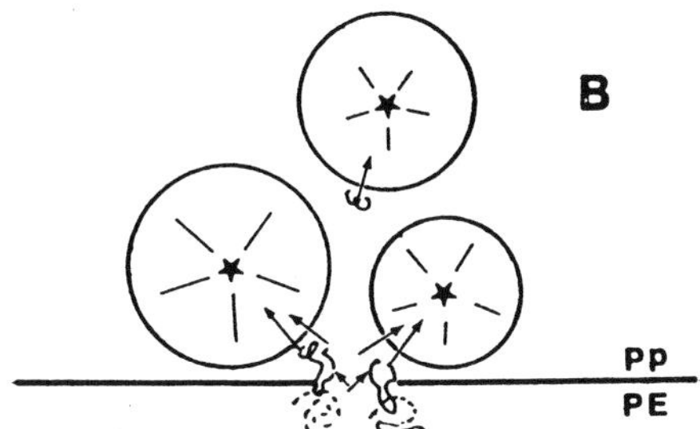

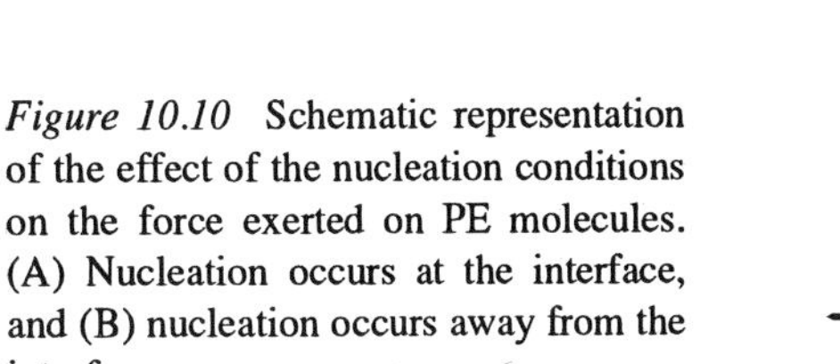

Figure 10.10 Schematic representation of the effect of the nucleation conditions on the force exerted on PE molecules. (A) Nucleation occurs at the interface, and (B) nucleation occurs away from the interface.

(different melt index) could develop similar interface structure provided they have sufficient time to flow.

The tensile strength of the interface obtained at 141 °C for 1 hour is about 5.5 MPa, and is lower than the value of that obtained at 136 °C (10 MPa). Considering the temperature dependence of the crystallization rate, the higher the crystallization temperature, the smaller the spherulite size at constant time, and in turn the smaller the dimension of PE influxes formed.

10.4.3 Nonisothermally Crystallized PP/HDPE

We also studied the structure and strength of quench-cooled PP/PE interfaces. The thermal history in this process better exemplifies that experienced by PP/PE blends during commercial melt processing. Although the melting temperature, T_m, of PP (160 °C) differs from those of HDPE (130 °C) and LLDPE (124 °C), the crystallization temperatures, T_c, are very close and overlap. For example, at a cooling rate of 10 °C/min, the crystallization temperatures are 119 °C (HDPE), 113 °C (PP), and 110 °C (LLDPE); the HDPE is expected to crystallize first at the PP/HDPE interface. The spherulite growth rate of PP is also less than that of HDPE or LLDPE at 119 °C.

The crystallization behavior of PP/PE interfaces at a constant cooling rate was studied using sections of the laminates on the Mettler hot stage with a polarizing microscope. A cross section of a PP/HDPE laminate was held at 190 °C for 10 minutes and then cooled at 10 °C/min. The PP nuclei first developed randomly at 126 °C over the entire area of the PP side. The HDPE nuclei at this point were invisibly tiny. As the temperature dropped to 121 °C, the HDPE nuclei spread rapidly and uniformly over the HDPE side, as shown in Figure 10.11(A). In a short time, the crystallization of the HDPE was completed, forming a solid boundary with the PP, which was not yet fully crystallized. As soon as the spherulites in the HDPE side appeared, a number of new PP nuclei occurred along the track of the interface; the latter can be seen clearly in Figure 10.11(B), where we remelted the same section to 150 °C after the crystallization had occurred at 121 °C.

The nucleation of the new PP spherulites near the interface may occur by several mechanisms, of which two are: (1) The crystallized boundary of the HDPE behaves as a solid surface, which acts to nucleate the molten PP. The PP and PE melts wet each other and epitaxial growth is possible. (2) Stresses developing at the interface may come from the difference in thermal expansion coefficients, 6.8×10^{-5} in/in·°C (PP) and 3.3×10^{-4} in/in·°C (PE), and volume contraction due to crystallization. In the case where the HDPE side crystallized first, the PP influxes were difficult to form since the HDPE spherulites were usually very small (Figure 10.11). Also, new PP spherulites formed near the interface may have made PP influx formation too difficult. These interfaces are subsequently weak, which is unfortunate since most polymer processing has melt histories like that discussed in this section.

10.4.4 Nonisothermally Crystallized PP/LLDPE

At a cooling rate of 10 °C/min, the spherulite nuclei first appeared in the PP side and fewer nuclei appeared near the interface. When the PP spherulites had spread over the entire PP area, including the boundary region of the PP, and the temperature of the section had reached 111 °C, the LLDPE still had not crystallized (Figure 10.12B). The PP side of the interface crystallized first, apparently without the inducing of nuclei in the LLDPE side that was observed for the PP/HDPE case; then the LLDPE began to crystallize. This behavior is similar in several respects to the isothermal crystallization case. The LLDPE influxes also formed, but not to the same extent as in the isothermal case. Since the temperature dropped so fast, the PP spherulites nucleated rapidly and could not form the proper boundary for accepting the LLDPE influxes. Thus, the PP/LLDPE interfaces are stronger than the PP/HDPE under constant cooling conditions, but weaker than the isothermally crystallized interfaces.

Several interesting phenomena were observed in the SEM micrographs of PP/PE interfaces. If the crystallization time is long, we can see concentric rings on the surface of the PP spherulites (Figure 10.13). The rings form continuous contour lines through several adjoining spherulites. Padden and Keith once showed the same phenomenon using polarized optical microscopy, and attributed the regular banding pattern to rhythmic fluctuations in the mode of crystallization in parallel with the ± 2 °C cycling of the thermostatic control of the heater [24]. Rings form because the lamellae are

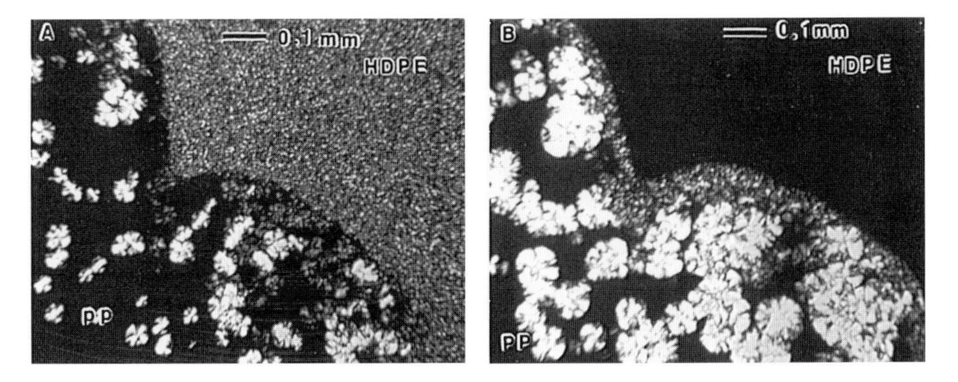

Figure 10.11 Crystallization of a PP/HDPE interface: (A) At a cooling rate of 10 °C/min, the HDPE solidified first (top right), shown at 121 °C; some spherulites are shown on the PP side (bottom left) near the interface. (B) The PP microspherulitic boundary near the interface is shown during remelting from 121 °C at 150 °C (Yuan and Wool).

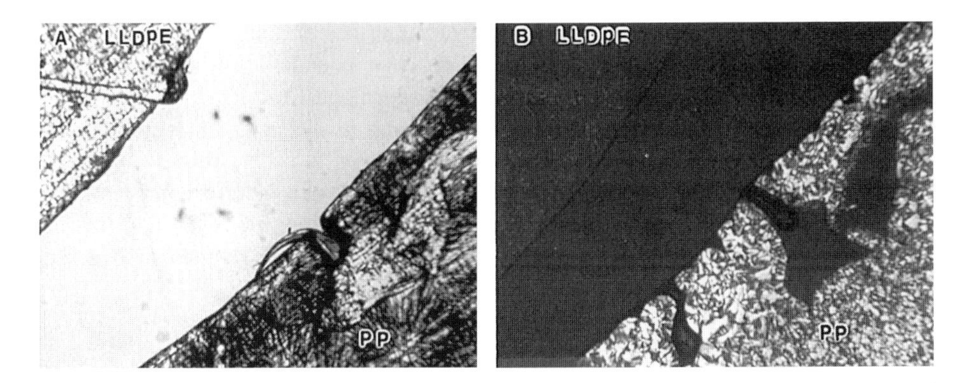

Figure 10.12 PP/LLDPE interface formed at a cooling rate of 10 °C/min. (A) Fractured PP/LLDPE interface; (B) heating (A) up to 150 °C, then cooling at 10 °C/min (the observation was made at 111 °C while LLDPE still had not crystallized). The black areas on the PP side are the PE influxes. Magnification of both photomicrographs is 80× (Yuan and Wool).

twisted and because the temperature fluctuates during spherulite growth. There are four types of spherulites formed during crystallization of polypropylene [24, 25]. They grow at different temperatures with different growth rates. The temperature range for formation of these spherulites is from approximately 128 °C to 138 °C. The temperature fluctuations of the laminate in the Carver press were in this range. The lines in Figure 10.13 therefore mirror the on/off cycle of the thermostat on the hot press. Given that the hot press had a ± 3 °C fluctuation in a 6-minute cycle, SEM analysis of the black rings in the solidified PP could be used to precisely determine local spherulite growth rates.

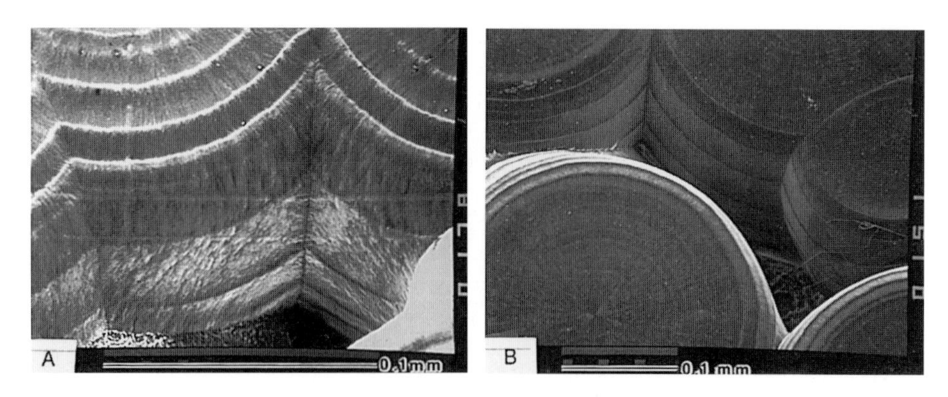

Figure 10.13 SEM micrographs of PP spherulites showing rings and contour lines on the adjoining spherulites. The spherulites were formed at a PP/HDPE interface under long-time isothermal crystallization conditions at 136 °C (Yuan and Wool).

Another interesting observation made with the SEM photos is that the spherulites in Figure 10.14 with clear black and white radial stripes on their tops are Type III spherulites. The Type III spherulite has a concave boundary toward the Type I spherulite, which is consistent with studies by Padden and Keith [24]. Since PP has so many structures at different levels corresponding to different temperatures, the interface strength could be affected in a complex manner by each of these structures.

In addition to the formation of influxes in the interstices of spherulites, intraspherulitic entrapment of melt is also possible. On several occasions, we noted that the PP spherulites on the fracture surface sported plastically deformed PE remnants. Thus, the

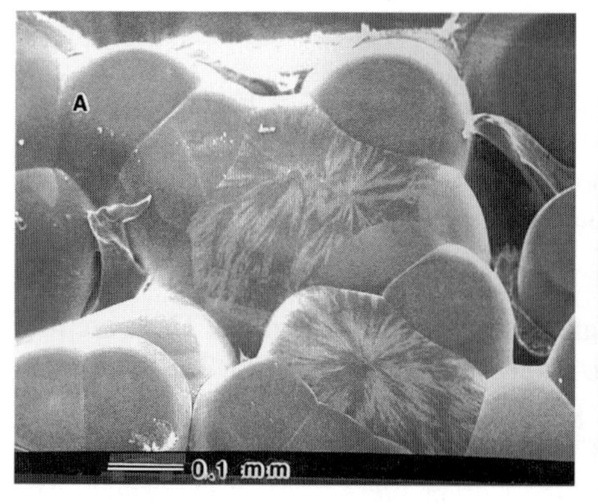

Figure 10.14 (A) SEM photo of PP spherulites obtained from a fractured PP/HDPE interface; (B) optical micrograph of Type III spherulites of PP showing characteristic bright/dark pattern (from P. H. Geil [25]).

polymer melt becomes entrapped in the growing lamellar crystallites. This provides another mechanism of mechanical interlocking of incompatible crystalline interfaces, and deserves further study since it is relevant to rapidly cooled interfaces.

10.5 Comments on Incompatible Semicrystalline Interfaces

The interface strength of laminated A and B crystallizable polymers is dependent on such factors as the nature of A and B, the crystallization temperature T_c, the crystallization time, the nucleation control, and the solidification conditions. The interface strength is related to its structure, which is affected by the crystallization conditions of the two contacting polymers. Several points are made in summary.

10.5.1 General Behavior of Influxes

1. For isothermal crystallization of incompatible A/B interfaces, the volume contraction associated with the nucleation of spherulites located away from the interface makes it possible for the A influxes to be formed between the B spherulites in the interface region. The magnitude of the volume contraction should be dependent on the degree of crystallinity and the density of the spherulites.
2. With increasing crystallization time, the size of the influxes increases and some pear-like influxes appear, resulting in an increase in the interface strength. The crystallization temperature and the solidification conditions control the size of the influxes.
3. With slow crystallization conditions, the higher crystallization temperature results in fewer and larger spherulites nucleating at random near the interface. Influxes form slowly and eventually grow to promote mechanical interlocking.
4. With fast crystallization conditions, a large number of nuclei form near the interface and tend to block off the influxes from the melt, which is also attempting to crystallize. Thus, rapid crystallization typically results in a very weak interface.
5. For the nonisothermal A/B case, one side crystallizes first and then induces crystals in the other side near the interface. Both events hamper the formation of influxes, and the interface strength is weak.

Suggestion to the Reader

A complete description of the influx formation process has not yet been formulated, and would make an excellent theoretical and experimental study. It would be useful to examine the incompatible interface formed with a crystallizable polymer and a noncrystallizable melt.

10.5.2 Comment on Recycled Plastics

Recycling of plastics is a useful method of addressing problems of plastic waste [30] and efficiently reusing valuable materials. Plastic garbage is separated and purified by a variety of techniques. For plastics such as PP and PE, the products are ground up and remelted, and often are poured into large mold sections to make such products as "wood" for park benches, soft playground surfaces, parking lot car wheel stops, etc. We observed an interesting phenomenon with these large sections.

The polymer melt containing many impurities is typically poured into a large mold (perhaps six inches in diameter and several feet in length). In the case of a circular cross section, as the polymer cools it crystallizes very slowly from the outer surfaces and pushes the incompatible impurities towards the center. This is similar to a radial zone refining process, and eventually all the impurities migrate into the center, from where they can be readily removed. Depending on the crystallization conditions, the impurities may be well separated to the center, or left in the form of concentric "tree rings" near the center of the section. The radial zone refining of recycled plastics offers an interesting new method of purifying and separating commingled plastics from municipal solid waste.

10.6 References

1. S. S. Voyutskii, *Autohesion and Adhesion of High Polymers*; Vol. 4 in series *Polymer Reviews*, H. F. Mark and E. H. Immergut, Eds.; Wiley–Interscience, New York; 1963.
2. H. H. Kausch, *Pure Appl. Chem.* ***55***, 833 (1983).
3. M. Tirrell, D. Adolf, and S. Prager, "Orientation and Motion at a Polymer–Polymer Interface: Interdiffusion of Fluorescent-Labelled Macromolecules"; chapter in *Orienting Polymers: Proceedings of a Workshop Held at the IMA, University of Minnesota, Minneapolis, March 21–26, 1983*, J. L. Ericksen, Ed.; No. 1063 in series *Lecture Notes in Mathematics*, A. Dold and B. Eckmann, Eds.; Springer-Verlag, Berlin; 1984; p 37; D. Adolf, M. Tirrell, and S. Prager, *J. Polym. Sci., Polym. Phys. Ed.* ***23***, 413 (1985); S. Prager and M. Tirrell, *J. Chem. Phys.* ***75***, 5194 (1981).
4. P.-G. de Gennes, *J. Chem. Phys.* ***55***, 572 (1971); .P.-G. de Gennes, *C. R. Hebd. Seances Acad. Sci., Ser. B*, ***291***, 219 (1980); .P.-G. de Gennes and L. Leger, *Ann. Rev. Phys. Chem.* ***33***, 49 (1982).
5. Y.-H. Kim and R. P. Wool, *Macromolecules* ***16***, 1115 (1983); H. Zhang and R. P. Wool, *Macromolecules* ***22***, 3018 (1989).
6. R. P. Wool, *J. Elastomers Plast.* ***17*** (Apr), 107 (1985); R. P. Wool and K. M. O'Connor, *Polym. Eng. Sci.* ***21***, 970 (1981); R. P. Wool, B.-L. Yuan, and O. J. McGarel, *Polym. Eng. Sci.* ***29***, 1340 (1989).
7. "Fracture Mechanics Studies of Crack Healing"; Chapter 10 in H. H. Kausch, *Polymer Fracture*; Springer-Verlag, Heidelberg; 2nd ed., 1987.

8. E. Helfand, "Polymer Interfaces"; chapter in *Polymer Compatibility and Incompatibility; Principles and Practices*, K. Šolc, Ed.; MMI Press Symposium Series; Harwood Academic Publishers, New York; 1982; T. A. Weber and E. Helfand, *Macromolecules* ***9***, 311 (1976); E. Helfand and Y. Tagami, *J. Chem. Phys.* ***56***, 3592 (1972).
9. S. Wu, H. Chuang, and C. D. Han, *J. Polym. Sci., Polym. Phys. Ed.* ***24***, 143 (1986).
10. B.-L. Yuan and R. P. Wool, *Polym. Eng. Sci.* ***30***, 1454 (1990); R. P. Wool, J. L. Willett, O. J. McGarel, and B.-L. Yuan, "Strength of Polymer Interfaces", *Polym. Prepr. (Am. Chem. Soc., Div. Polym. Chem.)* ***28*** (2), 38 (1987).
11. K. L. Foster and R. P. Wool, *Macromolecules* ***24***, 1397 (1991); J. L. Willett and R. P. Wool, *Macromolecules* ***26***, 5336 (1993).
12. D. R. Paul and S. Newman, Eds., *Polymer Blends Vol. 1*; Academic Press, New York; 1978; p 25.
13. R. P. Wool, *Rubber Chem. Technol.* ***57*** (2), 307 (1984).
14. A. Galeski and Z. Bartczak, *Polymer* ***27***, 544 (1986).
15. A. Galeski and Z. Bartczak, *Polymer* ***25***, 1323 (1984).
16. A. Galeski and E. Piorkowska, *J. Polym. Sci., Polym. Phys. Ed.* ***21***, 1313 (1983).
17. H.-W. Kammer, *Z. Phys. Chem. (Leipzig)* ***258***, 1149 (1977).
18. V. V. Shilov, V. V. Tsukruk, and Y. S. Lipatov, *Polym. Sci. USSR (Engl. Transl.)* ***26***, 1503 (1984).
19. J. Rault, *Crit. Rev. Solid State Mater. Sci.* ***13***, 57 (1986).
20. S. Wu, *Polymer Interface and Adhesion*; Marcel Dekker, New York; 1982.
21. D. G. Gray, *J. Polym. Sci., Polym. Lett. Ed.* ***12***, 509 (1974).
22. P. J. Flory, *Principles of Polymer Chemistry*; Cornell University Press, Ithaca, NY; 1953.
23. R. J. Young, *Introduction to Polymers*; Chapman and Hall, New York; 1983.
24. F. J. Padden, Jr., and H. D. Keith, *J. Appl. Phys.* ***30***, 1479 (1959).
25. P. H. Geil, *Polymer Single Crystals*; Robert E. Krieger Publishing Company, New York; 1973.
26. H. M. Zupko, *J. Appl. Polym. Sci.* ***18***, 2195 (1974).
27. C. C. Hsu, P. H. Geil, H. Miyaji, and K. Asai, *J. Polym. Sci., Part B: Polym. Phys.* ***24***, 2379 (1986).
28. G. Natta and P. Corradini, *Nuovo Cimento (Suppl.) Ser. 10*, **Vol. *15***, 40 (1960).
29. D. W. Van Krevelen, *Properties of Polymers: Their Estimation and Correlation with Chemical Structure*; Elsevier Scientific, New York; 1976.
30. *Facing America's Trash—What Next for Municipal Solid Waste*; Office of Technology Assessment, U.S. Congress Publication, October 1989.

11 Craze and Damage Healing*

11.1 Damage and Healing

In this chapter we examine microscopic damage and healing phenomena in multicomponent polymers, crystalline polymers, amorphous polymers, and some block copolymers. In previous chapters we were primarily interested in welding at a single plane or interface in the material. Here we are concerned with the evolution of cracks and microvoids throughout the body of a mechanically deformed material and the potential healing of such damage. This chapter is a prelude to the next and final chapter on crack healing where the damage is focused on a single crack.

11.1.1 Damage in Polymers

The response of a polymer body to an external mechanical or thermal stimulus is a function of cooperative events occurring at three material structural levels, namely, (1) single-chain, (2) multichain, and (3) microstructural. For example, the ultimate strength of a fibrous material in uniaxial deformation is bounded by the strength of a single chain or chain segment. Macroscopic fracture initiation and other damage processes depend on multichain interactions in crystalline, amorphous, or phase surface domains; these include time-dependent molecular stress distributions (spatial and numerical), orientation changes, conformational and configurational changes, and diffusion processes. The alignment of microstructural elements with respect to the deformation field determines the mechanical response and subsequent damage; the microstructures comprise aggregates of lamellar crystals, fibrils, filler particles (polymeric or nonpolymeric), sandwich structures of amorphous and crystalline regions, thermodynamically incompatible phase-separated domains, blend aggregates, and so forth.

The three material structural levels are common to the wide variety of polymer materials, the principal classes of which are (i) semicrystalline polymers, (ii) amorphous glassy polymers, (iii) elastomeric rubbery polymers, (iv) block copolymers, (v) filled elastomers, (vi) rigid composites, and (vii) blends. Material damage and fracture in these major classes have received considerable attention, and a few important reviews of the subject are in the references [1–7]. We will briefly discuss damage in single-chain, multichain, and microstructural material domains.

* Dedicated to J. E. Fitzgerald

Single-Chain Damage

Most theories on chain fracture utilize variations on Eyring's absolute reaction rate theory [8], which describes the kinetics and time of fracture, τ, in terms of the applied stress, σ, temperature, T, and several material constants as

$$\tau = \tau_0/2 \exp(\Delta H/RT) \operatorname{csch}(\tfrac{1}{2}\gamma\sigma/RT) \tag{11.1.1}$$

where ΔH is the enthalpy of activation, γ is the activation volume term, τ_0 is the oscillation period of atoms, and R is the gas constant. This theory tells us that the activation energy surface is linearly modified by the applied stress, and that the probability of chain rupture is related to the Boltzmann probability of encountering a high-amplitude oscillation (phonon) of sufficient energy to push the stressed bonds apart. The net rate of chain rupture is the result of the forward and reverse jumps over the activation energy barrier. At high stress, the reverse rate is negligible and Eq 11.1.1 can be expressed as

$$\tau = \tau_0 \exp\left(\frac{\Delta H - \gamma\sigma}{RT}\right) \tag{11.1.2}$$

Zhurkov *et al.* [9] have applied this equation to explain both single-chain fracture and total macroscopic fracture with the view that single-chain fracture is the controlling step. Many polymers exhibit macroscopic fracture behavior in which stress is a linear function of the log of the time, as predicted by Eq. 11.1.2, but many others do not, particularly those undergoing microstructural and multichain changes with time, as observed in plastic and viscoelastic deformation [6]. Applications of this equation to many polymer systems are reviewed by Kausch [2].

Multichain Damage

Multichain damage involves a deleterious change in the properties of groups of chains. It is a complex problem and no general treatment of it exists. One must consider interpenetration of chains, diffusional mixing and demixing, loss of entanglements (Chapter 7), configurational changes through conformational isomer population variation, orientation distribution changes, and many other perturbations of interchain reference configurations. Time-dependent redistribution of stress among chains controls changes in both conformation and orientation, and is a very important factor controlling the linear or nonlinear viscoelastic response of the polymer as well as the generation of internal fracture surfaces. For example, Table 11.1 highlights the qualitative differences in the molecular response in a viscoelastic polypropylene sample during uniaxial stress relaxation and creep [10].

Using dynamic infrared spectroscopy, we could deduce time-dependent rearrangements in the molecular stress distribution, orientation, and helical conformations. Obviously, creep and stress relaxation reflect very different multichain events that contribute to different levels of molecular damage and nonlinear viscoelastic behavior.

Table 11.1 Major Molecular Differences Resulting from Stress Relaxation and Creep of Viscoelastic Polypropylene

Stress relaxation	Creep
The number of highly stressed bonds decreases with time	The number of highly stressed bonds increases with time
The number of intermediately stressed bonds increases with time	The number of intermediately stressed bonds decreases with time
The number of helical conformations decreases	The number of helical conformations increases
The extent of orientation decreases	The extent of orientation increases
The number of bonds fractured is small	The number of bonds fractured is large

Microstructural Damage

The most important aspect of microstructural damage in multicomponent polymers is microvoid formation, particularly in the interfaces between contiguous microstructural elements. In semicrystalline polymers, filled elastomers, composites, and block copolymers, the interfaces are created by several processes, which include impingement growth during solidification (spherulite boundaries), heterogeneous formulation (rubber molecule on carbon black particle), and domain segregation (hard–soft phases in thermoplastic elastomers). In glassy polymers, crazing is the most important form of damage and has been reviewed by Kambour [12]. Voids, crazes, cracks, and so forth can range in size from a few angstroms to macroscopic dimensions. For example, Figure 11.1 shows the network of microcracks that develops in a hard elastic polypropylene fiber subjected to uniaxial strain (courtesy of L. Konapasek, Georgia Institute of Technology). This type of crack formation dominates the mechanical response of the fiber and gives it elasticity comparable to that of rubber (Section 11.5). Similar microvoids and crazes form in many other multicomponent polymers. We demonstrate in this chapter that the majority of these cracks can heal, so that the original mechanical strength of the material is restored.

11.2 Craze Growth and Healing

Craze nucleation and growth cause brittle fracture in amorphous polymers, so an understanding of these phenomena is important for the production of materials with optimal failure resistance. Several studies since 1940 have defined a variety of parameters governing crazing [12, 13].

Craze nucleation has been shown to be a complex function of the principal stresses. Sternstein *et al.* [14] proposed criteria for craze formation in a glassy polymer, the requirements being (1) a dilatational stress field and (2) a maximum principal stress greater than a threshold value. Craze growth, unlike nucleation, is controlled only by

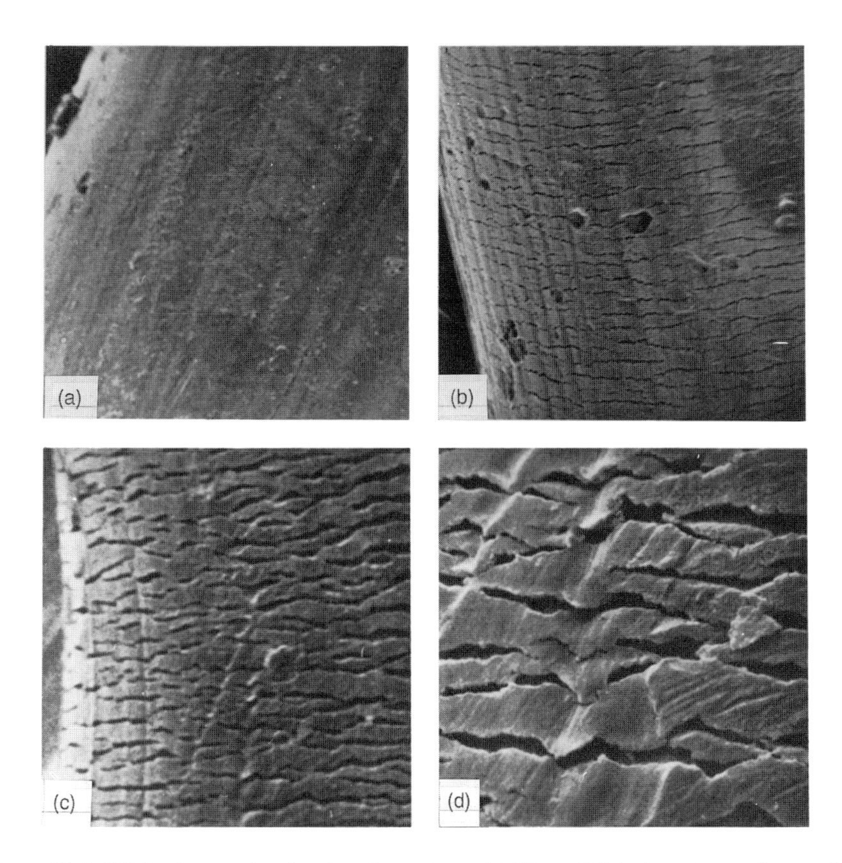

Figure 11.1 SEM micrographs showing the surface of 2.78 tex (25 den) springy polypropylene monofilament after extensions of: (a) up to 50%; (b) up to 100% (note the appearance of the surface cracks; (c) up to 200% (note the widening of the cracks); (d) the surface of (c) at higher magnification. Final magnifications are (a), (b), and (c), 1500×; (d), 3900× (courtesy of L. Konapasek).

the magnitude of the maximum principal stress. Argon and Salama [15] showed that craze growth rates were unaffected by the magnitude of the deviatoric stress (shearing stress with no volume change) but increased with the magnitude of the maximum principal stress.

The effect of molecular weight on crazing characteristics was investigated. It was found that, in order for stable crazes to form, the molecular weight of the sample must exceed the critical molecular weight, M_c [16–19]. Fellers and Kee [20] and Koltisko *et al.* [21] found that the stress at which crazes appeared was independent of molecular weight. In studies of craze density (number of crazes per unit area), Fellers and Kee [20] and Lainchbury and Bevis [19] found that it decreased for molecular weights below 100,000, while Koltisko *et al.* [21] found no effect of molecular weight on craze density. Argon and Hannoosh [22] have shown that the craze density for a surface under a given set of stress conditions becomes saturated with time. Mills [23] suggested that crazes grow past each other to a limited extent because craze interactions reduce the energetic driving force at the craze tips. The effects of craze density and craze interaction on growth rates have not been extensively studied.

11.2.1 Craze Healing Cases

Craze healing was studied by Wool and O'Connor [24] as part of a larger study of interfacial healing [25, 26, 27]. We grew crazes in polydisperse polystyrene, PS, at a constant stress, and measured the nucleation time, τ_1, and growth rate, $G_1 = 2da/dt$, where a is the half-length of the craze. The samples were unloaded, and healed at a constant healing temperature, T_h, for a measured healing time, t_h. The samples were then restressed to the same level as used initially and the nucleation time, $\tau_{2,i}$, and the growth rate, $G_{2,i}$, of the healed craze were measured. We identified five healing cases, displayed schematically in Figure 11.2.

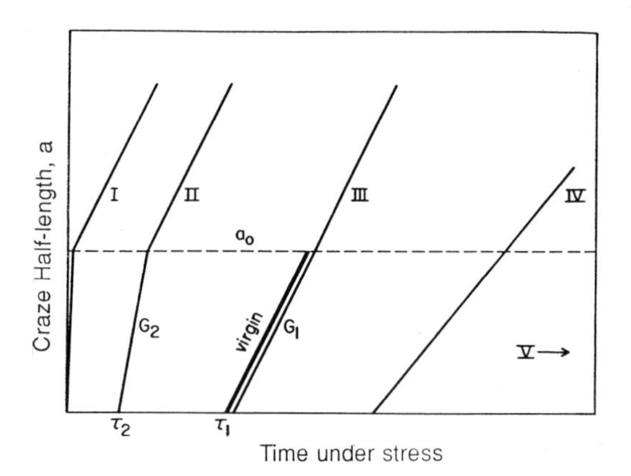

Figure 11.2 Nucleation and growth of crazes in the second loading after healing, compared schematically with the first loading (heavy line) where the craze grew to an initial half-length a_0. The five cases of healing are depicted. Case I—immediate reopening; Case II—partial healing; Case III—similar growth; Case IV—slower growth; Case V—no reappearance. The nucleation time, τ_1, and growth rate, G_1, for the virgin case, and the nucleation time, τ_2, and growth rate, G_2, for the Case II result are shown (Wool and O'Connor [24]).

Case I

At the lowest healing times the craze immediately renucleates and reopens to the final value of the original half length, a_0, and continues its growth at approximately the same rate, $G_{2,I} = G_1$, as that observed initially. Little or no healing has occurred.

Case II

At longer healing times the craze renucleates in a shorter time, $\tau_{2,\mathrm{II}} < \tau_1$, and grows at a faster rate, $G_{2,\mathrm{II}} > G_1$, until the original craze length is reached, and then resumes the original growth rate. Partial healing has occurred.

Case III

The nucleation times and growth rates are similar for the healed and virgin crazes. The craze is considered to be healed.

Case IV

Nucleation times are longer, $\tau_{2,\mathrm{IV}} > \tau_1$, and growth rates are slower, $G_{2,\mathrm{IV}} < G_1$, for the healed samples.

Case V

No renucleation of the craze is observed for the healed craze.

In Case I, healing of the initial craze in a mechanical sense is assumed not to have occurred (except for a small change in the wetting energy of about 1 J/m^2) even though the craze has disappeared optically. In Cases III, IV, and V complete healing of the initial craze has apparently taken place. In Case II, the initial craze is considered to have partially healed. Craze healing was found to occur both above and below T_g to a temperature of 70 °C, and to occur uniformly along the craze interface.

11.2.2 Craze Test Methods

The crazes were grown under a uniaxial stress around a circular hole in a PS film of monodisperse molecular weight (with molecular weight ranging from 88,000 to 1,334,000) as shown in Figure 11.3, following the method of Sternstein *et al.* [14] and O'Connor [24]. The stress distribution is well known from classical elasticity theory; this allows calculation of the major principal stresses for the individual crazes [28]. Craze healing and growth were observed on a microscope (Aus Jena) equipped with a tensile stage and dark-field accessories, and recorded on videotape using closed-circuit TV.

In dark-field microscopy, images of sharp contrast are produced as the crazes and other defects scatter the convergent hollow-cone illumination. Crazes appear white against a black background, as shown in Figure 11.3; the contrast facilitates the recording of data. The initial stress loading rate was approximately 8 MPa/min until the predetermined local stress level was reached, and crazes were then allowed to grow at a constant level of local stress, in the range 14–28 MPa. The crazes studied (except for those in samples with M_n of 88,000) were chosen to be one or more hole radii away from the hole, where the craze density was not great and the stress field did not vary greatly. When the crazes reached a total length of 80 to 120 μm, the samples

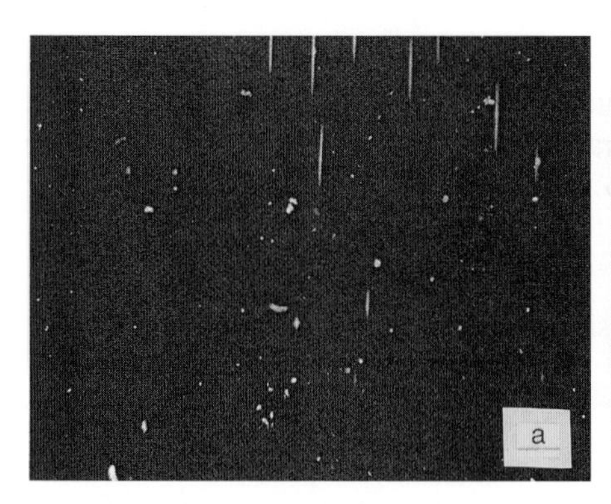

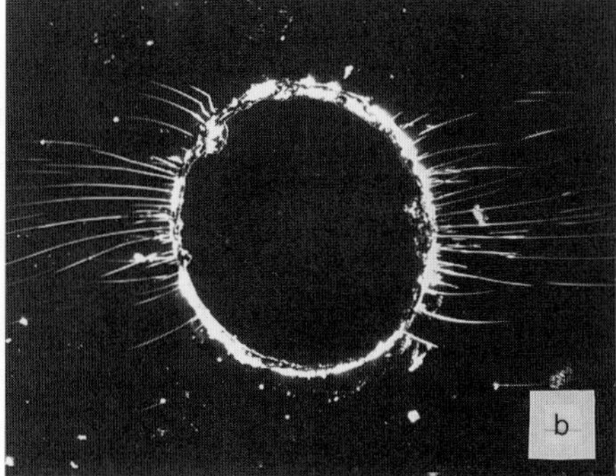

Figure 11.3 (a) Craze growth as seen by dark-field microscopy. (b) Craze growth as seen by dark-field microscopy around a circular hole of diameter 1 mm. Magnification of both is approximately 22×.

were removed and placed in a nitrogen-filled oven at a temperature T_h, for a healing time t_h. Crazes were healed at one of two temperatures, 100 °C or 107 °C. The samples were then reloaded to the initial stress level, the original site of crazing was located, and the subsequent craze growth was recorded in the same manner as before. By this procedure, individual craze growth and healing could be monitored and recorded. The television field of view usually allowed several crazes to be observed simultaneously. Very little creep (less than 1%) occurred in the samples during the crazing process.

11.3 Kinetics of Craze Growth

11.3.1 Growth Rate as a Function of Principal Stress

A typical craze growth pattern is shown in Figure 11.4; the half-length of the craze, a, is plotted as a function of time [11]. The crazes that were used to determine the effect of principal stress on growth rate were sufficiently separated from other crazes so that growth was not affected by craze interactions (discussed in Section 11.3.4). In addition, the craze density was always less than 2500 cm^{-2} in the area being analyzed. The craze growth in Figure 11.4 shows an initial nonlinear growth region until the half-length, a, reaches 20 μm; then the craze growth becomes linear. The nonlinear region was not always apparent. The linear region was used to determine the slopes for the growth rate, $G = 2\mathrm{d}a/\mathrm{d}t$. The nucleation times, τ, were determined by extrapolation of the initial nonlinear region to $t = 0$ on the time axis. We found the craze growth rates to be unaltered when the thickness of the craze approached the thickness of the sample.

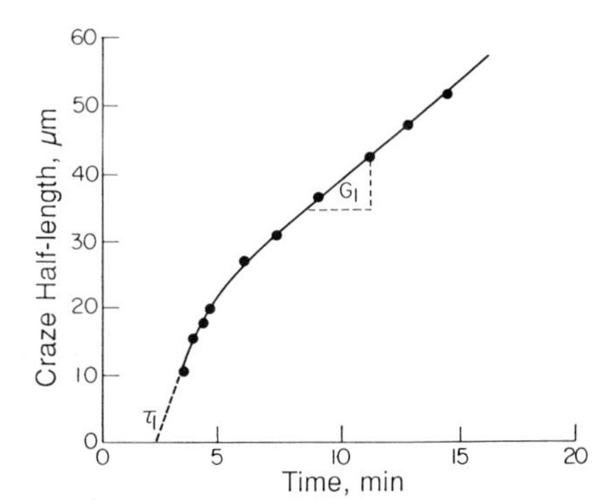

Figure 11.4 Growth behavior of a craze in polystyrene of M_n 217,000. The local maximum principal stress, σ, was 21.2 MPa (McGarel and Wool, [11]).

The growth rate, G, for monodisperse polystyrene samples of M_n 217,000 as a function of the major principal stress is shown in Figure 11.5. Our values are similar in magnitude to those reported by Argon and Salama [15]. If the growth rate, $G = 2\mathrm{d}a/\mathrm{d}t$, is proportional to σ^n, a log–log plot of Argon's data gives

$$G \sim \sigma^{9.4} \tag{11.3.1}$$

with a correlation coefficient r^2 of 0.99. A log–log plot of our data in Figure 11.5 for M_n of 217,000 yields a value of n of 5.99. However, for stresses of 19–23 MPa, the dependence of growth rate on stress for both sets of data is approximately linear as

$$G \approx 3.5\,(\sigma - \sigma_0) \qquad (\mu\mathrm{m/Pa}) \tag{11.3.2}$$

with a slope in the range of 3–4 μm/MPa and an intercept σ_i of about 18 MPa. This empirical approach seems reasonable since it also suggests that crazes do not grow below the stress level σ_0. Similar results were obtained for other molecular weights with M_n greater than 160,000, but the slope depended on molecular weight, as discussed in the next section.

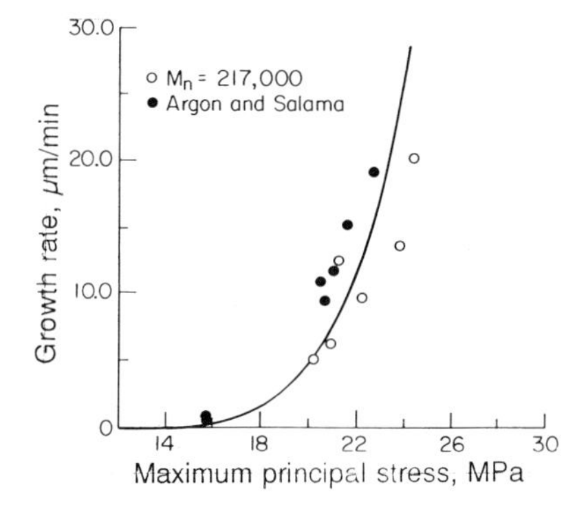

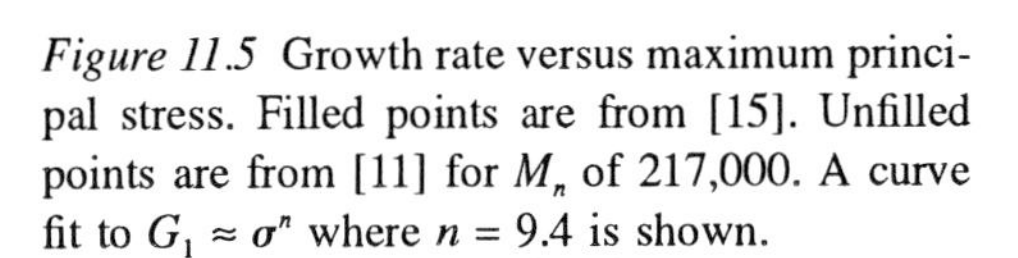
Figure 11.5 Growth rate versus maximum principal stress. Filled points are from [15]. Unfilled points are from [11] for M_n of 217,000. A curve fit to $G_1 \approx \sigma^n$ where $n = 9.4$ is shown.

11.3.2 Craze Growth Rate as a Function of Molecular Weight

Growth rates for crazes grown under local stresses ranging from 18.7 to 22.7 MPa, in polystyrene with molecular weights, M_w, in the range 88,000–1,334,000, are given in Table 11.2 [11]. As in the previous section, the crazes chosen to determine the effect of molecular weight on growth rate were adequately distant from other crazes and were taken from samples where the craze density was less than 2500 cm^{-2}. Individual craze growth data for which the local stress was approximately 21.3 MPa are shown in Figure 11.6 for three molecular weights. A strong dependence of growth rate on molecular weight is observed, but the nucleation times appear to be little affected.

Table 11.2 Craze Growth in Polystyrene as a Function of Molecular Weight

M_w $\times 10^{-3}$	Growth rate, μm/min	Maximum principal stress, MPa	Craze density, cm^{-2}
88	1082	21.7	150
88	980	18.7	150
160	26.2	21.0	2350
160	18.8	20.4	2350
160	21.0	22.3	290
160	15.4	21.7	290
217	5.22	20.3	2350
217	12.6	21.3	2350
217	9.84	22.3	2350
217	6.38	21.0	2350
330	13.1	18.0	1760
330	8.10	19.2	1320
330	6.98	19.0	1760
523	14.2	18.8	2200
523	10.0	20.4	440
523	5.26	19.9	440
1334	18.8	20.2	1170
1334	11.5	20.6	2350
1334	5.66	20.2	2350
1334	13.7	20.7	2350

The growth rates at each molecular weight for the crazes listed in Table 11.2 were averaged and are plotted against molecular weight in Figure 11.7. The growth rate decreased rapidly with increasing molecular weight until M_n was approximately 200,000, and then remained constant.

The craze growth behavior in Figure 11.7 is consistent with observations on the molecular weight dependence of fracture properties presented in Chapter 8. The critical stress intensity factor in PS, K_{1c}, is known to increase with molecular weight until M_n

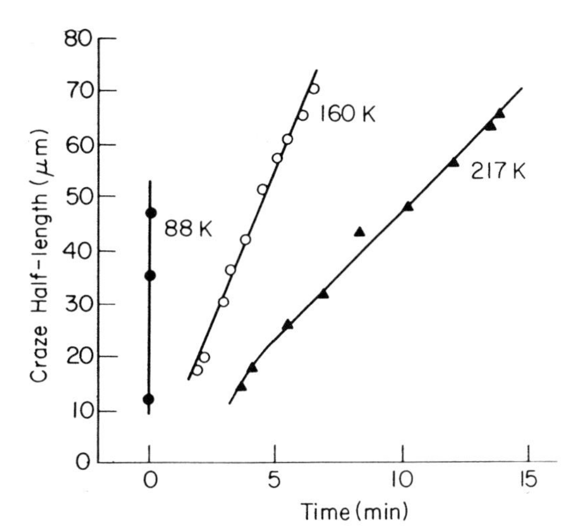

Figure 11.6 Growth behavior of crazes in PS is shown for three molecular weights, M_n, of 88,000, 160,000, and 217,000. The local maximum principal stress was approximately 21.3 MPa (McGarel and Wool).

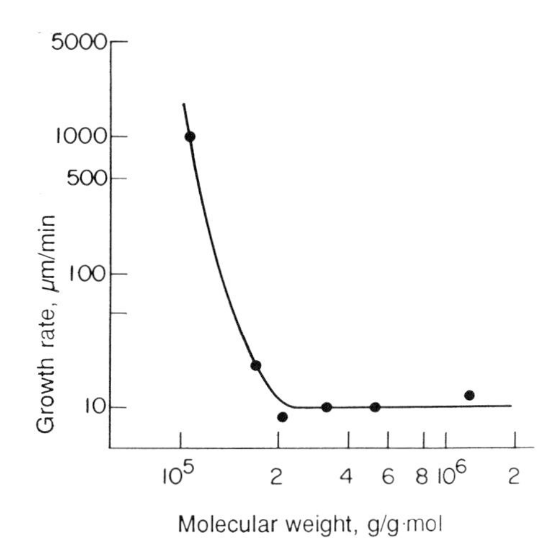

Figure 11.7 Growth behavior of crazes in PS of three different M_n; the local maximum principal stress was approximately 19.9 MPa (McGarel and Wool).

is approximately 200,000 [27, 29, 30]. For PMMA, it was found that K_{1c} is constant for viscosity average molecular weights greater than 200,000 [31, 32]. Since craze nucleation and growth are controlling factors in the fracture process for amorphous polymers, it is not surprising that K_{1c} increases with molecular weight for samples where M_n is below 200,000. This is particularly evident in the behavior of the samples with M_n of 88,000, where only one craze longer than 50 μm propagated and resulted in failure.

In related studies, Koltisko *et al.* [21] observed longer crazes for specimens of M_n of 100,000 compared to samples with M_n of 1,800,000, suggesting a higher craze growth rate for the samples of lower molecular weight. Chau *et al.* [33] grew crazes using a spherical stretching method so that a craze mesh was obtained. The size of the mesh was found to decrease with increasing molecular weight until M_n was approximately 200,000. If one assumes that the crazes propagate until they intersect another craze, and that nucleation is independent of molecular weight, one concludes that higher growth rates should result in larger sizes of mesh for the lower molecular weight samples.

Kramer [13] and Argon and Salama [15] have modeled the growth of crazes with the Saffman–Taylor meniscus instability mechanism discussed in Chapter 7. Using TEM, Kramer demonstrated that the craze tip behaved as an amorphous fluid undergoing an instability, to produce characteristic Saffman–Taylor fibrillar patterns, as shown in Figure 11.8. He derived the following equation for the steady-state velocity of the craze tip, G,

$$G = \sqrt{3}/2\,(\epsilon_f h/n+2)\,[\sqrt{3}h(\beta^* S_t)^2/(8\sigma_f \Gamma)]^n (1-(2\Gamma/\beta^* S_t h)^{2n}] \tag{11.3.3}$$

where Γ is the effective polymer surface energy, h is the height of the craze, and $\beta^* S_t$ is the hydrostatic stress midway between the void fingers. The fluid is considered to have a non–Newtonian fluid law of the form

$$\epsilon = \epsilon_f (\sigma/\sigma_f)^n \tag{11.3.4}$$

where ϵ_f and σ_f are material parameters and n is the non-Newtonian exponent. The polymer surface energy is given by

$$\Gamma = \Gamma_0 + 1/(4 d \nu_e U) \tag{11.3.5}$$

where Γ_0 is the surface energy of the crazes, d is the craze diameter, ν_e is the entanglement density, and U is the energy to break a chain. The hydrostatic stress midway between the void fingers is assumed to be proportional to the applied stress, σ_{ap}, so that Eq 11.3.3 can be simplified to determine its stress dependence as follows

$$G \sim \sigma_{ap}^{2n} (1 - 2\Gamma/\sigma_{ap} h)^{2n} \tag{11.3.6}$$

With applied stresses, σ_{ap}, of the order of 20 MPa (Figure 11.5), craze heights h of about 10^{-6} m, and 2Γ of about 1 J/m^2, then $2\Gamma/\sigma_{ap}h \approx 1/20$ and Eq 11.3.6 reduces to

$$G \sim \sigma_{ap}^{2n} \tag{11.3.7}$$

For non–Newtonian flow, the value of $2n$ is typically between 10 and 20 [13]; this strong dependence of craze growth upon applied stress is consistent with Eq 11.3.1.

If we speculate that pseudo-Newtonian flow characteristics dominate the craze fibril drawing process, then n = 1 and Eq 11.3.3 reduces to the approximation

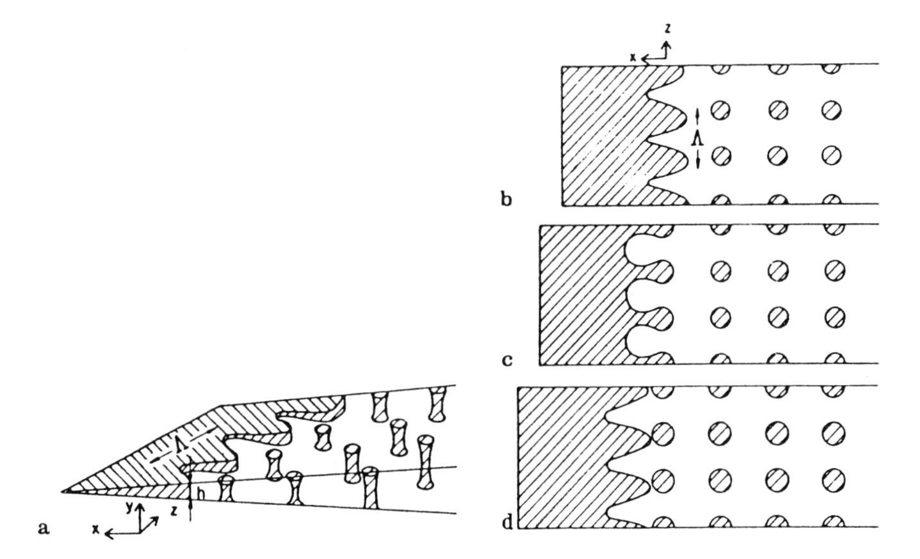

Figure 11.8 Craze formation and growth by the Saffman–Taylor meniscus instability mechanism (courtesy of E. J. Kramer).

$$G \approx \sigma_{ap}^2 h^2/(16\eta\Gamma) \tag{11.3.8}$$

Note that the growth rate in this scenario behaves as $G \sim 1/\eta$, which could help explain the large molecular weight dependence observed in Figure 11.7. The Newtonian flow assumption may be questionable in light of dominant non–Newtonian flow characteristics demonstrated for the Saffman–Taylor mechanism in Chapter 7. However, the fibril drawing process from a glass is different from the high shear rate effects on a fluid above its glass transition temperature.

There is a similarity between Eq 11.3.8 and the expression for the fatigue crack propagation rate, $da/dN \sim K^2/M^{5/2}$ (Chapter 8), via $G \sim \sigma^2/\eta$. Kramer also observes that the craze thickening velocity v_z behaves as $v_z \sim M^{-5/2}$ at craze deformation temperatures near 75 °C. This velocity is the speed at which the craze fibril neck propagates into the undrawn glass in a direction perpendicular to the direction of craze length propagation. He makes the interesting point that the time, τ_d, for the neck to propagate through a chain of end-to-end vector R, at temperatures 10 °C below T_g, is about equal to the relaxation time, T_r, of the melt at (T_g + 130 °C).

The apparent viscosity of the craze material, η_d, is estimated from Eq 11.3.8 as

$$\eta_d \approx \sigma_{ap}^2 h^2/(16 G\Gamma) \tag{11.3.9}$$

If G is 10 μm/min at M of 200,000 and σ_{ap} is 21 MPa (Figure 11.7), and if we let h be about 1 μm and Γ be about 1 J/m^2, then $\eta_d \approx 10^9$ poise (1 poise is defined as 0.1 N·s/m^2). The viscosity/molecular weight relation for PS was determined by Allen and Fox [39] at 217 °C as

$$\eta_{217} = 1.7 \times 10^{-14}\ M^{3.47} \qquad \text{(poise)} \tag{11.3.10}$$

For PS with M of 200,000, $\eta_{217} = 4 \times 10^4$ poise. If we relate the craze viscosity to the zero-shear viscosity, the craze viscosity of $\eta_d \approx 10^9$ poise corresponds to a local craze tip temperature, T_c, of about 170 °C, using an activation energy of 98 kcal/mol. This order of magnitude for the temperature agrees well with a William–Landel–Ferry (WLF) analysis for PS, which gives T_c of about 173 °C.

The energy per unit area of craze, G_f, due to fibril formation is obtained from the Saffman–Taylor analysis in Chapter 8 as

$$G_f = 4\,(\alpha_f\,\eta\,\Gamma\,G)^{1/2}/\pi \tag{11.3.11}$$

where $\alpha_f \approx 4$ is the craze fibril draw ratio. Substituting for the craze velocity G from Eq 11.3.8 above, we obtain the simple and interesting result

$$G_f \approx \sigma\,h \tag{11.3.12}$$

which is similar to the Dugdale model for fracture. When σ is 20 MPa and h is 1 μm, $G_f = 20$ J/m^2. If σ is constant, the value of h would have to increase to about 50 μm in order for G_f to be equal to the virgin fracture energy, G_{1c}, of about 1000 J/m^2. Since values of h on the order of 50 μm are not observed experimentally, we conclude, as we did for the Saffman–Taylor experiments (Chapter 8), that the viscous dissipation contribution to the fracture energy far outweighs the surface energy contribution, G_f.

The dependence of G on σ^2 discussed above suggests a modification of the craze growth equation (Eq 11.3.2) in the form

$$G \approx 0.28\,(\sigma - \sigma_i)^2 \tag{11.3.13}$$

with σ_i of about 16 MPa. The latter value gives a critical strain for crazing as $\epsilon_c = \sigma_i/E$, so that for PS with M_n greater than M_c, when E is 3 GPa, $\epsilon_c = 0.53\%$.

11.3.3 Growth Rate as a Function of Craze Density

We found that the craze density, ϕ_c, has a significant influence on the growth rate of individual crazes [11]. This effect is important in the production of high-impact-resistance plastics and must be considered when individual craze growth rates are evaluated. Averaged values for craze growth for M_n of 160,000 and 1,334,000 as a function of craze density are shown in Figure 11.9. In this figure, the growth rate

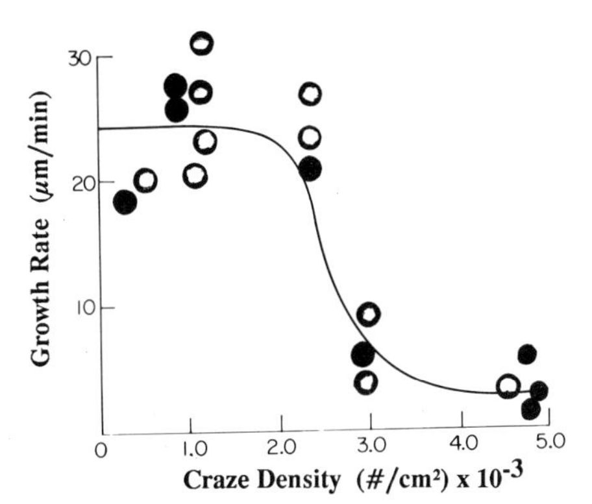

Figure 11.9 Growth rate as a function of craze density for two molecular weights. At high craze density, growth rates are dominated by craze density effects. Unfilled points are for M_n of 160,000; filled points are for M_n of 1,334,000 (McGarel and Wool).

decreases rapidly above a craze density of approximately 2500 cm^{-2}. All samples displayed a similar reduction in growth rates with increasing craze density.

The pronounced effect of craze density on growth rate is currently not well understood. If relaxation of the strain energy density is the driving force for craze formation, it is reasonable to assume that crazes in a region of high craze density have slower propagation rates than those in a similar area with a lower craze density. However, those who seek a solution to this problem should keep several points in mind:

a. craze growth rates decrease with increasing craze density;
b. the individual growth rates are constant with constant density, so that crazes appear to experience the effect of the other crazes even at short times, and
c. growth rates depend on both stress and craze density, but the latter appears to dominate at high density.

We suspect that the craze density effect on growth rate is determined by craze interactions, so that $\phi_c \approx 2/w_c$, where w_c of approximately 10 μm is the critical interaction distance between growth paths of crazes (Figure 11.10), as discussed in the next section.

11.3.4 Craze Interactions

The effect of interaction between individual crazes on growth rate was examined by McGarel and Wool [11]. For crazes separated by a distance w, shown schematically in Figure 11.10, the distance between approaching craze tips, d, was measured as a function of time. Negative d values correspond to craze tips that have grown past one another. We used interaction data for crazes of half-length greater than 30.0 μm with constant growth rates, since growth rates can be nonlinear when the craze half-length is less than 20 μm (see Figure 11.4).

In Figure 11.11, the craze tip distance, d, is plotted as a function of time (growth-rate normalized) for several interacting crazes whose separation distances, w, ranged

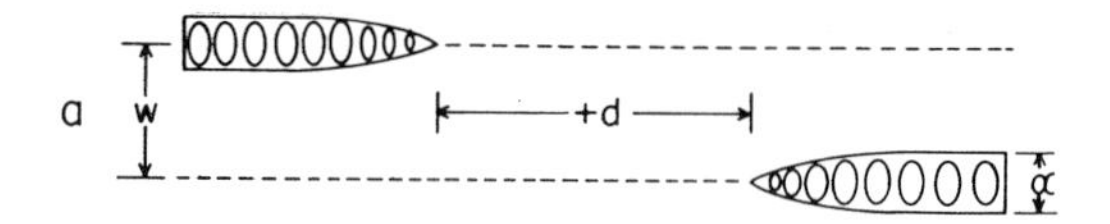

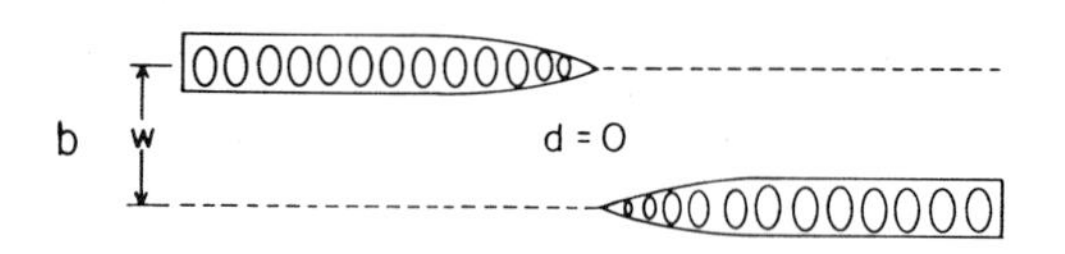

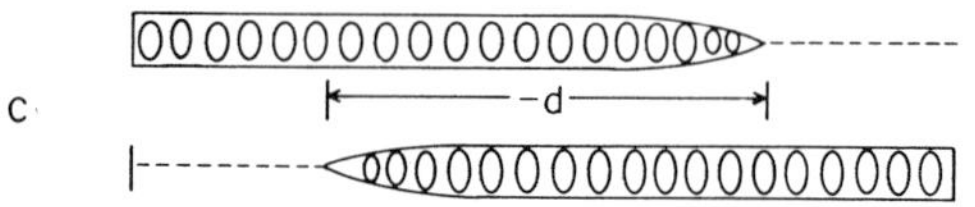

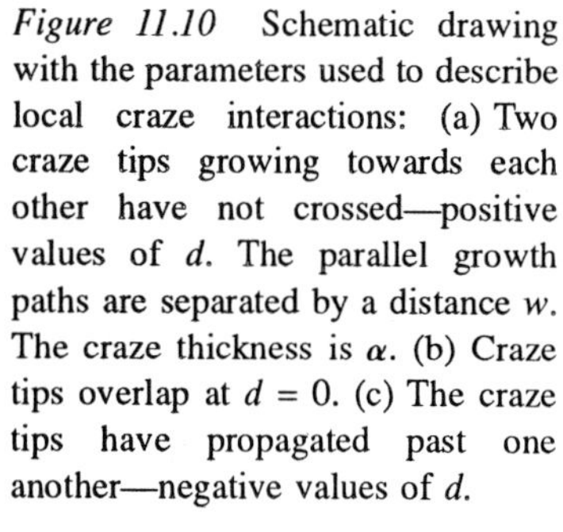

Figure 11.10 Schematic drawing with the parameters used to describe local craze interactions: (a) Two craze tips growing towards each other have not crossed—positive values of *d*. The parallel growth paths are separated by a distance *w*. The craze thickness is α. (b) Craze tips overlap at $d = 0$. (c) The craze tips have propagated past one another—negative values of *d*.

Figure 11.11 Effect of the separation distance, *w*, of parallel crazes on the interaction between approaching craze tips. For *w* greater than 40 μm, no significant change is noted in the growth rates of the approaching craze tips (McGarel and Wool).

from 9 to 64 μm. We normalized the growth data by plotting *d* versus $t(G/G_0)$, where *G* was the growth rate for each craze and G_0 was chosen as 1 μm/min. The purpose of this was to demonstrate the effect of *w* on the behavior of *d* versus *t* independently of growth rate, applied stress, and molecular weight. Several points can be made about the craze interactions in Figure 11.11.

1. The growth rates decrease when craze tips begin to overlap, at *d* approximately equal to 0.
2. Crazes separated by distances *w* of about 40 μm are not significantly affected by craze overlap.

3. The growth directions after overlap are unchanged despite the sometimes large reduction in growth rate. This is consistent with the craze fibrillar material being capable of transmitting the stress across the craze.
4. For w smaller than 40 μm, the growth rates following overlap (negative d) appear to decrease with decreasing w. In fact, the growth rate for the crazes separated by 9.4 μm eventually went to zero, as shown in Figure 11.11.

One expects that when the craze opens by an amount h, the local strain on the material between crazes drops from an initial value of $\epsilon = \sigma_{ap}/E$ to $\epsilon \approx h/w$. When ϵ approaches $\epsilon_i = \sigma_i/E$, the craze no longer grows. For PS, ϵ_i is about 0.006 and crazes that are 10 μm apart require a craze-opening displacement, h, of only about 600 Å to critically slow their growth. The craze interaction effects also offer an explanation for the effect of craze density on craze growth (Figure 11.9).

Exercise

Explain why reaching the critical craze density of 2,500 cm^{-2} causes a rapid drop in the growth rate of crazes.

11.4 Craze Healing

11.4.1 Kinetics of Craze Healing

We first grew crazes at a known stress, healed them at temperatures T_h, of about 100 °C or 107 °C, for various healing times t_h, and then regrew them at the same stress. Four (I, II, III, V) of the five craze healing cases described earlier (Figure 11.2) were observed in this study. Case IV was not observed by McGarel and Wool [11], and it occurred only rarely in the study by Wool and O'Connor [24].

An example of Case I healing is shown in Figure 11.12 for a sample with M_n of 217,000 where the craze was propagated at a maximum principal stress of 23.9 MPa. The craze nucleated at τ_1 of about 1.5 minutes and grew to a half-length, a_0, of about 35 μm at a growth rate $G_1 = 2da/dt = 13.8$ μm/min. The sample was then healed at T_h of about 100 °C for a healing time t_h of 380 minutes, at which point the craze was no longer visible in a dark-field optical microscope. Upon being reloaded to its original stress level, the craze immediately reopened to its original length, a_0, of 35 μm. That is, τ_2 was 0, and G_2 was approximately infinite when a was less than a_0. When a was greater than a_0 the growth rate of the craze was similar to the initial growth rate ($G_2 \approx G_1$).

An example of Case II is shown in Figure 11.13 for the same molecular weight and healing temperature as discussed in Case I but for a longer healing time. The craze initially nucleated at τ_1 of about 1 minute, grew at a rate, G_1, of 2.2 μm/min at σ of 20.5 MPa until a_0 was 25 μm. The craze was healed at $T_h = 100$ °C for a time t_h of

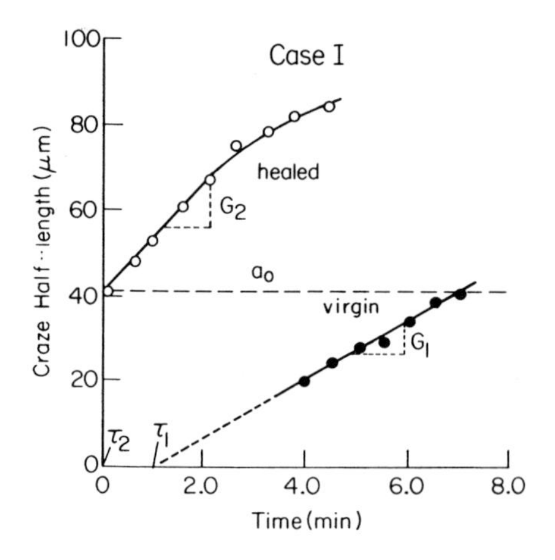

Figure 11.12 Kinetics of craze growth for Case I with M_n = 217,000, σ = 23.9 MPa, T_h = 100 °C, t_h = 380 min, and $a_0 \approx$ 35 μm (McGarel and Wool).

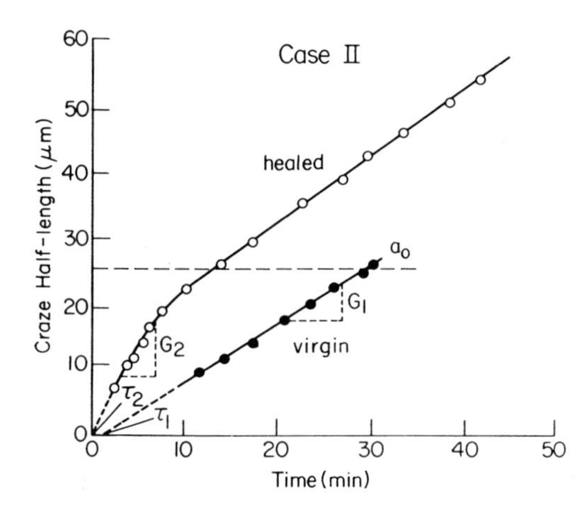

Figure 11.13 Kinetics of Case II craze growth for M_n = 217,000, σ = 20.5 MPa, T_h = 100 °C, t_h = 1405 min, and $a_0 \approx$ 25 μm [11].

1405 minutes. Upon being reloaded, the craze nucleated in a shorter time, τ_2, of about 0 and regrew at a faster rate, G_2, of 4.5 μm/min until it approached the final length of the original craze, a, of about 25 μm. The craze then continued its growth at a rate similar to that observed initially, with $G_2 = G_1$ for a greater than a_0.

In terms of craze growth kinetics, the ratio of the growth rates is related to the viscosities from Eq 11.3.9 by

$$G_2/G_1 = \eta_1/\eta_2 \tag{11.4.1}$$

in which η_2 is the apparent viscosity of the healed craze. This relation assumes that no change has occurred in h, Γ, and σ.

Case III is shown in Figure 11.14 for a sample with M_n of 712,000. This sample was healed at T_h of 107 °C for 1810 minutes. The nucleation times τ_1 and τ_2 were both

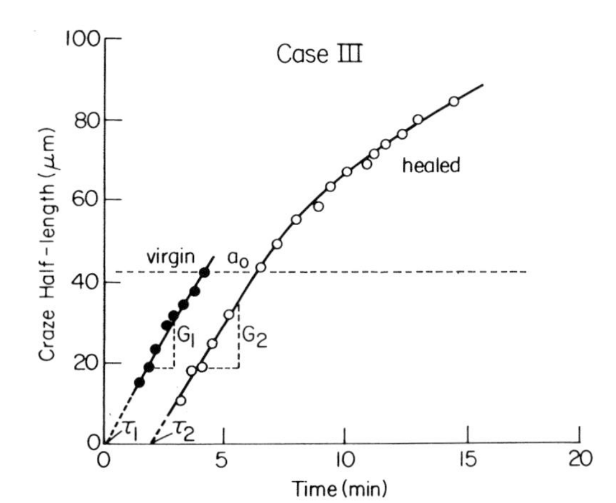

Figure 11.14 Kinetics of Case III craze growth for M_n = 712,000, σ = 24.9 MPa, T_h = 107 °C, t_h = 1810 min, and $a_0 \approx 43$ μm [11].

about 2 minutes, and the growth rates G_1 and G_2 were both about 19 μm/min for all values of a at σ of 24.9 MPa. From the standpoint of the craze kinetics, the sample shown in Figure 11.14 can be considered to be completely healed, but this may not mean that the sample is completely healed with regard to other mechanical testing methods. In the welding of symmetric interfaces, we can see that interdiffusion is necessary to restore the virgin material. The craze was formed from virgin material by a fibril drawing process, with tightening of slack between entanglements and creation of surface area with a high fibril content, possibly determined by some bond rupture. The relaxation of the slack between entanglements should occur at times corresponding to the Rouse relaxation time, $\tau_{RO} \sim M^2$. Both the restoration of the original molecular weight (which had changed because of fracture) and the healing of surfaces between craze fibrils require relaxation times of the order of the reptation time, $T_r \sim M^3$, similar to symmetric welding. In Figure 11.14, the reptation time is orders of magnitude greater than the observed healing time, namely, about 1,800 minutes.

The craze fibril surface area A_f (per unit of craze area) can be determined as follows. The craze fibril diameter d_f is determined from the Saffman–Taylor relation (Chapter 7) and Eq 11.3.8 as

$$d_f = \pi\Gamma/(\alpha_f \sigma) \tag{11.4.2}$$

where Γ is the surface tension, $\alpha_f \approx 4$ is the fibril draw ratio, and σ is the stress required to make the fibrils. With Γ = 0.08 J/m^2 and $\sigma \approx 16$ MPa, then $d_f \approx 40$ Å. This is of the same order of magnitude as observed by Berger and Kramer [38]. The craze fibril surface area per unit area of craze, A_f, is

$$A_f = 4h/(\sqrt{\alpha_f}\, d_f) \tag{11.4.3}$$

When $h = 1$ μm, then A_f = 500 m^2/m^2, which is a considerable increase in surface area. It is not clear how this large surface area heals itself.

Case IV was not observed. This may be because of several factors, including the fact that monodisperse-molecular-weight samples were used in this study rather than the polydisperse samples used by Wool and O'Connor. Moreover, relatively few examples of Case IV were noted in the earlier study.

An example of Case V occurred in a sample where M_n was 714,000, τ_1 was 0.5 minute, and G_1 was 22.6 μm/min at σ of 25.5 MPa and a_0 of 40 μm. Following a healing time of 1810 minutes at 107 °C as in Case III above, the craze when restressed to its original value did not renucleate and grow. However, other crazes nucleated and grew in the vicinity of the healed craze. The individual crazes discussed for Cases III and V were both measured on the same sample in the same experiment. The fact that a healed craze does not renucleate could also be explained by a change in the physical attitude of the nucleating particle at the original nucleation site, or a change in the local stress distribution at the nucleation site.

11.4.2 Molecular Weight Effects on Craze Healing

Several significant differences exist between this craze healing study and previous experiments on healing of symmetric glassy interfaces. Crazes are composed of fibrils of high plasticity; strains of 500% have been observed in the midrib region [13]. The relaxation of the highly oriented fibrillar material can be expected to contribute to the healing of the interface, as do also the interfacial diffusion processes. The molecular weight of the crazed material may undergo significant reduction due to chain scission, and could be of a highly polydisperse nature. The healing time of the interface would then be expected to depend upon the molecular weight of both crazed and uncrazed material. Craze healing should also be helpful in understanding crack healing in amorphous polymers, as reported by Kausch and co-workers [35] and discussed in Chapter 12.

In the craze healing study, we found that higher molecular weight samples required longer healing times to reach similar stages of healing. For example, the samples with M_n of 712,000 required 1810 minutes at T_h of 107 °C before Case V was observed, while a sample with M_n of 217,000 that was healed at the same temperature required only 700 minutes to achieve the same degree of healing. These healing times are orders of magnitude smaller than those predicted from stress relaxation and diffusion data.

The onset of Case II healing at 100 °C for three molecular weights is depicted in Figure 11.15. A slope of 3 is shown for comparison purposes. The higher molecular weight samples required longer healing times to obtain Case II results, but the data are too limited to make comparisons with the prediction, $T_r \sim M^3$. If bond rupture occurs, we expect the healing times to be shorter than the relaxation times for the unbroken chains of the virgin material and we expect deviations to occur from an M^3 healing-time law, as discussed in the next section.

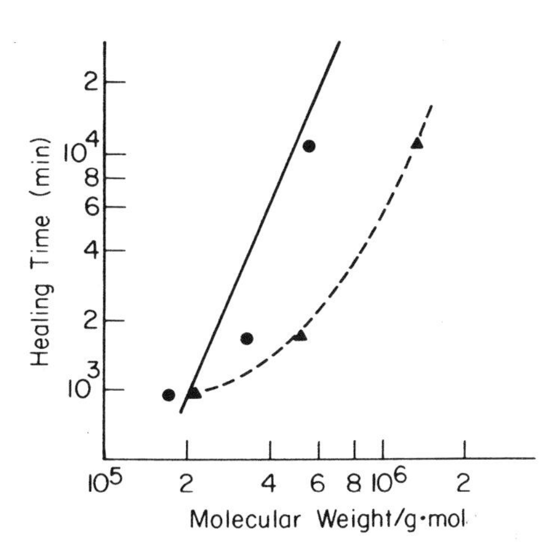

Figure 11.15 Onset of Case II healing at 100 °C as a function of molecular weight. The molecular weights plotted as triangular points (and dashed line) are based on the molecular weight of the virgin material. Estimates of the molecular weight of the craze material (circular points, solid line) were generated using data from [37]. The solid line has a slope of 3.

11.4.3 Molecular Weight Changes During Crazing

Recently several research groups have investigated whether appreciable chain scission occurs during the crazing process. Popli and Roylance used gel permeation chromatography (GPC) to measure the molecular weight distribution of manually crazed samples and discovered some reduction in molecular weight [36]. It was unclear, however, whether the bond breakage was due to craze formation or to craze breakdown brought on by the manual mechanical working of the polymer films. Kramer derived an equation to predict the molecular weight changes occurring during fibril formation [37]. He describes the network of chains in a craze fibril by a series of vectors, d, that connect the entanglement points. The random orientation of each vector is described by a sphere centered at the midpoint of the vector. The craze fibril is considered to be formed from a "phantom fibril" of diameter D. The probability of a chain remaining intact depends on the percentage of the sphere residing in the phantom fibril D. Assuming that all the entanglement loss is accomplished by chain scission, Kramer derives

$$1/M_n' = 1/M_n + (1-q)/M_e \qquad (11.4.4)$$

where M_n' is the molecular weight of the craze fibrils and M_e is the entanglement molecular weight.

Experiments by the Urbana group (Chapter 8) in which monodisperse polystyrene was sliced into 1- and 2-μm slices also provide insight into the amount of chain scission that occurs during the crazing process. In this experiment, the microtome blade was preceded by a craze as it sliced through the sample. The very large surface area per gram generated permitted detailed measurement of the changes in molecular weight distribution due to the craze–fracture process. The slices of PS were analyzed by GPC and the molecular weight distribution was determined before and after slicing. The molecular weight change measured was about 6.5×10^{13} scissions/cm^2 and was largely independent of molecular weight at room temperature. This number is of the same order of magnitude as the number of molecular bridges traversing a unit area.

The molecular weight change observed with 1-μm slices is shown in Table 11.3 and compared to the calculated molecular weight changes predicted by Eq 11.4.4. Assuming a craze width of 0.1 μm, Eq 11.4.4 can be used to calculate the molecular weight of a 1-μm slice if chain scission occurs only during crazing and not during fibril breakdown. Equation 11.4.4 predicts values for M_n in the crazed regions that are less than or equal to M_n of the lowest molecular weight chains observed by GPC. However, as noted by Kramer [37], craze fibrils are not strictly cylindrical; numerous ties exist between fibrils. Kramer also acknowledges that the chains may disentangle without breaking to achieve the required entanglement loss. Both of these factors increase the molecular weight of the fibrils. Thus, we feel that bond rupture occurs during a craze–fracture process but is not as extensive as predicted by Eq 11.4.4.

Table 11.3 Comparison of Molecular Weight Changes Observed and Predicted for Crazing and Fracture

Initial M_n	M_n' from Eq 11.4.4 for crazes	M_n predicted using Eq 11.4.4 for 1-μm slice	M_n' observed [34]	Cutoff M_n [34]
324,000	24,000	144,000	208,000	24,000
712,000	25,000	192,000	352,000	33,000
1,335,000	26,000	221,000	546,000	46,000

Bond rupture during the crazing process may affect craze healing in several ways. If the craze growth rate depends largely on the number average molecular weight and not so much on the details of the distribution, then the craze growth rate as a function of molecular weight depicted in Figure 11.7 can be used to predict growth rates of crazes with bond rupture. We note in Figure 11.7 that the growth rate is independent of molecular weight for M_n greater than about 200,000. Thus, complete craze healing appears to have occurred for samples whose average molecular weight due to fracture in the interface remains above 200,000. However, the healing rates should depend on the lower average molecular weight of the healing interface. In Table 11.3, the molecular weight after fracture is observed to be higher than the threshold molecular weight of about 200,000. Figure 11.15 also shows the healing time as a function of the expected molecular weight with slice or fracture and we observe a better agreement with the M^3 prediction. When bond rupture occurs, the diffusion rate of the lower molecular weight species can be much faster than that of the original unfractured chains. Thus, the symmetric interface model (Chapter 8) may be appropriate to describe craze healing once molecular fracture is considered.

If the molecular weight of the craze falls below the threshold molecular weight of 200,000, the growth rate during healing may be severely affected. For example, samples with M_n of 160,000 that were healed at 100 °C failed to produce Case II results even for healing times that produced Case II results for samples with weights of 1,334,000. This observation may also be explained by chain scission occurring during crazing. In terms of the crazing kinetics, the effects of chain scission would be the most pronounced on the lower molecular weight samples. The slicing data predict that for an initial M_n of 160,000 the molecular weight after crazing/fracture should be about 120,000. The growth rate for the latter molecular weight would be approximately 20 times larger than that of the virgin material and would appear as a Case I result since the healing temperature would be near the glass transition temperature of the virgin sample and complete healing or motion of the chains would not be expected. However, limited healing of the craze region could occur due to local reductions in T_g. Similar calculations would produce a growth rate for the 217,000 sample that is only about twice the original growth rate, so that a Case II result would occur when sufficient healing had occurred. In contrast to the previous study using polydisperse PS, complete healing below T_g was not observed for monodisperse PS.

11.4.4 Summary of Craze Healing

The kinetics of craze growth for virgin and healed specimens was examined in this chapter and the following results were found.

1. The craze growth rates decreased with increasing molecular weight until M_n of about 200,000, after which the growth rates reached a constant value. This behavior was also consistent with the molecular weight dependence of fracture energies. The Saffman–Taylor meniscus instability mechanism gave useful predictions on the effects of stress and viscosity on the growth rate.
2. The craze growth rate was affected by the craze density. When the craze density was greater than about 2500 cm^{-2}, the growth rate of the crazes was slower than in samples of lower craze density.
3. The growth rate of individual interacting crazes depended on the separation distance measured normal to each craze and parallel to the applied stress. Crazes less than about 40 μm apart propagated more slowly when the craze tips overlapped.
4. Four cases of craze healing were observed: Case I at short healing times, when no healing was expected to occur; Case III and Case V at longer healing times, assumed to represent complete healing; Case II at intermediate healing times, assumed to represent partial healing.
5. Higher molecular weight specimens generally required longer healing times to reach similar stages in healing, compared to lower molecular weights. This was explained in terms of the reptation model for healing and the role of bond rupture affecting the molecular weight dependence of growth.
6. Significant bond rupture occurred during crazing. Taking this into account was important to understanding the craze healing data.

11.5 Void and Damage Healing

11.5.1 Microvoid Damage and Healing

Microvoid formation in the interfaces of microstructural elements is an important aspect of mechanical deformation in multicomponent polymer systems [29]. In semicrystalline polymers, filled elastomers, block copolymers, and composites, the interfaces are created by several processes, including impingement growth during solidification, heterogeneous formulation, and domain segregation. Cracks or voids may form during deformation, especially in the weak interfaces, and can vary in size from a few angstroms to macroscopic dimensions. When the stress is removed, the cracks can be permanent, instantaneously reversible (instant healing), or time-dependent reversible (slow healing).

Figure 11.16 shows the generic stress–strain behavior resulting from crack healing in multicomponent polymer systems. The microstructural damage is created by the first cycle of the triangular strain input. The stress response for the second cycle depends on the time, temperature, and humidity or other environmental factors of the sample in the rest state between cycles. As the cracks heal in the rest state, the mechanical properties of the sample are restored partially or fully to the virgin state depending on the extent of permanent damage.

Crack healing can be studied by a variety of experimental methods. Direct observation of crack closure modes and healing processes can be done with a transmission electron microscope using thin films that have been subjected to mechanical damage [51]. Block copolymers with segregated domain structures, such as styrene–butadiene–styrene (SBS) or styrene–isoprene–styrene (SIS), were found to be useful for this study. Similar work has been done on polyethylene by Miles *et al.* [41]; they observed closure of voids in films with a stacked lamellar (hard elastic) morphology. Photooptical methods can also be used to measure the relative amount of crack development and healing [51]. Transmitted light scattered by cracks can be investigated using a photoelectric cell in the ocular of an optical microscope. This method was used by Wool and O'Connor to study the contribution of voids during stress relaxation at constant strain in semicrystalline and block copolymers. When the sample is unloaded, the "stress-whitened" films heal in a temperature-dependent manner, as judged by the gradual restoration of the initial clarity and mechanical strength.

The degree of void damage that develops at constant strain is found to depend on the magnitude and duration of strain. In addition to direct observation of crack healing by optical and electron microscopy, a number of scattering techniques can be used to monitor crack history. These include small angle X-ray scattering (SAXS), light scattering, and neutron scattering. SAXS complements photooptical studies by monitoring growth or healing of voids before they reach the critical size necessary to scatter visible light. Other spectroscopic techniques can be used in conjunction with the above methods to evaluate mechanisms of healing. Broad-line proton NMR has

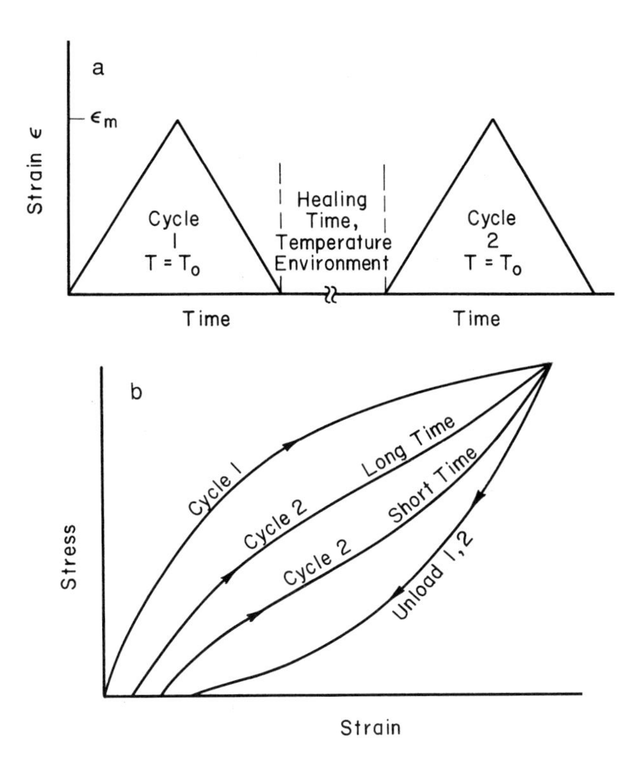

Figure 11.16 (a) The 2-cycle strain and thermal history used to evaluate mechanical recovery and crack healing in polymers. (b) The generic stress–strain behavior resulting from crack healing in multicomponent polymers.

been used to determine the behavior of the amorphous or fluid-like chains during deformation and healing of hard elastic polypropylene [42]. Infrared spectroscopy can also be used to investigate conformation, orientation, and other molecular changes during healing [43, 44]. Other useful techniques include mass spectroscopy, electron paramagnetic resonance (EPR), and luminescence methods. The success of the experimental method rests on the microscopic nature of the healing process.

Healing rarely involves re-formation of primary bonds associated with chain scission processes, although radical recombination can occur in crystal "cages" [45, 46] and to a certain extent in disordered molecular regions. Healing readily occurs by restoration of entanglements and secondary bonding between chains or microstructural components, so van der Waals or London dispersion forces play a very important role in healing. Other kinds of bonding that may be important in healing processes include hydrogen bonding and chemisorption of chains on filler particles. The rate of healing depends on the transport of the separated molecular components into each others' vicinities and the subsequent re-formation of the original or equivalent bonding states. Transport can be achieved by diffusional processes (Chapter 2) through Brownian motion and by bulk translation of microstructural components as strain energy is released. Since molecular mobility is necessary for achieving time-dependent healing, it is expected that healing rates are appreciably affected at temperatures near the glass transition temperature(s) of the healing component(s) in the material.

In the sections that follow, the general phenomena of crack healing are investigated in terms of mechanical properties. Mechanical recovery due to crack healing is examined as a function of time, temperature, and strain for semicrystalline polypropylene, SBS block copolymers, and carbon-black-filled natural rubber. An empirical theory of healing is developed that can be applied to all multicomponent polymer systems. It becomes apparent from the physical behavior of damage and healing processes that there is a microstructural analogy to molecular behavior that explains viscoelastic behavior of polymers. The microstructural contribution is very important for understanding nonlinear viscoelastic behavior in addition to bulk or "noncrack" constitutive behavior. While damage healing is advanced as a point of study herein, it is recognized that other noncrack mechanisms also contribute to the mechanical behavior described in Figure 11.16. However, the importance of healing is clearly demonstrable.

11.5.2 Mechanical Recovery

We can define an adiabatic energy balance for polymer mechanical deformation in terms of three mechanistic quantities as

$$\dot{W} = \dot{U} + \dot{D} + \dot{S} \tag{11.5.1}$$

where $\dot{W}$ is the rate of working (mechanical), $\dot{U}$ is the rate of change of strain energy, $\dot{D}$ is the rate of dissipation due to viscous and plastic processes, and $\dot{S}$ is the rate of change of damage or surface energy due to fracture mechanisms at the molecular or microstructural level.

For the strain history shown in Figure 11.16, the mechanical recovery R can be defined in several ways using parameters from Eq 11.5.1. For example, R can be defined in terms of the total work W_1 and W_2 done in cycle 1 and cycle 2, respectively, as R_W

$$R_W = \frac{W_2}{W_1} \tag{11.5.2}$$

The use of Eq 11.5.2 requires few assumptions but it does not provide information regarding specific contributions of the deformation mechanisms during recovery. More detailed information can be obtained by analysis of the stress–strain data in terms of the energetic parameters, U, D, and S. In the elastic fracture case, D is 0 and we can define recovery as

$$R = \frac{S_2}{S_1} \tag{11.5.3}$$

where S_1 and S_2 are the total fracture work to create voids, break bonds, etc., in the first and second deformation cycles, respectively, and can be obtained from Eq 11.5.1 as

$$S = W - U \tag{11.5.4}$$

The stored elastic energy, U, is the integral below the unload tensile curve. Thus, it is assumed that during healing, all mechanical recovery is expressed in changes of surface energy as the cracks heal and bonds reunite, and no change occurs in the strain energy function of the material. The latter requirement means that the tensile unload curves of the two cycles be superimposed (Figure 11.16). This procedure works well with materials that exhibit a predominantly elastic–fracture mechanical response but it may be difficult to apply to highly viscous systems. With these considerations in mind, we will use Eq 11.5.3 as a working definition of mechanical recovery, R.

11.5.3 Mechanical Recovery in Polypropylene

Hard elastic polypropylene (HPP) fibers with stacked lamellar morphologies were studied. These materials have been the subject of many studies [47–50]. Their tensile properties are unusual and are shown in Figure 11.17. HPP is highly crystalline (the density is 0.92 g/cm^3) with the molecular chain axes preferentially oriented along the fiber axis. Large deformations are accommodated by void formation between lamellae and aggregates of the microstructure [47–50] (Figure 11.1). The internal crack surface area varies approximately linearly with strain and is about 100 m^2/g at 50% elongation. Thus, this material is excellent for the study of crack healing. As shown in Figure 11.17, the second-cycle stress response depends on the healing time of the fibers between deformation cycles. At short healing times, considerable stress softening is observed, but at longer times the original mechanical properties of the fibers are restored as the cracks heal.

We investigated recovery in HPP fibers (provided by H. Noether of Celanese) quantitatively as a function of time, temperature, and strain, using the strain history shown in Figure 11.16. A strain rate of 10/min was applied on an Instron tensile tester with a gage length of 2.0 in and a crosshead speed of 20 in/min. The fibers were tested at ambient room temperature and humidity for both first and second cycles. To study the effect of temperature on healing, we placed the fibers in a constant temperature box, −17 °C to 80 °C, between deformation cycles for the desired healing time, t. The healing time was measured as the time between the unloading at zero strain in the first cycle and the beginning of the second cycle. The maximum strain, ϵ_m, was the same for both cycles.

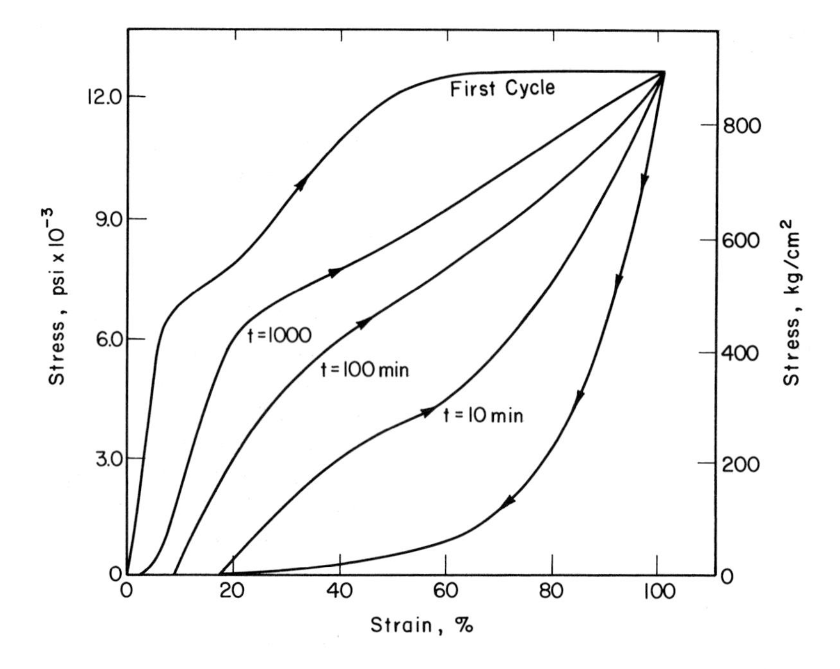

Figure 11.17 The effect of healing time on the stress–strain behavior of hard elastic polypropylene fibers at room temperature.

The effect of healing time and strain maximum on recovery of HPP fibers at room temperature is shown in Figure 11.18. From Eq 11.5.3, the recovery R is plotted as the ratio of the fracture energies. R is approximately linear with log t, and faster recovery occurs at lower strain maxima. The data did not obey any simple healing law of the type $R \sim t^{\alpha}$ observed for symmetric interfaces. The data are qualitatively similar to recovery data obtained by Cannon *et al.* [48], who used the fiber length ratio as a definition of recovery, that is, the fraction of the initial set recovered with time.

The effect of temperature on recovery is shown in Figure 11.19 for strain maxima of 50, 75, 125, and 150%. In each case, the recovery increases with temperature, as might be expected from the enhanced molecular mobility at higher temperatures. At −17 °C (T_g is about −10 °C), no recovery was observed for any of the strain series. In each case the initial or instantaneous recovery occurred at room temperature but further recovery did not occur below T_g. We were able to restart recovery by heating the sample above T_g and stop it by allowing the temperature to fall below T_g. We also observed that at 150% strain, where considerable damage had been created, almost complete mechanical recovery was obtained at higher temperatures.

At temperatures much higher than 80 °C, annealing rather than healing processes become important and the constitutive microstructural character of the material may change. In such a case crack healing arguments for the interpretation of the data are

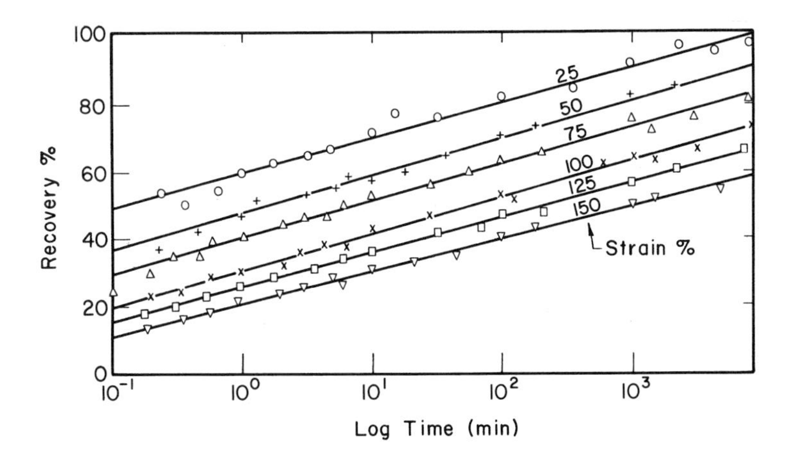

Figure 11.18 The mechanical recovery as a function of log healing time for HPP samples subjected to several strain maxima at room temperature.

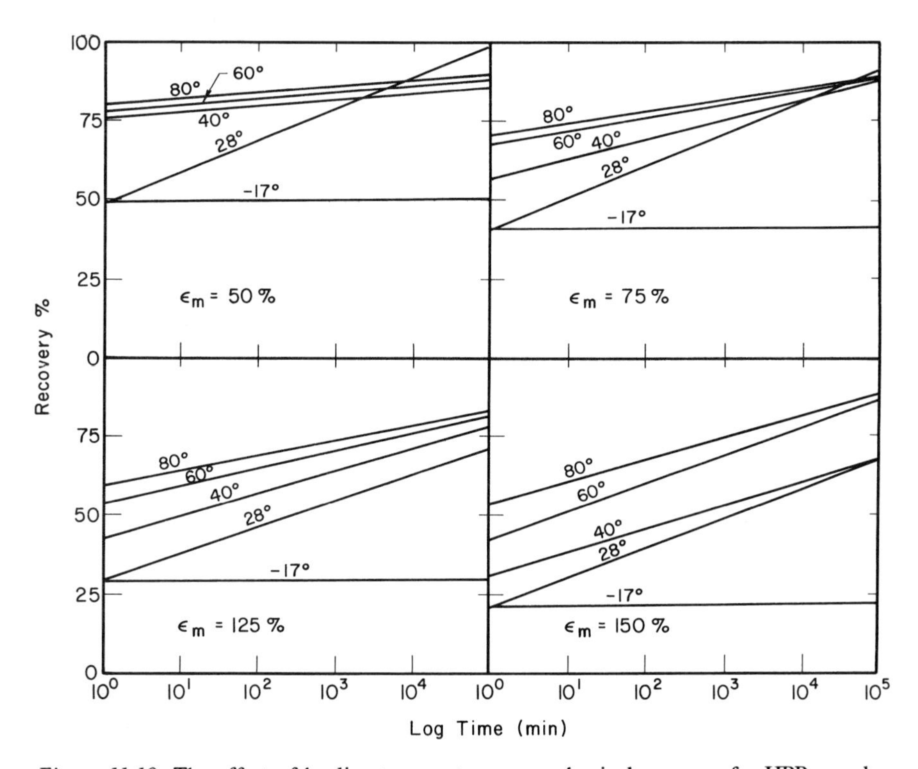

Figure 11.19 The effect of healing temperature on mechanical recovery for HPP samples.

rendered inappropriate. It should also be noted that despite the thermal treatments between deformation cycles, the unload curves were superimposed fairly accurately, which allowed the subtraction of the strain energy component from the total work input. At strains less than 15%, little hysteresis was observed even though cracks formed. Microcracks can contribute elastically to the mechanical response without hysteresis effects, that is, cracks can open and close in a reversible elastic manner especially at low strains [40].

In HPP fibers, crack healing involves molecular motion in the amorphous regions between the crystalline lamellar surfaces as well as healing of internal crystal defects. Broad-line proton NMR studies [42] of HPP fibers indicate that a large increase in the central component of the derivative absorption spectrum associated with the more mobile or fluid-like chains occurs with increasing strain. In the rest state between deformation cycles the intensity of this fluid-like central component decreases with healing time in a manner closely paralleling the mechanical recovery. The increased mobility with strain can be attributed both to the increase in internal surface area providing additional configurational freedom to the amorphous chains on the lamellar surface, and to the generation of new amorphous material. During healing, the debonded lamellar surfaces come together and the intervening amorphous chains intermingle and readsorb on the surfaces, assisted by van der Waals interactions. The increasing restriction of mobility of chain segments during healing causes the observed decrease in the fluid-like central component of the NMR broad-line spectrum.

Density measurements taken on HPP fibers during healing show an increase in density from 0.86 to 0.93 g/cm^3. If one assumes little density change of the lamellar crystals, then very large density changes must occur in the amorphous component. An increase in density must contribute to a decrease in the chain mobility, which is consistent with the observed NMR behavior. Stress infrared studies [44] of HPP films indicate a decrease in both molecular orientation and helical conformation of the PP molecules with deformation. During healing in the rest state, the molecular c-axis orientation and conformational regularity are restored. The increase in orientation is attributed to the deflected lamellae on the void surfaces returning to their initial stacked morphology, with a subsequent restoration of the chain stem alignment. This interpretation is supported by X-ray studies of lamellar bending in hard elastic polymers [52, 53].

Crack healing in semicrystalline polymers can also be interpreted in terms of fracture mechanics, as a process that restores the adhesive fracture energy, Γ, at each point on the interface to its original value. When we applied the peel adhesion model discussed in Chapter 7 to HPP, we found that Γ varied along the crack surface during healing. We thus observed that healing occurred simultaneously at all points on the crack surface but at different rates. As the crack propagates by lamellar debonding, the molecular damage or defects on the crack surfaces vary with position in such a manner that the relative damage is least in the vicinity of the crack tip and greatest towards the center of the crack away from the crack tip. This concept of surface damage is useful for interpreting modes of healing in molecular and macroscopic models [54].

In the case of HPP, the adhesive fracture energy for lamellar debonding was calculated to be about 0.1 J/m^2. Kaelble [55] suggests that Γ values for interfacial debonding in polymers can be interpreted as the sum of the reversible thermodynamic contribution plus a diffusional demixing component. In HPP, the diffusional demixing of amorphous chains represents about 70% of the total adhesive fracture work. From this viewpoint healing involves a rapidly reversible or instantaneous component due to the thermodynamic surface tension effect, followed by a slower healing rate involving diffusional mixing of chains in the amorphous interface. The latter effect is consistent with the T_g dependence of healing and is also useful for understanding deformation-rate-sensitive adhesive fracture energies.

11.5.4 Block Copolymers

Mechanical recovery experiments were conducted on SBS block copolymers (Kraton 1101) by O'Connor and Wool. Film samples 3 × 0.5 × 0.01 in were cast from a solution of THF/MEK (tetrahydrofuran/methyl ethyl ketone) (90/10). Under controlled evaporation conditions, the styrene component segregates into domains embedded in the polybutadiene matrix. The resulting material is a thermoplastic elastomer in which the hard styrene domains act as cross-links for the rubbery butadiene matrix [56, 57]. During mechanical deformation, voids form in the interfaces between domains and domain clusters. Healing of these voids can be observed by photooptical techniques or TEM [51].

Results on healing in SBS films are shown in Figure 11.20 [51]. The mechanical recovery, R, is plotted as a function of log time for healing temperatures ranging from −70 °C to 43 °C. Again, we obtained R by separating the fracture energy component from the total work input using the strain history shown in Figure 11.16 and a maximum strain of 300%. At the higher temperature of 43 °C, complete recovery is obtained in about 12 hours. At T of −70 °C, time-dependent recovery is not observed.

The effect of T_g on healing is interesting in this material, since we have two discrete T_g values for the domains and the matrix as well as a range of possible T_g values in the interfacial regions due to chain mixing. T_g is 100 °C for polystyrene and −85 °C for polybutadiene. Since complete recovery is observed below 100 °C, the polystyrene chains do not appear to play a major role in the healing process. However, healing ceases at some temperature above the T_g of −85 °C for polybutadiene, so it is probable that the mobile chains that are responsible for healing are those in the interfaces. These interfacial regions are mixtures of styrene and butadiene segments and thus have a resultant effective T_g higher than that of polybutadiene.

Recovery in SBS is more rapid than in HPP (Section 11.5.3). This is due to the higher mobility of the butadiene segments compared to polypropylene and to the cross-linked morphology providing a strong elastic memory [56]. Also, while SBS was subjected to much higher strains than the HPP fibers, the degree of microstructural damage was considerably lower in the block copolymer. The mechanical hysteresis and

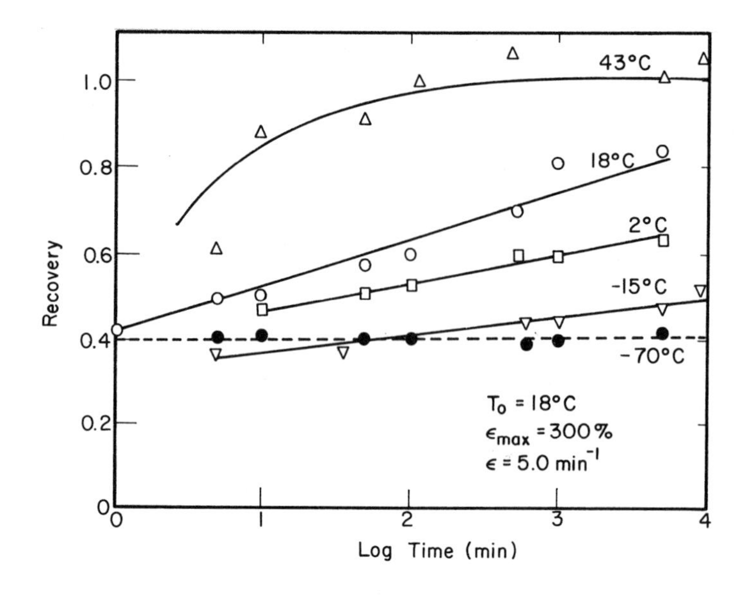

Figure 11.20 Recovery of SBS block copolymers as a function of healing time and temperature (Wool and O'Connor).

healing observed in these experiments are in general agreement with results of other investigators [55–60].

The generation of voids and their disappearance is readily observable by optical methods [51]. Figure 11.21 shows a sequence of dark-field photomicrographs of a styrene–isoprene–styrene (SIS) Kraton 1107 sample. The samples were cast from a mixture of THF and MEK, to produce films about 75–125 μm thick. The dark-field images were obtained under the following conditions (a–e):

a. The dark-field optical image of the virgin as-cast SIS film shows no defects. The light transmittance, T (%), is essentially zero. The featureless image is due to the lack of any source of scattering in the virgin film.
b. The sample is deformed to 250% constant strain and held for 1 hour at 22 °C. Voids eventually form (after about 1 hour) and scatter light in the dark-field image. The amount of scattering depends on the wavelength of the light. Here the transmittance, T, is 95% at 405 nm.
c. When the sample in (b) is unloaded to zero stress, the voids immediately close up and the image becomes dark instantly. The voids may not have closed completely, but they are small enough to no longer scatter light. However, little or no void healing has occurred, since the voids immediately reopen when the sample is reloaded right away to the same strain, and the bright image in (b) returns.
d. When the sample is healed in the rest state at 22 °C for 7 hours, and then re-strained to 250% for 10 minutes, the image shown in (d) is obtained. A significant number of voids are missing compared to (b). Most of the voids in (b) have healed,

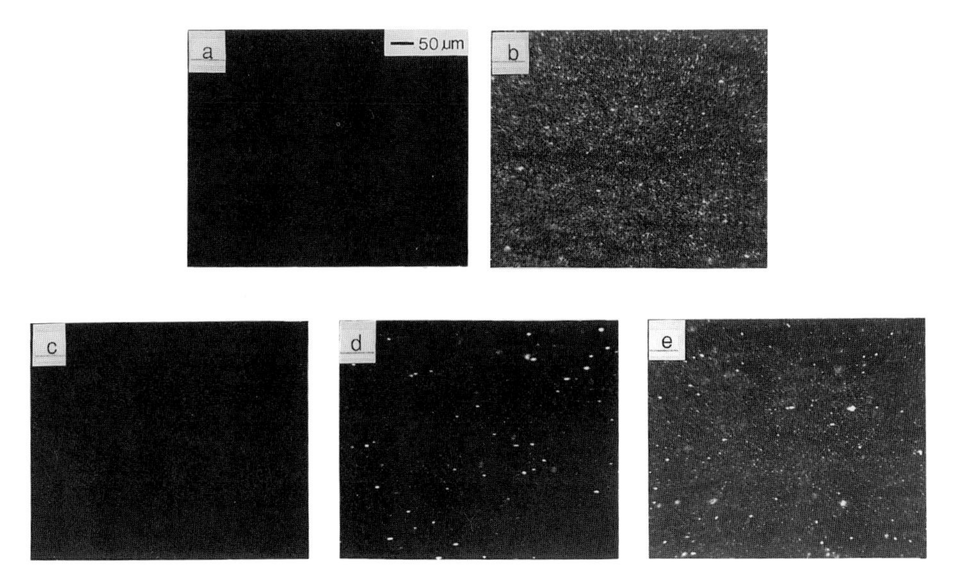

Figure 11.21 Dark-field photomicrographs of the same sample showing (a) the virgin sample; (b) the voids present after 1 h at 22 °C and 250% strain; (c) the immediate disappearance of voids upon unloading to zero stress; (d) the voids present after 10 min at 250% strain and 22 °C, following 7 h of healing at 48 °C at zero stress; (e) the voids present after an additional 50 min at 250% strain and 22 °C. The stretch direction is horizontal. The brighter spots in (d) and (e) are dust particles that became attached to the film surface during handling (O'Connor and Wool).

at least to undetectably small sizes. The brighter spots in (d) are not voids but are dust particles picked up by the sample during the experiment.

e. After an additional 50 minutes at 250% strain, the voids return and the transmittance is the same as that in (b).

The kinetics of void formation is shown in terms of the wavelength-dependent optical transmittance in Figure 11.22, where the sample discussed above was held at a constant strain of 250%. This experiment was done in a Brice–Phoenix light scattering photometer [51]. The transmittance was normalized to the value obtained immediately after application of the strain step function. After an apparent initiation time, the transmittance decreased monotonically for all wavelengths. The initiation time probably corresponds to the time required for voids to grow to a size at which they begin to scatter light, rather than to a sudden appearance of voids at about 30 minutes. The initiation time was generally observed to decrease with increasing strain level.

The stress relaxation response for the constant strain experiment in Figure 11.22 cannot be interpreted in terms of simple viscoelastic functions, since most of the relaxation is facilitated by void formation. This is a good example of where the initial stored elastic energy is dissipated primarily by damage rather than by purely viscous

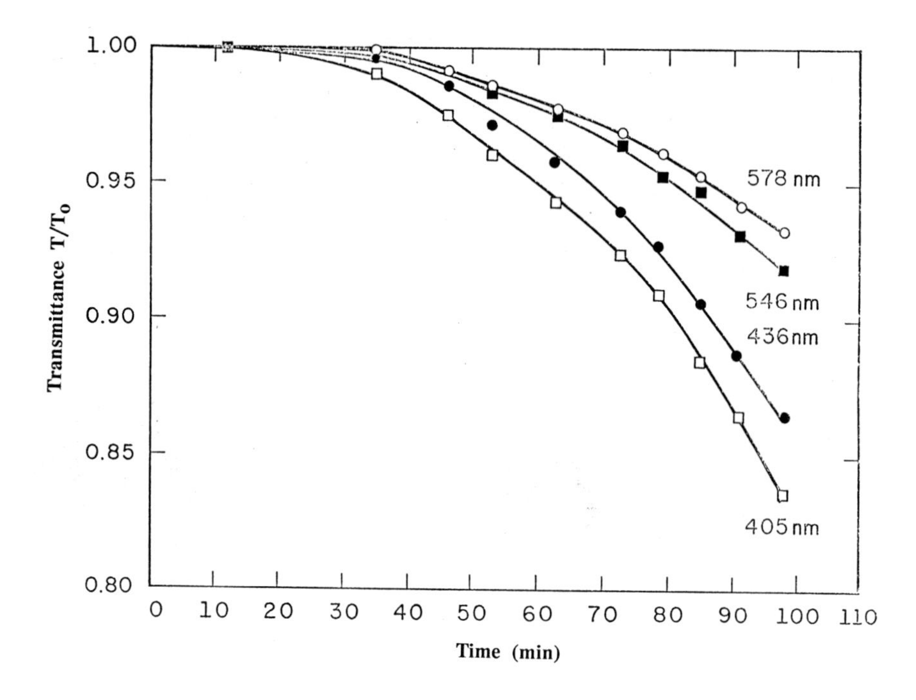

Figure 11.22 Wavelength dependence of optical transmittance during void formation at 250% strain and 22 °C. The transmittance was normalized to T_0, the value obtained immediately after the constant strain was applied.

processes, and is a commonplace occurrence in multicomponent polymer systems. Thus, spring–dashpot viscoelastic models may not be the best approach to understanding or describing this behavior.

The isothermal optical kinetics of void healing is shown in Figure 11.23. The voids were first formed at the reference damage condition, 7 hours at 250% strain at 22 °C. The sample was unloaded and allowed to heal for a given time and temperature. The samples were then re-strained to the reference strain and the transmittance at 546 nm immediately recorded. The optical recovery exhibits a sigmoidal recovery on a log time scale, with little recovery occurring at short times and full recovery possible at long times. Using Mie scattering theory we could determine that the small voids healed preferentially with time, often leaving some large voids [51].

An increase in the healing temperature shifts the recovery response to shorter times, similar to the time–temperature correspondence observed for molecular relaxations. The data can be superimposed if they are shifted along the time axis, with an apparent activation energy of 9 kcal/mol.

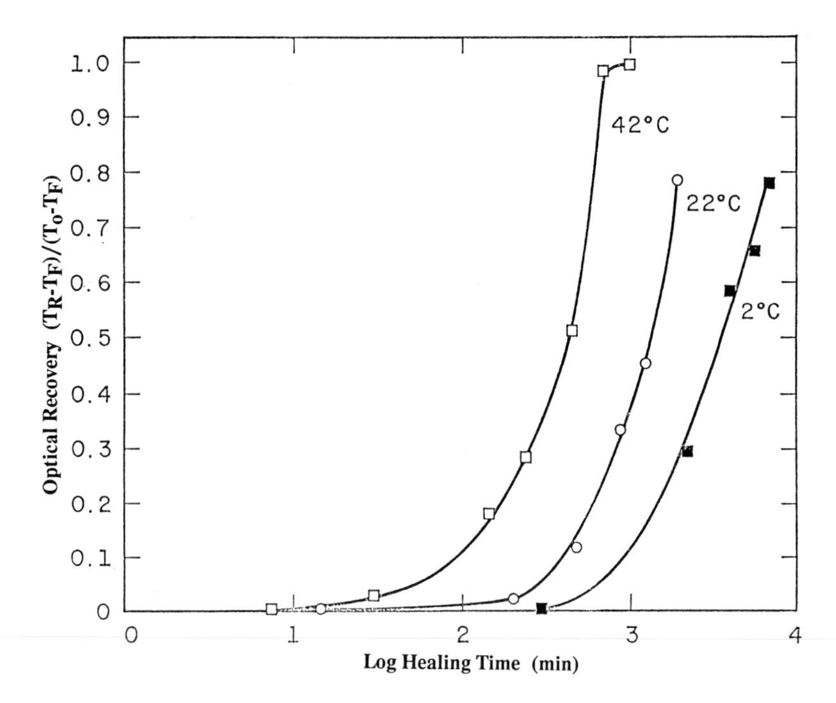

Figure 11.23 Optical recovery versus healing time at the healing temperatures indicated. Voids were formed at reference conditions of 250% strain and 22 °C.

11.5.5 Filled Elastomers

The stress softening effect in filled natural rubber was investigated by Mullins [61, 62], Bueche [63], and others [64]. Many processes contribute to this effect but the dominant molecular process is considered to be debonding of matrix chains from the filler particles and subsequent disentanglement. Mullins observed that this effect was reversible, that is, healing would occur in the rest state after deformation [62]. He determined recovery values, R, as a function of time and temperature from stress ratios at the same strain level in the first and second deformation cycle and found the healing rate increased with temperature with an approximately linear log time dependence.

Upadhyayula and Wool examined mechanical damage and healing in carbon-black-filled vulcanized natural rubber (provided by Dr. A. Aggarwal of General Tire and Rubber) using the strain history shown in Figure 11.16 [65]. The samples contained 34 and 74 parts per hundred (phr) of carbon black. Strains of 100% and 200% were used with healing temperatures of 18 °C and 40 °C. The samples were 3.0 in long,

0.5 in wide, and 0.084 in thick. The results on the effect of strain level and temperature on R are shown in Figure 11.24 and Figure 11.25, respectively (the solid lines are based on a healing theory developed in the next section). Again, we observe an approximately linear recovery with log time. The lower strain causes the least damage and thus higher recovery (in some cases, the recovery can be independent of strain) and, as Mullins observed [62], faster recovery occurs at the higher temperatures.

Volume measurements were taken of the filled rubber materials. A sample strained to 200% and unloaded had an initial volume increase of 0.5%. This volume increment decreased to zero after about 30 minutes of healing at 18 °C and there was no difference in density between the healed and a fresh virgin sample.

Healing in filled elastomers is considered to involve readsorption of chains on the filler particle surfaces and closure of voids in the polymer matrix. Carbon black used in materials such as auto tires is considered to be a very active filler, so that debonded chains would have little difficulty reattaching once they diffused to the particle surface. Recovery in these materials is rapid, similar to recovery rates in SBS, and is presumably controlled by the high mobility of the elastomeric chains.

11.5.6 Solid Rocket Propellants

Solid rocket propellants, such as used in Space Shuttle engines, missiles, lunar modules, and space stations, behave similarly to auto tire rubber. Rocket propellant typically consists of a highly filled polymer matrix. The filler is largely (about 80% by volume) ammonium perchlorate oxidizer, with additives such as burning rate catalysts; the polymer phase consists of a partially cross-linked polymer, such as hydroxy-terminated polybutadiene. The viscoelastic constitutive relations for these materials are complex and nonlinear because of damage and healing processes [68–71].

We examined the mechanical recovery of "live" and "inert" solid rocket propellants (SRP). [An inert SRP is one in which the explosive oxidizing agent has been replaced by salt (NaCl)]. SRP "dog bones" were subjected to the sawtooth strain cycles, using a strain maximum of 12%. Figure 11.26 shows the characteristic stress–strain response, where the second cycle followed healing treatments at 24 °C. The response is typical for a filled elastomer. Eventually, the SRP recovers its original mechanical properties. The rate of recovery increases with temperature.

Live propellants behave similarly (R. P. Wool, unpublished data) and it may be reassuring to astronauts that damage inflicted to their engines in the form of microcracks has the opportunity to heal completely. Such microcracks increase the internal surface area available for combustion in the motor and are potentially disastrous. In space environments, controlled solar annealing of SRP motors could improve their reliability.

J. E. Fitzgerald reports on experiments (about 1963 at Lockheed Propulsion Company, Redlands, CA) with a highly filled PBAN (polybutadiene acrylonitrile copolymer) SRP [72]. Samples were held at a 10% uniaxial strain in air at 50 °C for 24 hours. The usually observed damage, a decrease in modulus caused by chain

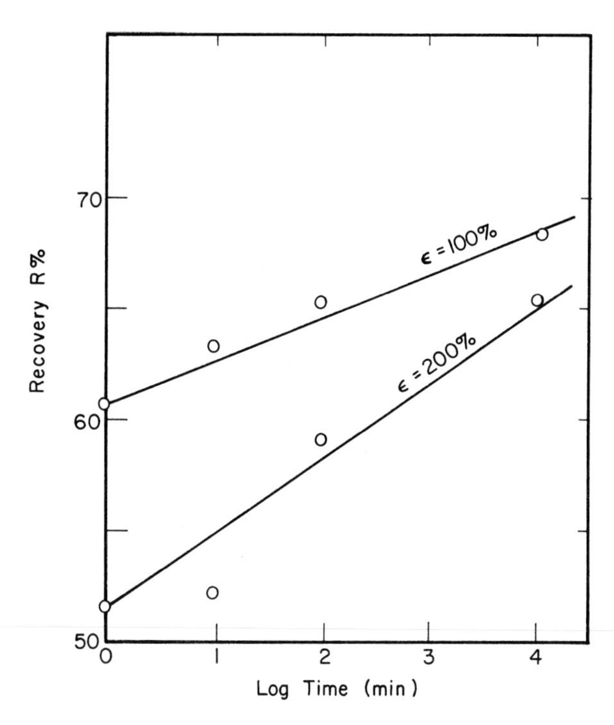

Figure 11.24 Recovery of carbon-black-filled natural rubber (74 phr) as a function of healing time at room temperature for strain maxima of 100% and 200%.

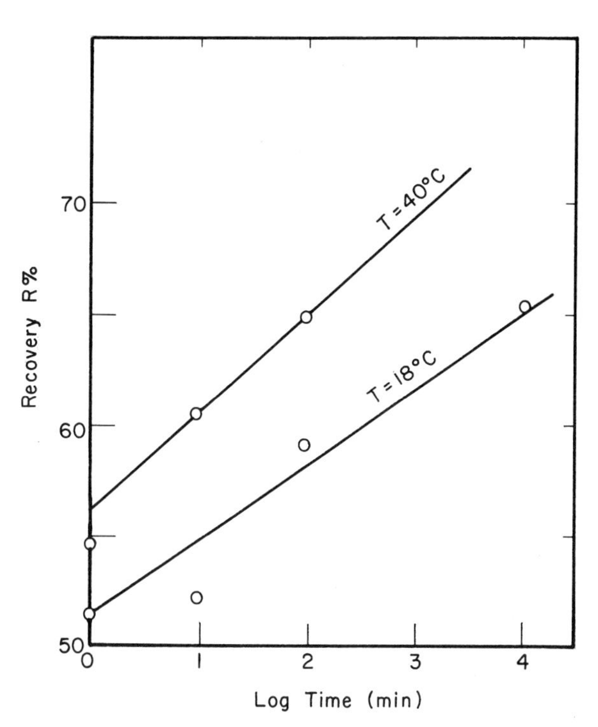

Figure 11.25 Recovery of carbon-black-filled natural rubber as a function of healing time at healing temperatures of 40 °C and 18 °C. The maximum strain was 200%.

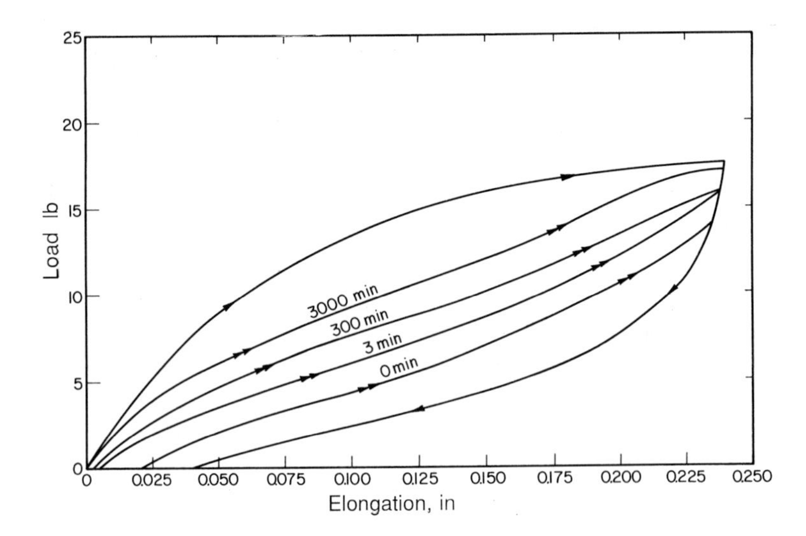

Figure 11.26 Mechanical recovery of a salt-filled solid rocket propellant as a function of healing time (Wool).

scission, was in this case negated by oxidative cross-linking, so that the samples when retested at room temperature exhibited a permanent set of 10% strain but gave stress–strain behavior identical to the virgin material, a rather unusual coincidence.

11.6 Kinetics of Damage Healing

11.6.1 An Empirical Theory of the Kinetics of Damage Healing

The generic character of crack healing and its influence on mechanical recovery have been demonstrated in a number of multicomponent polymers in Section 11.5. In this section, a general kinetic theory of crack or damage healing is presented to describe mechanical recovery as a function of time, temperature, and strain level. For each material, the healing mechanisms are complex and unknown. Therefore, we resort to an empirical theoretical approach that proves to have universal descriptive ability.

The mode of crack closure and healing is an important factor in the determination of recovery kinetics. Cracks can heal by a line mode, a point mode, or a combination of modes. The point mode involves a zip-up mechanism in which healing occurs only at the crack tip. The line mode involves simultaneous healing at all points on the crack surfaces. These healing modes are mechanically and kinetically distinguishable. Crack healing in brittle inorganic materials such as NaCl [66] or LiF [67] occurs preferentially by the point closure mode but the line mode also contributes, especially at higher temperatures. Our observations of crack healing in polymers indicate a line mode of healing is favored. The following analysis is based on an averaged line mode of healing but is also applicable to point mode when geometric factors at the crack tip are taken into consideration.

We propose that healing depends on the rate of disappearance of damage or defects in the debonded interface between the crack surfaces. The defects were created during the debonding process to make the voids; they result from molecular rearrangement and dislocations on or off the crack surfaces. Healing occurs as the molecular configuration at the interface is restored to its original or equivalent bonding state in the virgin material. This theory assumes that the complex process of healing is controlled by the slowest step, namely defect or damage disappearance via molecular rearrangement, and that sufficient driving forces exist to allow the debonded surfaces to approach each other. The driving force for crack closure prior to healing may arise from a combination of stored strain energy in the bulk material surrounding the crack, nonbonded potentials, connecting chains or fibrils between surfaces, and so forth.

Let D be the concentration of damage or defects per unit volume of crack interface. It is assumed that healing occurs at the same rate at all points on the crack surfaces. Therefore D is a parameter averaged over the damage distribution on the crack surface. The rate law for damage disappearance is not known, so we propose the following general kinetic expression

$$-\frac{\mathrm{d}D}{\mathrm{d}t} = kD^{n} \tag{11.6.1}$$

where k is a temperature-dependent rate constant and n is the order or power of the healing process. Ideally for crack healing involving multiple cooperative molecular motion, k and n are multiples and sums, respectively, of contributions from individual kinetic processes, but preferably should be regarded as averaged quantities reflecting a convolution of nonsimultaneous molecular mechanisms with varying rate laws.

The mechanical recovery R is related to D by

$$D = S_1 \beta (1 - R) \tag{11.6.2}$$

where β is a proportionality constant and S_1 is the initial work due to fracture (Eq 11.5.3). Equation 11.6.2 reasonably assumes that the time-dependent damage is linearly related to the difference between the fracture energies of the first and second deformation cycles.

The initial damage existing at $t = 0$ in the rest state is

$$D_0 = S_1 \beta (1 - R_0) \tag{11.6.3}$$

where R_0 is the instantaneous recovery component. Integrating Eq 11.6.1 and substituting for D and D_0 using Eq 11.6.2 and Eq 11.6.3, respectively, we obtain the mechanical recovery as a function of time, temperature, and R_0 as

$$R = 1 - (1 - R_0)/[1 + D_0^{n-1}(n-1)\, k_0\, e^{-E/k_B T}\, t]^{1/(n-1)} \tag{11.6.4}$$

where $n \neq 1$ and $k = k_0 \exp(-E/k_B T)$ has been defined as an Arrhenius rate constant, with k_0 the pre-exponential term, E the activation energy, k_B Boltzmann's constant, and T the temperature.

R_0 is a parameter that depends on the magnitude of the strain, which can be obtained experimentally and for HPP fibers has the form

$$R_0 = J\epsilon^{-m} \qquad \text{for } R_0 \leq 1 \tag{11.6.5}$$

where J and m are constants and ϵ is the maximum strain shown in Figure 11.16. The value of ϵ for which R_0 is equal to 1 reflects the strain range for perfectly reversible cracks or viscoelastic behavior in each system.

To obtain the kinetic constants from mechanical recovery studies, Eq 11.6.4 is simplified and rearranged as

$$(1 - R_0/1 - R) = (1 + Kt)^{\alpha} \tag{11.6.6}$$

where

$$K = D_0^{n-1}(n-1)\, k_0 \exp -E/k_B T \tag{11.6.7}$$

and $\alpha = 1/n - 1$.

For Kt much greater than 1, we obtain from Eq 11.6.6

$$\log(1 - R_0/1 - R) \cong \alpha \log K + \alpha \log t \tag{11.6.8}$$

Thus, a plot of the left side of Eq 11.6.7 versus log t, if linear, provides α from the slope, and K from the intercept (α log K). Once K is known at different temperatures, the other kinetic constants can be calculated.

When K and α are known, R can be expressed simply as

$$R = 1 - (1 - R_0)/(1 + Kt)^{\alpha} \tag{11.6.9}$$

which is the general expression proposed for healing in multicomponent systems.

For a first-order healing process not considered above, $n = 1$, we obtain from Eqs 11.6.1, 11.6.2 and 11.6.3

$$\ln(1 - R_0/1 - R) = kt \tag{11.6.10}$$

for which the kinetic constants can be readily determined experimentally.

The recovery half-life, τ, is measured at R of 0.5 and determined from Eq 11.6.9:

$$\tau = 1/K\{[2(1 - R_0)]^{1/\alpha} - 1\} \qquad (11.6.11)$$

These equations predict that R increases with time and temperature, and that faster recovery occurs at lower strains (large R_0). If R_0 is equal to 0, the recovery rate is independent of strain. The permanent damage contribution, D_p, can be determined either by a modification of the definition of R in terms of energy contributions for permanent damage or experimentally from the relation

$$D_p = 1 - R_\infty \qquad (11.6.12)$$

where R_∞ is the observed recovery at very long times.

The recovery behavior predicted by this kinetic approach is shown in Figure 11.27. From Eq 11.6.9, R is plotted as a function of log t with $K = 1$, $\alpha = 0.2$, $R_0 = 0.2$, and a reference temperature T_0. At higher temperatures (larger values of K), the curve is shifted to the left, and at lower healing temperatures, it is shifted to the right along the log time axis. The sigmoidal recovery curve is in fact the shape of a master recovery curve that describes healing of cracks and microstructural damage for most polymeric materials subjected to any particular strain history, not necessarily that outlined in Figure 11.16. At short times, the lower plateau can be extrapolated to the instantaneous recovery contribution, and at long times the upper plateau describes either full recovery or residual permanent damage. The master curve is associated with a reference testing temperature, T_0, and strain, ϵ, and can be generated experimentally by time–temperature superposition of recovery half-life, R, data at different temperatures [51].

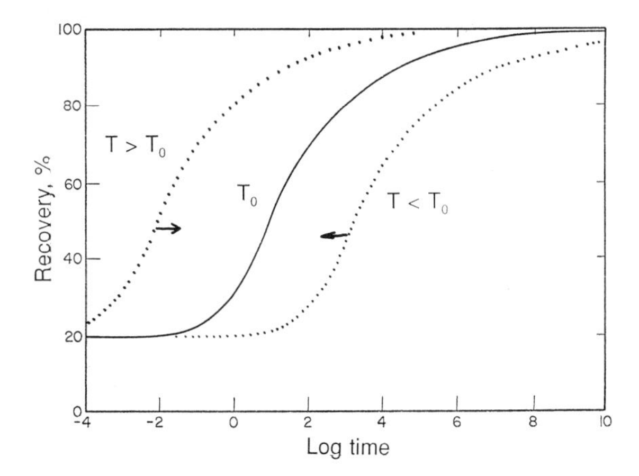

Figure 11.27 The theoretical master recovery curve.

11.6.2 Applications of Kinetic Theory

Hard Elastic Polypropylene

We evaluated the kinetic theory of crack healing by applying Eq 11.6.7 to the experimental data. Figure 11.28 shows a plot of $\log(1 - R_0)/(1 - R)$ versus $\log t$ for HPP samples subjected to a maximum strain of 150% (as in Figure 11.16) and several healing temperatures. At long times, when Kt is much greater than 1, the data are linear, from which we obtain the slope, and calculate $\alpha = 0.09 \pm 0.01$ and $n \approx 12$. The extrapolated intercept, $\alpha \log K$, is temperature dependent and yields an apparent activation energy, E, of about 28 kcal/mol. Similar behavior is obtained for other strain levels over the same temperature range.

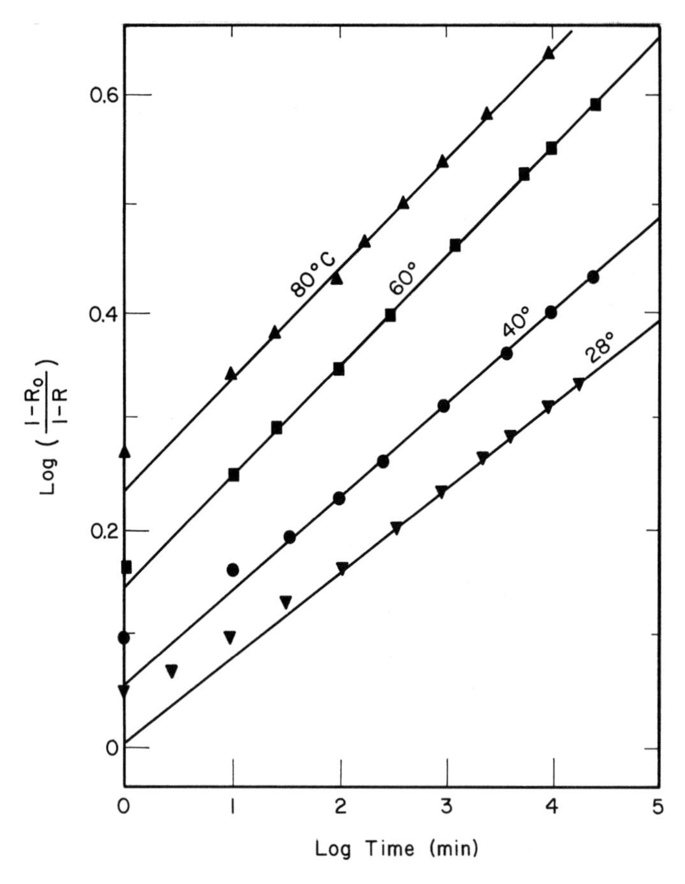

Figure 11.28 Log $(1 - R_0)/(1 - R)$ plotted as a function of log healing time for HPP fibers at several healing temperatures. The strain maximum was 150%.

R_0 was determined as a function of strain at room temperature, and can be described by Eq 11.6.5 as

$$R_0 = 0.19\, \epsilon^{-0.635} \tag{11.6.13}$$

At R_0 of 1, ϵ is 7.31%, which can be interpreted as the maximum strain limit for rapid crack reversibility in these hard elastic polypropylene fibers.

Using the kinetic constants and the value for R_0 at each strain level, we determined the recovery R as a function of time and temperature from Eq 11.6.9. Figure 11.29 shows a comparison of theory with experiment for the recovery at room temperature at several strain maxima. The theoretical plots are slightly curved but fit the data better than the linear log time plots used in Figure 11.18.

Figure 11.30 shows the comparison of theory with the experimental data for the 125% and 150% strain series at different temperatures. A reasonably good correlation is observed. The effect of T_g is not intrinsically incorporated in the theory, although little recovery is predicted near T_g at −17 °C. The temperature dependence can also be introduced using a WLF approach [51].

Block Copolymers

The data in Figure 11.20 for healing of SBS at different temperatures were analyzed with the kinetic theory. Figure 11.31 shows the master curve for SBS, where the data were shifted along the log time axis so that they were superimposed, at a reference temperature of 18 °C. No vertical shifts were used. The healing data are well

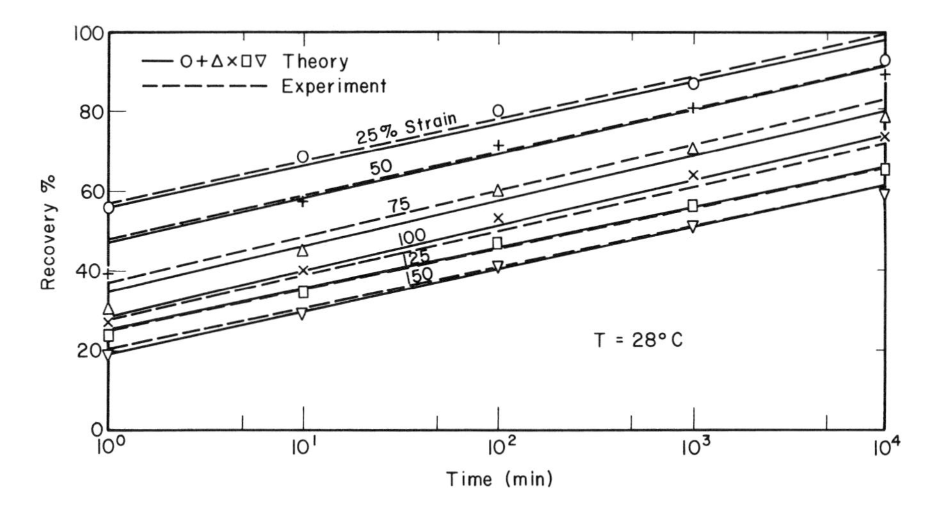

Figure 11.29 A comparison of theory with experimental recovery data for HPP fibers. The strain maxima ranged from 25% to 150% at 28 °C.

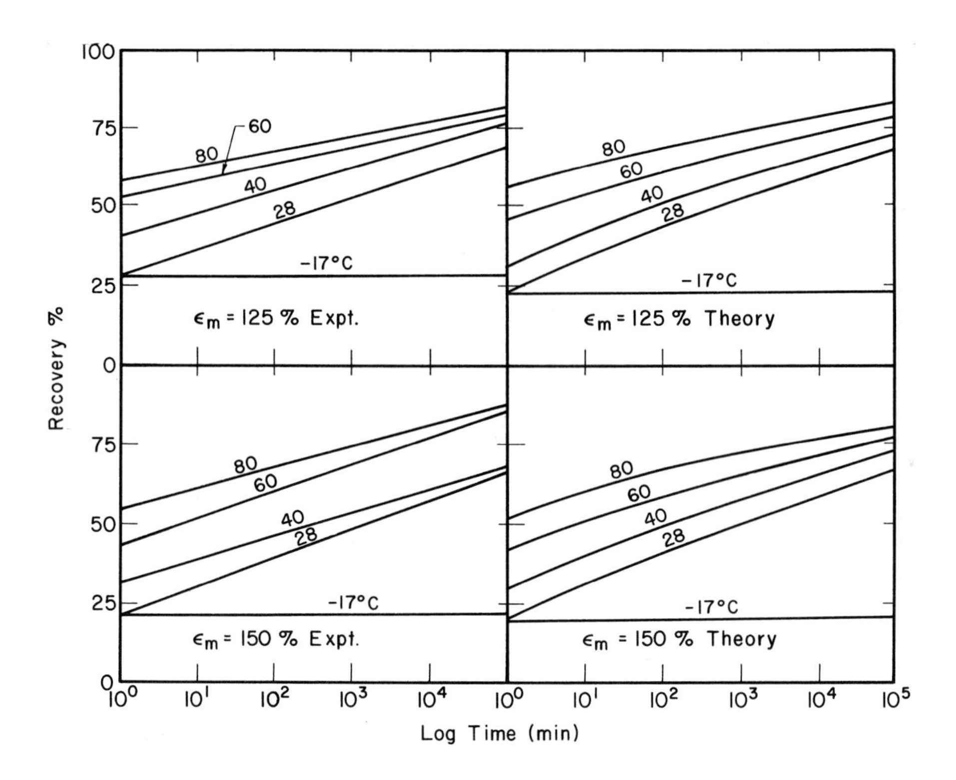

Figure 11.30 The temperature-dependent experimental recovery data for HPP fibers compared with the theoretical prediction for strain maxima of 125% and 150%.

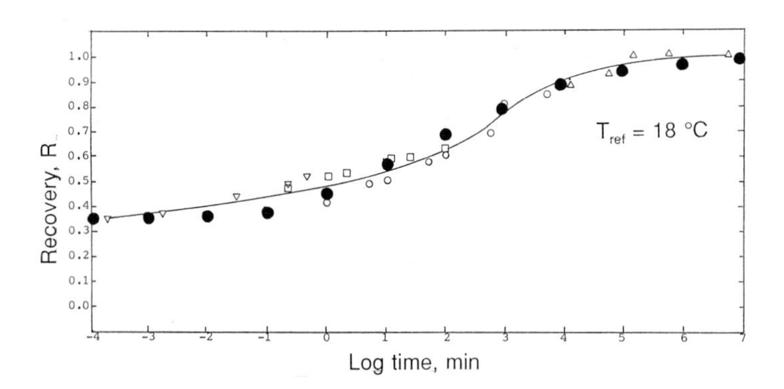

Figure 11.31 Master curve for mechanical recovery. Points of Figure 11.20 were horizontally shifted relative to 18 °C, the temperature of deformation. No vertical shift was used. The −70 °C data are not included.

represented by the kinetic theory using $K = 1$, $R_0 = 0.35$, and $\alpha = 0.18$. The shift factor used an activation energy $E_a = 48$ kcal/mol.

Filled Elastomers

The kinetic constants for the carbon-black-filled rubber were found to be $K = 24.9$/min and $\alpha = 0.224$ at 18 °C; at 40 °C, $K = 936.8$/min and $\alpha = 0.042$. The solid lines in Figures 11.24 and 11.25 are the theoretical lines using Eq 11.6.9.

Solid Rocket Propellants

Figure 11.32 shows the healing master curve for the salt-filled SRP. The kinetic constants are $K = 18.443$/min, $\alpha = 0.071$, $R_0 = 0.26$ at 12% strain, and $T_0 = 24$ °C.

The kinetics of healing in a live propellant (Space Shuttle motor) can be well described by $\alpha = 0.053$, $R_0 = 0.66$, and $T_0 = 20$ °C. The temperature dependence of K was

$$K(T) = K_0 \exp - (E_a / RT) \tag{11.6.14}$$

where $K_0 = 18 \times 10^{36}$/min and $E_a = 50$ kcal/mol. The latter value is consistent with those for most rubber materials.

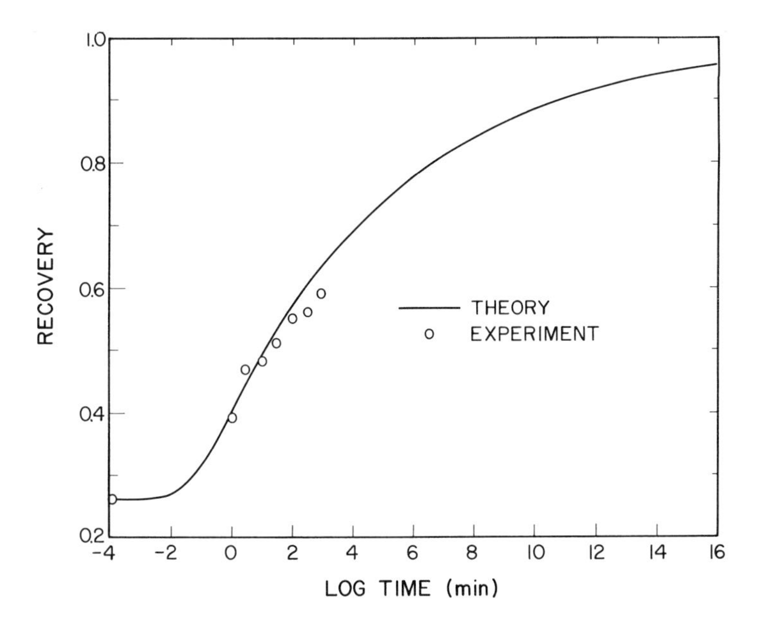

Figure 11.32 The theoretical master healing curve for a salt-filled solid rocket propellant, with experimental data points.

Despite the simplicity of the kinetic healing model, it is very useful for quantitatively describing mechanical recovery data and is in excellent qualitative agreement with the known physical processes of crack healing in all materials investigated.

Exercise

Use the kinetic theory of damage healing to analyze and compare both mechanical data (Figure 11.20) and optical data (Figure 11.23) for block copolymers.

11.6.3 Summary of the Concepts of Damage Healing

The concept of damage healing was discussed and shown to play an important role in determining mechanical properties of semicrystalline polymers, block copolymers, and filled elastomers. Important aspects and consequences of crack healing are summarized below.

1. Crack and void healing is a general phenomenon that occurs in many polymer systems. It affects mechanical behavior and is an important consideration in the mechanistic evaluation of constitutive relations.
2. Healing rates are controlled by temperature, strain history, environment, and the nature of the microstructural damage. Nonpermanent cracks can heal either instantaneously or in a time-dependent manner.
3. The concept of cracks opening and closing in a time-dependent manner provides an alternative microstructural interpretation of viscoelastic behavior in addition to pure molecular relaxation processes. Rate-dependent mechanical properties that are controlled by crack healing can be subjected to time–temperature superposition analyses similar to the WLF approach with viscoelastic polymers.
4. A semiempirical kinetic theory of crack healing based on the rate of disappearance of damage adequately describes mechanical recovery due to crack healing and provides the rationale for constructing recovery master curves.

11.7 References

1. W. Brostow, and R. C. Corneliussen, Eds., *Failure of Plastics*; Hanser Publishers, New York; 1986.
2. H. H. Kausch, *Polymer Fracture*; Springer-Verlag, Berlin; 2nd ed., 1987.
3. E. H. Andrews and P. E. Reed, "Molecular Fracture in Polymers"; in *Failure in Polymers; Molecular and Phenomenological Aspects*; Vol. 27 in series *Advances in Polymer Science*, Springer-Verlag, Berlin; 1978; p 1.
4. H. H. Kausch, J. A. Hassell, and R. I. Jaffee, Eds., *Deformation and Fracture of High Polymers*; Plenum Press, New York; 1973.
5. E. H. Andrews, *Fracture in Polymers*, Oliver and Boyd, Edinburgh, UK; 1968.
6. F. Bueche, *J. Appl. Phys.* ***26***, 1133 (1955).

7. H. Liebowitz, *Fracture: An Advanced Treatise*; Academic Press, New York; 1972.
8. A. Tobolski and H. Eyring, *J. Chem. Phys.* ***11***, 125 (1943).
9. S. N. Zhurkov, E. E. Tomashevskii, and V. A. Zakrevskii, *Sov. Phys.—Solid State (Engl. Transl.)* ***3***, 2074 (1962).
10. R. P. Wool, *Polym. Eng. Sci.* ***20***, 805 (1980).
11. O. J. McGarel and R. P. Wool, *J. Polym. Sci., Part B, Polym. Physics* ***25***, 2541 (1987).
12. R. P. Kambour, *Macromol. Rev.* ***7***, 1 (1973).
13. E. J. Kramer, "Microscopic and Molecular Fundamentals of Crazing"; chapter in Vol. 52/53 in series *Advances in Polymer Science: Crazing*, H. H. Kausch, Ed., Springer-Verlag, Berlin; 1983; p 1.
14. S. S. Sternstein, L. Ongchin, and A. Silverman, "Inhomogeneous Deformation and Yielding of Glasslike High Polymers"; chapter in *Polymer Modification of Rubbers and Plastics*, H. Keskkula, Ed.; Vol. 7 in series *Applied Polymer Symposia*; Wiley–Interscience, New York; 1968; p 175.
15. A. S. Argon and M. M. Salama, *Philos. Mag.* ***36***, 1217 (1977).
16. A. N. Gent and A. G. Thomas, *J. Polymer Sci. Part A-2* ***10***, 571 (1972).
17. T. E. Brady and G. S. Y. Yeh, *J. Mater. Sci.* ***8***, 1083 (1973).
18. S. Wellinghoff and E. Baer, *J. Macromol. Sci., Phys.* ***B11***, 367 (1975).
19. D. L. G. Lainchbury and M. Bevis, *J. Mater. Sci.* ***11***, 2222 (1976).
20. J. F. Fellers and B. F. Kee, *J. Appl. Polym. Sci.* ***18***, 2355 (1974).
21. B. Koltisko, N. Balashin, A. Hiltner, and E. Baer, "Irreversible Deformation Mechanisms in Thin Polymeric Films", paper presented at 41st Annu. Tech. Conf. Soc. Plast. Eng., *Soc. Plast. Eng. Tech. Papers* ***29***, 440 (1983).
22. A. S. Argon and J. G. Hannoosh, *Philos. Mag.* ***36***, 1195 (1977).
23. N. J. Mills, *J. Mater. Sci.* ***16***, 1332 (1981).
24. R. P. Wool and K. M. O'Connor, *Polym. Eng. Sci.* ***21***, 970 (1981).
25. R. P. Wool and K. M. O'Connor, *J. Appl. Phys.* ***52***, 5953 (1981).
26. Y.-H. Kim and R. P. Wool, *Macromolecules* ***16***, 1115 (1983).
27. R. P. Wool, "Strength Development at a Symmetric Polymer–Polymer Interface"; In *Advances in Rheology 3. Polymers*, B. Mena, A. Garcia-Rejon, C. Rangel-Nafaile, Eds.; Proc. IXth Int. Congr. Rheology; Univ. Nacional Autónoma de México, Mexico, 1984; p 573; R. P. Wool, *Rubber Chem. Technol.* ***57*** (2), 307 (1984).
28. S. Timoshenko, *Theory of Elasticity*; McGraw–Hill, New York; 1934; p 76.
29. R. P. Wool, "Crack Healing in Semicrystalline Polymers, Block Copolymers and Filled Elastomers"; chapter in *Adhesion and Adsorption of Polymers*, L.-H. Lee, Ed.; Vol. 12A in series *Polymer Science and Technology*; Plenum Press, New York; 1980; p 341.
30. Robertson, R. E., "The Fracture Energy of Low Molecular Weight Fractions of Polystyrene"; chapter in *Toughness and Brittleness of Plastics*, R. D. Deanin and A. O. Crugnola, Eds.; Vol. 154 in Advances in Chemistry Series; American Chemical Society, Washington, DC; 1976; p 89.
31. R. P. Kusy and D. T. Turner, *Polymer* ***18***, 391 (1977).
32. P. Prentice, *J. Mater. Sci.* ***20***, 1445 (1985).
33. C. C. Chau, L. C. Rubens, and E. B. Bradford, *J. Mater. Sci.* ***20***, 2359 (1985).
34. J. L. Willett, K. M. O'Connor, and R. P. Wool, *J. Polym. Sci., Part B: Polym. Phys.* ***24***, 2583 (1986).
35. K. Jud, H. H. Kausch, and J. G. Williams, *J. Mater. Sci.* ***16***, 204 (1981); H. H. Kausch, *Pure Appl. Chem.* ***55***, 833 (1983).
36. R. Popli and D. Roylance, *Polym. Eng. Sci.* ***22***, 1046 (1982).

37. E. J. Kramer, *Polym. Eng. Sci.* ***24***, 761 (1984).
38. L. L. Berger and E. J. Kramer, *Macromolecules* ***20***, 1980 (1987).
39. Y. R. Allen and T. G. Fox, *J. Chem. Phys.* ***41***, 337 (1964).
40. R. P. Wool, *Polym. Eng. Sci.* ***18***, 1057 (1978).
41. M. Miles, J. Petermann, and H. Gleiter, *Progr. Colloid Polym. Sci.* ***62***, 478 (1977).
42. R. P. Wool, M. I. Lohse, and T. J. Rowland, *J. Polym. Sci., Polym. Lett. Ed.* ***17***, 385 (1979); M. I. Lohse, T. J. Rowland, and R. P. Wool, "NMR Study of Void Healing in Hard Elastic Polypropylene", *Bull. Am. Phys. Soc.* ***24*** (3), 348 (1979).
43. R. P. Wool and W. O. Statton, "Dynamic Infrared of Polymers"; chapter 12 in *Applications of Polymer Spectroscopy*, E. G. Brame, Jr., Ed.; Academic Press, New York; 1978; p 185.
44. R. P. Wool, *J. Polym. Sci., Polym. Phys. Ed.* ***14***, 1921 (1976).
45. K. L. DeVries and D. K. Roylance, *Prog. Solid State Chem.* ***8***, 283 (1973).
46. L. A. Davis, C. A. Pampillo, and T. C. Chiang, *J. Polym. Sci.* ***11***, 841 (1973).
47. I. K. Park and H. D. Noether, *Colloid Polym. Sci.* ***253***, 824 (1975).
48. S. L. Cannon, G. B. McKenna, and W. O. Statton, *J. Polym. Sci., Macromol. Rev.* ***11***, 209 (1976).
49. B. S. Sprague, *J. Macromol. Sci., Phys.* ***8***, 157 (1973).
50. E. S. Clark, "A Mechanism of Energy-Driven Elasticity in Crystalline Polymers"; In *Structure and Properties of Polymer Films*, R. W. Lenz and R. S. Stein, Eds.; Vol. 1 in series *Polymer Science and Technology*; Plenum, New York; 1973; p 267.
51. K. M. O'Connor and R. P. Wool, *J. Appl. Phys.* ***51***, 5075 (1980).
52. H. D. Noether and W. Whitney, *Kolloid Z. Z. Polym.* ***251***, 991 (1973).
53. H. Ishikawa, H. Numa and M. Nagura, *Polymer* ***20***, 516 (1979).
54. R. P. Wool, *Int. J. Fract.* ***14***, 597 (1978).
55. D. H. Kaelble and E. H. Cirlin, *J. Polym. Sci.: Polym. Symp.* ***43***, 131 (1973).
56. S. L. Aggarwal, *Polymer* ***17***, 938 (1976).
57. M. Morton, *J. Polym. Sci. C* ***60***, 1 (1970).
58. J. L. LeBlanc, *J. Appl. Polym. Sci.* ***21***, 2419 (1979).
59. E. Pedemonte, *et al.*, *Polymer* ***16***, 531 (1975).
60. Y. D. M. Chen and R. E. Cohen, *J. Appl. Polym. Sci.* ***21***, 629 (1977).
61. L. Mullins, *Rubber Chem. Technol.* ***21***, 281 (1948).
62. L. Mullins and N. R. Tobin, *Rubber Chem. Technol.* ***30***, 555 (1957).
63. A. M. Bueche, *J. Polym. Sci.* ***25***, 139 (1957).
64. A. F. Blanchard and D. Parkinson, *Ind. Eng. Chem.* ***44***, 799 (1952).
65. S. K. Upadhyayula and R. P. Wool, "Mechanical Damage and Healing in Filled Elastomers", *Bull. Am. Phys. Soc.* ***24***, 259 (1979).
66. S. M. Park and D. R. O'Boyle, *J. Mater. Sci.* ***12***, 840 (1977).
67. R. Raj, W. Pavinich, and C. N. Ahlquist, *Acta Metall.* ***23***, 399 (1975).
68. J. E. Fitzgerald, "Analysis and Design of Solid Propellant Grains"; chapter in *Mechanics and Chemistry of Solid Propellants*, A. C. Eringen, H. Liebowitz, S. L. Koh and J. M. Crowley, Eds.; Proceedings of the Fourth Symposium on Naval Structural Mechanics, 1985; Pergamon Press, New York; 1967; p 19.
69. M. H. Quinlan, *On the Theory of Materials with Permanent Memory*; Ph.D. Thesis, University of Utah, Salt Lake City, UT; 1974.
70. W. Noll, *Arch. Ration. Mech. Anal.* ***2***, 197 (1958).
71. B. D. Coleman and W. Noll, *Arch. Ration. Mech. Anal.* ***6***, 335 (1960).
72. J. E. Fitzgerald, private communication.

12 Crack Healing*

12.1 Introduction to Crack Healing

In this chapter we consider the healing of single cracks formed by fracture of virgin test specimens. The specimens can have DCB, wedge cleavage, single-edge notch, compact tension, or impact fracture mechanics configurations. The sample is first fractured or fatigued, its fracture properties K_{1c}, G_{1c}, or da/dN are determined, the surfaces are brought back into contact, the crack is healed for a time t at temperature T, and the piece is then refractured. Crack healing provides a new "back-door" approach to the understanding of fracture of virgin materials. At the instant of critical fracture in virgin materials, many events are occurring simultaneously and it is difficult to separate the important contributions. However, in crack healing, we slowly converge on the virgin strength, with control of time, temperature, and molecular weight, and this permits us to investigate the fracture mechanism in more detail.

We examine crack healing in polymers of different molecular architectures. The bulk of this chapter focuses on healing of fracture specimens composed of linear chains, such as PS and PMMA, and that is followed by studies of branched or partially cross-linked polybutadiene and highly cross-linked epoxies. The potential of materials with infinite fatigue lifetimes is investigated, followed by a concluding section on self-repair in smart composites.

In Section 12.2, we comment on the elegant healing experiments conducted by Kausch's group at Lausanne [1–5]. Related experiments done by our Urbana group are presented in Section 12.3.

12.2 Crack Healing at Lausanne

12.2.1 Crack Healing of PMMA and PSAN

In seminal papers on crack healing, Jud, Kausch and Williams [1, 2] used a compact tension (CT) method to determine the K_{1c} values of fractured and healed PMMA and PSAN (25 mole % AN). The CT specimens were made by compression-molding pellets and were cut to dimensions of 26 mm length, 26 mm width, and 3 mm depth.

---

* Dedicated to H. H. Kausch

A precrack was introduced with a saw and sharpened with a razor blade. The CT pieces were fractured in tension and gave smooth surfaces.

PMMA samples with M_w of 1.2×10^5 ($M_w/M_n = 2$) and the PSAN samples with M_w of 1.2×10^5 were fractured at room temperature, healed above the glass transition temperature (100 °C), and refractured at room temperature. Healing was conducted in a hot press at constant temperature. The surfaces were kept in contact during healing with the light pressure of a rubber band.

The main result is that the critical stress intensity factor, $K_{1c}(t)$, could be described by

$$K_{1c}(t) = K_{1c\infty}\,(t/\tau_0)^{1/4} \tag{12.2.1}$$

where $K_{1c\infty}$ is the virgin strength and τ_0 is the time to achieve the virgin strength. Since the fracture energy G_{1c} is equal to K_{1c}^2/E, it also follows that

$$G_{1c}(t) = G_{1c\infty}\,(t/\tau_0)^{1/2} \tag{12.2.2}$$

where $G_{1c\infty}$ is the virgin fracture energy. Figure 12.1 shows $K_{1c}(t)$ versus t for PMMA/PMMA, PSAN/PSAN, and PMMA/PSAN crack interfaces at several temperatures. Each point represents an average of about 20 samples, as considerable data scatter occurs with these experiments, but a good linear relation is observed. They found that the activation energy for healing PMMA, E_a, was 65 kcal/mol in a range of $T_g < T < (T_g + 15$ °C). Note that little difference is seen in the healing response of the different polymer pairs.

The apparent diffusion coefficient, D, for PMMA can be determined from Figure 12.1 as

$$D = R^2/(3\pi^2\tau_0) \tag{12.2.3}$$

where R is the end-to-end vector. For example, $\tau_0 \approx 100$ min, $M = 1.2 \times 10^5$, and $R = 243$ Å, at 378 K (5 °C above T_g), and hence we obtain

$$D_{378} = 3.3 \times 10^{-17} \quad (\text{cm}^2/\text{s}) \tag{12.2.4}$$

This analysis assumes that the time to achieve complete healing is the reptation time T_r. Since the molecular weights used were of the order of the upper molecular weight, M^*, of about 200,000, this approach is reasonable. If M is greater than M^*, then τ^0 is less than T_r, since the molecules do not have to diffuse a distance equal to their radius to be completely healed with respect to G_{1c} measurements (Chapter 7). This may not be the case for subcritical crack propagation, or for fatigue measurements based on da/dN measurements (Chapter 8).

Kausch and co-workers [2, 3] compared their healing data to diffusion coefficients for PMMA determined from viscoelastic data proposed by Graessley [6], using

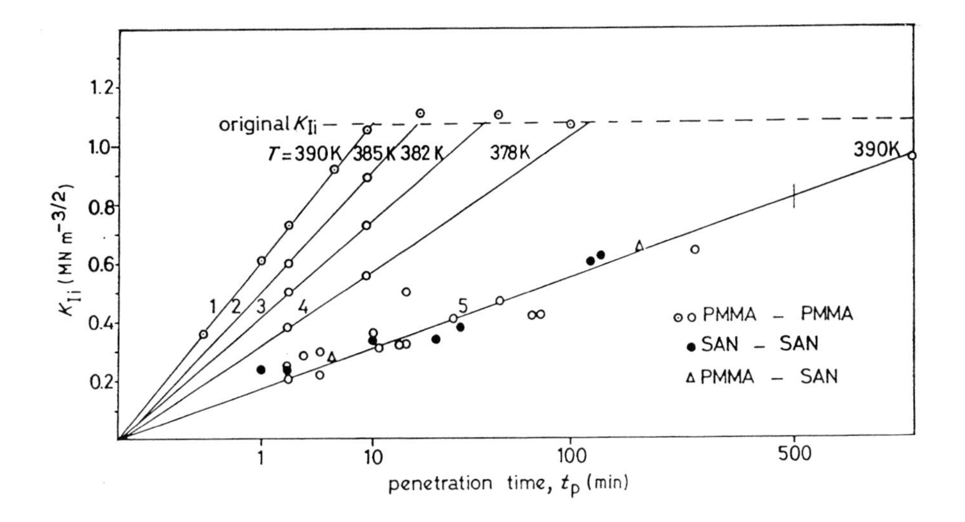

Figure 12.1 Plot of fracture toughness against t_p [2].

$$D = G_N^0(\rho RT/G_N^0)^2 \ (R_g^2/M) \ [M_c/M^2 \eta (M_c)] \qquad (12.2.5)$$

where $G_N{}^0 = 6.36 \times 10^4$ N/m^2 is the plateau modulus and $\eta(M_c) = 3.78 \times 10^8$ N·s/m^2 (at 378 °C) is the zero-shear viscosity evaluated at the critical entanglement molecular weight, M_c. Eq 12.2.5 gives D values that are about one order of magnitude lower than the D value derived from the healing data. This difference may be due to bond rupture lowering the molecular weight on the crack surfaces so that the apparent diffusion coefficient from healing appears to be higher than the bulk value.

12.2.2 Surface Rearrangement Effects on Healing in PMMA

When a fresh fracture surface is created, significant surface rearrangement can occur. This can involve relaxation of fibrillar material, chemical reactions, diffusion of low-molecular-weight material, or other processes. Petrovska-Delacrétaz and Kausch considered the effects of various surface treatments on the crack healing process in PMMA, as follows.

A Surface: Fractured surfaces were healed at (T_g + 5 °C).

B Surface: Fractured surfaces were annealed at (T_g + 17 °C) for varying times t_s, and then healed at (T_g + 5 °C).

C Surface: Fractured surfaces were polished and then healed at (T_g + 5 °C).

The results of the surface treatments on the healing curve K_{1c} versus $t^{1/4}$ are shown in Figure 12.2. The freshly fractured surface (A) heals most rapidly. When the surfaces

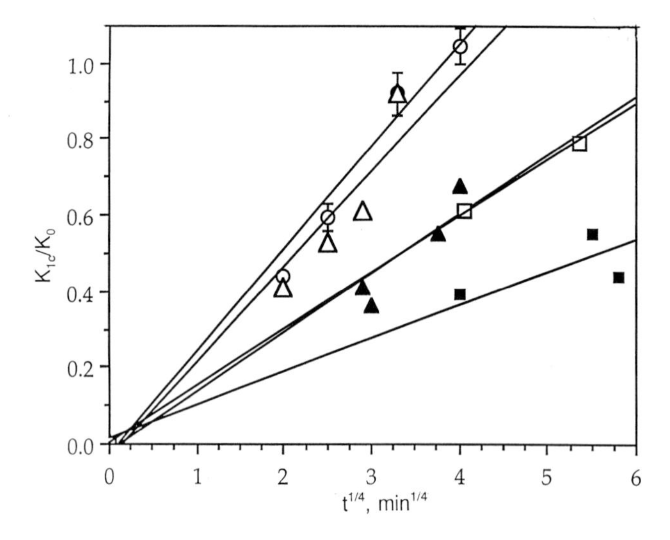

Figure 12.2 Healing curves for PMMA with different surface states. Key to symbols (see text for fuller explanation): ○, Surface A; Δ, Surface B annealed for 10 min; ▲, Surface B annealed for 65 min; ■, Surface B annealed for 256 min; □, Surface C [4].

are annealed before healing (state B), the slope decreases with increasing annealing time; the longer the sample is annealed, the more slowly it heals. The polished surface (state C) also heals more slowly than surface A, but with a slope almost the same as that of the B surface that was annealed for 65 min. Annealing prior to healing allows the lower molecular weight fractured chains to diffuse away from the interface. These chains contribute to rapid wetting and diffusion during normal healing. We expect that in their absence the slope of the healing curve is decreased. Polishing removes the fibrillar material on the fracture surfaces and creates a new surface with reduced molecular weight. However, it appears that the new surface due to polishing is comparable to the partially annealed surface, which causes us to think that polishing does less damage to the surface molecular weight distribution than does fracture. This is reasonable, since the fibrillar material is rich in low-molecular-weight fragments (see discussion in Chapter 11), which polishing reduces. However, polishing also both destroys a surface layer of chains and alters the surface molecular weight.

12.2.3 Effect of Molecular Weight on Healing of PMMA

Kausch and co-workers studied the effect of molecular weight on the crack healing rate of PMMA using molecular weights ranging from 88,500 to 727,000 [4]. The healing was done at (T_g + 5 °C) and the results are shown in Figure 12.3. There is a clear effect of molecular weight: the healing curves can be described by

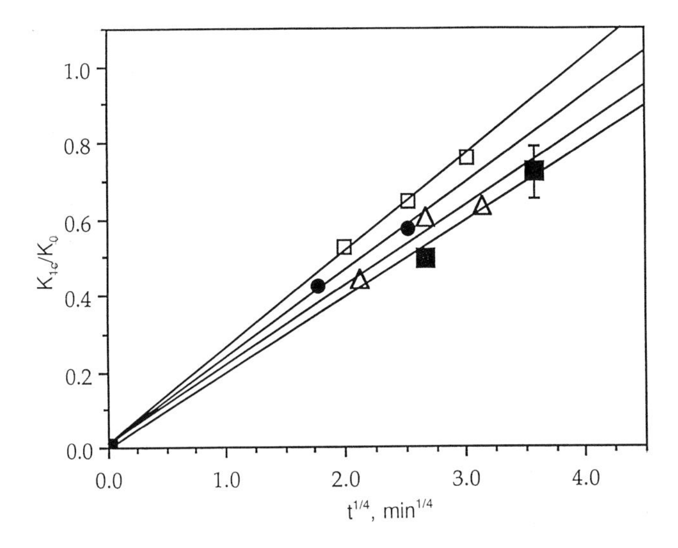

Figure 12.3 Healing curves for PMMA with different molecular weights. Key to symbols: ▫, M = 88,500; •, M = 162,000; Δ, M = 404,000; ▪, M = 727,000 (Kausch and co-workers).

$$K_{1c}(t) \sim t^{1/4} M^{-0.14} \tag{12.2.6}$$

The exponent of 0.14 is not readily predictable from simple reptation theory and is smaller than that observed for symmetric welding (exponents in the range 0.25–0.4). We interpret the exponent of 0.14 as follows.

The number of broken bonds per unit area, N_f, created in an interface of thickness X during fracture is given by (Chapter 8)

$$N_f = \rho N_a X/2 \,(1/M_f - 1/M) \tag{12.2.7}$$

where M_f is the molecular weight after fracture. Appropriate values are $X \approx 1000$ Å, $N_f = 6 \times 10^{13}/\text{cm}^2$. Thus the molecular weight on the surface layer is estimated as

$$M_f = \beta/(N_f + \beta/M) \tag{12.2.8}$$

where $\beta = \rho N_a X/2$. For example, with molecular weights, M, of 88,500, 162,000, 404,000, and 727,000, we obtain values of M_f of 34,000, 41,700, 49,000, and 52,000, respectively. Thus, fracture significantly reduces the molecular weight on the surface but also shrinks the differences between molecular weights, causing a lowering of the exponent for the molecular weight dependence of healing.

The chain fracture effect discussed above also helps us understand the surface rearrangement data in Section 12.2.2. For example, in Figure 12.2, a sample with initial molecular weight, M, of 727,000 and a surface molecular weight, M_f, of 52,000 after fracture would be expected to undergo a slope change of about 50%, if the molecular weight exponent is ¼, that is, if $K_{1c} \sim t^{1/4}M_f^{-1/4}$. This implies that as annealing of the surfaces proceeds, the lower molecular weight species diffuse into the bulk and restore the initial higher molecular weight distribution. As M_f approaches M, the slope decreases to a final value, which should be the same as that of the symmetric interface weld (no chain fracture) discussed in Chapters 7 and 8.

12.3 Crack Healing at Urbana

12.3.1 Crack Healing Methods for Glassy Polymers

We conducted crack healing experiments with PS, PMMA, PSAN, PC, and cross-linked epoxies [7–11]. Pellets of these materials were molded into plates and cut into compact tension (CT) fracture mechanics specimens. To initiate a crack, a razor blade was tapped lightly into the slot. In the case of PMMA, the razor-notched specimens were simply loaded at a constant extension rate of 2.1 mm/min at room temperature until the single crack advanced approximately 10 mm into the uncracked region of the specimen. However, loading of PS in a similar manner resulted in a diffuse zone of crazes, rather than a single crack advancing from the tip of the razor notch. In order to create a sharper single crack with a minimal craze zone, we loaded the CT specimens cyclically from 0 to 20 lb at a frequency of 20 Hz. This treatment resulted in slow crack growth at a rate of approximately 1.2 mm/min. Figure 12.4 shows SEM micrographs of the fracture surfaces created in this manner. Fatigue striations, which are related to the crack advance per load cycle, are clearly seen. We will meet these surfaces again when we discuss fatigue healing in Section 12.7.

After fracture, the cracks were healed isothermally in a nitrogen atmosphere at several temperatures above T_g (T_g is 100 °C for PS and 105 °C for PMMA). After much experimentation, we devised a simple procedure that enabled us to achieve reproducible healing results. Each specimen was held in a brass clip, as shown in Figure 12.5. The role of the clip was to promote surface approach and wetting as the specimen expanded upon heating. The crack generally stopped being visible after about 10 minutes or more in the oven. At times of high ambient humidity, bubbles of trapped gas often remained in the interface for long times. Preliminary healing experiments on PS indicated that the crack and the bubbles could completely disappear in 3 minutes if the specimens were stored in vacuum at room temperature for 24 hours before healing. Presumably, this treatment removed adsorbed water vapor from the freshly created fracture surfaces.

For PMMA, the blade was forced in sufficiently to reopen approximately 3 mm of the healed interface. For PS, a razor notch of 1 mm length was introduced, followed by cyclic loading at 20 Hz until about 3 mm of the healed interface was reopened. The strength of the interface was measured via (Chapter 7)

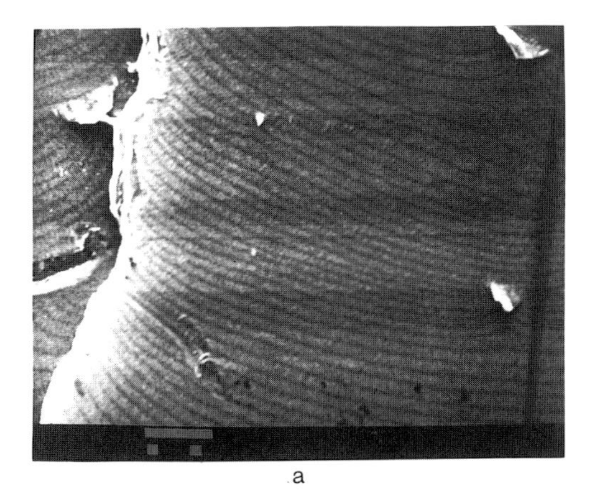

a

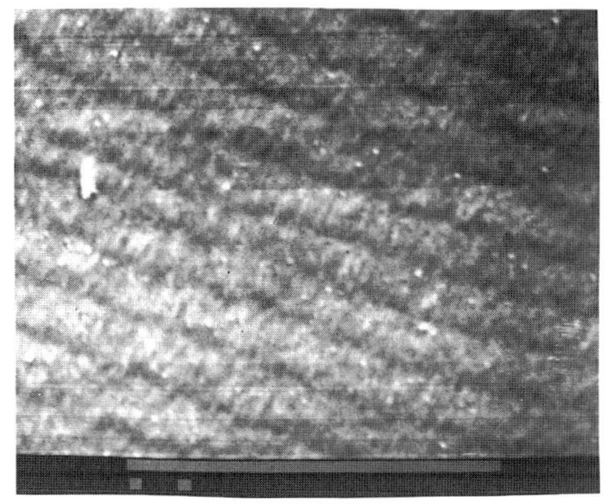

b

Figure 12.4 SEM micrographs of the PS surfaces prior to healing, showing the characteristic fatigue striations. The bar below each micrograph represents 100 μm (O'Connor and Wool).

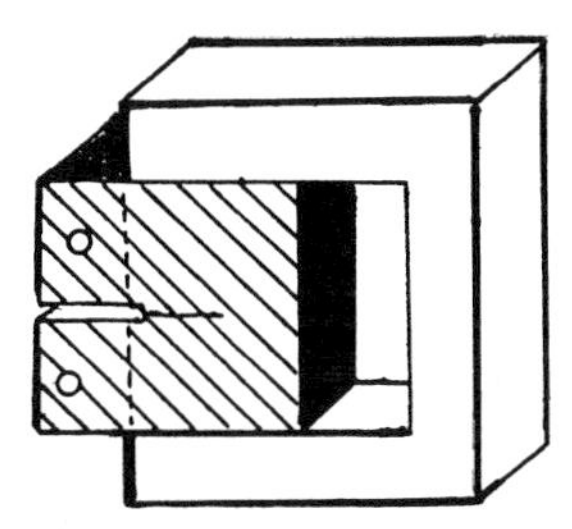

Figure 12.5 Constraining clamp for experimental specimens.

$$K_{1c} = P_c / B W^{1/2} f(a/W) \tag{12.3.1}$$

where P_c is the critical load required to separate the partially healed interface, B is the specimen thickness, a is the crack length as measured from a line connecting the loading pins to the crack tip, and W is the specimen width as measured from the loading pins to the opposite edge of the specimen. The function f is given by

$$f(\frac{a}{W}) = (2+\frac{a}{W})\frac{[0.886+4.64\frac{a}{W}-13.32(\frac{a}{W})^2+14.72(\frac{a}{W})^3-5.6(\frac{a}{W})^4]}{(1-\frac{a}{W})^{3/2}} \quad (12.3.2)$$

Equation 12.3.2 represents an approximation to the full solution that is accurate within 0.5% for $0.2 < a/W < 1$. The critical load, P_c, was determined from the drop or abrupt change in slope of the load versus displacement behavior that corresponded to crack growth. The crosshead speeds were 2.1 mm/min for PMMA and 0.25 mm/min for PS.

12.3.2 Crack Healing of PS and PMMA

The typical load versus displacement behavior of PS (M_n of 142,000 and M_w of 260,000) compact tension specimens after various degrees of healing is shown in Figure 12.6. The initial slope of these curves is the inverse of the specimen

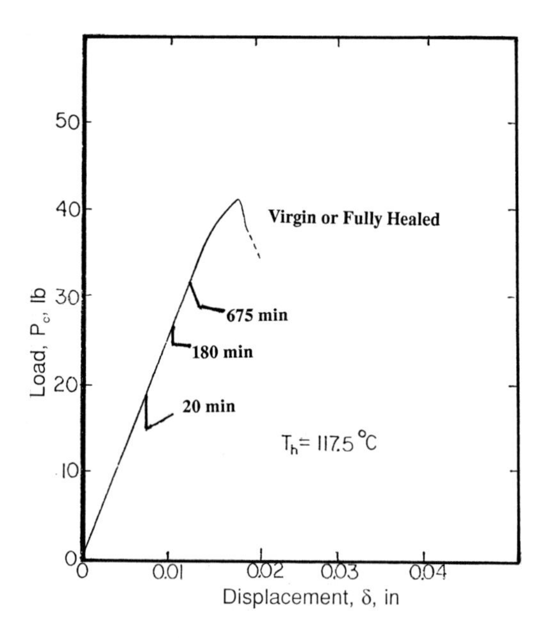

Figure 12.6 Load versus displacement behavior for PS fractured in the compact tension geometry, for virgin behavior and at various healing times. At a given crack length, the compliance is independent of healing time (Wool and O'Connor).

compliance, and represents the effective modulus, E. For a given crack length, we see from the Figure 12.6 that the modulus behaves as

$$E \sim t^0 M^0 \tag{12.3.3}$$

That this modulus is independent of both time and molecular weight is not trivial and helps to simplify the healing models. The critical load used for calculation of K_{1c} in Eq 12.3.1 is that value at which the load drops sharply, corresponding to rapid crack advance. As healing approaches completion in PS, crack advance is slower and the critical load is that value at which the load versus displacement behavior stops being linear. Crack growth causes a change in compliance and therefore a change in slope.

The healing results for PS are shown in Figure 12.7, where K_{1c} is plotted against $t^{1/4}$ for healing at 110 °C and 125 °C, and each data point is the average of readings from at least 10 specimens [10]. The data are described in terms of K_{1c} as

$$K_{1c} = K_{1c}^0 + Jt^{1/4} \tag{12.3.4}$$

where $K_{1c}{}^0$ is the extrapolated intercept at zero time. The slope J was 0.140 at 110 °C and 0.161 at 125 °C. The intercept is most likely a consequence of polydispersity effects on healing as the lowest molecular weight species diffuse rapidly across to give

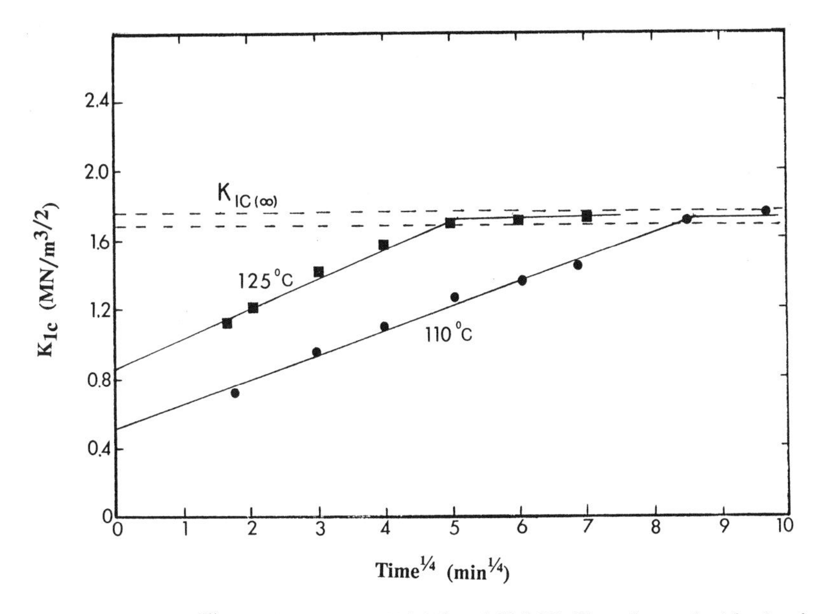

Figure 12.7 K_{1C} versus $t^{1/4}$ for PS healed at 110 °C and 125 °C. $K_{1C(\infty)}$ is reached in the time t_∞ (Wool and O'Connor).

some initial strength in the first 15 minutes of healing (Chapter 7). Thus we hesitate to derive too much quantitative information, such as activation energies (about 28 kcal/mol), and so forth, from the slopes. However, the time to achieve complete strength can be used to determine the diffusion coefficient for this PS sample with M_w of 260,060. Substituting into Eq 12.2.3 the value $R^2 = 344.4^2$ Å^2, we can write

$$D = 4 \times 10^{-13}/\tau \qquad (\mathrm{cm}^2/\mathrm{s}) \tag{12.3.5}$$

where τ is the healing time. At 125 °C, $\tau^{1/4} \approx 4.8$ min$^{1/4}$, $\tau \approx 530$ min (from Figure 12.7) and $D = 1.25 \times 10^{-17}$ cm^2/s. This agrees with experimental D values (from Chapter 5) of $D_{260K} \approx 3 \times 10^{-17}$ cm^2/s for the same molecular weight at 125 °C.

Healing results for PMMA at two different molecular weights (M_n of 110,000 and M_w of 240,000 for one, and M_n of 74,400 and M_w of 157,000 for the other) are shown in Figure 12.8; the healing temperature was 118 °C, and readings from at least 10 specimens are averaged for each data point. Again, the data are described by $K_{1c} \sim t^{1/4}$ with $K_{1c}{}^0 \approx 0$. We predict that the ratio of slopes J_1/J_2 for M_1 of 110,000 and M_2 of 74,400 is $(M_1/M_2)^{-1/4} = 0.907$, which agrees well with the (least squares) experimental slopes of Figure 12.8, $J_1/J_2 = 0.914$.

Our relatively large initial $K_{1c}{}^0$ value for PS [8–10] (Figure 12.7) points up a striking difference in the intrinsic healing behavior of PS and PMMA. For PMMA, our results are similar to those of Kausch and co-workers [1–5] where $K_{1c}{}^0 \approx 0$. The

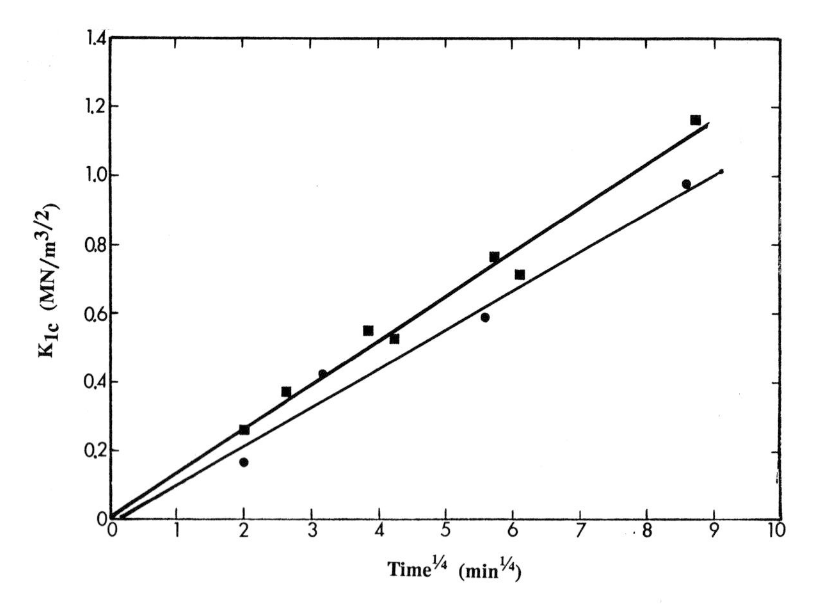

Figure 12.8 K_{1C} versus $t^{1/4}$ for PMMA of two different molecular weights; ▪, M_n = 74,400; • , M_n = 110,000. Healing temperature was 118 °C (Wool and O'Connor).

large $K_{1c}{}^0$ may be attributed to the nature of the fracture surface in the PS specimens, in terms of both chain fracture and molecular orientation. However, our healing rates for PMMA are much longer than those reported by Kausch. For example, compare data in Figure 12.1 and Figure 12.8 at 390 K (117 °C) and similar molecular weights, M_w, of the order of 1.2×10^5. Kausch obtained complete healing in about 10 minutes; our material required about 4,000 minutes. This may be due to material differences and methods of crack healing. Commercially available "PMMA" comes in many varieties and is often a copolymer of MMA with other acrylates (Chapter 7), resulting in polymers with different M_c, C_∞, R_g, and related dynamical properties.

12.4 Healing of Elastomers

In this section, we describe experiments in which we investigated aspects of crack healing in elastomers. Amorphous, noncrystallizable elastomers have a random-coil chain conformation in the bulk state. The fact that these materials can be cross-linked to different extents could be used to study molecular dynamics in terms of linear versus nonlinear chains. A relatively low T_g allows healing experiments to be conducted conveniently at room temperature, and the low modulus of elastomers results in rapid surface wetting under modest contact forces. Thus, the complications of convoluted wetting and diffusion are largely avoided. A further consequence of the low T_g is that fracture tends to be governed by the chain pullout mechanism rather than by chain fracture. We are also interested in these HTPB pure resins because this is the binder phase used in many solid rocket propellants (discussed in Chapter 11).

12.4.1 Sample Preparation and Characterization

Experiments were conducted on a hydroxy-terminated polybutadiene (HTPB) that was lightly cross-linked with isophenyl diisocyanate, which acts as a chain extender and end-linking agent. Equilibrium swelling in cyclohexane resulted in weight and volume swelling ratios of 4.62 and 5.30, respectively. Extraction in boiling benzene for 30 days resulted in an extractable fraction of 15%, presumed to be linear chains. Hence, this material represents an interesting blend of linear, branched, and partially cross-linked materials that we expect to manifest differences in healing behavior.

Well-designed healing experiments have at least two clear requirements. First, the surface formation and contacting procedures must be reproducible, so that subsequent mechanical property measurements of the interface reflect the healing process only. Second, time-dependent wetting must be minimized, because of its convolution with the intrinsic healing behavior. The strength development associated with diffusion is most directly observed when wetting is nearly instantaneous. Rapid wetting is facilitated not only by smooth surfaces, but also by temperatures that are high relative to T_g, so that viscoelastic response to the contact forces is rapid.

We used an edge-notched specimen (shown schematically in the inset of Figure 12.10) for the tests on HTPB [9]. A triangular notch, 6.3 mm deep with a 45° included angle, was cut into the center of the specimen edge with a razor blade. Fracture of such a specimen in uniaxial tension results in a rough, slip–stick fracture initiated at the root of the notch. This type of fracture surface is characteristic of elastomers. Although rapid wetting is still possible because of the low T_g (T_g is about −70 °C), the rough surfaces do not permit the two pieces to be brought back into contact in a reproducible manner. A more satisfactory method for creating the surfaces to be healed was to cut the specimens in two with a single pass through the notch with a sharp razor. After they were cut apart, the two halves were immediately brought back into contact with a small force applied by hand. After it had healed for the desired time at room temperature under no external forces, the interface was refractured at a constant uniaxial strain rate of 0.15 s^{-1}. The fracture stress was calculated from the load at separation and the interfacial contact area; the fracture energy or work was the integrated area under the stress–strain curve. Similar tests are often conducted in the elastomer industry, where they are termed *butt tack tests*.

12.4.2 HTPB Stress–Strain Healing Behavior

Figure 12.9 shows the stress–strain behavior of HTPB in the virgin state and at various healing times, with fracture and healing all done at room temperature [9]. From such data, one may obtain independent measurements of fracture stress, strain, and energy

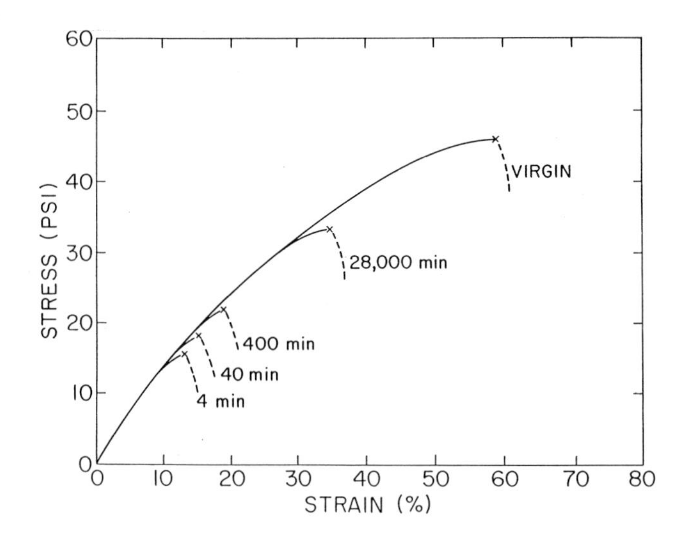

Figure 12.9 The stress–strain behavior of HTPB for the virgin state and for various healing times (Wool and O'Connor).

as functions of healing time. As with other crack healing experiments (for example, Figure 12.6) we observe that the reload curves are superimposed on the original stress–strain curves, which implies that the modulus remains unchanged by the presence of the interface.

Figure 12.10 shows the fracture stress, σ, plotted against $t^{1/4}$, where each data point is the average of readings from at least three tested samples. The linearity of the data supports the healing relation

$$\sigma = \sigma_0 + K t^{1/4} \tag{12.4.1}$$

where K is the slope of a least squares fit of the data, and σ_0 is the initial "wetting stress" obtained by extrapolation to t of 0, and is most likely a combination of contact forces and low-molecular-weight diffusion. The solid line is drawn with $\sigma_0 = 12.76$ psi and $K = 1.69$ psi/min$^{1/4}$. At very long times (240 days), the data level off at $\sigma_\infty = 38$ psi [10], which is less than the virgin stress obtained without cutting the sample ($\sigma_\infty = 46$ psi). An apparent conclusion is that interpenetration of chain segments, rather than chemical cross-links, plays the dominant role in the load-bearing mechanism of cross-linked polymers. A slow diffusion process is evident in Figure 12.10, since several months is required for full healing. This points out the dramatic effect of branches and cross-links on the reptation process. For a similar system of linear chains, healing would occur orders of magnitude faster.

The fit of the data in Figure 12.10 to the intrinsic healing prediction indicates that rapid wetting is achieved, which is consistent with the rapid optical disappearance of the interface during the contacting process. However, data taken at very short times

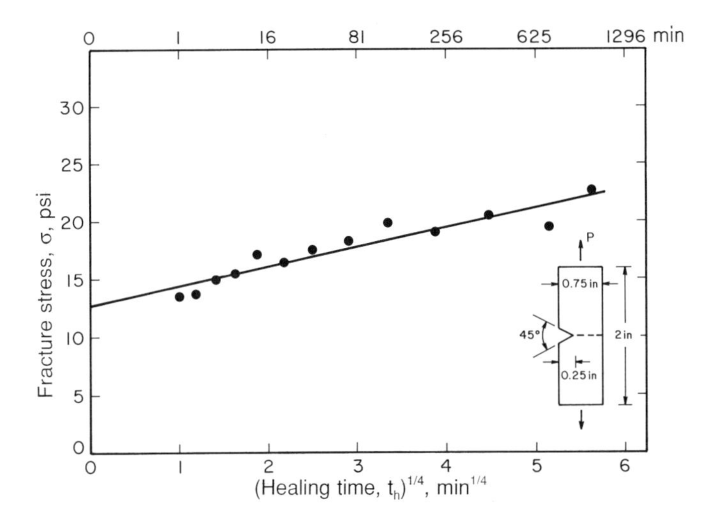

Figure 12.10 Fracture stress σ versus $t^{1/4}$ for HTPB fractured as shown. The line is a least squares fit (Wool and O'Connor).

(t less than 60 seconds) can be used to observe time-dependent wetting. The observed stress should be the convolution product of the intrinsic stress and wetting functions, as

$$\sigma(t) = \int_{-\infty}^{t} [\sigma_0 + K(t-\tau)^{1/4}] \frac{\partial[1-(1-\sigma_0)e^{-kt^m}]}{\partial\tau} d\tau \tag{12.4.2}$$

where m is a constant. Since the diffusion rate is so small in this material, a good assumption is that σ_0 is much greater than $Kt^{1/4}$ for very short times. Then Eq 12.4.2 reduces to the "wetting solution",

$$\sigma = \sigma_0 [1-(1-\phi_0)e^{-kt^m}] \tag{12.4.3}$$

Figure 12.11 shows the healing behavior over the first 60 seconds. The solid line is the extrapolation of Eq 12.4.1, with σ_0 and K as determined in Eq 12.4.1. The instantaneous fractional wetted area, ϕ_0, is determined from $\sigma(t{=}0)/\sigma_0$. The dashed line through the data in Figure 12.11 was plotted from Eq 12.4.3 with $\phi_0 = 0.73$, $k = 0.19\ s^{-1}$, and $m = 1$. This value of m corresponds to instantaneous nucleation of wetted areas and radial spreading of these areas according to $t^{1/2}$. As is evident in Figure 12.10, complete wetting ($\sigma = \sigma_0$) is achieved in about 30 seconds, after which diffusion begins to contribute.

In retrospect, an alternative explanation to the initial rapid gain in strength, discussed as wetting above, involves the diffusion of the linear chain fraction in the lightly cross-linked HTPB sample. The linear chains diffuse much more rapidly than the branched or cross-linked chains and are expected to contribute to the initial strength gain σ_0. In Chapter 8 we saw that elastomeric chains with M of about 300,000 could attain their optimal strength contribution in about 60 seconds, which is

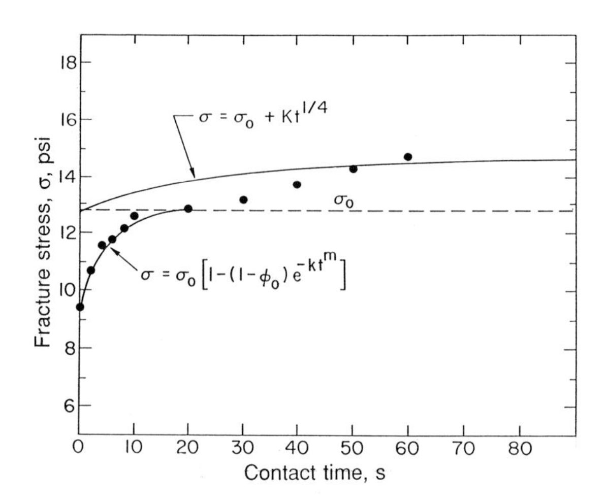

Figure 12.11 Time-dependent wetting for HTPB at times less than 1 minute (Wool and O'Connor).

comparable to the wetting times in Figure 12.11. If the extractable fraction of linear chains, f, is 0.15, we have the contribution to the fracture stress as σ_0(linear) $\approx f^{1/2}\sigma_\infty$. With $\sigma_\infty = 38$ psi, this predicts that σ_0(linear) ≈ 14 psi, which is comparable to σ_0 in Figure 12.11. This interesting result however, carries with it the assumption that the linear chains contribute as much (on a contour–length basis) to the strength as do the branched and cross-linked materials.

Comment on Healing with Mixtures of Linear, Branched, and Cross-Linked Chains

It would be interesting to examine healing in model polymer mixtures where we have a known fraction of linear, branched, and cross-linked chains contributing to the diffuse structure of the interface. Since the different chains can be made with well-separated relaxation times, one should be able to observe the contributions of each component from the resulting plateaus in the healing behavior.

12.4.3 Fracture Energy for Crack Healing

The intrinsic healing behavior for fracture energy, W, is shown in Figure 12.12. The data points represent the area under the stress–strain curve at each healing time. The solid line is given by

$$W = \tfrac{1}{2} E(\sigma_0 + K t^{1/4})^2 \tag{12.4.4}$$

where $E = 94.6$ psi (an average value over the deformation range), and σ_0 and K are as determined previously from the stress data of Figure 12.10. The data deviate from

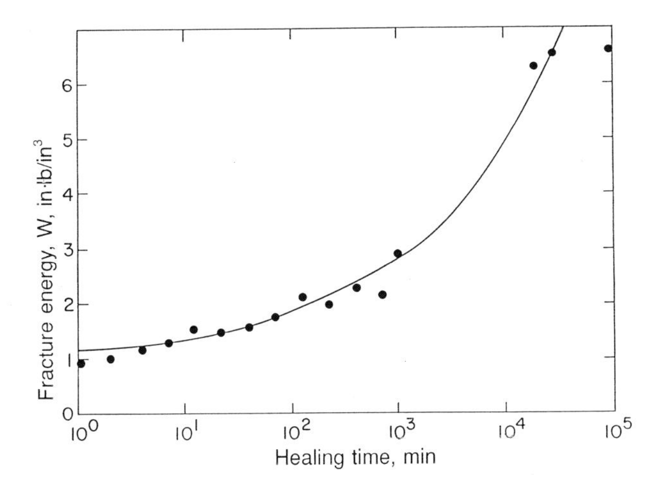

Figure 12.12 Fracture energy, W, versus log time. The line is plotted using Eq 12.4.4 with E of 94.6 psi (Wool and O'Connor).

Eq 12.4.4 at short and long times for the same reasons as mentioned above for the stress data. At long times ($Kt^{1/4}$ greater than σ_0), a dependence on $t^{1/2}$ is more apparent.

12.4.4 Surface Rearrangement Studies

We investigated the effect of surface rearrangement prior to contact and healing in fractured specimens of HTPB [10]. Samples were fractured and held apart under dry N_2 for various times, t_{sr}, and then healed for a constant time of 10 minutes. Figure 12.13 shows the fracture stress after the 10-minute healing as a function of $t_{sr}^{1/4}$; each data point is an average of readings from at least two tested samples. As is evident from Figure 12.13, the data can be represented by

$$\sigma = \sigma^* - K_{sr} t_{sr}^{1/4} \tag{12.4.5}$$

where σ^* of 18.27 psi is the intercept at t_{sr} of 0, and the slope, K_{sr}, is 0.845 psi/min$^{1/4}$. The value of σ^* correctly matches that of Figure 12.10 and Eq 12.4.1 at a healing time of 10 minutes, since that experiment was conducted at t_{sr} of 0. At long rearrangement times, σ approaches a plateau equal to σ_0, the wetting stress in Eq 12.4.1.

A possible explanation of the dependence on $t^{1/4}$ and the plateau value of σ_0 is that surface rearrangement involves the diffusion of chain ends and linear chain segments into the bulk, so that they are not available to initiate healing after contact is established. After a rearrangement time corresponding to the plateau of Figure 12.13, virtually no chain segments diffuse across the interface over the 10-minute healing

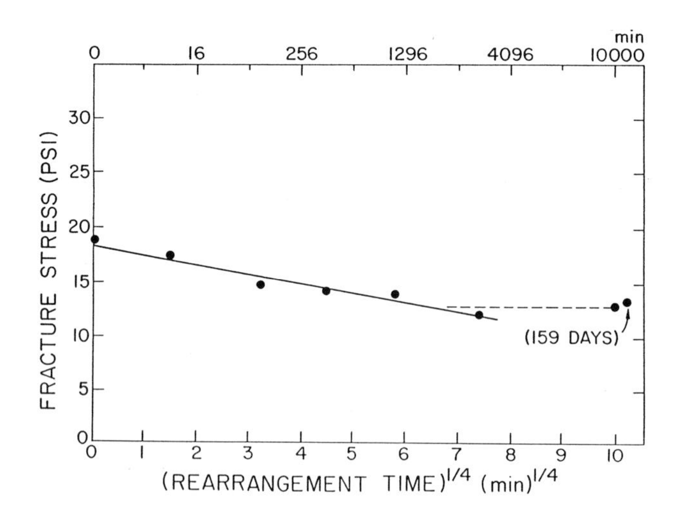

Figure 12.13 Fracture stress versus $t^{1/4}_{sr}$. The sample surfaces were held apart for a time t_{sr}, after which the interfaces were healed for a constant 10 minutes (Wool and O'Connor).

period. It should be noted that the slope K_{sr} is exactly one-half the magnitude of the slope K (no surface rearrangement) in Figure 12.10 and Eq 12.4.1.

Several explanations are possible to account for the decrease in σ with rearrangement time. The exposed surfaces may have been affected by chemical or physical interactions with atmospheric gases, water vapor, or dust. These interactions would not necessarily lead to a $t^{1/4}$ decrease in σ, and would likely result in σ approaching zero rather than σ_0 at long times. To resolve this question, a set of samples was fractured and held apart under dry N_2 for a constant time t_{sr} of 4320 minutes, which corresponds to the plateau region of Figure 12.13. The subsequent healing behavior of such identically conditioned samples is shown in Figure 12.14, where each data point is an average of readings from at least three samples. The same wetting stress σ_0 is obtained at $t = 0$ as was obtained in Figure 12.11. However, this value persists up to approximately 40 minutes of healing, after which the stress increases with $t^{1/4}$. Clearly, the surface rearrangement effect is reversible and may be associated with chain ends pulled out from the surface at fracture, or the surface concentration of linear chains.

The slope of the time-dependent data in Figure 12.14 is identical to that of Figure 12.10. The prior surface rearrangement introduced an apparent diffusion initiation time τ_i so that, for t greater than τ_i

$$\sigma = \sigma_0 + K(t - \tau_i)^{1/4} \tag{12.4.6}$$

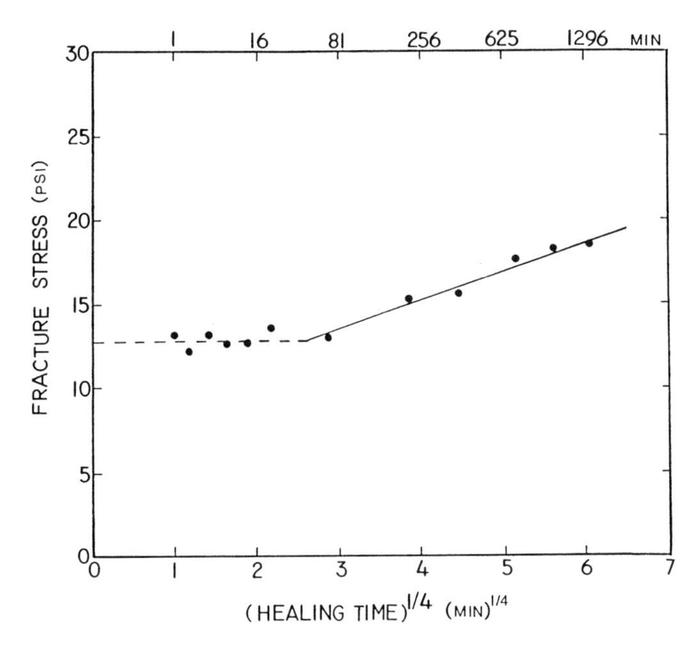

Figure 12.14 Fracture stress versus $t^{1/4}$ for sample surfaces held apart for 4320 min under dry N_2 before being brought into contact (Wool and O'Connor).

where τ_i is 40 minutes, and σ is equal to σ_0 for t less than τ_i. It is expected that τ_i depends on the time allowed for rearrangement.

It would be useful to reexamine these data with other dynamics models (for tethered molecules, branched chains, cilia, and partially cross-linked chains). Our suggestion would be to calculate the average contour length for each molecule and derive a strength law with methods similar to those used for the linear chains.

12.5 Impact Healing

12.5.1 Izod Impact Healing

Izod impact experiments were conducted on PMMA and polystyrene (PS) notched samples using Method C in ASTM D 256 [20]. The sample configuration is shown in the inset of Figure 12.15. The results on both materials were the same qualitatively in terms of slope arguments, and only the PS data are reported here. The sample preparation and characterization are similar to work reported above. Each sample was fractured, brought back into contact at 20 °C, and healed in a vacuum oven for times in the range 10^1–10^4 minutes.

The assumption that wetting is instantaneous is extremely poor for all such experiments, due to the large surface area of the fracture and its roughness. In the range of healing temperatures from T_g to (T_g + 40 °C), the crack interface slowly and uniformly disappears by multiple nucleation of wetting points in the interface, propagation of wetted pools, and eventual total coalescence of wetted areas. For example, at T of 125 °C, partial wetting has occurred at 10 minutes, with complete

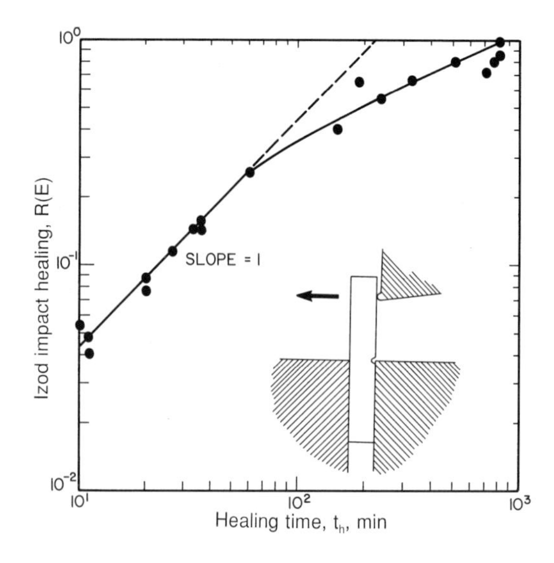

Figure 12.15 Plot of log R versus log t for healing of PS fractured in the Izod impact test, where $R = E/E_\infty$ (Wool and O'Connor).

disappearance of the interface at 60 minutes. From scanning electron micrographs we observe that the fracture surface adjacent to the notch is usually rougher than the fracture surface that is farther away; this further complicates the healing problem because of changes in the notch geometry with time and nonuniform healing near the notch tip.

We can write a simple approximate function for time-dependent uniform wetting of the interface as

$$\phi(t) = k_s\ U(t) \tag{12.5.1}$$

where k_s is a constant and $U(t)$ is the Heaviside step function. Convoluting Eq 12.5.1 and Eq 12.4.4, we obtain the time-dependent impact work, $W(t)$, as

$$W(t) = \tfrac{1}{2}\, k_s E(\sigma_0^2 t + 8/5\, \sigma_0\, K t^{5/4} + \tfrac{2}{3} K^2\, t^{3/2}) \tag{12.5.2}$$

Thus, during the wetting stage, we have a time dependence with a mixture of exponents, 1, $^5/_4$, and $^3/_2$. We see in Figure 12.15 that a slope of about unity reasonably well describes the impact energy recovery data during the 60-minute interval while the crack interface wets. At later times, when the wetting stage is completed, a new slope develops, with a value of about ½, as expected for the diffusion-controlled fracture energy. If wetting took place instantaneously and the initial wetting energy were very low, we would expect a dependence of the impact work on $t^{1/2}$ over the entire healing range, similar to results obtained for compact tension samples.

12.6 Crack Healing in Cross-Linked Polymers

12.6.1 Crack Healing of Epoxy

Given the prevalence of crack healing in polymers under the right conditions, one might wonder whether a polymer exists that does not heal. We once conjectured as a naive response to that question that a highly cross-linked system such as an epoxy would not be expected to heal. A paper by Outwater and Gerry in 1969 [11] demonstrated that Epon 826 (100 parts) cured with methyl anhydride (90 parts) and benzyl dimethyl amine (1 part) at 121 °C, did crack heal. They measured the fracture energy of the cured epoxy using a double torsion method on a plate sample with a central crack. When the crack grew halfway through the plate, they healed it at 148 °C and 204 °C. Complete healing was observed, and the original fracture energy, $G_{1c} \approx 200$ J/m^2, and crack propagation velocity were recovered. Healing would occur only at temperatures above the cure temperature of 121 °C. They also examined the effect of post-cure on epoxy healing, since it was possible that the epoxy interface healed by post-curing reactions. They found that healed samples when refractured and rehealed also recovered full strength.

Encouraged by the experiments of Outwater and Gerry, we examined crack healing in Epon 828 (bisphenol A) cured with TETA. We conducted exploratory Izod impact fracture and healing tests. We found that the crack interface optically disappeared and that partial (about 50%) impact energy recovery was obtained. Some samples did not heal and others broke upon cooling. We attribute the latter results to surface roughness effects associated with the Izod method, but conclude that epoxies can heal, under the right conditions.

The mechanism of healing in cross-linked materials in unclear. We speculate that the fracture process creates many new chain ends that did not exist in the virgin material and are now available for interdiffusion at the interface. Also, the freshly created fracture surfaces when rejoined may permit additional curing to occur by providing a new spatial arrangement of functional groups. A useful test would be to determine whether nonfractured (virgin) epoxy surfaces heal.

12.6.2 Rubber Toughening in Epoxies

Rubber and glass particles are known to toughen epoxy matrices [12]. We wish to know the effect of the rubber fraction on G_{1c} of the composite. The following is a novel analysis based on simple percolation concepts of connected damage zones. Consider an epoxy matrix containing a volume fraction p of rubber particles. Rubber toughening occurs by additional energy dissipation through crazing, plastic shear yielding of matrix ligaments between particles, and deformation of the rubber particle. Figure 12.16 shows the schematic effect of a low-modulus rubber particle in a higher modulus matrix. The stress field at the rubber particle is similar to that described in

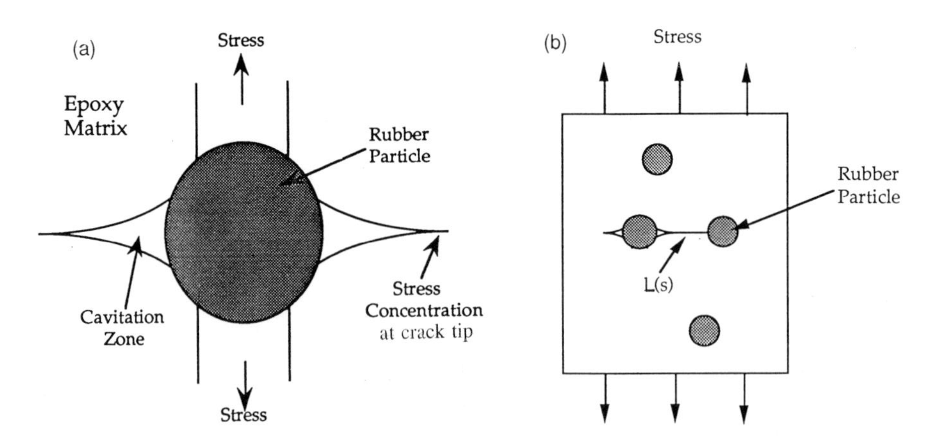

Figure 12.16 (a) Damage and cavitation at a rubber particle in a stressed epoxy matrix. (b) The mean free damage length, L, between rubber particles in an epoxy matrix. (Figure kindly drawn by Michael Gustek, Univ. of Illinois, 1993.)

Chapter 11 for crazes around a hole (Figure 11.3) in a uniaxially stressed PS film. The stress concentration normal to the equatorial plane causes crazes to nucleate and propagate perpendicular to the direction of the applied stress. We let the contribution of the rubber particles to the fracture energy, G_r, be expressed as

$$G_r \sim N_c L \tag{12.6.1}$$

where N_c is the effective number of particle clusters per unit volume and L is the average mean free path length of the craze or damage zone emanating from each cluster (Figure 12.16). The crazes grow to a length L, until they reach another particle or the surface of the composite. The rubber particles act as stress concentrators, either as individual particles at low concentration, or as clusters at higher concentration.

The number of damage nucleation clusters is given by

$$N_c = p/(Vs) \tag{12.6.2}$$

where $V = 4/3\pi r^3$ is the volume of a particle of radius r, and s is number of particles per cluster, determined from scalar percolation (Chapter 4) as

$$s = |p - p_c|^{-v} \tag{12.6.3}$$

in which p_c is the percolation threshold and v is the critical exponent. At p_c, the rubber particles form a continuous phase of touching particles with very little strength, and we assume that the matrix becomes very weak, or $G_r \approx 0$. Thus, we have the number of clusters per unit volume as

$$N_c = p\,|p - p_c|^{v} / V \tag{12.6.4}$$

The number of clusters decreases with increasing p since their size becomes larger. At p_c, N_c approaches zero because the sample contains essentially only one large cluster, the percolating structure. The average damage length L is determined by the distance between clusters as $L \sim N_c^{-1/3}$, so that $G_r \sim N^{2/3}$, or

$$G_r \sim p^{2/3}\,|p - p_c|^{2v/3}\;V^{-2/3} \tag{12.6.5}$$

This function goes through a maximum value of G^* when p is equal to p^*, which is itself determined by

$$p^* = p_c/(1 + v) \tag{12.6.6}$$

In three dimensions, $p_c \approx 31\%$, $v = 0.43$, and the fracture energy reaches a maximum at a rubber particle fraction of $p^* \approx 21\%$. At p^*, the average cluster size is 2–3 particles, which act as the nucleus for single crazes. Normalizing G_r with respect to G^*, we have the resulting ratio of fracture energies

$$G_r / G^* = \{(p/p^*)\,[(p_c - p)/(p_c - p^*)]^{\nu}\}^{2/3} \tag{12.6.7}$$

in which $0 \leq p \leq p_c$.

The behavior of this function is shown in Figure 12.17. The position of the maximum at 21% is independent of individual particle size. Figure 12.17 shows experimental data for rubber-filled epoxies, reviewed by Garg and Mai [12], and analyzed in terms of Eq 12.6.7. Garg and Mai find that the initial fracture energy is very low (about 120 J/m^2) but increases to a maximum of about 35,000 J/m^2 at about 16–17 wt% rubber. We obtain the volume fraction, p, from the weight fraction by correcting for density differences as follows:

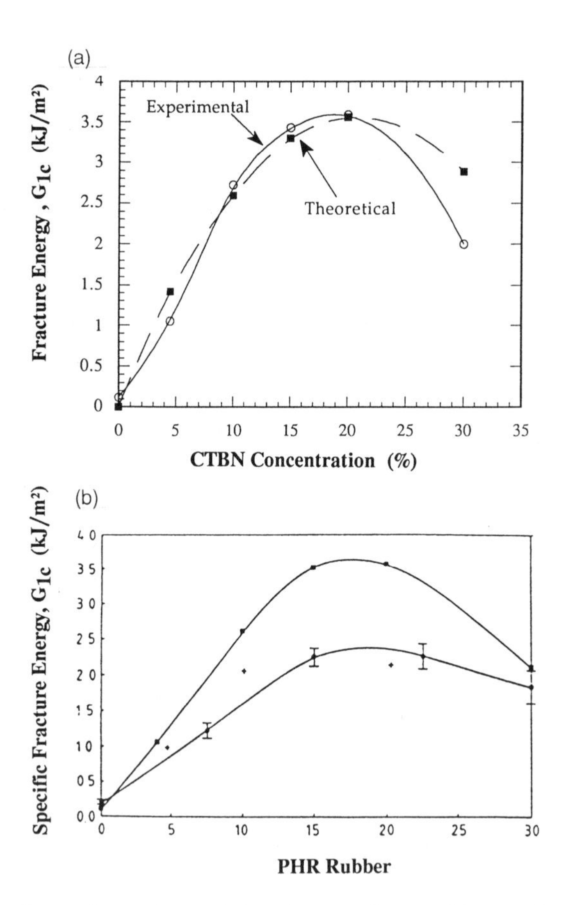

Figure 12.17 (a) Comparison of percolation damage theory with experimental data for G_{1c} versus rubber particle fraction in an epoxy matrix. (b) Data of several investigators for the fracture energy, G_{1C}, of epoxy versus rubber fraction [12]. (Figure kindly drawn by Michael Gustek, Univ. of Illinois, 1993.)

$$p = (w_r / d_r) / [w_r / d_r + w_e / d_e] \quad (12.6.8)$$

in which w_r, d_r, w_e, and d_e are the weight of the rubber, density of the rubber, weight of the epoxy, and density of the epoxy, respectively. From Figure 12.17, $w_r \approx 17\%$ and $w_e \approx 83\%$; and substituting $d_r = 1.2$ and $d_e = 1.6$, we obtain $p^* \approx 21\%$. The agreement between the values obtained from the theory and from experiment is interesting considering the simplicity of the theory.

The amplitude of the fracture energy at p^* depends on the particle radius r (from Eq 12.6.5) as

$$G_r \sim 1/r^2 \quad (12.6.9)$$

Smaller particles result in higher fracture energy. The lower limit of the particle size is determined by its mechanistic role in toughening. For example, a rubber particle has to be large enough to span a craze width and allow the craze to form while stabilizing it through bridging. This also requires that the rubber particle adhere sufficiently to the matrix. If poor adhesion occurs, a critical crack may propagate through the craze and cause fracture of the sample.

Some exceptions

The above analysis is based on energy dissipation per unit volume. There are several important situations where this analysis does not apply.

a. Fatigue: In subcritical cyclic fatigue of a sharp crack, the damage is largely confined to a single propagating crack with a craze at its tip. The fatigue stress may generate many crazes within the body of the material. However, these may have little effect on da/dN other than to cause some crack blunting on occasion. As long as the craze breaks down at the same rate as in the virgin material, there should be little change in da/dN with p. Matrices that undergo extensive shear banding are expected to deviate from this behavior.
b. Sharp Crack: In critical propagation of a sharp crack through a line deformation zone, the fracture energy should be proportional to the number of particles per unit area, and hence

$$G_r \sim p \qquad \text{for } p \leq p_c \quad (12.6.10)$$

provided that good adhesion exists between the particles and the matrix. This function should reach a maximum at p_c.
c. Poor Adhesion: If we have poor adhesion with the matrix, then for a sharp propagating crack

$$G_r \sim G_0 - \alpha p \quad (12.6.11)$$

where α is a constant determined by $\alpha = G_0/p_c$, and G_0 is the matrix fracture energy.

d. Crack Healing: If the rubber particles rupture during fracture, crack healing is adversely affected by the presence of the particles at the healing interface and the weld is likely to be very weak.
e. Fiber-Reinforced Composites: The presence of the fibers reduces the propagation length L of the crazes and undermines their stability, resulting in lower fracture energies compared to the bulk.

For glass-particle-reinforced epoxies, the damage occurs at the particle poles parallel to the applied stress, since the glass modulus is much higher than that of the epoxy matrix. Glass and rubber are not expected to contribute additively to the toughening of epoxies, since their energy-absorbing mechanisms can interact with each other.

12.7 Fatigue Healing

We asked whether cracks formed in subcritical cyclic growth could be healed. An affirmative answer offered the potential of treating composites and similar plastic components with fatigue-healing treatments on a periodic basis and essentially giving them an infinite fatigue life. This is important for the design of smart composites, which are discussed in Section 12.8. The answer to this question happens to be very different from that we found in the normal crack healing studies.

12.7.1 Fatigue Crack Healing

Polystyrene compact tension samples were fatigued in constant displacement tension–tension control at room temperature at a frequency of 20 Hz by Klosterman [13]. The number of cycles was recorded each time the crack length was incremented by 0.544 mm, and the virgin da/dN slope was determined. The maximum starting load was 22.5 ± 1.5 lb. Fatigued samples were healed in a vacuum oven for times of 15 to 180 minutes. After healing, the crack growth rate da/dN was again monitored and compared with the virgin crack growth response. Figure 12.18 shows the histogram of virgin da/dN data (180.3 crack length units = 1 mm). Crack growth rates, da/dN, for virgin samples ranged between 5.32×10^{-4} mm/cycle and 2.97×10^{-3} mm/cycle, with an average crack growth rate of 1.05×10^{-3} mm/cycle.

Healing at T_h of 123 °C (T_g of PS is 100 °C) in a vacuum produced complete recovery of the crack growth rate at relatively short times (as low as 15 min). We found that fatigue cracks would wet uniformly and optically disappear within 6 to 10 minutes. The healed crack growth rate data did not significantly differ from virgin crack growth rate data (Figure 12.19). Samples healed at 123 °C for as little as 15 minutes were observed to be completely recovered. Complete healing is observed (with respect to da/dN) at times much shorter than the time to achieve either complete interpenetration (at R_g) or fracture strength, G_{1c}. Figure 12.20 summarizes the observations for fatigue crack healing.

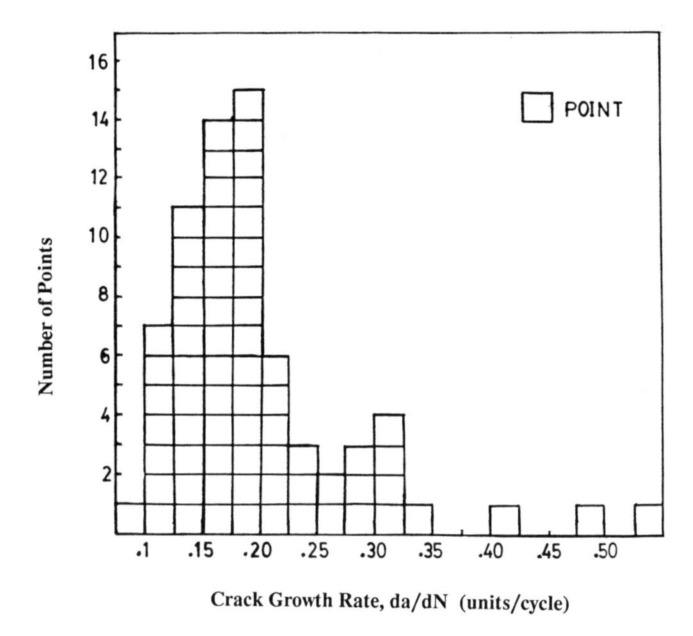

Figure 12.18 Histogram of virgin crack growth rate data (Klosterman and Wool).

We believe that the lack of a significant time dependence for fatigue crack healing is due to the ability of the slowly moving crack to select a path away from the original interface. Similar effects were observed for some tests of symmetric interfaces (Chapter 8). However, fatigue healing of welded parts with DCB specimens showed a much greater time dependence, and the fact that CT specimens were used here may have contributed to the above results. Weak incompatible interfaces show relatively high strength when cracks migrate into the virgin material on either side. Also, low-molecular-weight species may assist in rapid healing in the first 10 to 30 minutes. These results are at least encouraging for promoting fatigue healing in composites.

12.7.2 Fatigue Healing of Short-Fiber-Reinforced Polystyrene

We examined da/dN healing in a more complex system, PS reinforced with chopped glass fiber. Rectangular specimens of 20% randomly oriented short-fiber-reinforced polystyrene with edge notches were fatigued in uniaxial tension. These samples were examined by transmitted light microscopy. The observed damage was a mixture of (a) matrix cracking, (b) matrix crazing, (c) fiber cracking, (d) fiber–matrix debonding, and (e) fiber pullout.

We found that when the samples were healed in a Mettler hot stage at 131 °C for 205 minutes, matrix and interfacial damage in front of and adjacent to the crack

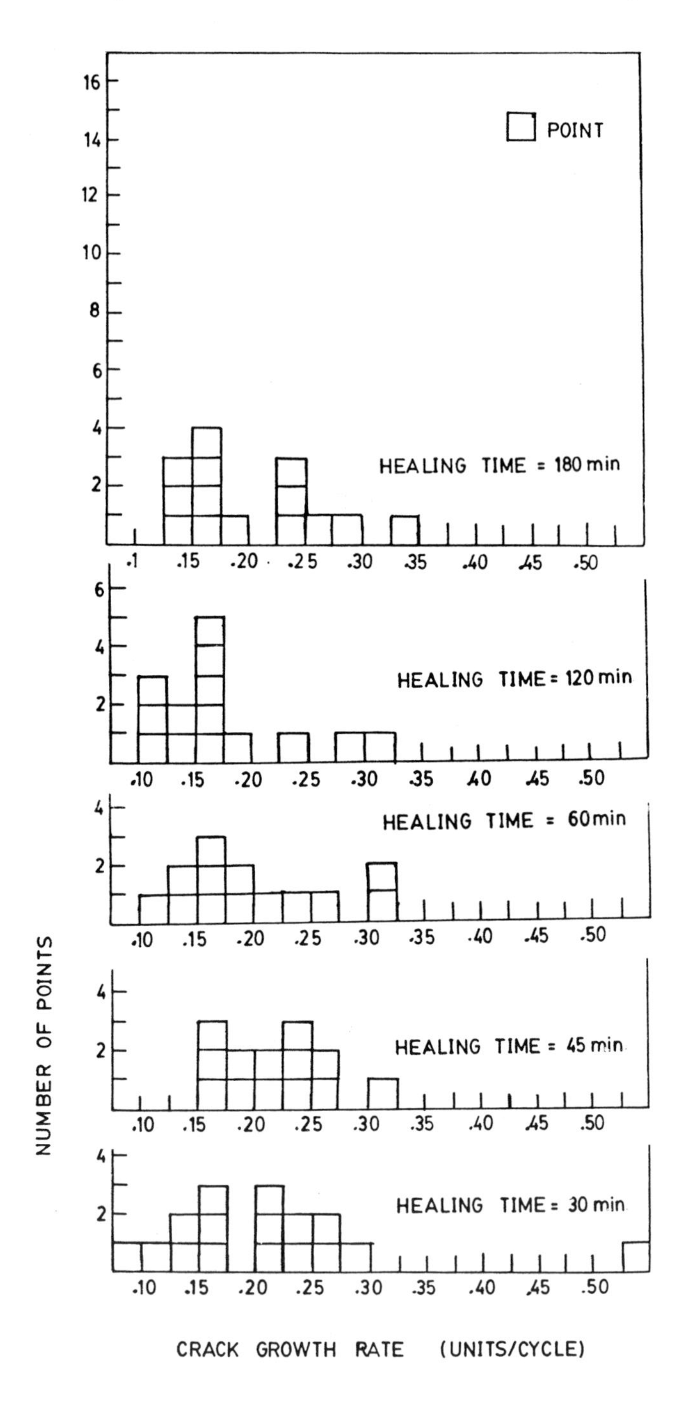

Figure 12.19 Fatigue growth rates at several healing times (Klosterman and Wool).

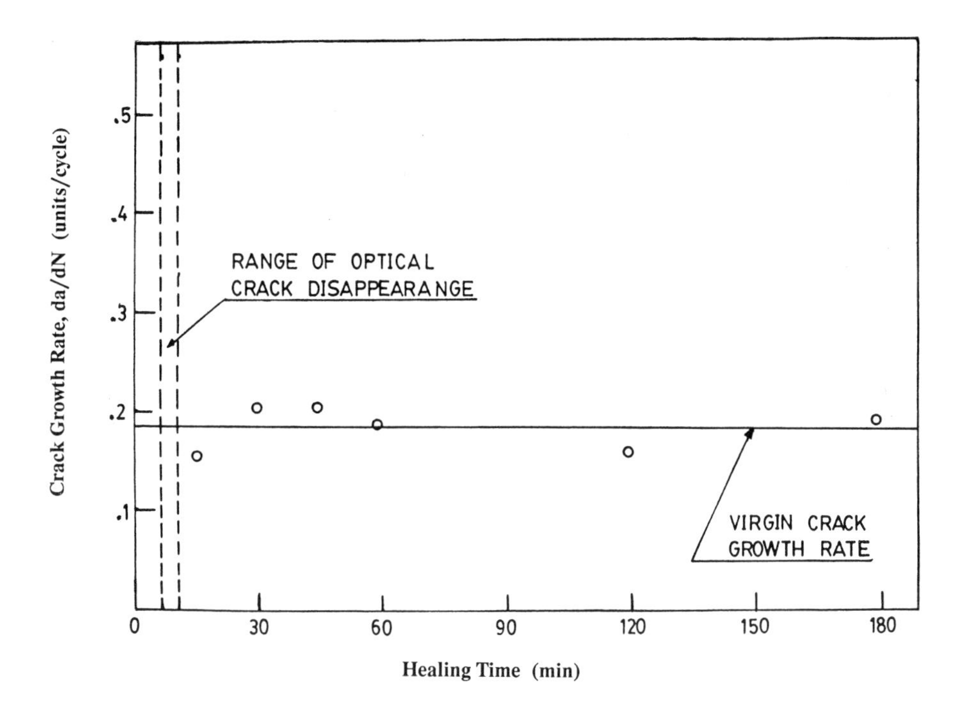

Figure 12.20 Observations of fatigue crack healing in polystyrene.

partially healed. The crack retreated to a length shorter than the initial crack length. The fatigue crack itself was unable to mend under the above conditions because of fiber debonding and fiber pullout.

In an attempt to obtain further healing, we constrained samples on three sides and applied a load of 6.5 lb perpendicular to the rectangular face. The constrained samples were healed for 720 minutes in a vacuum oven at 131 °C. Under these conditions, the crack and all other damage became optically undetectable. After complete optical healing was attained, the samples were refatigued. The following results were obtained.

Initial crack growth rate: 6.63×10^{-5} mm/cycle

Healed crack growth rate: 1.32×10^{-4} mm/cycle

We found that in the fiber-reinforced case, the recovery ratio equals about 50%. Thus, even though samples looked completely healed (through the microscope), they had fatigue healed only partially. Fiber breakage and fiber–matrix debonding reduce the load-carrying capability of the polystyrene matrix. Since fiber reinforcements as a general rule increase fatigue life, we concluded that the fiber breakage and fiber–matrix debonding resulted in a lower fatigue life. Thus, even if the polystyrene matrix fully recovers, the permanent fiber damage and fiber–matrix debonding leave additional sites for cracks and crazes to form upon refatiguing. Solutions to this problem are discussed in Section 12.8.

12.8 Self-Repair of Smart Materials

12.8.1 Smart Materials

We can bring about the healing of damage in materials by applying healing treatments, as discussed above, or by having the healing treatment built into the material so that it is available when needed. A material that could heal itself when damaged would be considered a "*smart*" *material*, in contrast to those materials that have to suffer the consequences. For example, in fiber–composite materials, we are concerned about mechanical and environmentally induced damage in the form of delamination, matrix cracking, fiber debond from the matrix, fiber fracture or pullout, and other processes such as distortion and shrinkage that might cause the material to become defective or fail catastrophically. A smart material is one that can somehow sense the damage process and correct it by initiating a self-repair action. This action could be the release of self-repair fluids that work by rebonding the fractured surfaces. Self-repair could have a profound effect on personnel safety and the lifetime of the composite.

Smart materials combine normal structural components, such as plastic, ceramic, metals, and concrete, with other components, which **sense** the environment of the structure and **actuate** a corrective measure. Some sensors may simply monitor a variable such as stress or temperature. For example, a mountain climber would appreciate a rope that changes color in places that are dangerously weakened. Piezoelectric polymers and ceramics such as poly(vinylidene chloride) and lead zirconium titanate have been used as both pressure sensors and actuators (reviewed in [14]). By virtue of their electrical polarity, they can turn mechanical force into electrical signals (sensors) and turn electrical signals into mechanical force (actuators). Shape–memory alloys such as nickel–titanium operate as actuators with respect to temperature. When the temperature goes above a certain transition point, a crystal phase transition (for example, martensitic) occurs and the entire metal changes shape reversibly. The resulting forces can be considerable and can be used to actuate a corrective measure in the material system. Electrorheological fluids, consisting of polarizable polymer or ceramic particles in a fluid, serve as actuators by changing from a liquid to a gel-like solid when electric fields are applied. Electrooptical materials and thermooptical materials change their optical properties and can be used as self-actuating radiation shields. Embedded optical fibers can be used to monitor local environments by spectroscopic methods and are an inexpensive sensor.

12.8.2 Self-Repair of Fiber-Filled Composites

Professor C. Dry [15–17] (University of Illinois) used hollow fibers filled with fluids in composites to actuate self-healing in the composites. The fibers, when broken, release their fluids, which act to heal the mechanically induced damage in the polymer or concrete matrix. Dry and Sottos [16] investigated the possibility of developing a smart polymer matrix composite that had the ability to self-repair internal microcracks resulting from thermal–mechanical loading. As shown in Figure 12.21, they examined controlled cracking of hollow repair fibers dispersed in a composite matrix. The fluids are released either by rupture, fiber pullout, or fiber debonding. The subsequent release of chemicals results in the sealing of matrix microcracks and the rebonding of damaged interfaces. Materials that are capable of passive, smart self-repair have several features, which they analyze in the following manner.

1. The material must be subject to gradual deterioration, for example, by dynamic loading that induces microcracks.
2. The fiber contains self-repair fluid.
3. The fibers require a stimulus to release the repairing chemical.
4. A coating or fiber wall must be removed in response to the stimulus.
5. The fluid must promote healing of the composite damage.

Several interesting self-repair tests were conducted by Dry and co-workers [15–17] on polymer matrices. Experiments were performed on polymer samples containing an embedded continuous metal fiber with two self-repair fibers in its vicinity, one containing a monomer and the other containing a cross-linking agent. They first mechanically debonded the metal fiber from the matrix without disturbing the self-repair fibers. At this point, the metal fiber could be readily removed with little additional stress. The entire composite was then stressed and the self-repair fluids were released. The fluids mixed, diffused to the debonded area and cured in place, which resulted in a considerable gain in the fiber pullout stress. They obtained similar results in controlled impact tests; when subcritical impact resulted in the formation of many microcracks, the self-repair fluid migrated into the cracks and interfaces and promoted healing.

This approach to self-repair is very flexible with regard to the design of fiber, number of fibers, healing fluid, fiber construction, fiber coatings, and so forth. The healing fluid could be a cross-linking epoxy or polyester, a chemical agent that reacts with the matrix, or one that reacts selectively with the damaged surface. For example,

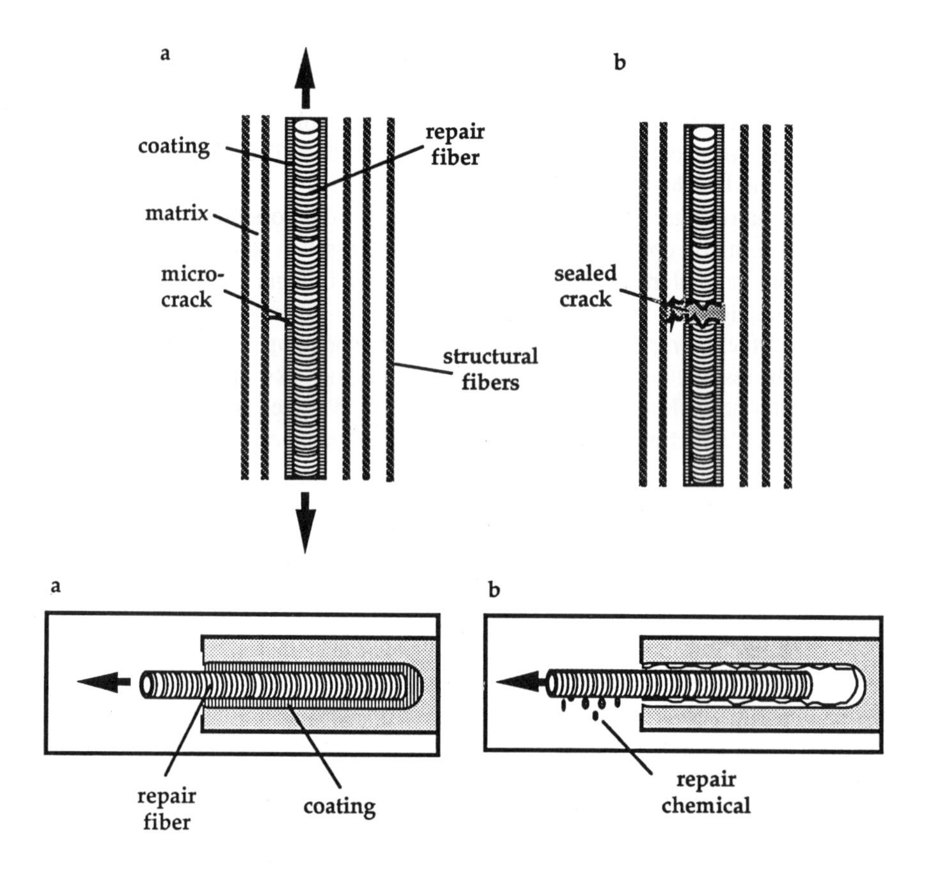

Figure 12.21 Schematic drawing of hollow-fiber fluid-release design to repair cracks in smart fiber-filled composites (courtesy of C. Dry).

one could use vinyl monomers in places where microcracks in the matrix generate free radicals. The free radicals would then polymerize the monomer fluid and help promote strength. Matrix solvents would also be useful to promote solvent bonding in microvoids. Thermally induced healing can also be induced by exothermic reactions of the self-repair fluid, either with itself or with damaged components of the composite.

12.8.3 Stages of Healing in Smart Materials

Discussions with C. Dry resulted in the following analysis of crack healing in smart materials by self-healing fluids, either solvents or reacting fluids [18]. The objective is to develop a framework for evaluating the healing of damage processes in composites with self-repair systems. This is a very complicated problem, given the many variables and interrelationships. We seek solutions that use methodologies similar to those that were used to solve crack-healing problems. Healing of cracks and microscopic damage has been described in terms of several stages of healing, involving surface rearrangement, surface approach, wetting, diffusion, and randomization (Chapter 8). This approach has been successfully used for thermal healing of damage. Here, we describe the stages of healing and how they should be used to promote self-repair by time-release systems involving solvents and adhesives.

Surface Rearrangement

Before bringing the damaged surfaces into contact, one should consider the roughness or topography of the surface and how it changes with time, temperature, and pressure following contact with the fluid. In fractured polymers, rearrangement of fibrillar morphology and other factors can affect the rate of crack healing. Chain-end distributions near the surface can change as molecules diffuse back into the bulk polymer. Spatial changes of the molecular weight distribution may also occur, for example, where the low-molecular-weight species preferentially migrate to the surface. In time-released solvents (or adhesives), surface rearrangement is affected by the solvent–polymer interaction. Chemical reactions, for example, oxidation and cross-linking, can occur on the surface and complicate the molecular dynamics of diffusion. Solvents can also generate additional damage, for example, by causing crazes to swell and allowing them to propagate further.

Each material and self-repair release technique possesses unique surface rearrangement processes that may need to be quantified. The use of solvent systems that promote surface segregation of chain ends, for example, would be highly conducive to rapid healing of damage. The relation between solvent concentration and surface rearrangement dynamics needs to be quantified, primarily through the effect of solvent on both the glass transition temperature, T_g, and the relaxation times, τ_s, of the surface molecules (Chapter 7). The dynamics of the surface layer rearrangement is similar to that in some bulk processes, but in general the surface molecules should have enhanced mobility due to higher degrees of freedom.

Surface Approach

Surface approach considers time-dependent contact of the different parts of the surfaces to create the interface. For example, in crack, craze, and void healing, contact may be achieved at different locations at different times in the interface depending on the closure mode. Slow closure of a double cantilever beam crack results in different extents of healing along the closed crack. This surface approach stage typically contributes as a boundary value problem to the other stages of wetting and diffusion for thermal healing processes; however, for self-repair systems, surface approach could be the critical factor. The nature of the damage evolution, for example in highly filled composites, could result in microcracks whose surfaces are held apart by damage debris or fibers. Solvent systems may force the surfaces together by pressure swelling. Also, adhesives may play a very important role where the microvoids can be filled by the low-viscosity fluid that then reacts and bonds the surfaces together.

Wetting

When the damaged surfaces approach each other, they need to wet each other and form an interface before the healing process can continue. With self-repair fluids, the wetting of the damage surfaces by the fluid must also be considered. Wetting at the interface can be time-dependent. Because of surface roughness and other factors, good contact and wetting are not achieved instantaneously at all locations. Typically, wetted pools are nucleated at random locations at the interface and propagate radially until coalescence and complete wetting are obtained. Contact theories suggest that complete strength may be obtained when the interface has been wetted. This is important for healing of fiber surfaces to the composite matrix. For symmetric interfaces, interdiffusion is necessary for strength development. Both contact and interdiffusion dynamics are sensitive to the solvent concentration. The time dependence of viscous flow to promote contact and that of interdiffusion may be comparable, since they are subject to the same molecular dynamic processes.

Diffusion

The diffusion stage discussed for thermal welding is expanded to bring into consideration the role of the diffusion of small solvent molecules into the damage zones and polymer matrix. The diffusion stage is the most important in terms of its intrinsic contribution to strength development for high-molecular-weight polymers. The intrinsic diffusion function for polymer interdiffusion (Chapter 2) is

$$H(t) = H_\infty (t/T_r)^{r/4} \qquad (12.8.1)$$

in which the reptation time has a dependence on polymer concentration

$$T_r(c) = T_r(1)\, c^{5/4} \qquad (12.8.2)$$

However, Eq 12.8.2 applies when the polymer is already in the melt. If the polymer exists below its glass transition temperature, as would be expected for most composite matrices, then the greatest effect on the diffusion process comes through the lowering of the T_g by the solvent. This problem is quite complex, but one could attempt to solve it in the following manner. The addition of a low-molecular-weight diluent sharply decreases T_g. The reduction is at first linear via [19]

$$T_g(c) = T_g(1) - k(1-c) \tag{12.8.3}$$

where k is a constant that ranges from 200 °C to 500 °C, depending on the solvent or plasticizer. For example, the addition of toluene to polystyrene at a concentration of about 0.7 (weight fraction) causes the glass transition temperature to fall from 100 °C to about 0 °C. Thus, such a PS–solvent mixture at room temperature would be expected to heal at a rate comparable to that of undiluted PS at about 120 °C. With the empirical Vogel–Fulcher relation for the diffusion coefficient (Chapter 8), the concentration dependence of the self-repair time for diffusion becomes

$$T_r(c) = (R^2/3\pi^2 T)\, 10^{\{B/[T - T_\infty(1) + k(1-c)]\} - A} \tag{12.8.4}$$

in which A and B are constants. This equation can be readily characterized for a given solvent matrix system. With increasing solvent concentration, the reduction in T_g is no longer linear, and more complex expressions must be used, as suggested by Ferry [19].

The interaction of the solvent with the matrix is greatest in the vicinity of the release fibers, and the flow or diffusion of the fluid throughout the matrix must be evaluated. Dry has suggested that d'Arcy's law could be used to describe the flow of fluid in capillary pores [17].

Randomization

The randomization stage refers to the equilibration of the nonequilibrium conformations of the chains near the surfaces, and, in the case of crack healing and processing, the restoration of the molecular weight distribution and random orientation of chain segments near the interface. The stages of crack healing and self-repair can have interactive time-dependent functions such that the repair processes involve five-way convoluted functions. Therefore, the experimenter must take great care in the conduct of experiments designed to critically explore theories of self-repair.

The use of healing processes to increase the safety and reliability of materials systems has enormous potential for future applications. The use of self-healing smart materials is especially important in cases where it may not be possible or practical to repair the material once it has been put into use, for example, in deep space exploration, satellites, artificial organs, rocket motors, and large-body constructions. May we take it on faith that those things coming closest to having eternal life will require the most healing?

Final Summary

The relations we have developed throughout this book are brought together in summary form in Table 12.1.

Table 12.1 Polymer Interface Adhesion Relations

Relation	Form of the equation	Applicability
Adhesion dynamics	$G_{1c} = G_{1c\infty}(t/T_r)^{1/2}$	$(t \leq T_r)$
Welding time	$T_r \sim M^3$	$(M \geq M_c)$
	$T_r \sim M^{*2}M$	$(M > M^*)$
Strength versus molecular weight	$G_{1c} \sim M/M_c[1 - (M_c/M)^{1/2}]^2$	$(M \leq 8M_c = M^*)$
	$G_{1c} \sim M$	$(M < M_c)$
Tack	$\sigma \sim t^{1/4}M^{-1/4}$	$(t \leq T_r)$
Green strength	$\sigma, \epsilon \sim (M^{1/2} - M_c^{1/2})$	$(M < 8M_c)$
Weld fatigue	$da/dN \sim t^{-5/4}M^{5/4}$	$(t \leq T_r)$
Fatigue versus molecular weight	$da/dN \sim M^{-5/2}$	$(\Delta G$ constant)
"Nail" solution	$G_{1c} = \frac{1}{2}\mu_0 V \Sigma L^2$	$(L \leq L_c)$
Interface structure	$H = H_\infty(t/T_r)^{r/4}$	$(t \leq T_r)$
Virgin structure	$H_\infty = M^{(3r-s)/4}$	$(r,s = 1,2,3)$
Structure versus strength	$G_{1c} \sim H^{2/r}$	$(r = 1,2,3)$
Fractal roughness	$N_f = H^{d/4}/M$	$(d = 2,3)$
Critical entanglement	$M_c \sim a^2C_\infty$	(random coil)
Incompatible interface	$G_{1c} = G_{1c}^*(d_\infty/R_g)^2$	$(d_\infty \leq 2R_g)$
Interface width	$d_\infty = 2b/(6\chi)^{1/2}$	$(\chi > \chi_c)$
Compatibilizer number	$G_{1c}(\Sigma) = G_\infty(\Sigma/\Sigma_\infty)^2$	$(N > N_c)$
Coupling agent number	$G_{1c} = G_\infty(\Sigma/\Sigma_\infty)$	$(N < N_c)$
Optimal number of chains	$\Sigma_\infty \sim N^{-1/2}$	(random coil)
Adhesive thickness	$G_{1c} \sim G_\infty(h/h_\infty)^2$	$(h \leq h_\infty)$
Optimal thickness	$h_\infty \sim N^{1/2}$	(random coil)

a = area of chain; b = bond length; C_∞ = characteristic ratio; d = dimension; da/dN = crack increase per cycle; d_∞ = thickness of incompatible interface; G_{1c} = critical fracture energy; $G_{1c\infty}$ = virgin fracture energy; G_{1c}^* = virgin fracture energy of side A or B of the incompatible A/B interface; G_∞ = optimal fracture energy; G^* = maximum fracture energy; H = molecular property; H_∞ = virgin property; h = thickness of adhesive interlayer; h_∞ = optimal interlayer thickness; L = nail length; L_c = critical nail length; M = molecular weight; M_c = critical entanglement molecular weight; $M^* = 8M_c$; N = degree of polymerization of compatibilizer/coupling agent; N_c = critical entanglement number of compatibilizer or coupling agent; N_f = interface roughness; R_g = radius of gyration; $r,s = 1,2,3$; T_r = relaxation time; t = welding time; V = velocity; ΔG = fatigue driving energy; ϵ = fracture strain; μ_0 = coefficient of friction; Σ = number of chains/unit area; Σ_∞ = saturation number of chains; σ = fracture stress; χ = Flory–Huggins parameter; χ_c = critical χ value.

12.9 References

1. K. Jud and H. H. Kausch, *Polym. Bull.* ***1***, 697 (1979).
2. K. Jud, H. H. Kausch, and J. G. Williams, *J. Mater. Sci.* ***16***, 204 (1981).
3. "Fracture Mechanics Studies of Crack Healing"; Chapter 10 in H. H. Kausch, *Polymer Fracture*; Springer-Verlag, Heidelberg; 2nd ed., 1987.
4. D. Petrovska-Delacrétaz and H. H. Kausch, "Interdiffusion of Macromolecules from Surfaces of Different States", *Proceedings of IBM Polymer Symposium, Florence, Italy*, May 10–13 (1989); paper based on the Ph.D. Thesis of D. Petrovska-Delacrétaz, "Effets Mécaniques de la Cicatrisation des Polymères de Structures Différentes", Thèse No. 866, École Polytechnique Fédérale de Lausanne (1990).
5. H. H. Kausch and M. Tirrell, *Annu. Rev. Mater. Sci.* ***19***, 341 (1989).
6. W. W. Graessley, "Entangled Linear, Branched and Network Polymer Systems—Molecular Theories"; chapter in *Synthesis and Degradation; Rheology and Extrusion*; Vol. 47 in series *Advances in Polymer Science*, H.-J. Cantow *et al.*, Eds.; Springer-Verlag, Berlin; 1982; p 67.
7. R. P. Wool, *Polym. Eng. Sci.* ***18***, 1057 (1978).
8. R. P. Wool and K. M. O'Connor, *J. Appl. Phys.* ***52***, 5953 (1981).
9. R. P. Wool and K. M. O'Connor, *J. Polym. Sci., Polym. Lett. Ed.* ***20***, 7 (1982).
10. K. M. O'Connor, *Crack Healing in Polymers*; Ph.D. Thesis, University of Illinois, Urbana, IL; 1984.
11. J. O. Outwater and D. J. Gerry, *J. Adhes.* ***10***, 290 (1969).
12. A. C. Garg and Y.-W. Mai, *Compos. Sci. Technol.* ***31***, 179 (1989).
13. D. H. Klosterman, *Fatigue Healing Studies in Polystyrene and Short Fiber Reinforced Polymers*; M.S. Thesis, Univ. of Illinois, Urbana, IL; 1984.
14. I. Amato, "Animating the Material World", *Science* ***255***, 284 (Jan. 17, 1992).
15. C. M. Dry, "Smart Building Materials Which Prevent Damage and Repair Themselves"; in *Smart Materials Fabrication and Materials for Micro-Electro-Mechanical Systems: Symposium Held April 28–30, 1992, San Francisco, CA;* Vol. 276 in series *Materials Research Society Proceedings*; Materials Research Society, Philadelphia, PA; 1992; p 311.
16. C. M. Dry, and N. Sottos, "Passive Smart Self-Repair in Polymer Matrix Composite Materials"; in *Smart Structures and Materials 1993: Smart Materials*, V. K. Varadan, Ed.; *Proceedings of 1993 North American SPIE Conference on Smart Structures and Materials*; Vol. 1916 in series *SPIE Proceedings*; SPIE, Bellingham, WA; 1993; p 438.
17. C. M. Dry, "Smart Materials which Sense, Activate and Repair Damage"; in *First European Conference on Smart Structures and Materials, Glasgow, Scotland*, B. Culshaw, P. T. Gardiner, and A. McDonach, Eds.; Vol. 1777 in series *SPIE Proceedings*; SPIE, Bellingham, WA; 1992; p 367.
18. R. P. Wool and C. M. Dry, work in progress (1993).
19. J. D. Ferry, *Viscoelastic Properties of Polymers*; John Wiley & Sons, New York; 3rd ed., 1980.
20. ASTM D 256 – 92, Standard Test Methods for Impact Resistance of Plastics and Electrical Insulating Materials. In *Vol. 08.01, Plastics (I), 1993 Annual Book of ASTM Standards*; American Society for Testing and Materials, Philadelphia, PA, 1993.

Appendix

Polymer Abbreviations Used in This Book

Abbreviated term	*Designation*
ABS	Acrylonitrile/butadiene/styrene
dPS*	Polystyrene-d8, fully deuterated polystyrene
DHD*	Polystyrene triblock, ends deuterated
DPS*	Deuterated polystyrene
EPDM	Ethylene/propylene/diene
EPR*	Ethylene–propylene rubber
HDH*	Polystyrene triblock, center deuterated
HDPE	High density polyethylene
HPP*	Hard elastic polypropylene
HPS*	Normal (protonated) polystyrene
HTPB*	Hydroxy-terminated polybutadiene
LCP	Liquid crystal polymer
LDPE	Low density polyethylene
LLDPE	Linear low density polyethylene
PA	Polyamide (nylon)
PAMS*	Poly(α-methyl styrene)
PAN	Polyacrylonitrile
PB	Polybutene-1
PBAN	Polybutadiene-acrylonitrile
PBMA*	Poly(butyl methyl acrylate)
PC	Polycarbonate
PDMS*	Polydimethylsiloxane
PE	Polyethylene
PEEK	Polyether-ether-ketone
PEI	Polyetherimide
PEO	Poly(ethylene oxide)
PET	Poly(ethylene terephthalate)
PI	Polyimide
PIB	Polyisobutylene
PIP*	Polyisoprene
PMMA	Poly(methyl methacrylate)

POMS*	Poly(*o*-methyl styrene)
PP	Polypropylene
PS	Polystyrene
PSAN*, SAN	Poly(styrene–co-acrylonitrile)
PSU	Polysulfone
PVAc*	Poly(vinyl acetate)
PVC	Poly(vinyl chloride)
PVOH*	Poly (vinyl alcohol)
PVP	Polyvinylpyrrolidone, poly(2-vinylpyridine)
SAN, PSAN*	Styrene–acrylonitrile
SBR*	Styrene–butadiene rubber
SBS*	Styrene–butadiene–styrene triblock copolymer
SI*	Styrene–isoprene copolymer
SIS*	Styrene–isoprene–styrene triblock copolymer
UHMWPE	Ultra-high molecular weight polyethylene

* Not in ASTM D 1600 (Chapter 1, [49])

Index

Notes: Entries beginning with a numeral precede the alphabetical entries. Those beginning with a Greek letter follow the alphabetical entries, in Greek alphabetical order. Quantities are indexed under their full name (*e. g.*, glass transition temperature), with a cross-reference under their symbol if the symbol is in frequent use (*e. g.*, T_g). Alphabetization is letter-by-letter. Full names corresponding to polymer abbreviations are in the Appendix.